Isotopes in the Physical and Biomedical Sciences

Volume 2

Isotopic Applications in NMR Studies

Isotopes in the Physical and Biomedical Sciences

Edited by E. Buncel and J.R. Jones

Isotopes in the Physical and Biomedical Sciences

Volume 2

Isotopic Applications in NMR Studies

Edited by

E. Buncel
Queen's University, Kingston, Ontario, Canada

and

J.R. Jones
University of Surrey, Guildford, England

ELSEVIER

Amsterdam — Oxford — New York — Tokyo   **1991**

ELSEVIER SCIENCE PUBLISHERS B.V.
Sara Burgerhartstraat 25
P.O. Box 211, 1000 AE Amsterdam, The Netherlands

Distributors for the United States and Canada:

ELSEVIER SCIENCE PUBLISHING COMPANY INC.
655, Avenue of the Americas
New York, NY 10010, U.S.A.

Library of Congress Cataloging-in-Publication Data
(Revised for volume 2)

Isotopes in the physical and biomedical sciences.

   Includes bibliographies and indexes.
   Contents : v. 1, pt. A. Labelled compounds -- v. 2.
Isotopic applications in NMR studies.
   1. Radioactive tracers. 2. Isotopes. 3. Radio-
pharmaceuticals. I. Buncel, E. II. Jones, J. R.
(John Richards) [DNLM: 1. Isotope Labeling. 2. Iso-
topes. QD 466.5 I85]
QD607.I86 1987   541.3'884   87-9159
ISBN 0-444-42809-7 (pt. A.)
ISBN 0-444-89090-4 (Vol. 2)

ISBN 0-444-89090-4

This book is printed on acid-free paper.

Printed in The Netherlands

ISOTOPES IN THE PHYSICAL AND BIOMEDICAL SCIENCES

Contributors to Volume 2

A.D. Bain

Department of Chemistry, McMaster University,
Hamilton, Ontario, L8S 4M1, Canada

K.M. Brière

The Ottawa-Carleton Chemistry Institute,
University of Ottawa Campus, Ottawa, Ontario,
K1N 9B4, Canada

J.H. Davis

Physics Department, University of Guelph,
Guelph, Ontario, N1G 2W1, Canada

C. Detellier

The Ottawa-Carleton Chemistry Institute,
University of Ottawa Campus, Ottawa, Ontario,
K1N 9B4, Canada

H.P. Graves

The Ottawa-Carleton Chemistry Institute,
University of Ottawa Campus, Ottawa, Ontario,
K1N 9B4, Canada

C.J. Jameson

Department of Chemistry, University of Illinois
at Chicago, P.O. Box 4348, Chicago, Illinois
60680, U.S.A.

K. Kamienska-Trela

Institute of Organic Chemistry, Polish Academy
of Sciences, 01-224 Warszawa, Poland

E. Kupče

Institute of Organic Synthesis, Latvian Academy
of Sciences, Riga, 226006, Latvia, U.S.S.R.

L. Lao

Department of Chemistry, McMaster University,
Hamilton, Ontario, L8S 4M1, Canada

E. Lukevics

Institute of Organic Synthesis, Latvian Academy
of Sciences, Riga, 226006, Latvia, U.S.S.R.

T.J. Simpson

School of Chemistry, University of Bristol,
Bristol, BS8 1TS, England

R.D. Thomas

Center for Organometallic Research and
Education, Department of Chemistry, North Texas
State University, Denton, Texas 76203, U.S.A.

P.G. Williams

National Tritium Labeling Facility, Lawrence
Berkeley Laboratory, University of California,
Berkeley, California 94720, U.S.A.

VOLUME 1. LABELLED COMPOUNDS (PART A)

PREFACE TO THIS SERIES OF MONOGRAPHS

Since the end of the Second World War there has been a tremendous increase in the number of compounds that have been synthesized with radioactive or stable isotopes. These have found application in many diverse fields, so much so, that hardly a single area in the pure and applied sciences has not benefited. Not surprisingly this has been reflected in the appearance of related publications. The early proceedings of the symposia on Advances in Tracer Methodology (ref. 1) were soon followed by a number of Euratom sponsored meetings in which the methods of preparing and storing labelled compounds featured prominently (refs. 2-4). In due course a resurgence of interest in stable isotopes, brought about by their greater availability (also lower cost) and partly by the development of new techniques such as gas chromatography-mass spectrometry (GC-MS), led to the publication of the proceedings of several successful conferences (refs. 5-7). More recently conferences dealing with the synthesis and applications of isotopes and isotopically labelled compounds have been established on a regular basis (refs. 8-10). Reflecting these trends, the Journal of Labelled Compounds was established in 1963, and published quarterly. Now it encompasses Radiopharmaceuticals and is produced monthly.

In addition to the proceedings of conferences and journal publication, individuals have left their mark by producing definitive texts, usually on specific nuclides. Only the classic two volume publication of Murray and Williams (refs. 11, 12), now over 30 years old and out of print, attempted to do justice to several nuclides. With the large amount of work that has been undertaken since then it seems unlikely that an up-dated edition could be produced.

The alternative strategy was to ask scientists currently active to review specific areas and this is the approach adopted in the present series of monographs. In this way it is intended to cover the broad advances that have been made in the synthesis and applications of isotopes and isotopically labelled compounds in different areas of the physical and biomedical sciences. Though this aim may be difficult to achieve with any given volume, from the viewpoint of selection of particular topics and availability of authors, it is hoped that the volumes will, nevertheless, prove to be useful to readers in general, as well as to researchers and practitioners in the use of isotopes.

E. Buncel

J.R. Jones

X

REFERENCES

1   S. Rothchild (ed.), Advances in Tracer Methodology, Plenum Press, New York. Vol.1, 1963; Vol.2, 1965; Vol.3, 1966; Vol.4, 1968.

2   J. Sirchis (ed.), Proc. Conf. Methods of Preparing and Storing Marked Molecules, Brussels, 1963. EUR 1625e, 1964.

3   J. Sirchis (ed.), Proc. 2nd. Int. Conf. on Methods of Preparing and Storing Labelled Compounds, Brussels, 1966. EUR 3746-d-f-e, 1968.

4   J. Sirchis (ed.), Proc. Symp. on Preparation and Bio-Medical Applications of Labelled Molecules, Venice 1964. EUR 2200e, 1966.

5   P.D. Klein and S.V. Peterson (eds.), Proc. First Int. Conf. on Stable Isotopes in Chemistry, Biology and Medicine, May 9-11, 1973, CONF-730525. U.S. Dept. of Commerce, Virginia.

6   E.R. Klein and P.D. Klein (eds.), Stable Isotopes, Proceedings of the 3rd Int. Conf., Oak Brook, Illinois, May 23-26, 1978, Academic Press, New York, 1979.

7   H.L. Schmidt, H. Förstell and K. Heinzinger (eds.), Stable Isotopes. Proceedings of the 4th Int. Conf. Jülich, March 23-26, 1981. Elsevier, Amsterdam, 1982.

8   W.P. Duncan and A.B. Susan (eds.), Synthesis and Applications of Isotopically Labeled Compounds. Proceedings of the First International Symposium, Kansas City, MO, USA 6-11 June 1982, Elsevier, Amsterdam, 1983.

9   R.R. Muccino (ed.), Synthesis and Applications of Isotopically Labeled Compounds. Proceedings of the Second International Symposium, Kansas City, MO, USA 3-6 September 1985, Elsevier, Amsterdam, 1986.

10 T.A. Baillie and J.R. Jones (eds.), Synthesis and Applications of Isotopically Labelled Compounds. Proceedings of the Third International Symposium, Innsbrück, Austria, 17-21 July 1988, Elsevier, Amsterdam, 1989.

11 A. Murray III and D.L. Williams, Organic Syntheses with Isotopes. Part I: Compounds of Isotopic Carbon. Interscience Publishers, New York 1958.

12 Part II: Organic Compounds Labeled with Isotopes of the Halogens, Hydrogen, Nitrogen, Oxygen, Phosphorus and Sulfur, Interscience Publishers, New York 1958.

CONTENTS

Isotopes in the Physical and Biomedical Science, Vol. 2, edited by E. Buncel and J.R. Jones
© 1991 Elsevier Science Publishers B.V., Amsterdam

Chapter 1

THE DYNAMIC AND ELECTRONIC FACTORS IN ISOTOPE EFFECTS ON NMR PARAMETERS

CYNTHIA J. JAMESON

Department of Chemistry, University of Illinois, Chicago

CONTENTS

INTRODUCTION

The other chapters in this volume involve observation of the NMR nucleus (such as ^{2}H, ^{3}H, or ^{13}C) which itself is the isotopic label. In this chapter we consider the mass effects of the isotopic label on the chemical shifts and coupling constants of the neighboring resonant nuclei, a secondary isotope effect. We also consider the mass effect on the reduced coupling constants when one of the coupled nuclei is the isotopic label, a primary isotope effect.

The physical picture is a simple one: When a heavier isotope replaces an atom in a molecule its effects are felt by every resonant NMR nucleus in the molecule because the changed mass causes changes in the rovibrational averaging of all electronic properties of the molecule. In particular, the chemical shift of each resonant nucleus and the spin-spin coupling between it and other nuclei are affected to a larger or smaller extent dependent on its location in the molecule relative to the site of the perturbation (mass change). What this means in a practical sense is that we can use the isotope effects on these electronic quantities as a way of arriving at the location (structure-wise) of the mass-label. All resonant nuclei linked to the label site by some efficient electronic transmission path will report on its location, providing a multiplicity of useful information. There are obvious applications of this to the determination of mechanisms of reactions.

There are two important factors affecting the magnitudes and signs of these isotope effects. One, the dynamic factor, has to do with the slight change in the rovibrationally-averaged geometry of the molecule upon substitution of an atom by a heavier isotope; the dynamic average bond lengths tend to become shorter. The other, an electronic factor, has to do with the sensitivity of the chemical shift or of the coupling constant to a change in molecular geometry. The dynamic factor depends on the potential energy surface of the molecule. Anharmonic bond stretching plays an important role in the mass-dependence of the mean bond length. On the other hand, the electronic factor depends on the chemical shift range of the resonant nucleus, which itself is a measure of the sensitivity of the nuclear magnetic shielding to molecular structure. The electronic factor depends on the electronic transmission pathway leading from the site of the isotopic substitution to the site of the resonant nucleus.

II. ISOTOPE EFFECTS ON CHEMICAL SHIFTS

NMR isotope shifts are useful in that, where an isotopic label is introduced, every neighboring resonant nucleus experiences a slight chemical shift. If labeling is less than 100%, the resonant nuclei in both the labeled and the unlabeled molecules are observed, with intensities according to statistical distribution, and positions dependent on the electronic and dynamic factors. The magnitude of the shift depends on the fractional change in mass at the substitution site, on the remoteness of the resonant nucleus from the substitution site and on the sensitivity of the nuclear magnetic shielding to bond lengthening at the substitution site. Every resonant nucleus in the isotopically labeled molecule experiences this isotope shift but the shift is only observable under favorable conditions such as when it is larger than the half-width of the peaks.

We consider in the following sections the intrinsic isotope shifts. That is, except for Sect. IIF, we limit our discussions to those observations in which the distinct isotopomers are intact molecules, non-exchanging, or at least the exchange (of ^2D for ^1H for example) is slow compared to the characteristic NMR time, the reciprocal of the isotope shift in Hertz. The shifts are observed in the same sample rather than two samples containing isotopically different solvents. Charged molecular species, in which the isotopomers have slightly different first sphere and second sphere solvation cages or form intramolecular or intermolecular hydrogen bonds, may have solvent effects which could obscure the true nature of the intrinsic isotope shifts when the latter are small. H_2O and D_2O have slightly different liquid structures due to differential hydrogen-bonding strengths. The isotopic solvent effect could be important when the observed nucleus is in a charged molecular species.

Isotope shifts depend on molecular structure since both the dynamic and electronic factors are dependent on structure. This observable is a convolution of electronic and dynamic terms of various orders of magnitude, all of which are specific to the molecule. Viewed in this light the quantitative interpretation or prediction of isotope shifts is molecule-specific and becomes extremely difficult for large molecules. However, there are several empirical trends which provide interesting insights into the separation of the electronic and dynamic factors.[1,2]

(A) General trends and some examples

The large body of data on isotope shifts have been collected and reviewed.[3-6] There are several general observations which have been made about magnitudes and signs of isotope shifts:

(1) Upon substitution with a heavier isotope the NMR signal of the nearby nucleus usually shifts towards lower frequencies (higher shielding). An example is shown in FIGURE 1. This trend is fairly general, although there are exceptions.

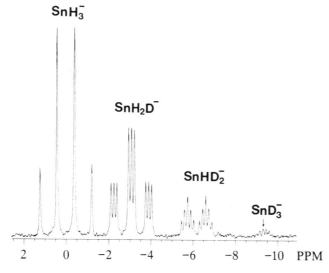

Fig. 1. Tin-119 NMR spectrum of $SnH_{3-n}D_n^-$ at -50°C in liquid ammonia, reproduced with permission from R.E. Wasylishen and N. Burford, *Can. J. Chem.*, 65 (1987) 2707.[7]

Among one-bond shifts, ^{13}C-induced ^{113}Cd and ^{199}Hg shifts in $M(CH_3)_2$,[8] and in fact all known ^{199}Hg isotope shifts, are of unusual sign.[9] Among 2-bond shifts the D-induced ^{13}C isotope shifts in the H-C-$^{13}C\oplus$, H-C-$^{13}C=O$, H-C-$^{13}C=S$ fragments are of unusual sign. So are the 3-bond D-induced ^{19}F shifts in the

$$H-\overset{\overset{\displaystyle F}{|}}{C}-C=O$$

fragment and the ^{17}O shift in the H-C-C=^{17}O fragment. Longer-range isotope shifts can be of unusual sign, and in some cases a continuum of values negative through positive have been observed in related molecules,[10] as shown in the examples in FIGURE 2.

(2) The magnitude of the isotope shift is dependent on how remote the isotopic substitution is from the observed nucleus. Although there are exceptions, one-bond isotope shifts are larger than two-bond or three-bond isotope shifts. An example is shown below [11]

$$\overset{\oplus}{D_3 N}\overset{0.33}{-CH_2}\overset{0.14}{-CH_2}\overset{0.01}{-CH_2}\overset{0.10}{-CH_2}\overset{0.29}{-CH}\overset{0.30}{-C}=O$$
$$\quad\quad\quad\quad\quad\quad\quad\quad\quad\quad\quad\quad\quad\quad\underset{\oplus ND_3}{|}\ \underset{OD}{|}$$

where the magnitudes of the D-induced ^{13}C shifts (in ppm) are indicated over each observed carbon. In para-difluorobenzene the ^{13}C- induced ^{19}F shifts are -0.090, -0.026, -0.005, and -0.005 ppm for ^{13}C substitution, one, two, 3, and 4 bonds away from the observed ^{19}F.[12]

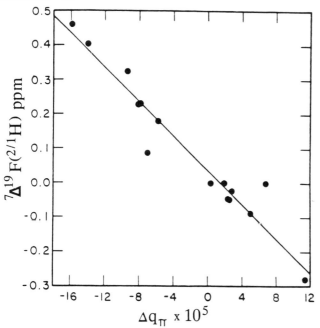

Fig. 2. The D-induced ^{19}F shift $^7\Delta^{19}F(^{2/1}H)$ in 4-fluorophenyl species vs. the change in pi electron density at fluorine induced by C-H bond shortening in MNDO calculations, reproduced with permission from D. A. Forsyth and J. R. Yang, *J. Am. Chem. Soc.*, 108 (1986) 2151.[10] Copyright 1986 American Chemical Society.

(3) The magnitude of the shift is a function of the observed nucleus and reflects its chemical shift range. In analogous molecules the isotope shifts very roughly scale as the chemical shift ranges of the observed nucleus: e.g., ^{31}P in PH_3 (-0.85 ppm per D) [13] compared to ^{15}N in NH_3 (-0.623 ppm per D) [14], ^{77}Se in H_2Se (-7.04 ppm per D) [15]

compared to ^{17}O in H_2O (-1.55 ppm per D) [14], and even in the case of the ^{13}C- induced ^{199}Hg and ^{113}Cd shifts in Me_2Hg (+0.4 ppm per ^{13}C) and Me_2Cd (+0.14 ppm per ^{13}C) [8] where the isotope shifts are of unusual sign.

(4) The magnitude of the shift is related to the fractional change in mass upon isotopic substitution. For a given resonant nucleus the largest shifts are induced by D or T substitution for H. For example, the one-bond ^{51}V shifts induced by $^{2/1}H$ and $^{13/12}C$ substitution in the $[CpVH(CO)_3]^-$ ion are respectively -4.7 ppm per D and -0.51 ppm per ^{13}C. [16]

(5) The magnitude of the shift is approximately proportional to the number of equivalent atoms which have been substituted by isotopes. In other words, isotope shifts exhibit additivity. This is quite evident in FIGURE 1. In $[Co(NH_3)_6]^{3+}$ complex ion in a H_2O-D_2O solvent, each of the 18 replaceable hydrogens in the complex can be H or D. Each of the 19 isotopomers (H_{18}, $H_{17}D$, ..., D_{18}) which gives rise to one ^{59}Co resonance signal can be observed in the spectra obtained in solvents of varying H : D ratios (FIGURE 3). The ^{59}Co shift relative to the H_{18} isotopomer is proportional to

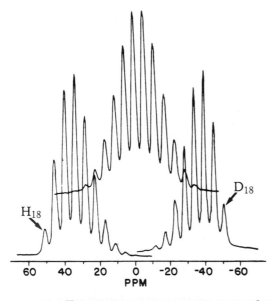

Fig. 3. The ^{59}Co NMR spectrum of $[Co(NH_3)_6]^{3+}$ ion in D_2O/H_2O solvent. The spectra, from left to right, are for 15%, 50%, 85% D_2O samples, reproduced with permission from J.G. Russell and R.G. Bryant, *Anal. Chim. Acta*, 151 (1983) 227.[17]

the number of H atoms replaced by D, so there are 19 peaks equally spaced.

These general trends have been explained by a theory based on rovibrational averaging [1,2,18] which we outline briefly in the next section. The interpretation of isotope shifts involves a consideration of the vibrational and rotational averaging of nuclear shielding. For this reason the isotope shift is intimately related to the observed temperature dependence of nuclear shielding in the gas phase in the zero-pressure limit. These two measurable properties share the same electronic factors - the change in shielding with bond extension or bond angle deformation.

(B) Rovibrational theory of isotope shifts

Nuclear magnetic shielding σ is dependent on the geometry of the molecule, and we are particularly interested in the geometries very close to the equilibrium geometry since vibration of the molecule involves only very small displacements away from this geometry. The average shielding can then be expressed as follows

$$\langle \sigma \rangle = \sigma_e + \sum_i \left(\frac{\partial \sigma}{\partial r_i} \right)_e \langle \Delta r_i \rangle + \sum_{ij} \left(\frac{\partial^2 \sigma}{\partial r_i \partial r_j} \right)_e \langle \Delta r_i \Delta r_j \rangle + \sum_{ij} \left(\frac{\partial \sigma}{\partial \alpha_{ij}} \right)_e \langle \Delta \alpha_{ij} \rangle + ... \tag{1}$$

Isotopic substitution changes all the dynamic averages $\langle \Delta r_i \rangle$, $\langle \Delta r_i^2 \rangle$, $\langle \Delta \alpha_{ij} \rangle$, etc., with the largest changes occurring in the bond displacements directly involving the substituted atom. The measured isotope shift is denoted by a symbol $^n \Delta A$ ($^{m'/m}X$) and is usually in units of ppm per $^{m'}X$ substitution, for the chemical shift of nucleus A in the isotopomer with the heavier isotope $^{m'}X$, relative to the isotopomer containing the lighter $^m X$ at a site n bonds away. This measured value, the isotopic chemical shift or the "isotope shift", is the difference $\langle \sigma \rangle - \langle \sigma \rangle^*$ where the * denotes the heavier isotopomer. For a one-bond isotope shift involving a group AX_n where the A-X bond in question is denoted by r_1

$$\langle \sigma \rangle - \langle \sigma^* \rangle = \left(\frac{\partial \sigma}{\partial r_1} \right)_e [\langle \Delta r_1 \rangle - \langle \Delta r_1 \rangle^*] + \left(\frac{\partial^2 \sigma}{\partial r_1^2} \right)_e [\langle (\Delta r_1)^2 \rangle - (\Delta r_1)^2 \rangle^*]$$
$$+ \left(\frac{\partial \sigma}{\partial \alpha_{12}} \right)_e [\langle \Delta \alpha_{12} \rangle - \langle \Delta \alpha_{12} \rangle^*] + ... + \sum_{j \neq 1} \left(\frac{\partial \sigma}{\partial r_j} \right)_e [\langle \Delta r_j \rangle - \Delta r_j \rangle^*] + ... \tag{2}$$

Δr_1 is the bond displacement at the substitution site and Δr_j are the other bond displacements. Thus, there is a primary dynamic effect which we shall denote by

$$\Delta \approx [\langle \Delta r_1 \rangle - \langle \Delta r_1 \rangle^*], \tag{3}$$

a change in the mean bond displacement at the site of substitution of $^m X$ by $^{m'}X$. The terms involving $\Delta\alpha_{12}$, the bond angle deformation at the substitution site, turns out to be unimportant in many cases where calculations have been carried out.[19,20] There are much smaller changes (secondary dynamic effect) in the bond lengths at sites remote from the substitution site; let us represent these smaller changes to be of magnitude δ,

$$\delta \approx [\langle\Delta r_j\rangle - \langle\Delta r_j\rangle^*] . \qquad (4)$$

A one-bond isotope shift is observed when the resonant nucleus is directly bonded to the substituted atom. In this case one type of term is much larger than the others because the nuclear magnetic shielding changes nearly linearly with bond extension at bond lengths close to the equilibrium geometry.[21-26]

$$\langle\sigma\rangle - \langle\sigma\rangle^* = (\partial\sigma/\partial r_1)_e \cdot \Delta \quad + \quad \text{smaller terms} \qquad (5)$$

The smaller terms include terms in the second and higher derivatives of nuclear shielding, terms involving bond angle deformations and derivatives of the shielding with respect to the bond angle, as well as the much smaller terms such as $(\partial\sigma/\partial r_j) [\langle\Delta r_j\rangle - \langle\Delta r_j\rangle^*]$, involving the shielding change due to the small secondary changes in bond length elsewhere in the molecule.

The factor $(\partial\sigma/\partial r_1)_e$ is the sensitivity of the shielding of the resonant nucleus to the lengthening of a bond to it (a primary electronic effect). The factor $(\partial\sigma/\partial r_j)_e$ may be exactly the same as $(\partial\sigma/\partial r_1)_e$, such as for ^{13}C shielding in CH_4. On the other hand, $(\partial\sigma/\partial r_j)_e$ could well be much smaller, as for ^{17}O in $^{17}O=C=^{16}O$ where r_j is the bond length involving the other oxygen. Bond angle deformation might also be important in some cases. Full calculations including the smaller terms have been carried out for diatomic molecules [21], H_2O,[19] and CH_4,[20] using ab initio theoretical values for the derivatives.[21,25,26]

Qualitatively, the mass change accompanying the replacement of ^mX by its heavier isotope $^{m'}X$ leads to a shorter average bond length. This happens because the potential energy surface of the molecule is an anharmonic one. Whereas the mean bond displacement, $\langle\Delta r\rangle = \langle r - r_e\rangle$, is zero for a harmonic oscillator, it is a mass-dependent non-zero value for the anharmonic oscillator. The amplitudes of motion of a lighter atom being larger than that of a heavier one, at each vibrational state the heavier isotopomer has a smaller average $\langle\Delta r\rangle$ than the lighter isotopomer. The vibrational frequencies being lower for the heavier isotopomer, the population of higher vibrational states are somewhat greater for the heavier isotopomer. Both effects constitute a mass dependence of the thermal average $\langle \Delta r \rangle$, the former being more important than the latter. Full vibrational calculations are possible for any molecule provided that enough spectroscopic constants are available to determine its potential surface, from which the quantities $\langle \Delta r_1 \rangle$, $\langle \Delta r_j \rangle$, $\langle \Delta \alpha_{12} \rangle$, $\langle (\Delta r_1)^2 \rangle$, etc., can be calculated.[18-20] For a diatomic molecule described by a Morse oscillator these vibrational averages take fairly simple forms

$$\langle \Delta r \rangle = \frac{3a}{2} \langle (\Delta r)^2 \rangle \tag{6}$$

$$\langle (\Delta r)^2 \rangle_v = 2 \left(\frac{B_e}{\omega_e} \right) r_e^2 \left(v + \frac{1}{2} \right) \tag{7}$$

where a is the Morse parameter describing the anharmonicity of the vibration. To obtain the thermal average we need to sum over all the states weighted by their populations.

If we use the harmonic oscillator density of states as an approximation, we find a very simple form for the thermal average

$$\langle v + \frac{1}{2} \rangle^T = \frac{1}{2} \coth (hc\omega_e/2kT) \ . \tag{8}$$

Thus, for a diatomic molecule the thermal average bond extension is given by

$$\langle \Delta r \rangle \approx \frac{3a}{2} \left(\frac{B_e}{\omega_e} \right) r_e^2 \coth \left(\frac{hc\omega_e}{2kT} \right) \tag{9}$$

and the thermal average shielding is given by

$$\langle \sigma \rangle \approx \sigma_e + \left[\left(\frac{d\sigma}{dr} \right)_e + \frac{1}{3a} \left(\frac{d^2\sigma}{dr^2} \right)_e \right] \langle \Delta r \rangle \text{ or } \langle \sigma \rangle \approx \sigma_e + \left[\frac{3a}{2} \left(\frac{d\sigma}{dr} \right)_e + \frac{1}{2} \left(\frac{d^2\sigma}{dr^2} \right)_e \right] \langle (\Delta r)^2 \rangle \tag{10}$$

and the isotope shift by

$$\langle \sigma \rangle - \langle \sigma \rangle^* \approx \left[\left(\frac{d\sigma}{dr} \right)_e + \frac{1}{3a} \left(\frac{d^2\sigma}{dr^2} \right)_e \right] \{ \langle \Delta r \rangle - \langle \Delta r \rangle^* \}. \tag{11}$$

In the above expression the first quantity (in square brackets) is a mass-independent purely electronic quantity, the second quantity (in curly brackets) is a mass-dependent dynamic factor.

In polytomic molecules the analogous equation, eq. (2), has a large number of terms even when only up to second derivatives are considered, due to the many types of internal displacements Δr_i, $\Delta\alpha_{ij}$, etc. Nevertheless, the general trends noted in Sect. IIA, which are easily explained by eq. (11) for diatomic molecules, are also found for the very large body of data on polyatomic molecules. The simplicity of eq. (11) can be preserved for one-bond isotope shifts in polyatomic molecules if we consider bond extension as the dominant contributor to isotope shifts. The simple relation (eq. (6)) for Morse anharmonic oscillators permits the discussions of the dynamic factor to be cast in terms of $\langle\Delta r\rangle$ alone and the electronic factor in square brackets in eq. (11) which includes both first and second derivatives will be represented in following discussions entirely by $(\partial\sigma/\partial r)_e$.

(C) The dynamic factors

(1) Mass Dependence

For a diatomic molecule, by making use of the implicit mass dependence of its spectroscopic constants B_e and ω_e in eq. (9), one can express the dynamic factors $[\langle\Delta r\rangle - \langle\Delta r\rangle^*]$ for the v=0 state as follows:

$$\langle\Delta r\rangle - \langle\Delta r\rangle^* = [1 - (\mu/\mu^*)^{1/2}]\langle\Delta r\rangle \tag{12}$$

and,

$$^1\Delta A\left(^{m'/m}X\right) = \langle\sigma\rangle - \langle\sigma\rangle^* = [1 - (\mu/\mu^*)^{1/2}] \bullet \left[\left(\frac{d\sigma}{dr}\right)_e + \frac{1}{3a}\left(\frac{d^2\sigma}{dr^2}\right)_e\right]\langle\Delta r\rangle + \ldots \tag{13}$$

The thermal average isotope shift is nearly the same as above except that coth $(hc\omega_e/2kT)$ is mass-dependent and this can be important when the vibrational frequency is low or the temperature is high, that is, when $(hc\omega_e/2kT)$ is less than about 3 or 4. Thus for several isotopomers, relative to the same parent molecule, the isotope shift is proportional to

$$[1 - (\mu/\mu^*)^{1/2}].$$

In the hydrogen fluorides this factor is 0.3936 for FT relative to FH and it is 0.2750 for FD relative to FH. In other words we expect

$$\frac{^1\Delta\ ^{19}F\ (^{3/1}H)}{^1\Delta\ ^{19}F\ (^{2/1}H)} \approx 1.43.$$

It has been shown [1] that upon substitution of heavier nuclei such as ^{13}C for ^{12}C, that is when $(\mu^* - \mu)/\mu^* \ll 1$, the above mass factor can be approximated by

$$(\mu^* - \mu)/2\mu^* \qquad \text{or} \qquad \frac{1}{2}\left(\frac{m' - m}{m'}\right)\left(\frac{m_A}{m_A + m}\right)$$

with an error of about 1% or less. For example, using this formula, we expect

$$\frac{^1\Delta\,^{31}P\,(^{36/32}S)}{^1\Delta\,^{31}P\,(^{34/32}S)} \approx 1.92.$$

It has been shown [27] that even for polyatomic molecules, $(m'-m)/m'$ appears in the dynamic factors $[\,\langle\Delta r\rangle - \langle\Delta r\rangle^* \,]$ and $[\,\langle\,(\Delta r)^2\,\rangle - \langle\,(\Delta r)^2\,\rangle^*\,]$ for the A^mX relative to A^mX. If our interpretation is valid, that the one-bond isotope shift can be considered as being dominated by a single product of an electronic and a dynamic factor, then indeed we should find the following:

(a) Where more than one isotope may be used for substitution we should be able to find a direct proportionality of $^1\Delta$ to $(m'-m)/m'$. Indeed we do. For the atomic masses, m=74, m'=76, 77, 78, 80, 82 of Se in the isotopomers of R_1SeSeR_2 (R_1, $R_2=CH_3$ and CF_3), the one-bond isotope shifts $^1\Delta^{77}Se\,(^{m'/74}Se)$ show straight lines for each set of R_1 and R_2 in FIGURE 4, reproduced from ref. 18. Similarly, the one-bond Se-induced and Te -induced ^{19}F isotope shifts are found to be proportional to $(m'-m)/m'$.[28] The former are shown in FIGURE 4. There are other examples involving fewer isotopes which also are consistent with this mass factor, as in m'=17 and18, m=16 in $^1\Delta^{13}C(^{m'/16}O)$ in CO molecule [31] and m'=34,36 for S in $^1\Delta^{31}P(^{m'/32}S)$ in thiophosphate anhydrides.[32]

(b) The factor $m_A/(m_A+m)$ is not so easily verified experimentally because this would mean comparing different resonant isotopes (A is the resonant nucleus) and the same set of m' and m. This factor should lead to an observed isotope shift which is larger for heavier resonant nuclei. A possible candidate is the ^{18}O- induced shift in ^{14}N and ^{15}N spectra of NO_3^-, for example. The $^1\Delta^{15}N(^{18/16}O)$ should be larger than the $^1\Delta^{14}N(^{18/16}O)$ by 3.7%. Only the ^{15}N data are available so far.[33] The $^1\Delta^{15}N(^{2/1}H)$ and the $^1\Delta^{14}N(^{2/1}H)$ in NH_4^+ ion have been observed. For D-substitution we cannot use the approximate form of the mass factor. The ratio of $[1-(\mu/\mu^*)^{1/2}]$ values in the dynamic factors of $^1\Delta^{15}N(^{2/1}H)$ relative to $^1\Delta^{14}N(^{2/1}H)$ is 1.005. The measured values of $^1\Delta^{15}N(^{2/1}H)$ [34] are indeed slightly greater than the measured values of $^1\Delta^{14}N(^{2/1}H)$ [14] but not by as much; the observed ratio is 1.002.

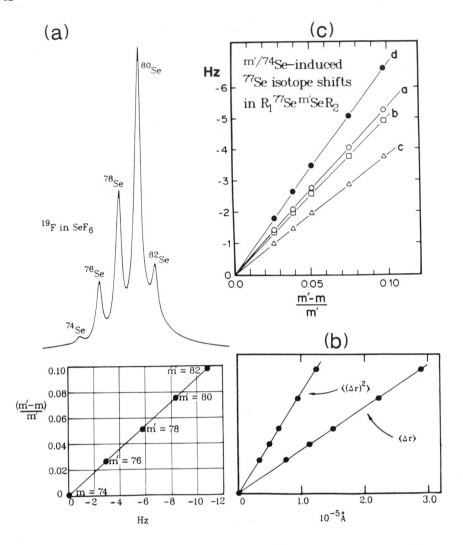

Fig. 4. (a) The effect of selenium isotopes on the ^{19}F nuclear shielding in SeF_6, reproduced from C.J. Jameson, A.K. Jameson, and D. Oppusunggu, *J. Chem. Phys.*, 85 (1986) 5480.[28] (b) The calculated mass dependence of the mean bond displacements and mean square amplitudes at 300K of the Se-F bond in SeF_6, reproduced from C.J. Jameson and A.K. Jameson, *J. Chem. Phys.*, 85 (1986) 5484.[29] (c) The effect of selenium isotopes on the ^{77}Se nuclear shielding in the diselenides $R_1\,^{77}Se\,^{m'}SeR_2$, where R_1, R_2 = CH_3, CF_3. Reproduced from C.J. Jameson and H.J. Osten, *J. Chem. Phys.*, 81(1984) 4293,[18] data taken from Ref.30. These illustrate the theoretically predicted mass factor $(m'-m)/m'$ in polyatomic molecules.

(c) Where several equivalent atoms ^mX may be replaced by the heavier $^{m'}X$ in the group AX_n, there will be a large change $(\partial\sigma/\partial r_i)_e [\langle\Delta r_i\rangle - \langle\Delta r_i\rangle^*]$ for each replaced atom, leading to the observed additivity of isotope shifts.[18] Even when all the terms in eq.(2) are included and calculated properly, the nearly strict additivity is preserved.[20] Although there are many smaller changes, such as in eq. (4), which are not quite the same as each ^mX is replaced by $^{m'}X$, it has been shown that these smaller changes lead to only slight deviations from additivity, which incidentally have been observed just as predicted.[7,18,34]

(d) The dynamic factor can make an important contribution to the differences in isotope effect in analogous molecules, such as in the comparison of $^1\Delta^{13}C(^{18/16}O)$ and $^1\Delta^{13}C(^{34/32}S)$. The former is -0.019 per ^{18}O in CO_2 [31] and the latter is -0.009 per ^{34}S in CS_2 [35] a ratio of 2.1. The dynamic factors in CO_2 and CS_2 have been calculated

$$\frac{\left[\langle\Delta r\rangle_{^{13}C\,^{16}O} - \langle\Delta r\rangle_{^{13}C\,^{18}O}\right]}{\left[\langle\Delta r\rangle_{^{13}C\,^{32}S} - \langle\Delta r\rangle_{^{13}C\,^{34}S}\right]} = 3.15.$$

Thus, a large part of the observed difference between the isotope shifts $^1\Delta^{13}C(^{18/16}O)$ and $^1\Delta^{13}C(^{34/32}S)$ is in the dynamic factors.

(2) *Mean bond displacement depends on* r_e

It has been shown that for ground state diatomic molecules the vibrational part of $\langle\Delta r\rangle$ can be described fairly well for closed or open shell molecules, for all types of bonds, electron deficient, single, double, triple bonds, between any two elements in the periodic table, by the relationship [27]

$$\langle\Delta r\rangle \approx \left(\frac{3h}{8\pi}\right) \mu^{-1/2}\, 10^{-D} \tag{14}$$

where $D \equiv \dfrac{r_e - a_3}{b_3} - \dfrac{3(r_e - a_2)}{2b_2}$ \hfill (15)

in which a_2, b_2, a_3, b_3 are the empirical parameters of Herschbach and Laurie characterizing any two rows of the periodic table.[36] The factor of fundamental constants $(3h/8\pi)$ is 19.35×10^{-3}. Similarly, in the same analysis

$$\langle(\Delta r)^2\rangle \approx \left(\frac{h}{4\pi}\right) \mu^{-1/2}\, 10^{+d} \tag{16}$$

where $d \equiv \dfrac{r_e - a_2}{2b_2}$. \hfill (17)

For a given bonded pair of atoms, $\langle\Delta r\rangle$ increases with increasing r_e. These are zero-point vibrational averages. The temperature dependence of $\langle \Delta r \rangle$ and $\langle (\Delta r)^2 \rangle$ are discussed elsewhere.[37] It has been shown that these equations can be used as well to estimate the dynamic parts for polyatomic molecules, except that the factor 19.35×10^{-3} is replaced by 22×10^{-3}. In comparing the dynamic parts for different bonding situations, it may be noted that the explicit r_e dependence in eq. (14-17) may play a role as r_e in turn depends on bond order, axial vs equatorial positions, and so on. However, as we shall see in the following sections, this is not as significant as the accompanying changes in the electronic factor.

The magnitude of the shift is a function of the observed nucleus and reflects its chemical shift range. Only a small part of this is due to the dynamic factor. For D-substitution in ^{13}C-H, ^{51}V-H, and ^{93}Nb-H bonds for example, the factor $[1-(\mu/\mu^*)^{1/2}]\langle \Delta r \rangle$ is 5.65, 5.54, and 5.05 respectively for bond lengths r_e=1.0871Å, 1.54Å, and 1.66Å, respectively. On the other hand the observed D-induced shifts in Me_3CH, $CpVH(CO)_3$, and $CpNbH(CO)_3$ are -0.472 [38] , -4.7 [16] , and -6 ppm [39]. The dynamic factors are in the opposite order compared to the observed isotope shifts, which confirms that the electronic factors are overwhelmingly larger for ^{93}Nb and ^{51}V compared to ^{13}C. This seems to indicate that the electronic factor in each case depends to a great extent on the general sensitivity of the nuclear magnetic shielding of each nucleus to changes in electronic environment, which sensitivity is indicated by its chemical shift range, which are roughly 680, 6000, and 5000 ppm respectively for ^{13}C, ^{51}V, and ^{93}Nb.

(D) Further trends in one-bond isotope shifts

We have noted the general trends in Sect. IIA. There are further correlations with indices of the chemical bond which have been observed in favorable cases to apply to related molecules. With other quantities more or less being equal, these correlations reveal important characteristics of the electronic factor, which is after all the chemical part of isotope shifts. Our ability to calculate the dynamic factor or at least determine a semi-quantitative estimate, allows us to extract from the measured isotope shift that part which reveals the sensitivity of the nuclear shielding to minor changes in the molecular geometry.

(1) Correlation with absolute shielding

In closely related compounds the least shielded nucleus exhibits the largest isotope shift. Of course this trend only becomes apparent when the dynamic terms are very similar. Examples are $^1\Delta^{13}C(^{18/16}O)$ in acetophenones [40], $^1\Delta^{19}F(^{13/12}C)$ in fluoromethanes [41], $^1\Delta^{13}C(^{2/1}H)$ in 1-D, 4-X- substituted benzenes.[42] This behavior is also observed for $^7\Delta F(^{2/1}H)$ in 4-fluorophenyl systems in FIGURE 2,[43] and two-bond isotope shifts $^2\Delta^{13}C(^{2/1}H)$ in H_3C ^{13}C R=X.[44] The last is shown in FIGURE 5.

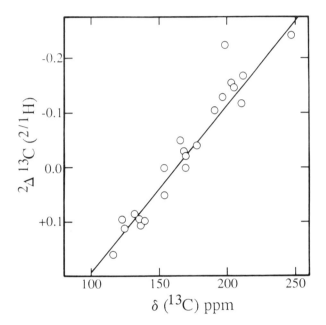

Fig. 5. The 2-bond deuterium isotope effects on a trigonal carbon vs. the ^{13}C chemical shift of the latter. Values plotted are $^2\Delta^{13}C(^{2/1}H)$ for replacement of CH_3 by CD_3, from C.H. Arrowsmith and A.J. Kresge, *J. Am. Chem. Soc.*, 108 (1986) 7918.[44] The largest negative isotope shifts are for the least shielded carbons.

This trend is consistent with the correlation discovered in studies of temperature-dependent chemical shifts, that the derivative of nuclear shielding with respect to bond extension is largely dependent on the paramagnetic term [28,45]. The larger the paramagnetic term the less shielded the nucleus, the greater the magnitude of the derivative, the greater the temperature dependence. Exactly the same electronic factors appear in both the temperature-dependent chemical shifts and the mass-dependent chemical shifts; only the dynamic factors are different.[2,37,46] Two examples are shown in FIGURE 6.

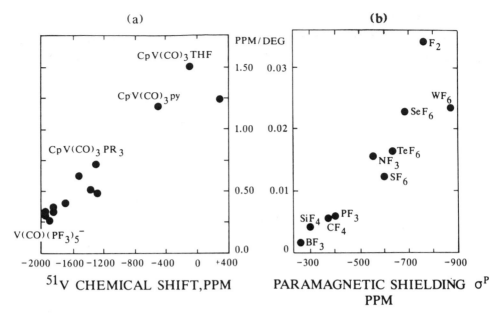

Fig. 6. Temperature coefficients of chemical shifts correlate with paramagnetic shielding (a) for ^{51}V in vanadium carbonyl complexes, and (b) for ^{19}F in binary fluorides, from C.J. Jameson, D. Rehder, and M. Hoch, *J. Am. Chem. Soc.*, 109 (1987) 2589,[47] and C.J. Jameson, A.K. Jameson and D. Oppusunggu, *J. Chem. Phys.*, 85 (1986) 5480.[28] The largest temperature coefficients are for the least shielded nuclei.

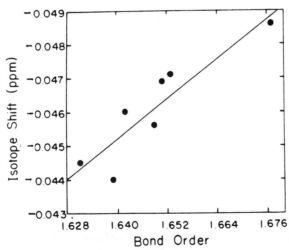

Fig. 7. The ^{18}O-induced ^{13}C shifts, $^{1}\Delta^{13}C(^{18/16}O)$ of carbonyl carbons in para-substituted acetophenones as a function of the carbonyl bond order, reproduced by permission of John Wiley & Sons, Ltd. from J. M. Risley, S. A. Defrees, and R. L. van Etten, *Org. Magn. Reson.*, 21 (1983) 28,[50]

(2) Bond order

The dependence of isotopic shifts on bond order was first noted in ^{18}O- induced ^{31}P shifts.[48] The nominal bond orders are shown to be in a monotonic correlation with $^1\Delta$.[49] Risley et al [50] found that the isotope shifts $^1\Delta C(^{18/16}O) = -0.0440$ to -0.04856 and the calculated bond orders in acetophenones give a linear plot shown in FIGURE 7.

The correlations indicate that we should expect to find theoretical derivatives of nuclear shielding to be greater upon stretch of a triple bond than upon stretch of a double bond. Indeed there are examples which indicate just this:

$$\text{empirical } (\partial\sigma^C/\partial r_{CO})_e \text{ ppm Å}^{-1} \text{ [31]}$$

CO	-456±15
O=C=O	-214±17

$$\text{theoretical } (\partial\sigma^C/\partial r_{CO})_e \text{ [51]}$$

CO	-573.9
$H_2C=O$	-406.6

Is there an underlying physical reason for this ? This is a very interesting question that can perhaps be answered by theoretical calculations such as by the IGLO method, in which the individual MO contributions to the shielding derivative can be examined. Unfortunately it has been found that theoretical calculations on nuclei in- volved in multiple bonds give less satisfactory agreement with experimental shield- ing. Answers to this question may require calculations including electron correlation.

(3) Effects of the net charge

It has been observed by Wasylishen and coworkers [7,53], that the magnitude of the isotope shift decreases with a net positive charge on the molecule and increases with a net negative charge on the molecule. An example is shown below [53]

	$^1\Delta^{119}Sn(^{2/1}H)$ ppm per D	$(\partial\sigma/\partial r)_e$ ppm Å^{-1}
^{119}Sn in SnH_4	-0.403	-92
^{119}Sn in $SnH_3{}^+$	-0.05 ± 0.03	-10

There are no lone pairs involved in the above. The empirical electronic factors shown above were deduced by Leighton and Wasylishen (using eq. (1) and the methods of es- timation of $\langle\Delta r\rangle$ described in the preceding sections). The trend is that the derivative increases algebraically with algebraically increasing net charge. This was also found in the theoretical ab initio studies by Chesnut and Foley.[22] Examples are shown below.

$$(\partial\sigma/\partial r)_e \text{ ppm } \overset{\circ}{A}^{-1} \text{ }^{22}$$

^{11}B in BH_3	-3.5
^{11}B in BH_4^-	-27.
^{27}Al in AlH_3	+84.2
^{27}Al in AlH_4^-	+11.6

It appears that in a molecule with a net negative charge the nuclear shielding is more drastically affected by a bond stretch. Except for the diamagnetic part, nuclear shielding and local charge density are not directly related. However, a net negative charge on the molecule will mean that, upon bond extension, there will be more electron density to follow or not follow the nucleus as it moves out. A net positive charge will have the opposite effect. The dependence of the shielding derivative on the net charge is clear in the comparison of theoretical and empirical derivatives in HCN versus CN⁻ :

$$(\partial\sigma/\partial r)_e \text{ ppm } \overset{\circ}{A}^{-1}$$

^{13}C in HCN	-263	theor,[51]	
^{13}C in CN⁻	-538.7	theor,[51]	-473 emp.[54]
^{15}N in HCN	-675.4	theor,[51]	
^{15}N in CN⁻	-892.2	theor,[51]	-872 emp.[54]

There are other data which could be used as examples to illustrate the above trends of isotope shifts and net charge. However, they also involve the presence or absence of lone pairs on the resonant nucleus, which we now consider.

(4) Lone pair effects

It had been predicted that the presence of a lone pair gives a larger negative isotope shift.[55] This was later demonstrated experimentally by Wasylishen et al. in the comparisons of related systems in TABLE 1. One possible explanation is that the observed differences are partly due to dynamic factors, i. e., there may be significant $\langle\Delta\alpha\rangle$ terms in the bent and pyramidal molecules where the resonant nucleus bears a lone pair or two. Full calculations on H_2O molecule show that the $^1\Delta^{17}O(^{2/1}H)$ is largely due to the $\langle\Delta r\rangle$ term although there are smaller other terms.[19] Thus, the differences between molecules with and without lone pairs likely comes from the electronic factor being a larger negative number when lone pairs are present. These are indeed found to be the case, as shown in TABLE 2.

TABLE 1

One-bond isotope shifts in small molecules with and without lone pairs

		number of lone pairs	Average $^1\Delta$ ppm per D	Ref
^{15}N	in NH_3	1	- 0.623	34
	NH_4^+	0	- 0.293	34
^{31}P	in PH_2^-	2	- 0.2760	7
	PH_3	1	- 0.846	7
	PH_4^+	0	very small	7
^{119}Sn	in SnH_3^-	1	- 3.281	56
	SnH_4	0	- 0.403	53
	SnH_3^+	0	- 0.05 ± 0.03	53
^{15}N	in NO_2^-	1	- 0.138	33,57
	NO_3^-	0	- 0.056	33
^{17}O	in H_2O	2	- 3.090	34
	H_3O^+	1	- 0.3	58

TABLE 2

The electronic factors $\dfrac{(\partial\sigma/\partial r)_e}{ppm\ \overset{o}{A}^{-1}}$

		Ab initio	theoretical	From expt
N in	NH_3	- 130.3 [22]	-144 (-70 from LP) [24]	-124 [34]
	NH_4^+	- 67.9 [22]		-60 [34], -65 [27]
P in	PH_2^-			-585 [7]
	PH_3	- 150.8 [22]	- 154 (-47 from LP) [24]	-180 [7]
	PH_4^+	- 52.9 [22]		very small [7]
Sn in	SnH_3^-			- 750 [56]
	SnH_4			- 92 [53]
	SnH_3^+			- 10 [53]

Theoretical calculations too show that the change in shielding with bond extension is greater for the systems with one or more lone pairs. In comparing NH_3 with NH_4^+ there is a factor of roughly 2 in the electronic factors (-130.3 vs -67.9 ppm $\overset{o}{A}^{-1}$) and also in the observed isotope shifts (-0.623 ppm vs. -0.293 ppm).

In particular, calculations by Fleischer and Kutzelnigg [24] for NH_3 and PH_3

provide separately the lone pair contributions (shown in parentheses in the second column of numbers in TABLE 2). Roughly 1/2 and 1/3 of the change in shielding upon bond extension in these molecules come from lone pair molecular orbitals. It would thus be predicted that with a lone pair centered on the resonant nucleus, larger isotope shifts may be expected, all other things being roughly similar. The nitrogen in NO_2^- has a lone pair that NO_3^- does not have, so $^1\Delta N(^{18/16}O)$ should be larger for NO_2^- than for NO_3^-, and it is (-0.138 compared to -0.056 ppm per ^{18}O).[33,57] Similarly for aniline, $^1\Delta N(^{2/1}H)$ = -0.714 ppm per D (neat)[34] while for anilinium + ion, $^1\Delta N(^{2/1}H)$ = -0.40 ppm per D.[59] In the above examples and those in TABLE 1 and 2 the lone pair and the net charge effects cannot be separately considered; those effects are in the same direction, exaggerating the observed differences.

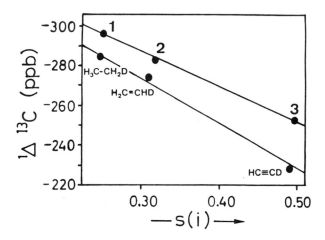

Fig. 8. Correlation between $^1\Delta^{13}C(^{2/1}H)$ and $^1J(CD)$ in H_3C-CH_2D, H_2C=CHD, and HC≡CD and also for Ph-CH_2-CH_2D (1), E isomer of PhHC=CHD (2), and PhC≡CD (3). $^1J(CD)$ are plotted in terms of the fractional s characters s(i) derived directly from the observed $^1J(CD)$. Reproduced with permission from J.R. Wesener, D. Moskau, and H. Gunther, *J. Am. Chem. Soc.*, 107 (1985) 7307.[38] Copyright (1985) American Chemical Society.

(5) Correlation with spin-spin coupling

Since the one-bond coupling constant is a purely electronic quantity, it must be the electronic factors in the one-bond isotope shifts which allow linear correlations with the coupling constant to become apparent. This was most clearly observed for $^1\Delta^{31}P(^{18/16}O)$ vs. $^1J(P^{17}O)$[60], for $^1\Delta^{19}F(^{13/12}C)$ vs. $^1J(CF)$ in fluoromethanes[41], and for $^1\Delta^{119}Sn(^{13/12}C)$ vs. $^1J(SnC)$ in $R_{4-n}Sn(C≡CH)_n$ (R=Me, Et, n=0-4)[61] and similarly for

$Me_{4-n}Sn(CH=CH_2)_n$.[61] In each case the magnitude of the isotope shifts increase with increasing magnitude of the coupling constant. Wesener et al. expressed the correlation of the isotope shift with the hybridization of the resonant carbon derived directly from the observed $^1J(CH)$ [38] in FIGURE 8.

   (6) Increments due to substituents

An interesting incremental effect observed in isotope shifts can be attributed to the electronic factor. The successive replacement of H by CH_3 or by Ph substituents at the observed nucleus leads to incremental changes in the isotope shift.[38] The effect of phenyl substitution is dramatic. Examples are shown below.

		$^1\Delta$ ppm per D	Ref
^{15}N	in NH_3	- 0.623	34
	in $PhNH_2$	- 0.715	34
^{31}P	in PH_3	- 0.846	13
	in $PhPH_2$	- 1.21	62
^{13}C	in CH_4	- 0.192	63
	in $PhCH_3$	- 0.28	64
^{77}Se	in SeH_2	-7.02	15
	in $PhSeH$	-7.96	65

Successive substitution shows the increments are nearly additive. For example, (to avoid confusion the replaced ^1H is written as a D in the following)

		$^1\Delta^{13}C(^{2/1}H)$ ppm per D	Ref
^{13}C	in CH_3D	- 0.187	38,63
	in $PhCH_2D$	- 0.2755	38
	in $(Ph)_2CHD$	- 0.342	38
	in $(Ph)_3CD$	- 0.4377	38

that is, an increment of about -0.081 ppm per Ph, more clearly shown in FIGURE 9. A somewhat larger increment is observed for CH_3 (about -0.095 ppm per CH_3) as shown below:

		$^1\Delta^{13}C(^{2/1}H)$ ppm per D
^{13}C	in CH_3D	- 0.187
	in H_3CCH_2D	- 0.284
	in $(H_3C)_2CHD$	- 0.3759
	in $(H_3C)_3CD$	- 0.4722

There is a factor of 2.5 in the magnitude of the isotope shift in $(CH_3)_3CD$ compared to CH_3D. There are no data on Bu_3CD compared to H_3CD but the $^1\Delta^{13}C(^{2/1}H)$ in BuH_2CD is -0.301 [38] , an increment of -0.114 ppm per Bu. If Bu substitution does follow an incremental pattern as CH_3 substitution does, Bu_3CD would have an isotope shift of -0.529 ppm, i.e., a factor of 2.8 in going from H_3CD to Bu_3CD. The large increase of $^1\Delta^{119}Sn(^{2/1}H)$ in going from H_3SnD (-0.463 ppm [53]) to Bu_3SnD (-1.62 ppm [66]), a factor of 4.0, is therefore not unexpected.

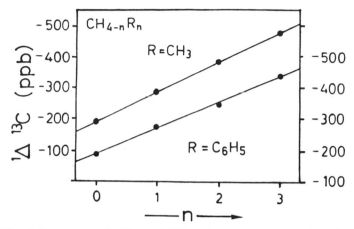

Fig. 9. Incremental effects of CH_3 substitution and of phenyl substitution on $^1\Delta^{13}C(^{2/1}H)$, reproduced with permission from J.R. Wesener, D. Moskau, and H. Günther, *J. Am. Chem. Soc.*, 107 (1985) 7307.[38] Copyright (1985) American Chemical Society.

It can be supposed that part of these observed increments may be due to incremental changes in [$\langle\Delta r_{CH}\rangle$ - $\langle\Delta r_{CD}\rangle$] upon substitution at C. It has been found that $\langle\Delta r\rangle$ is related to bond length in a general way as shown in eq. (14)-(15) in Sect. IIC2, which provides a general relationship for fairly broad comparisons. Although we do not expect very accurate predictions from this, the form of the relationship leads to larger Δr for longer r_e. Therefore, electronic factors being equal, a larger dynamic term for a longer r_e would lead to a larger isotope shift for longer bonds. It is well known that substitutions have an incremental effect on r_e of CH [67] and therefore also on [$\langle\Delta r_{CH}\rangle$ - $\langle\Delta r_{CD}\rangle$]. For example, from Table 2 of Ref 67 we have

H_3C-H	$r_{C-H} = 1.081Å$
MeH_2C-H	1.083
Me_2HC-H	1.085
Me_3C-H	1.087

These small increments ($\sim 2 \times 10^{-3}Å$) in r_e correspond to even smaller increments in $\langle\Delta r\rangle$ which are much too small ($\sim 2 \times 10^{-5}$ Å) to account for the observed increments in isotope shifts, even though the changes are in the right direction. Therefore, the electronic factor clearly must be responsible for these increments. Once again, in view of the well-known incremental substituent effects on nuclear shielding, it is not surprising that there appear to be incremental substituent effects on shielding derivatives.

(7) Bond length

An interesting trend is the effect of chain prolongation in the isotope shift of a terminal $^{13}CH_2D$ as shown in FIGURE 10a. This is entirely parallel to the change in r_e for the terminal CH, as seen in FIGURE 10b, the sharp rise in going from methane to ethane, the flattening out at pentane and longer, even including the slight dip at butane. These values of r_e are the ab initio calculated bond lengths which have been found to have a clearly established excellent correlation with the reported isolated C-H stretching frequencies.[67,69] The latter were found to have excellent correlation with spectroscopically determined r_0 values for a wide range of organic compounds: [67]

$$r_0^{expt}(CH)/\text{Å} = 1.3982 - 1.023\text{x}10^{-4}(v_{CH}^{iso}/cm^{-1})$$

in which differences in length of 0.0005Å could be distinguished. The correlation with calculated r_e is [69]

$$r_e^{calc}(CH)/\text{Å} = 1.2719 - 0.639\text{x}10^{-4}(v_{CH}^{iso}/cm^{-1})$$

in which differences in length of 0.0001Å could be distinguished. The measured r_e are known to be longer by ~ 0.01Å than calculated by ab initio SCF methods, and the slope is known to be larger than calculated by nearly a factor of two. Snyder et al.[69] have found a simple relation between the above and the local structure in the immediate vicinity of the C-H bond, such that the isolated C-H frequency or the C-H bond length is not significantly influenced by structure beyond the next nearest neighbors of the carbon of a given Ċ-Ḣ bond, and can be predicted from simple additivity relations that reflect local structure, i.e.,can be predicted by how many atoms (H or C) are trans or gauche to the Ḣ across the Ċ-C bond.[69] Here again, the r_e(C-H) differences are very small, $3.2\text{x}10^{-3}$Å over-all, so the slight changes in $\langle \Delta r \rangle$ with chain length can not possibly account for the observed increments in isotope shifts. Therefore, the electronic factor must be changing precisely as shown in FIGURE 10a or 10b. The parallel behavior of r_e and $^1\Delta$ with increasing chain length almost surely indicates an electronic factor that reflects the r_e behavior, since r_e is after all a purely

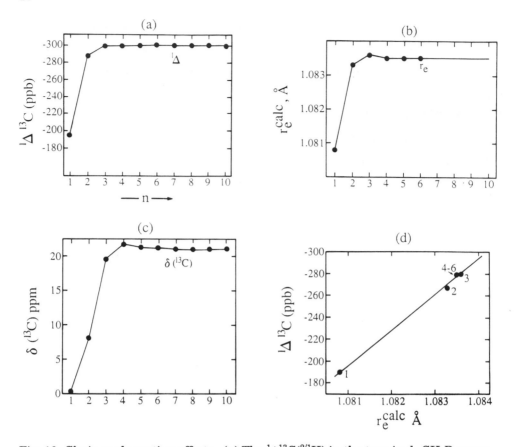

Fig. 10. Chain prolongation effects. (a) The $^1\Delta^{13}C(^{2/1}H)$ in the terminal -CH_2D group varies with number of carbons in the normal alkanes, with permission from J.R. Wesener, D. Moskau, and H. Günther, *J. Am. Chem. Soc.*, 107 (1985) 7307.[38] (b) r_e in the terminal -CH_3 group varies with number of carbons in the normal alkanes. (c) ^{13}C chemical shift of the terminal-CH_3 group. Data taken from Ref. 68. (d) The one-bond isotope shift for the terminal -CH_3 in the normal alkanes correlate with the r_e (CH).

electronic phenomenon. There are smaller known contributions to $^1\Delta^{13}C(^{2/1}H)$ in CH_4 from second order and second derivative terms.[18,20] It is just simpler to think of such an electronic factor as $(\partial\sigma^C/\partial r_{CH})_e$ rather than some composite including second-order and second derivative contributions. These molecule-specific smaller contributions may change in going from CH_4 to n-decane, but it is difficult to see how these contributions could be responsible for the precise similarities in shape of FIGURE 10a and 10b. Finally a comparison of FIGURE 10a, 10b and 10c gives convincing proof

that the electronic factors are responsible for the changes in FIGURE 10a. The observations in FIGURE 10c have been described in terms of increments due to "α-, β-,
and γ- effects" of methyl substitution on ^{13}C- chemical shifts in alkanes, where γ- effects lead to shielding and α-, β- effects lead to deshielding.[68,70] Comparison of
FIGURE 10c with 10a offers convincing evidence that the well-known incremental α-,
β-, γ- effects which apply to the ^{13}C shielding also apply to its derivatives with respect
to bond extension. A plot of $^1\Delta$ vs r_e from FIGURE 10a and 10b is shown in FIGURE
10d. This is probably the simplest relation between $^1\Delta$ and bond length. In these
homologous compounds there are no accompanying changes in hybridization or bond
order which could be modifying both the bond length and $^1\Delta$. What we find is that
the magnitude of the isotope shift increases with increasing bond length. In
cyclohexane the comparison of $^1\Delta^{13}C(^{2/1}H_{ax}) = -0.445$ ppm and $^1\Delta^{13}C(^{2/1}H_{eq}) = -0.395$
ppm [38] is consistent with the bond length for the axial CH being longer than the
equatorial by 0.0017Å.

In contrast, the dependence of one-bond isotope shifts on bond length which
have been noted in the literature have been in the opposite direction to that just discussed. That is, shorter bond lengths have been associated with larger isotope shifts.
FIGURE 11 shows the linear dependence of $^1\Delta F(^{34/32}S)$ on the SF bond length in
a wide variety of compounds.[71] Further examples are the $^1\Delta^{31}P(^{18/16}O)$ in
oxyphosphoranes [32,72], and $^1\Delta P(^{15/14}N)$ in phosphorinanes [73], and $^1\Delta Se(^{13/12}C)$ in a
variety of organoselenium compounds.[74] In each of these examples the electronic environments are drastically different for the compounds being compared, in particular,
the bond orders are different. What is apparently an increase of isotope shift with
decreasing bond length is actually more appropriately attributed to an increase in
isotope shift with increasing bond order. Again, this can not be interpreted primarily
in terms of the dynamic factors. First of all, as already discussed, the dependence of
the dynamic factor on the bond length is in the opposite direction. Secondly, it is far
too small to account for the observed changes in the isotope shift in any of these systems. Therefore, we can only attribute these correlations to changes in the electronic
factor accompanying these bond order changes which incidentally are also reflected
by bond length changes. The dependence of the shielding derivative on bond order
has already been discussed in Sect. II D2 for cases in which the bond order is fairly

well defined and it was found that theoretical carbon shielding derivatives with respect to C-O stretch were indeed larger for CO than CO_2, for CO than H_2CO. Therefore, there is theoretical support for the concluson that the correlations such as shown in FIGURE 11 provide relations between two purely electronic quantities: $(\partial\sigma/\partial r)_e$ and r_e in these molecules.

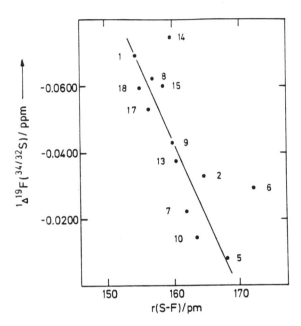

Fig. 11. Correlation of the [34]S-induced ^{19}F shifts with the S-F bond length in a wide variety of compounds such as FSSF (10), SF_4(eq.) (1), and SO_2ClF (18), reproduced with permission from W. Gombler, *Z. Naturforsch.* B, 40 (1985) 782.[71]

(E) Long-range isotope shifts

These are generally smaller than one-bond shifts; nevertheless a large number of them have been observed, some over a distance of seven bonds. The question is, what information is contained in these long range shifts? What we propose to show here is that these long-range shifts are a property of the electronic transmission path from the site of substitution up to the resonant nucleus.

(1) Dominance of the secondary derivative

When the observed isotope shift is over a distance of 2 or more bonds then the magnitudes can only be explained if terms beyond the primary ones are included in the discussion [75]. For one-bond shifts, both the electronic and the dynamic factors are the primary terms: the change in the mean bond length when the mass of one of the atoms in the bond is changed, and the response of the nuclear shielding to a stretch of a bond directed to the resonant nucleus. Since 2-bond shifts involve a remote bond (i.e., an atom is substituted at a bond remote from the resonant nucleus), then secondary factors have to be considered, that is, the response of the shielding to a stretch of a remote bond, or else the change in the mean bond length when the mass of an atom in a remote bond is changed.

For example, for the resonant A nucleus in A-Y-X, the primary dynamic factor is of course the change (call it Δ) in the Y-X bond upon substitution of ^mX by $^{m'}X$. The primary electronic factor is $(\partial\sigma^A/\partial r_{AY})_e$. The secondary electronic factor is $(\partial\sigma^A/\partial r_{YX})_e$ and the secondary dynamic factor is the much smaller change (call it δ) in the A-Y bond upon substitution of ^mX by $^{m'}X$.

The secondary dynamic factor can be thought of in terms of interactions between local vibrational modes. Take as local modes the Y-X stretch and the A-Y stretch. The small change δ in the bond length A-Y upon isotopic substitution of X is dependent on the coupling between the Y-X and the A-Y stretches. In short, the leading terms in a 2-bond isotope shift can be written in the form

$$(\partial\sigma^A/\partial r_{AY})_e \, \delta + (\partial\sigma^A/\partial r_{YX})_e \, \Delta + \ldots \tag{18}$$

where
$$\delta \equiv \langle r_{AY}\rangle_{AY^{m'}X} - \langle r_{AY}\rangle_{AY^mX} \tag{19}$$

$$\Delta \equiv \langle r_{YX}\rangle_{AY^{m'}X} - \langle r_{YX}\rangle_{AY^mX} \tag{20}$$

For example, the ^{17}O isotope shift due to ^{18}O substitution in CO_2 is the ^{17}O shielding difference between $^{17}O_1-^{12}C-^{16}O_2$ and $^{17}O_1-^{12}C-^{18}O_2$, which can be written approximately as

$$\langle\sigma^O - \langle\sigma^O\rangle\rangle^* \approx \left(\frac{\partial\sigma^{O_1}}{\partial r_{CO_1}}\right)_e \delta + \left(\frac{\partial\sigma^{O_1}}{\partial r_{CO_2}}\right)_e \Delta \tag{21}$$

Note that $(\partial\sigma^{O_1}/\partial r_{CO_2})_e$ is a measure of the change in the ^{17}O shielding upon a change in the length of the remote CO bond, a secondary electronic factor which depends on the transmission of the electronic information along the O-C-O path. Δ is of course the $\langle \Delta r \rangle$ - $\langle \Delta r \rangle^*$ that we have been able to easily estimate as discussed in Sect. IIC, whereas δ is a much smaller long-range dynamic factor, a molecule-specific quantity dependent on the over-all potential energy surface of the molecule. There are indications that even in long-range isotope shifts the terms of the order of Δ are dominant compared to the terms in δ.

We have stated that the signs and magnitudes of isotope shifts across 2 or more bonds reflect the signs and magnitudes of the secondary derivatives.[75,76] What is the evidence for secondary derivatives being important for isotope shifts over 2 bonds and longer paths? The observed additivity of long-range isotope shifts is completely consistent with this. Isotope shifts from equivalent substitution sites such as a remote CH_3 group involve the same secondary derivative and one Δ term for each C-D replacing a C-H. There are other clues. One is that the mass dependence of the 2-bond isotope shift is the same as the 1-bond. Vibrational calculations showed [75] that

$$\frac{\langle \Delta r_{CH} \rangle_{CH_4} - \langle \Delta r_{CT} \rangle_{CT_4}}{\langle \Delta r_{CH} \rangle_{CH_4} - \langle \Delta r_{CD} \rangle_{CD_4}} = 1.426.$$

In the one-bond isotope shift, we have seen in Sect. IIB that the leading term is

$$\left(\frac{\partial \sigma}{\partial r} \right)_e [\, \langle \Delta r_{CH} \rangle - \langle \Delta r_{CD} \rangle \,] \quad \text{for deuterium substitution and}$$

$$\left(\frac{\partial \sigma}{\partial r} \right)_e [\, \langle \Delta r_{CH} \rangle - \langle \Delta r_{CT} \rangle \,] \quad \text{for tritium substitution.}$$

The observed ratio of tritium and deuterium one-bond isotope shifts in $^{13}CH_3C(O)CH_3$ is [77]

$$\frac{^1\Delta^{13}C\,(^{3/1}H)}{^2\Delta^{13}C\,(^{2/1}H)} = 1.424 \pm 0.025$$

That this ratio is indistinguishable from 1.426 serves to reassure us that the leading term is dominant in one-bond shifts. The two-bond isotope shifts have also been observed in $CH_3{}^{13}C(O)CH_3$, and these have the ratio

$$\frac{^2\Delta^{13}C\,(^{3/1}H)}{^2\Delta^{13}C\,(^{2/1}H)} = 1.41 \pm 0.12.$$

This ratio too is indistinguishable from 1.426, which indicates that in the leading terms the dominant contribution is

$$\left(\frac{\partial \sigma^{C(O)}}{\partial r_{CH}} \right)_e [\, \langle \Delta r_{CH} \rangle - \langle \Delta r_{CT} \rangle \,] \quad \text{for tritium substitution and}$$

$$\left(\frac{\partial \sigma^{C(O)}}{\partial r_{CH}}\right)_e \quad [\,\langle \Delta r_{CH} \rangle - \langle \Delta r_{CD} \rangle\,] \quad \text{for deuterium substitution.}$$

The minor changes in the C-C and C=O bond lengths upon deuterium substitution are apparently not important although the electronic factors associated with them may be large.

Another clue is that the long-range isotope shifts correlate with indicators of electronic transmission paths:

(1) dihedral angle dependence of 3-bond shifts

(2) stereospecificity (cis vs. trans vs. gauche) of isotope shifts parallel to that of spin-spin coupling

(3) correlation of long-range isotope shifts with electron-withdrawing / donating ability of substituents, even across a path traversing 7 bonds.

The negative one-bond isotope shift $^1\Delta^{13}C$ ($^{2/1}H$) in $(CH_3)_2C=O$ means that $(\partial \sigma^{C(H)}/\partial r_{CH})_e$ is negative (the usual sign). Incidentally, $(\partial \sigma^{C(O)}/\partial r_{CO})_e$ is also negative, as observed in the $^1\Delta^{13}C(^{18/16}O) = -0.050$ ppm [78] in this molecule. What about the sign of the secondary derivative $(\partial \sigma^{C(O)}/\partial r_{CH})_e$? C-H bond lengthening increases the gross atomic charge at $^{13}C(=O)$ in $(CH_3)_2C=O$ and at $^{13}C\oplus$ in $(CH_3)_2C\oplus$, according to theoretical self-consistent field (STO-3G level) calculations.[79] Since an increase in gross atomic charge usually (though not always) means an increase in nuclear shielding, the $(\partial \sigma^{C(O)}/\partial r_{CH})_e$ is probably positive. Indeed, in both $(CH_3)_2{}^{13}CO$ and $(CH_3)_2{}^{13}C\oplus$ the 2-bond ^{13}C shift upon deuteration is positive, $+ 0.054$ [77,79-81] and $+ 0.133$ ppm per D [82] respectively. Thus, the important term must be

$$(\partial \sigma^{C(O)}/\partial r_{CH})_e \bullet \Delta$$

and not

$$(\partial \sigma^{C(O)}/\partial r_{CO})_e \bullet \delta.$$

This is very encouraging because the secondary dynamic factor δ is so dependent on the entire molecular framework and potential energy surface in which vibrations take place that it is difficult to estimate. On the other hand Δ has a very straightforward dependence on the r_e and masses in the local fragment and can be estimated,[27] as discussed in Sect. IIC2. This means that if the δ terms are unimportant, the sign and magnitude of the long-range isotope shifts are a direct measure of the sign and magnitude of $(\partial \sigma/\partial r_{remote})_e$. This derivative is stereospecific, dependent on electron-withdrawing / donating abilities of substituents, and has the usual electronic-transmission-path-dependence of various observables such as long range spin-spin coupling, substituent effects on chemical shifts, etc.

How well are these ideas supported by calculations of secondary derivatives? Unfortunately the shielding derivatives $(\partial\sigma^C/\partial r_{CH})_e$ across the fragments $H\text{-}C\text{-}^{13}C\oplus$ have not been calculated. On the other hand the following are known: [51]

path	$(\partial\sigma^C/\partial r_{CH})$ / ppm $\overset{\circ}{A}^{-1}$
$H\text{-}C\equiv^{13}C$ in $HC\equiv CH$	+12.5
$H\text{-}C=^{13}C$ in $H_2C=CH_2$	+5.3
$H\text{-}C\text{-}^{13}C$ in $H_3C\text{-}CH_3$	-12.4

and for $H\text{-}C\text{-}^{17}O$ in $H_2C=O$, $(\partial\sigma^O/\partial r_{CH})_e$=+94.0 ppm $\overset{\circ}{A}$.[51] So indeed, both + and - signs are possible for these secondary derivatives.

In $CH_3C(O)X$ too, the observed $^2\Delta^{13}CO(^{2/1}H)$ is positive and the magnitude appears to be related to the electron withdrawing / donating ability of X, as shown in the examples below.

X in $CH_3{}^{13}C(O)X$	$^2\Delta^{13}CO(^{2/1}H)$, ppm per D [83]
CH_3COO	+0.008
CH_3O	+0.010
OH	+0.013
F	+0.014
Ph	+0.039
CH_3	+0.054
H	+0.072

Long-range isotope shifts are usually large enough to be observed only when the electronic transmission pathway involves a pi system.[84] For example, the D-induced ^{19}F isotope shifts in 4-fluorophenyl systems correlate linearly with the calculated changes in pi electron density at the F atom upon a C-H bond shortening, as shown in FIGURE 2. It appears that the electronic transmission across the ring determines the sign of the effectively 7-bond isotope shift, and that $(\partial\sigma^F/\partial r_{CH})_e$ is determined largely by changes in the pi charge density at the F atom, $(\partial q_\pi/\partial r_{CH})_e$. The change in the pi charge densities at F atom upon shortening the C-H bond in the X group in X-⟨O⟩-F calculated by MNDO range from +11.4 x 10^{-5} (electron-donating X) to -15.9 x 10^{-5} (electron-withdrawing X) [10] and correlate with the isotope shift

$^7\Delta^{19}F$ $(^{2/1}H)$ of -0.28 ppm to +0.461 ppm.[10,43] The signs of the observed isotope shifts imply that the long-range derivative $(\partial\sigma^F/\partial r_{CH})_e$ is positive when X is an electron-withdrawing group in $X=\overset{\oplus}{C}\text{-}(CD_3)_2$, and negative when X is an electron-donating group as in $X=\overset{\oplus}{C}\text{-}(CHD_2)(CD_3)$.

Although charge densities do not tell the whole story about nuclear shielding, in these specific examples the shielding changes are mimicked by the changes (upon remote bond extension) of the pi charge densities or the gross atomic charges centered at the resonant nucleus. In FIGURE 2 the signs and relative magnitudes are mimicked well enough that in these cases we may assume that the secondary derivative is directly related to the observed long-range isotope shift.

(2) Stereospecificity

Once we accept the notion that isotope shifts across 2 or more bonds reflect secondary (long-range) electronic factors and primary (local) dynamic factors, then the dependence of 3-bond shifts on dihedral angle is not surprising. For example, there is a quantitative correlation of vicinal isotope effects with dihedral angles ϕ of the bonding pathway $^{13}C\text{-}C\text{-}C\text{-}D$ in the isotopomers of norbornane.[85] A Karplus-type relationship like that for the vicinal spin-spin coupling constant and ϕ fits the data well. This is also observed in exo-1,6 trimethylene norbornan-3-one and the related alcohol,[86] as well as for ^{119}Sn isotope shifts.[66] An explanation which is consistent with our previous analysis (Sect. E1) is that the isotope shift is determined primarily by a shortened average bond length at the D-substitution site combined with a long-range derivative which describes the change in shielding at ^{13}C upon a remote bond stretching. This derivative describes the change in shielding along the same electron-transmission pathway as for vicinal coupling constants. Thus, it is expected that,

as $\quad |\,^3J_{trans,cis}\,| \quad > \quad |\,^3J_{gauche}\,|$

so do $\quad |\,^3\Delta^{13}C(^{2/1}H)_{trans,cis}\,| \quad > \quad |\,^3\Delta^{13}C(^{2/1}H)_{gauche}\,|$

and as $\quad ^3J = a + b\cos\phi + c\cos 2\phi$

so do $\quad ^3\Delta A(^{2/1}H) = a_0 + a_1\cos\phi + a_2\cos 2\phi \quad$ where ϕ is the dihedral angle.

This is indeed found to be the case for $^3\Delta^{13}C$ in adamantane [87], proadamantanes [88], and $^3\Delta^{19}F$ in deuterated acetyl and propionyl fluorides,[83] and for $^3\Delta^{19}F$ in deuterated fluorocyclohexane and bicyclo [2.2.1] heptane [89]. Similarly, there is a general correlation of $^{2,3}\Delta^{19}F(^{2/1}H)$ with $^{2,3}J(HF)$ coupling in fluoroethenes.[76] Unlike ^3J however, a detailed angle dependence for isotope shifts that includes all known examples cannot be formulated using the same set of a_0, a_1, and a_2. Nevertheless, there is a clear angular dependence in each class of related examples. The stereospecificity of long-range isotope shifts confirm the dominance of the long-range electronic factor rather than the long-range dynamic factor, a fortunate occurrence since the former is far more interesting and useful than the latter.

(F) Isotope exchange equilibria

A common means of preparing partly deuterated chemical species in solution is by dissolving a solute in water of known deuterium / hydrogen ratio and waiting until the sample has come to equilibrium with respect to D/H exchange. Examples are shown in FIGURE 3 for solutions with 0.15, 0.50, and 0.85 D isotopic fraction.[17] Depending on the solvent isotope composition (call the deuterium fraction d, the proton fraction $(1-d)$, the relative intensities of the ^{59}Co peaks of the various

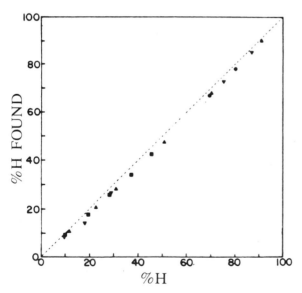

Fig. 12. The proton fraction found from analysis of intensities in Fig. 3 as a function of the D_2O/H_2O solvent composition. The deviations from the dotted line are due to isotope effects on the chemical equilibria involving H/D exchange between the hexammine sites and the water sites. This is a plot of $(1-f)$ on the ordinate and $(1-d)$ on the abcissa. Reproduced with permission from J.G. Russell and R.G. Bryant, Anal. Chim. Acta, 151 (1983) 227.[18]

isotopomers H_{18}, H_pD_{18-p}, through D_{18} will be given strictly by the statistical formula,

$$I = [\, 18! \,/\, p! \,(18-p)! \,] \,(1-d)^p \,d^{18-p}$$

provided that there are no isotope effects on the chemical equilibrium . Indeed, in this particular case the found fraction f of deuterated species slightly differs from d as shown in FIGURE 12. This figure shows explicitly the proton fraction $(1-f)$ expressed in percent, found by fitting intensities of the several peaks in the spectrum of each prepared solvent composition $(1-d)$ expressed in percent. The precision and accuracy of the data are sufficient to clearly indicate that there are isotope effects on the exchange equilibria. In this case, the so-called isotopic fractionation factor for H-D exchange in aqueous solution is larger than 1.0, that is the D prefers to attach to the N in the hexammine complex rather than the O in water.

Isotope effects on chemical equilibria have been studied since 1933 (H.C. Urey).[90-93] NMR spectroscopy has provided quantitative experimental information.[94-96] In this section we will discuss only the use of isotope shifts to determine the isotopic fractionation factors in OH, NH, SH, etc... groups exchanging hydrogen and deuterium with hydroxylic solvents, H_2O/D_2O in particular.[97,98] Many of the systems studied involve multiple equilibria. Nevertheless, each of the many equilibrium constants involved can be expressed in the same form. Let us consider a ^{15}N site (such as aniline) with two exchangeable protons as an example and for the purpose of this illustration, limit the chemical species to the following:

$$RNH_2 + D_2O \overset{K_1}{\rightleftharpoons} RNHD + HDO \tag{22}$$

$$RNHD + HDO \overset{K_2}{\rightleftharpoons} RND_2 + H_2O \tag{23}$$

$$RNHD + D_2O \overset{K_3}{\rightleftharpoons} RND_2 + HDO \tag{24}$$

$$RNH_2 + HDO \overset{K_4}{\rightleftharpoons} RNHD + H_2O \tag{25}$$

$$H_2O + D_2O \overset{K_w}{\rightleftharpoons} 2HDO \tag{26}$$

$$RNH_2 + RND_2 \overset{K_{self}}{\rightleftharpoons} 2RNHD$$

Let f_1 = fraction present as RNHD

f_2 = fraction present as RND_2

f_0 = fraction present as RNH_2

such that $f_1 + f_2 + f_0 = 1$. Ordinarily the individual fractions f_1, f_2, and f_0 can be observed directly from the spectrum as in aniline and aniline derivatives [99]. The spectra appear as in the example of FIGURE 1 for ^{119}Sn in $SnH_{3-n}D_n^-$ when the isotope exchange is slow relative to the NMR time scale, that is, when the rate of exchange is less than the isotope shifts in hertz. The equilibrium constants can then be expressed as:

$$K_1 = \frac{f_1}{f_0}\frac{(HDO)}{(D_2O)} \qquad\qquad K_2 = \frac{f_2}{f_1}\frac{(H_2O)}{(HDO)} \qquad\qquad (27)$$

$$K_3 = \frac{f_2}{f_1}\frac{(HDO)}{(D_2O)} \qquad K_4 = \frac{f_1}{f_0}\frac{(H_2O)}{(HDO)} \qquad K_w = \frac{(HDO)^2}{(H_2O)(D_2O)}$$

K_w is known theoretically as 3.85 at 298 K. the experimental value is about 3% lower.[92] From this K_w the ratios of $(HDO) : (H_2O) : (D_2O)$ can be calculated at each solvent composition d. Even when the isotopic exchange is fast, resulting in only an average resonance signal, it is still possible to obtain the fractionation factor in the method proposed by Jarrett and Saunders [97,98]. The frequency of the average resonance signal of the probe nucleus is measured in a D_2O/H_2O mixture of known isotopic composition $D/H = d/(1-d)$. The shift is measured relative to a separate H_2O solution. The measured chemical shift between the probe nucleus in the H_2O solution and in the separate D_2O solution includes the intrinsic isotope shift and solvent isotope shifts. For example the D-induced intrinsic ^{19}F isotope shift in HF is -2.5±0.5 ppm [100] whereas the measured shift between HF in H_2O and DF in D_2O was around -6 ppm [91]. In the case of $^{19}F^-$ ion in KF solutions or of $^{37}Cl^-$ ion, the entire shift observed is the solvent isotope shift. The experimental quantity f is then obtained as

$$f = \frac{V_{ave} - V_{in\ H_2O}}{V_{in\ D_2O} - V_{in\ H_2O}} \qquad (28)$$

In the particular example of two equivalent exchangeable protons, f can be related to the fractions of the individual isotopomers if the isotope shifts are taken to be <u>strictly additive</u> (which is a very good approximation in most cases) and if the self exchange is strictly statistical. Thus, in the above example,

$$f = \frac{1}{2}f_1 + f_2 \qquad\qquad \text{and } 1 - f = \frac{1}{2}f_1 + f_0 \qquad\qquad (29)$$

Thus, we can write

$$K_1 = \frac{2f}{1-f}\frac{(HDO)}{(D_2O)} \qquad\qquad K_2 = \frac{1}{2}\frac{f}{1-f}\frac{(H_2O)}{(HDO)} \qquad (30)$$
$$K_3 = \frac{1}{2}\frac{f}{1-f}\frac{(HDO)}{(D_2O)} \qquad\qquad K_4 = 2\frac{f}{1-f}\frac{(H_2O)}{(HDO)}$$

It turns out that

$$\frac{(HDO)}{(D_2O)} \approx \frac{2(1-d)}{d}, \qquad \frac{(H_2O)}{(HDO)} \approx \frac{(1-d)}{2d}, \qquad \text{and } \frac{(H_2O)}{(D_2O)} \cong \frac{(1-d)^2}{d^2} \qquad (31)$$

The above ratios of water isotopomers are exactly true only if $K_w = 4$ exactly, but for

K_w = 3.85 they are close enough. In other words, using the proper (HDO)/(D$_2$O), etc., ratios calculated from K_w = 3.85 change the results only minimally and all changes are a small percent of the experimental error. Thus, each of the above equilibrium constants can be written in the following simple form:

$$K \approx (\text{sym. nos.}) \left(\frac{f}{1-f} \right) \left(\frac{1-d}{d} \right) \tag{32}$$

The factors involving the symmetry numbers which accounts for the number of indistinguishable sites (the symmetry number of a molecule is defined as the number of different values of rotational coordinates which correspond to one orientation of the molecule) are respectively 4, 1/4, 1, 1 for K_1 through K_4 respectively (and is 4 for K_w). What this means is that no matter how many equivalent sites are involved, the equilibrium constants are all related to

$$K_{frac} \approx \frac{f}{1-f} \frac{1-d}{d} \tag{33}$$

within the approximations already stated above. The individual equilibrium constants will have different factors containing symmetry numbers, of course, and therefore differ by these factors from one another. The studies by Jarrett and Saunders used a wide variety of probe nuclei including ^{31}P, ^{19}F, ^{37}Cl, ^{14}N, ^{13}C, to witness isotope exchange involving O-H, N-H, S-H, and F-H sites in competition with H$_2$O/HDO/D$_2$O sites. Their tabulated values of frequencies at various values of d have been used to calculate f for the systems shown in FIGURE 13. The best value of K_{frac} is probably that value around $d=0.5$ where the observed chemical shifts are largest. They reported an average value of K_{frac} for each system. The method of analysis presented in FIGURE 13 allows an examination of systematic errors in the assumptions or the experiments. Although the experimental scatter is worse in pyrrolidine, the results all show that this empirical measure of isotopic fractionation, $K_{frac} \approx \frac{f}{1-f} \frac{1-d}{d}$, tends to decrease with increasing deuterium fraction in the solvent, to an extent which depends on the system, whereas a true thermodynamic equilibrium constant does not. Using the properly calculated (HDO)/(D$_2$O) ratios from K_w = 3.85 makes very slight corrections in the right direction but still well inside experimental error. Activity coefficients will depend on isotopic composition of the solvent but that variation is probably not large enough to account for the systematic trend in FIGURE 13. The latter could well be due to systematic experimental errors.

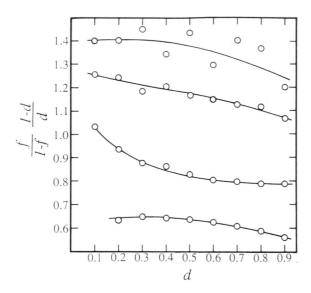

Fig. 13. Isotopic fractionation factor $K_{frac}=[f/(1-f)][(1-d)/d]$ versus d for H/D exchange between the water sites and the N-H site in pyrrolidine, the N-H site in NH_4^+, the O-H site in H_3PO_4, and the S-H in 2-mercaptoethane sulfonic acid Na^+ salt, respectively. Calculated from the experimental data of Ref.97 and 98.

An interesting general result in Jarrett and Saunders's studies which are reflected in these 4 examples (and also, incidentally in Russell and Bryant's example in FIGURE 12) is that N-H sites have $K_{frac} > 1.0$ whereas S-H, and F-H sites have $K_{frac} < 1.0$ in competition with the O-H site in water.

There is a simple theoretical explanation for this. The equilibrium constant for isotopic exchange can be written in terms of the molecular partition functions [90,91]. The potential functions being assumed the same for isotopic molecules (Born-Oppenheimer approximation), the ΔE of the reaction from the zero of energy of reactants to the zero of energy of products is zero in isotopic exchange. The rotational parts are replaced entirely by reciprocals of symmetry numbers and mass factors when the sum over the rotational states is approximated by the classical limit. Since the symmetry numbers have already been separated from K_{frac} all we have to include for a simple exchange such as

$$AH + DB = AD + HB \tag{34}$$

are the following terms:

$$K_{frac} = \frac{v_{AD}}{v_{DB}} \cdot \frac{v_{HB}}{v_{AH}} \cdot e^{(v_{DB}-v_{AD}-v_{AH}-v_{HB})/2kT} \cdot \frac{(1-e^{-v_{AH}/kT})(1-e^{-v_{DB}/kT})}{(1-e^{-v_{AD}/kT})(1-e^{-v_{HB}/kT})} \tag{35}$$

with kT = 208.5 cm^{-1} at 300K. In harmonic vibration the first factor is simply

$$\left(\frac{\nu_{AD}\,\nu_{HB}}{\nu_{DB}\,\nu_{AH}}\right)_{harm} = \left[\frac{(m_H+m_B)\,(m_A+m_D)}{(m_D+m_B)\,(m_A+m_H)}\right]^{1/2} \tag{36}$$

Right away we can see that in competition with an O-H site, A being lighter than oxygen atom (such as in N-H) will give rise to a factor greater than 1, whereas A being heavier than O (such as in F-H, Cl-H, S-H) will give rise to a factor less than 1. Actually the second factor $e^{(\nu_{DB}-\nu_{AD}+\nu_{AH}-\nu_{HB})/2kT}$ is the most important one. As an example we can put in some typical numbers. Using $\nu_{AH} \approx 3495$ cm^{-1}, $\nu_{AD} \approx 2491$ cm^{-1}, $\nu_{HB} \approx 3657$ cm^{-1}, $\nu_{DB} \approx 2672$ cm^{-1}, we find

$K_{frac} \approx 1.05$ for NH.

Using $\nu_{AH} \approx 2615$ cm^{-1} and $\nu_{AD} \approx 1892$ cm^{-1}, we find

$K_{frac} \approx 0.53$ for SH.

Of course the use of the harmonic vibrational partition function is not quite correct. Nevertheless, the general comparison between N-H vs O-H on the one hand and S-H vs O-H on the other hand, is well-reproduced. Force constants are in general smaller for the longer A-H bonds involving heavier A, such as sulfur, leading to $K_{frac} < 1$. In general "the heavy isotope tends to concentrate (relative to the light isotope) in that species where it is more tightly bound " (i.e., largest force constants) [92]. That $K_{frac} \neq$ 1.0 is indicative of a change of force constant between the two hydrogen sites is probably uninteresting in itself since vibrational spectroscopy is a more direct measure of such changes. However, in solution these isolated molecule partition functions are probably inappropriate so that experimental direct measures of K_{frac} such as provided by the NMR method are necessary. For an O-H site competing with the O-H site in water, the results are more interesting, K_{frac} tends to be < 1 when the hydrogen is involved in strong internal hydrogen bonding and K_{frac} increases with increasing hydrogen bond strength relative to water. The important result is that even when resonance signals from the individual isotopomers are not observable, the NMR isotope shift still provides useful information.

III. ISOTOPE EFFECTS ON COUPLING CONSTANTS

These isotope effects are analogous to the effects on nuclear shielding. A review of available data and the rovibrational theory applied to J has been presented by us [101] and will not be repeated here. These isotope effects are of importance to the fundamental understanding of spin-spin coupling constants as a molecular electronic property. The available experimental and theoretical information on J is not as extensive as for nuclear shielding. There are fewer calculations of the isotropic value of J and even fewer calculations of the components of the J tensor. There is meager experimental information on the J tensor components, largely due to the fact that the traceless direct dipolar coupling tensor D which does not contribute to the isotropic average spin-spin coupling, has components which are usually much larger than the components of J. Since the observed components in oriented molecules are those of (D+J), J tensor components have to be obtained by difference. The D tensor for nuclei N and N′ depends only on geometry, goes roughly as $R^{-3}_{NN'}$, and can explicitly be written in terms of the cartesian positions of the coupled nuclei relative to the external magnetic field. Thus, the J tensor can only be obtained if the rovibrationally averaged nuclear positions can be determined independently with sufficient precision. Experiments which are directly related to isotope effects on an electronic property are the measurements of the temperature-dependence of the property in the absence of intermolecular effects.[102] There are far fewer temperature-dependent studies of J and even fewer studies in the gas in the zero-pressure limit. The indirect spin-spin coupling constant J is a useful index of the chemical bond and isotope effects on it provide information about the sensitivity of J to bond extension. Since the isotope effects on the isotropic average J can be determined more precisely than the individual J tensor components or the anisotropy, the isotope effects can be considered as the major source of experimental information for assessing the quality of ab initio theoretical calculations of indirect spin-spin coupling.

The working definitions of the isotope effects on J are: $\Delta_s \, ^nJ(AB)[^{m'}/^mX]$ is the secondary isotope effect on the coupling constant between nucleus A and B due to the substitution of ^mX by the heavier isotope $^{m'}X$ somewhere in the molecule.

$$\Delta_s \, ^nJ(AB) \, [^{m'}/^mX] = | \, ^nJ(AB) \, |^* - | \, ^nJ(AB) \, | \tag{37}$$

$\Delta_p \, ^nJ(A^{2/1}H)$ is the primary isotope effect on the coupling constant between nucleus A and H due to the substitution of H by D, the coupled nuclei being separated by n bonds.

$$\Delta_p \, {}^nJ(A^{2/1}H) = |\, {}^nJ(AD)\, |* \frac{\gamma_H}{\gamma_D} - |\, {}^nJ(AH)\, | \qquad\qquad (38)$$

where $\gamma_H/\gamma_D = 6.514\ 398\ 04\ (120).$[103] $\qquad\qquad\qquad\qquad\qquad\qquad (39)$

Of course if the purely electronic quantities, the reduced coupling constants are compared,

$$^nK(AH) \equiv 4\pi^2 \, {}^nJ(AH)/h\gamma_A\gamma_H \qquad\qquad (40)$$

then the isotope effect is just a difference:

$$\Delta_p \, {}^nK(A^{2/1}H) = |\, {}^nK(AD)\, |* - |\, {}^nK(AH)\, |. \qquad\qquad (41)$$

Only the absolute values are compared, for practical reasons, since the absolute sign of spin-spin coupling constants are not always known. As in isotope shifts the asterisk * denotes the heavier isotopomer. Where both secondary and primary isotope effects are observed, the primary isotope effect should be obtained from $J(SnD)$ in SnH_2D^- and $J(SnH)$ in SnH_3^- for example, rather than $J(SnD)$ and $J(SnH)$ in the same isotopomer SnH_2D^-. The latter difference includes both primary and secondary isotope effects.

Isotope effects on nuclear shielding are usually visually obvious in a high-resolution NMR spectrum since the peaks of the isotopomer are shifted from the parent species. On the other hand, isotope effects on coupling constants are only observed as very slightly different multiplet splittings for each isotopomer (secondary isotope effects) or as a slightly smaller (usually) or larger $J(AD)$ than the expected $J(AH)$ / 6.514 398 04 (primary isotope effects). Because of this factor of 6.514... the isotope effects on J are less precisely determined than the isotope effects on shielding. According to these working definitions a positive isotope effect on J means that the reduced coupling is larger in the deuterated species.

(A) General trends

The general trends and their theoretical explanation have been presented by Jameson and Osten:[101]

(1) The sign of the primary or secondary isotope effect on the coupling constant is not directly related to the absolute sign of the coupling constant. Whether it is related to the absolute sign of the reduced coupling constant is not yet established since one-bond coupling isotope effects are available presently for positive reduced coupling constants only.

(2) Primary isotope effects are negative or positive, the positive signs being

found only in molecules involving one or more lone pairs of the coupled nuclei. For example, primary isotope effects are positive in H_2Se,[15] in PH_2^-,[7] and in PH_3,[13] and in other 3-coordinate phosphorus, but negative in PH_4^+ [7] and in 4- and 5-coordinate phosphorus.[62] It is positive in SnH_3^- (one lone pair) [7] and negative in SnH_4 and SnH_3^+ (no lone pairs) [53].

(3) Secondary isotope effects can have either sign, but are often negative. Positive signs have been observed where triple bonds are involved (as in $HC\equiv CH$ and $HC\equiv N$) but also in some of the same systems with positive primary isotope effects.

(4) Secondary isotope effects are roughly additive upon substitution of several equivalent sites neighboring the coupled nuclei. For example, in NH_4^+ ion, each D substitution decreases $^1J(^{14}NH)$ by 0.05 ± 0.02 Hz,[14] and in PH_3, each D substitution decreases $^1J(PH)$ by 2.5 Hz.[13] Small deviations from additivity have been observed.

(5) The magnitudes of isotope effects are small, the largest primary effect being about 10% of the coupling constant in SnH_3^- and the largest secondary effects being about 3% in SnH_3^-, so that only the effects of deuterium (and tritium) substitution, where the largest fractional changes in mass are involved, have been observed. Some of the larger primary isotope shifts are +11.5 Hz for Δ_p $^1J(PH)$ in PH_3 [13] and +10.5 Hz in SnH_3^-.[7] Secondary isotope shifts as large as -2.0 Hz per D (for $^1J(SiF)$) [104] and +3.0Hz per D (for $^1J(SnH)$) [7] have been observed.

(6) The magnitudes of the isotope effects are roughly proportional to the fractional change in mass, in the very few instances where effects of isotopic substitution of ^1H by ^2H and ^3H have been reported.

(B) Theory and selected examples

Just as in isotope shifts, the isotope effect on J can be written in terms of products of electronic and dynamic factors as follows. For a diatomic molecule,

$$\langle J \rangle^* - \langle J \rangle = (\partial J/\partial r)_e[\langle \Delta r \rangle^* - \langle \Delta r \rangle] + \frac{1}{2}(\partial^2 J/\partial r^2)_e[\langle (\Delta r)^2 \rangle^* - \langle (\Delta r)^2 \rangle] +... \qquad (42)$$

The dynamic factors are the same as we have already discussed, (and indeed the same for all rovibrational averaging of any molecular electronic properties).[105] Only the electronic factors need to be discussed here. Of course, γ of some nuclei are negative so we really should compare derivatives of the reduced coupling $\left(\dfrac{\partial K}{\partial r}\right)_e$ etc..., where

$$\left(\frac{\partial J}{\partial r}\right)_e = \frac{h\gamma_N\,\gamma_{N'}}{4\pi^2}\left(\frac{\partial K}{\partial r}\right)_e \qquad (43)$$

One important difference between J and σ is that the latter is a one-center property whereas J is a two-center property. In polyatomic molecules there are usually several electronic factors of comparable size which contribute to the isotope effects on J.

For polyatomic molecules, if ^1J is most sensitive to the bond length, the one-bond shifts can also be written in the same way as for diatomic molecules, the terms such as $(\partial J/\partial\alpha)_e \bullet [\langle\Delta\alpha\rangle^* - \langle\Delta\alpha\rangle] + ...$ being less important. Let us consider SnH_4 as an example of a polyatomic molecule. If we consider only the terms in the first derivatives as the leading terms in the spin-spin coupling then,

$$\langle K\rangle_{bond\,1} \approx K_e + \left(\frac{\partial K}{\partial r}\right)_e \Delta r_1 + \left(\frac{\partial K}{\partial r'}\right)_e (\Delta r_2 + \Delta r_3 + \Delta r_4) + ... \qquad (44)$$

If we further neglect the secondary effects on an Sn-H bond length due to substitution of a remote H by D, then we may represent $\langle\Delta r_{SnH}\rangle$ in all the isotopomers by d and all $\langle\Delta r_{SnD}\rangle$ by (d-Δ) since the average length of the Sn-D bond is shorter than the SnH bond. Then we may write the reduced coupling constants in the following simple forms:

SnH_4 :

$$K(SnH) \approx K_e + \left[\left(\frac{\partial K}{\partial r}\right)_e + 3\left(\frac{\partial K}{\partial r'}\right)_e\right] d + ... \qquad (45)$$

SnH_3D :

$$K(SnH) \approx K_e + \left(\frac{\partial K}{\partial r}\right)_e d + \left(\frac{\partial K}{\partial r'}\right)_e [2d + d - \Delta] + ...$$

$$K(SnD) \approx K_e + \left(\frac{\partial K}{\partial r}\right)_e (d-\Delta) + 3\left(\frac{\partial K}{\partial r'}\right)_e d + ...$$

SnH_2D_2 :

$$K(SnH) \approx K_e + \left(\frac{\partial K}{\partial r}\right)_e d + \left(\frac{\partial K}{\partial r'}\right)_e [d + 2(d-\Delta)] + ...$$

$$K(SnD) \approx K_e + \left(\frac{\partial K}{\partial r}\right)_e (d-\Delta) + \left(\frac{\partial K}{\partial r'}\right)_e [(d-\Delta) + 2d] + ...$$

$SnHD_3$:

$$K(SnH) \approx K_e + \left(\frac{\partial K}{\partial r}\right)_e d + \left(\frac{\partial K}{\partial r'}\right)_e [3(d-\Delta)] + ...$$

$$K(SnD) \approx K_e + \left(\frac{\partial K}{\partial r}\right)_e (d-\Delta) + \left(\frac{\partial K}{\partial r'}\right)_e [2(d-\Delta) + d] + ...$$

SnD_4 :

$$K(SnD) \approx K_e + \left[\left(\frac{\partial K}{\partial r}\right)_e + 3\left(\frac{\partial K}{\partial r'}\right)_e\right] (d-\Delta) + ...$$

We can further estimate Δ which is literally

$[\langle \Delta r_{SnH} \rangle - \langle \Delta r_{SnD} \rangle]$ as $\{1-(\mu/\mu^*)^{1/2}\} \langle \Delta r_{SnH} \rangle$ or $0.290 \langle \Delta r_{SnH} \rangle$.

Estimating $\langle \Delta r_{SnH} \rangle$ by the approximate method used in isotope shifts,[27] using $r_e \approx$ 1.70Å, we get

$\langle \Delta r_{SnH} \rangle \approx 17.80 \times 10^{-3}$ Å   or $\Delta = 5.161 \times 10^{-3}$ Å.

The observed coupling constants can be converted to the reduced coupling constants

$K = (4\pi^2/h\ \gamma_{Sn}\ \gamma_{H\ or\ D})\ J$

in reduced units of $10^{19} J^{-1} T^2$. Thus converted, the experimental data are shown below:

	Expt [53]		Calc	
	K(SnH)	K(SnD)	K(SnH)	K(SnD)
SnH_4	429.19±.02		429.20	
SnH_3D	428.81±.02	428.21±.14	428.81	428.16
SnH_2D_2	428.43±.02	427.78±.14	428.42	427.77
$SnHD_3$	428.08±.04	427.49±.14	428.04	427.39
SnD_4		426.91±.29		427.02

Using the parameters

$\left(\dfrac{\partial K}{\partial r}\right)_e \approx 200$ reduced units per Å, $\left(\dfrac{\partial K}{\partial r'}\right) \approx 75$ reduced units per Å,

the leading terms in the vibrational correction for K(SnH) in SnH_4 is +7.565 reduced units so that $K_e \approx 421.63$ reduced units. Using these parameters we are able fit all 8 experimental numbers to within experimental error, as shown in the right side of the above table.

An analogous treatment of SnH_3^- leads to the estimates

$\left(\dfrac{\partial K}{\partial r}\right)_e \approx -580$ reduced units per Å $\left(\dfrac{\partial K}{\partial r'}\right)_e \approx -130$ reduced units per Å.

These provide a rough vibrational correction of -14.95 reduced units for K(SnH) in SnH_3^- so that K_e is estimated as 39.04 reduced units. Comparisons of the calculated and experimental values are shown below:

	Expt[7]		Calc	
	K(SnH)	K(SnD)	K(SnH)	K(SnD)
SnH_3^-	24.09±0.01		24.09	
SnH_2D^-	24.76±0.02	27.09±0.14	24.76	27.08
$SnHD_2^-$	25.40±0.03	27.72±0.22	25.43	27.75
SnD_3^-		28.53±0.43		28.42

The agreement with the 6 experimental numbers is very good. We have neglected differential intermolecular effects on the coupling constants of the isotopomers of SnH_3^-, which could be significant. These quantities can be translated back to Hz $\overset{\circ}{A}^{-1}$

	SnH_4	SnH_3^-
$\left(\dfrac{\partial J(SnH)}{\partial r}\right)_e$	-900 Hz $\overset{\circ}{A}^{-1}$	+2600 Hz $\overset{\circ}{A}^{-1}$
$\left(\dfrac{\partial J(SnH)}{\partial r'}\right)_e$	-340 Hz $\overset{\circ}{A}^{-1}$	+580 Hz $\overset{\circ}{A}^{-1}$
$\langle J(SnH)\rangle - J_e$	-34.1 Hz	+67.3 Hz
J_e	-1899.3 Hz	-175.8 Hz

It is interesting that although the magnitude of the SnH coupling constant is about one order of magnitude smaller in SnH_3^- compared to SnH_4, the sensitivity of the spin-spin coupling to bond extension is almost 3 times as great. The lone pair (on Sn in SnH_3^-) is known to be responsible for negative contributions to the reduced coupling,[106,107] leading to a much smaller magnitude of the spin spin coupling. Apparently it is also the lone pair which is responsible for the greater sensitivity to bond extension. By similar procedures, the change in the reduced coupling constant with extension of the bond $(\partial K/\partial r)_e$, or of an adjacent bond $(\partial K/\partial r')_e$, can be estimated in other systems such as SnH_3^+, PH_4^+, PH_3, PH_2^-, and SeH_2 from experimental values of J(AH) and J(AD) combined with values of the dynamic factor calculated by methods described here. These values are compared in TABLE 3 with some published theoretical values. Since all are expressed as γ-free reduced coupling constants these are purely electronic quantities which can be directly compared with each other in sign and magnitude. Theory has shown that the Fermi contact term is most sensitive to bond stretch.[108,109] Thus,it is not surprising that the magnitude of the empirical derivatives in TABLE 3 increase with increasing spin density at the nucleus just as the reduced coupling constants are known to do.[106] The earlier prediction by Jameson and Osten that $(\partial K(AH)/\partial r)_e$ will generally be negative when A has lone pairs but will be positive otherwise [101] appears to be correct. Other interesting trends which also emerge in TABLE 3 are that the dependence of the coupling on the extension of a neighboring bond $(\partial K/\partial r')_e$, is usually smaller, and usually depends on whether A has a lone pair or not (PH_3 is an exception , however). The values in TABLE 3 for molecular ions should be considered with caution for there may be differential intermolecular effects for the isotopomers, which we have ignored.

In summary, theory supports the following general behavior: where no lone pairs are involved on atom A, the magnitude of the AD coupling is smaller than expected from the observed AH coupling. Further, the magnitude of the AH coupling becomes smaller with increasing mass of neighboring atoms. Where there is one or more lone pairs on A, the magnitude of the AD coupling is larger than expected from the observed AH coupling. The magnitude of the AH coupling (except for PH_3) becomes larger with increasing mass of neighboring atoms.

TABLE 3

Changes in reduced spin-spin coupling upon extension of the bond r and upon extension of a neighboring bond r′

Molecule	$(\partial K/\partial r)_e$ [a]	$(\partial K/\partial r')_e$ [a]	Ref(expt)	Ref(theor)
Without lone pairs				
HD	+149.79			108
CH_4	+ 41.175	+26.8 [b]		110
CH_4	$[(\partial K/\partial r)_e + 3(\partial K/\partial r')_e] = 121.8$		102	
PH_4^+	+104±20	+7.5±2	7	
$H_2P(O)OH$	+50			101
SnH_4	+200±30	+75±7	53	
SnH_3^+	+600±400	~0	7	
With lone pairs				
$^{13}C^{17}O$	negative			109
$^{14}N^{15}N$	negative			109
HF	-51.56			111
PH_3	-230			101
PH_3	-422	+46.5	13	
PH_2^-	-125	-26	7	
SeH_2	-190±60	-71±9	15	
SnH_3^-	-580±40	-130±10	7	

[a] Error estimates are based only on the quoted errors in experimental values of J and do not include errors associated with rovibrational averaging or failure to take into account intermolecular effects and second and higher derivatives.
[b] From the observed temperature dependence of J(CH) in CH_4.

IV. CONCLUSIONS

The isotope effects on nuclear shielding and on spin-spin coupling can be measured with precision and are found to have many useful applications. The very large collection of these observations have been systematically discussed here in terms of general trends and specific dependences on the chemist's usual concepts (net charge on the molecule, presence or absence of lone pairs, bond order, hybridization, etc...).

We have found that the observed isotope effects on shielding and on the spin-spin coupling can be interpreted with a theoretical model within the Born-Oppenheimer approximation. The observed isotope effect is therefore, by its nature, a convolution of electronic and dynamic parts which can be individually calculated by ab initio theoretical methods and combined in the way described by eq. 2, each term being a product of an electronic factor and a dynamic factor, up to second derivatives or even up to sixth derivatives, as desired. In the very few cases in which this full calculation has been done (H_2O, CH_4, and a few diatomic molecules), it has been found that beyond the largest single term there are other smaller but significant terms. Therefore, in accurate work it is obviously necessary to do the complete calculations. On the other hand, short of doing such complete calculations, we have seen that important information , albeit semi-quantitative, can be obtained from the measured isotope effects.

The trends which emerge from the measured isotope effects become apparent when one of the terms in the above equation dominates. In particular, for one-bond shifts we considered cases in which the dominant term is

$$^1\Delta A\,(^{m'/m}X) \approx \left(\frac{\partial\sigma^A}{\partial r_{AX}}\right)_e \left[\,\langle\Delta r_{A^mX}\rangle - \langle\Delta r_{A^{m'}X}\rangle\,\right] \tag{46}$$

and in long-range isotope shifts we considered cases in which the dominant term is

$$^n\Delta A\,(^{m'/m}X) \approx \left(\frac{\partial\sigma^A}{\partial r_{YX}}\right)_e \left[\,\langle\Delta r_{Y^mX}\rangle - \langle\Delta r_{Y^{m'}X}\rangle\,\right] \tag{47}$$

where the Y-X bond is n bonds away from resonant nucleus A.

We have shown the mass-dependence of the dynamic term in the square brackets, and elicited the $[(m'-m)/m']\,[m_A/(m_A+m)]$ factors in it by using a diatomic model. We have also shown the dependence of $\langle\Delta r\rangle$ itself on the bond length r_e at the equilibrium geometry, on reduced mass, and on the general constants which characterize the harmonic and anharmonic force constants and which largely depend only on the rows in the periodic table to which the bonded atoms belong. Those isotope shift observations which exhibit the mass dependence of the dynamic factors do indeed exhibit the mass factors in this particular form. Many trends which have been noted and which provide the means with which the very large body of isotope shift observations can be sorted out have been explained with the rovibrational model.

Our goal was to elicit the electronic part directly from experiment by separating out and calculating or estimating the dynamic part wherever possible. If we can do this then we can examine the electronic part which provides us with direct invaluable information about the shielding of a nucleus in a molecule. Thus, we can view the isotopic substitution at some site as a minor perturbation of the electronic shielding at the site of the resonant nucleus. The electronic transmission path between the substitution site and the nuclear site can thus be probed.

The early questions asked of ab initio calculations were, how well do theoretical methods of calculating shielding reproduce the isotropic chemical shifts (^{13}C relative to some reference environment such as CH_4 for example)? More recently we have asked, how well do the calculations reproduce the shielding tensor components? This is a more stringent test of ab initio theory. In the isotope shift experiments we can pose an even more stringent test, of how well do the calculations reproduce the partial derivatives of shielding with respect to various coordinates, e.g., $(\partial\sigma/\partial r)_e$, $(\partial\sigma/\partial\alpha)_e$, etc...? At the outset we compared theoretical derivatives in very small molecules (where calculations with large basis sets are feasible) to the empirical ones that we have obtained by separately calculating the dynamic part of the isotope shift. This way we checked how well our model holds up. As we found good agreement between the empirical derivatives elicited by our model from experiment and the derivatives calculated by ab initio theoretical methods, we now have confidence in using the model to elicit other empirical derivatives from experiment in larger systems where ab initio calculations are less feasible. In so doing, then the very large collection of isotope shifts can provide insight into the shielding function (a mathematical surface) in terms of its dependence on net charge of the molecule, bond order, presence of lone pairs, hybridization, etc.

The following are our major findings about these shielding derivatives:
The electronic factor in the one-bond shift $(\partial\sigma^A/\partial r_{AX})_e$ has been found to have the fol-

lowing attributes which explain the observed trends.

1. It is nearly always negative. In a linear molecule it is dominated by that component of the shielding tensor perpendicular to the bond.

2. For the same bond in homologous compounds it correlates with r_e, which after all is itself an electronic property.

3. For the same two bonded atoms its magnitude increases with bond order.

4. For the same bond in molecules related by presence or absence of a lone pair on A, $(\partial\sigma/\partial r)_e$ decreases algebraically with the presence of a lone pair.

5. It increases algebraically with net charge on the molecule.

6. It is easily visualized in terms of depletion of electron density in the immediate vicinity of nucleus A as nucleus A is moved away from X, but this is a simplistic view.

7. For the same nucleus, it correlates with the paramagnetic part of shielding as in comparisons of $(\partial\sigma^F/\partial r_{MX})_e$ in various MX_n, M being B, C, N, S, Se, Te, W, etc. atoms.

8. For various nuclei, it correlates with the chemical shift range, that is, $(\partial\sigma_A/\partial r)_e$ correlates with $\langle r^{-3}\rangle$ for atomic A. This is a consequence of the paramagnetic part of the shielding being more sensitive to bond stretching than the diamagnetic part.

The electronic factor in the long-range isotope shifts $(\partial\sigma^A/\partial r_{YX})_e$ is found to have the following attributes:

(1) The magnitude decreases (usually) as the Y-X bond becomes more remote from nucleus A.

(2) It is stereospecific. For n=3, cis, trans $\rangle$ gauche, and in some cases the dihedral angle dependence has been clearly observed and it is not unlike the Karplus relation of 3-bond spin-spin coupling with angle ϕ. This stereospecificity is purely electronic in nature.

(3) It correlates with other purely electronic quantities which are dependent on the same electronic transmission path such as substituent effects on chemical shifts by electron-withdrawing or donating groups.

(4) It is easily visualized in terms of the easily calculated change in electron density (gross atomic charge, pi electron density, etc.) at A upon extension of the

remote Y-X bond. This is a simplistic view, although several linear correlations bear out this simple relationship. The electronic transmission path is most easily explored by the isotope shift because substitution of an atom by a heavier one provides the least complications compared to other methods of investigation such as replacement by atoms or groups of different electronegativity or size.

The electronic factor in the one-bond coupling constant $(\partial K \, (AH) \, / \partial r_{AH})_e$ has been found to have the following attributes which explain the observed trends:

1. For A without lone pairs it is positive, that is, the magnitude of the coupling increases as the bond becomes stretched.

2. It is largely dominated by the Fermi contact term which is very sensitive to the distance between the two nuclei, much more so than the spin-dipolar and orbital contributions to coupling.

3. Probably due to the dominance of the Fermi contact term it is larger for nucleus A having larger atomic number.

4. For A with lone pairs this electronic factor is negative. It is known that lone pairs make negative contributions to the spin-spin coupling. The sign of this electronic factor confirms that the lone pair contributions are very sensitive to bond extension.

The electronic factor $(\partial K(AH)/\partial r_{AX})_e$ has been found to have the following attributes:

1. It is generally 2-5 times smaller than the primary electronic factor and is a measure of the sensitivity of the coupling across one bond to the lengthening of another bond involving one of the coupled atoms.

2. This is negative when A has lone pairs (PH_3 is an exception) and positive otherwise, once again an indication that the lone pair contribution to the coupling constant is sensitive to bond extension.

Going to higher and higher magnetic fields will not increase the magnitude of spin-spin splittings so the ease of observing the isotope effects on these will not improve, unlike isotope effects on chemical shifts. Nevertheless, these quantities are of fundamental interest as a crucial source of experimental data which provide critical tests of our quantitative theoretical understanding of J. Since the anisotropy of the J tensor is nearly always inextricable from the direct dipolar tensor in oriented molecules, only the $(\partial K/\partial r)_e$ and $(\partial K/\partial r')_e$ from isotope effects on shielding provide the critical tests, which are not otherwise available beyond the magnitude and absolute sign of the isotropic J observed in freely tumbling molecules, against which any ab initio theoretical calculations may be compared.

In summary, the dynamic factors and electronic factors in isotope effects on NMR parameters have been characterized. The theoretical model is general for any molecular electronic property. The effects can be completely understood within the context of the Born-Oppenheimer approximation. Complete calculations can be carried out for individual molecules within this theoretical framework. Where such calculations have been carried out, the agreement with experiment is good. Under conditions where the leading terms are dominant, the molecule-specific higher order terms are neglected so as to find explanations for and make predictions of sweeping generalities which have surfaced empirically in isotope effects on NMR parameters. The dynamic factors which have been deduced empirically under this approximation predicted the mass-factor dependence observed later. The electronic factors deduced from this model are found to be consistent in sign, in magnitude, in correlations with various quantities such as net charge, presence or absence of lone pairs, bond order, etc., with electronic factors obtained by ab initio calculations. By using estimated dynamic factors and the discovered dependence of the shielding derivatives on various chemical indices such as hybridization, bond order, bond length, net charge, presence of lone pairs, it is possible to estimate the magnitude of isotope effects in new systems for future applications.

REFERENCES

1. C.J. Jameson, *J. Chem. Phys.*, 66 (1977) 4983.

2. C.J. Jameson and H.J. Osten, *Ann. Reports NMR Spectrosc.*, 17 (1986) 1.

3. H. Batiz-Hernandez and R.A. Bernheim, *Prog. NMR Spectrosc.*, 3 (1967) 63.

4. P.E. Hansen, *Ann. Reports NMR Spectrosc.*, 15 (1983) 105.

5. D.A. Forsyth, in E.Buncel and C.C. Lee (Eds.), *Isotopes in Organic Chemistry*, Vol.6, chapter 1, Elsevier, Amsterdam, 1984, p.1.

6. P.E. Hansen, *Progr. NMR Spectrosc.*, 20 (1988) 207.

7. R.E. Wasylishen and N.Burford, *Can. J. Chem.*, 65 (1987) 2707.

8. J. Jokisaari and K. Raisanen, *Mol. Phys.*, 36 (1978) 113; J. Jokisaari, K. Raisanen, L. Lajunen, A. Passoja and P. Pyykkö, *J. Magn. Reson.*, 31 (1978) 121.

9. A. Sebald and B. Wrackmeyer, *J. Magn. Reson.*, 63 (1985) 397.

10. D.A. Forsyth and J.R. Yang, *J. Am. Chem. Soc.*, 108 (1986) 2157.

11. H.K. Ladner, J.J. Led and D.M. Grant, *J. Magn. Reson.*, 20 (1975) 530.

12. A. Pulkkinen, J. Jokisaari and T. Väänänen, *J. Mol. Struct,*, 144 (1986) 359.

13. A.K. Jameson and C.J. Jameson, *J. Magn. Reson.*, 32 (1978) 455.

14. R.E. Wasylishen and J.O. Friedrich, *J. Chem. Phys.*, 80 (1984) 585.

15. H.J. Jakobsen, A.J. Zozulin, P.D. Ellis and J.D. Odom, *J Magn Reson.*, 38 (1980) 219.

16. M. Hoch and D. Rehder, *Inorg. Chim. Acta*, 111 (1986) L13.

17. J.G. Russell and R.G. Bryant, *Anal. Chim. Acta*, 151 (1983) 227.

18. C.J. Jameson and H.J. Osten, *J. Chem Phys.*, 81 (1984) 4293.

19. P.W. Fowler and W.T. Raynes, *Mol. Phys.*, 43 (1981) 65.

20. W.T. Raynes, P.W. Fowler, P. Lazzeretti, R. Zanasi and M. Grayson, *Mol. Phys.*, 64 (1988) 143.

21. R. Ditchfield, *Chem. Phys.*, 63 (1981) 185.

22. D.B. Chesnut, *Chem. Phys.*, 110 (1986) 415.

23. D.B. Chesnut and C.K. Foley, *J. Chem. Phys.*, 85 (1986) 2814.

24. U. Fleischer, M. Schindler and W. Kutzelnigg, *J. Chem. Phys.*, 86 (1987) 6337.

25. P.W. Fowler, G. Riley and W.T. Raynes, *Mol. Phys.*, 42 (1981) 1463.

26. P. Lazzeretti, R. Zanasi, A.J. Sadlej and W.T. Raynes, *Mol. Phys.*, 62 (1987) 605.

27. C.J. Jameson and H.J. Osten, *J. Chem Phys.*, 81 (1984) 4300.

28. C.J. Jameson, A.K. Jameson and D. Oppusunggu, *J. Chem Phys.*, 85 (1986) 5480.

29. C.J. Jameson and A.K. Jameson, *J. Chem Phys.*, 85 (1986) 5484.

30. W. Gombler, *J. Magn. Reson.*, 53 (1982) 69.

31. R.E. Wasylishen, J.O. Friedrich, S. Mooibroek and J.B. Macdonald, *J. Chem. Phys.*, 83 (1985) 548.

32. C. Roeske, P. Paneth, M.H. O'Leary and W. Reimschüssel, *J. Am. Chem. Soc.*, 107 (1985) 1409.

33. K.K. Andersson, S.B. Philson and A.B. Hooper, *Proc. Natl. Acad. Sci. U.S.A.*, 79 (1982) 5871.

34. R.E. Wasylishen and J.O. Friedrich, *Can. J. Chem.*, 65 (1987) 2238.

35. S.A. Linde and H.J. Jakobsen, *J. Magn. Reson.*, 17 (1975) 411.

36. D.R. Herschbach and V.W. Laurie, *J. Chem. Phys.*, 35 (1961) 458.

37. C.J. Jameson, *J. Chem Phys.*, 66 (1977) 4977.

38. J.R. Wesener, D. Moskau and H. Günther, *J. Am. Chem. Soc.*, 107 (1985) 7307.

39. F. Naümann, D. Rehder and V. Pank, *J. Organomet. Chem.*, 240 (1982) 363.

40. J.R. Everett, *Org. Magn. Reson.*, 19 (1982) 86.

41. C.J. Jameson and H.J. Osten, *Mol. Phys.*, 56 (1985) 1083.

42. S. Berger and B.W.K. Diehl, *Tetrehedron Lett.*, 28 (1987) 1243.

43. J.W. Timberlake, J.A. Thomson and R.W. Taft, *J. Am. Chem. Soc.*, 93 (1971) 274.

44. C.H. Arrowsmith and A.J. Kresge, *J. Am. Chem. Soc.,* 108 (1986) 7918.

45. C.J. Jameson, *Mol. Phys.*, 54 (1985) 73.

46. C.J. Jameson, *J. Chem Phys.*, 67 (1977) 2771.

47. C.J. Jameson, D.Rehder and M. Hoch, *J. Am. Chem Soc.*, 109 (1987) 2589.

48. M. Cohn and A. Hu, *J. Am. Chem. Soc.*, 102 (1980) 913.

49. C.J. Jameson, in G.A. Webb (Ed.), *Nuclear Magnetic Resonance*, Specialist Periodical Reports, Vol.10, Royal Society of Chemistry, London, 1981, p.1.

50. J.M. Risley, S.A. Defrees and R.L. Van Etten, *Org. Magn. Reson.*, 21 (1983) 28.

51. D.B. Chesnut and C.K.Foley, *J. Chem. Phys.*, 84 (1986) 852.

52. M. Schindler and W. Kutzelnigg, *J. Chem. Phys.*, 76 (1982) 1919; *J. Am. Chem. Soc.*, 105 (1983) 1360; *Mol. Phys.*, 48 (1983) 781.

53. K.L. Leighton and R.E. Wasylishen, *Can. J. Chem.*, 65 (1987) 1469.

54. R.E. Wasylishen, *Can. J. Chem.*, 60 (1982) 2194.

55. H.J. Osten and C.J. Jameson, *J. Chem Phys.*, 82 (1985) 4595.

56. R.E. Wasylishen and N. Burford, *J. Chem. Soc. Chem. Commun.*, (1987) 1414.

57. R. L. VanEtten and J.M. Risley, *J. Am. Chem. Soc.*, 103 (1981) 5633.

58. G.D. Mateescu, G.M. Benedikt and M.P. Kelly, in W.P. Duncan and A.B. Susan (Eds.), *Synthesis and Applications of Isotopically Labeled Compounds*, Elsevier, Amsterdam, 1983.

59. A. Lycka and P.E. Hansen, *Magn. Reson. Chem.*, 23 (1985) 973.

60. R.D. Sammons, P.A. Frey, K. Bruzik and M.D. Tsai, *J. Am. Chem. Soc.*, 105 (1983) 5455.

61. S. Kerschl, A. Sebald and B. Wrackmeyer, *Magn. Reson. Chem.*, 23 (1985) 514.

62. A.A. Borisenko, N.M. Sergeyev and Y.A. Ustynyuk, *Mol. Phys.*, 22 (1971) 715.

63. M. Alei and W.E. Wageman, *J Chem. Phys.*, 68 (1978) 783.

64. T. Schaefer, J. Peeling and R. Sebastian, *Can. J. Phys.*, 65 (1987) 534.

65 R.E. Wasylishen and G. Facey, private communications.

66. J.P. Quintard, M. Degueil-Castaing, G. Dumartin, B. Barbe and M. Petraud, *J. Organomet. Chem.*, 234 (1980) 27.

67. D.C. McKean, J.E. Boggs and L. Schäfer, *J. Mol. Struct.*, 116 (1984) 313.

68. D.M. Grant and E.G. Paul, *J. Am. Chem. Soc.*, 86 (1964) 2984.

69. R.G. Snyder, A.L. Aljibury, H.L. Strauss, H.L. Casal, K.M. Gough and W.F. Murphy, *J. Chem. Phys.*, 81 (1984) 5352.

70. D.M. Grant and B.V. Cheney, *J. Am. Chem. Soc.*, 89 (1967) 5315.

71. W. Gombler, *Z. Naturforsch. B*, 40 (1985) 782.

72. P.B. Kay and S. Tripett, J. Chem. Res. (S), (1985) 156; *J. Chem. Soc. Chem. Commun.*, (1985) 135.

73. E. Kupce, E. Liepins and E. Lukevics, *Angew. Chem.*, 97 (1985) 588.

74. W. Gombler, *J. Am. Chem. Soc.*, 104 (1982) 6616.

75. H.J. Osten and C.J. Jameson, *J. Chem Phys.*, 81 (1984) 4288.

76. H.J. Osten, C.J. Jameson and N.C. Craig, *J. Chem Phys.*, 83 (1985) 5434.

77. C.H. Arrowsmith, L. Baltzer, A.J. Kresge, M.F. Powell and Y.S. Tang, *J. Am. Chem. Soc.*, 108 (1986) 1356.

78. J.M. Risley and R.L.VanEtten, *J. Am. Chem. Soc.*, 102 (1980) 4609.

79. K.L. Servis and R.L. Domenick, *J. Am. Chem. Soc.*, 108 (1986) 2211.

80. G.E. Maciel, P.D. Ellis and D.C. Hoper, *J. Phys. Chem.*, 71 (1967) 2161.

81. A. Allerhand and M. Dohrenwend, *J. Am. Chem. Soc.*, 107 (1985) 6684.

82. K.L. Servis and F.F. Shue, *J. Am. Chem. Soc.*, 102 (1980) 7233.

83. P.E. Hansen, F.M. Nicolaisen and K. Schaumburg, *J. Am. Chem. Soc.*, 108 (1986) 625.

84. S. Berger and H. Kuenzer, *Angew. Chem.*, 95 (1983) 321.

85. R. Aydin, W. Frankmölle, D. Schmalz and H. Günther, *Magn. Reson. Chem.*, 26 (1988) 408.

86. J.L. Jurlina and J.B. Stothers, *J. Am. Chem. Soc.*, 104 (1982) 4677.

87. R. Aydin and H. Günther, *Z. Naturforsch. B*, 34 (1979) 528.

88. Z. Majerski, M. Zuanic and B. Metelko, *J. Am. Chem. Soc.*, 107 (1985) 1721.

89. J.B. Lambert and L.G. Greifenstein, *J. Am. Chem. Soc.*, 96 (1974) 5120.

90. H.C. Urey, *J. Chem. Soc.*, (1947) 562.

91. H.C. Urey and D. Rittenberg, *J. Chem. Phys.*, 1 (1933) 137.

92. M. Wolfsberg, *Acc. Chem. Res.*, 7 (1972) 225.

93. M. Wolfsberg, *Ann. Rev. Phys. Chem.*, 20 (1969) 449.

94. M. Saunders and M.R. Kates, *J. Am. Chem. Soc.,* 99 (1977) 8071; M. Saunders, S. Saunders and C.A. Johnson, *J. Am. Chem. Soc.,* 106 (1984) 3098; M. Saunders, M.R. Kates and G.E. Walker, *J. Am. Chem. Soc.,* 103 (1981) 4623; M.Saunders, L.A. Telkowski and M.R. Kates, *J. Am. Chem. Soc.*, 99 (1977) 8070; F.A.L. Anet, V.J. Basus A.P.W. Hewett and M. Saunders, *J. Am. Chem. Soc.*, 102 (1980) 3945; M.Saunders and M.R. Kates, *J. Am. Chem. Soc.*, 105 (1983) 3571; M.Saunders in R.Sarma (Ed.), *Stereodynamics of Molecular Systems*, Pergamon Press, New York, 1979 p. 171.

95. S.I. Chan, L. Lin, D. Clutter and P. Dea, *Proc. Nat. Acad. Sci. U.S.A.*, 65 (1970) 816.

96. P.E. Hansen, F. Duus and P. Schmitt, *Org. Magn. Reson.*, 18 (1982) 58.

97. R.M. Jarrett and M. Saunders, *J. Am. Chem. Soc.*, 108 (1986) 7549.

98. R.M. Jarrett and M. Saunders, *J. Am. Chem. Soc.*, 107 (1985) 2648.

99. J. Reuben, *J. Am. Chem. Soc.*, 109 (1987) 316.

100. C.K. Hindermann and D.C. Cornwell, *J. Chem. Phys.*, 48 (1968) 2017.

101. C.J. Jameson and H.J. Osten, *J. Am. Chem Soc.*, 108 (1986) 2497.

102. B. Bennett and W.T. Raynes, *Mol. Phys.*, 61 (1987) 1423.

103. B. Smaller, E. Yasaitis and H.L. Anderson, *Phys. Rev.*, 81 (1951) 896.

104. M. Murray, J. *Magn. Reson.*, 9 (1973) 326.

105. C.J. Jameson in Z.B. Maksic (Ed.), *Theoretical Models of Chemical Bonding Part 3. Molecular Spectroscopy, Electronic Structure, and Intramolecular Interactions*, Springer-Verlag, Berlin, 1990, p.

106. C.J. Jameson in J. Mason (Ed.), *Multinuclear NMR Spectroscopy*, Plenum, New York, 1987, p.89.

107. C.J. Jameson in J.G. Verkade and L.D. Quin (Eds.), *^{31}P NMR Spectroscopy in Stereochemical Analysis*, Verlag Chemie International, Weinheim, 1987, p.205. Methods in Stereochemical Analysis, Vol. 8.

108. J. Oddershede, J. Geertsen and G.E. Scuseria, *J. Phys. Chem.*, 92 (1988) 3056.

109. J. Geertsen, J. Oddershede and G.E. Scuseria, *J.Chem. Phys.*, 87 (1987) 2138.

110. P. Lazzeretti, R. Zanasi and W.T. Raynes, *J. Chem. Soc. Chem. Commun.*, (1986) 57.

111. J.N. Murrell, M.A. Turpin and R. Ditchfield, *Mol. Phys.*, 18 (1970) 271.

Isotopes in the Physical and Biomedical Science, Vol. 2, edited by E. Buncel and J.R. Jones
© 1991 Elsevier Science Publishers B.V., Amsterdam

Chapter 2

^{3}H NMR STUDIES OF HYDROGEN ISOTOPE EXCHANGE REACTIONS

Philip G. WILLIAMS

National Tritium Labeling Facility, Lawrence Berkeley Laboratory

Berkeley, CA 94720, U.S.A.

CONTENTS

I.   INTRODUCTION

A.   General

A large number of publications now exist where ^{3}H NMR spectroscopy has been used as an analytical tool, but only a small proportion of these papers reflects an interest in the catalytic exchange method by which the subject compound was tritiated. Conversely, the field of isotope exchange is a large one, but only a small proportion of the results reported include ^{3}H NMR analyses. Perhaps this is a stage in the development of the ^{3}H NMR technique, from its initial discovery,[1] through its methodological development,[2] to its application in areas of greatest relevance.[3] The point is clear, that very few techniques offer the power of ^{3}H NMR spectroscopy for characterising exchange mechanisms - by clearly showing the labelling pattern in the substrate of interest. It is also clear that the majority of reports of ^{3}H NMR studies concern compounds labelled by synthetic techniques such as hydrogenation or tritio-dehalogenation rather than exchange methods, and the latter techniques have already been well reviewed.[3]

Previous reviews of the exchange literature with respect to ^{3}H NMR analysis have been brief overviews.[4-6] This review will bring together the published NMR results and emphasize the new information that the availability of the technique has afforded. In the case of several labelling techniques there are only a small number of results, and these have been tabulated. Other techniques are much better represented in the literature, and a selection of data will be presented in these instances. Many results confirm theories proposed from mass spectrometry data, but there are subtleties that this analytical technique could not reveal.

The positional information available from ^{3}H NMR study is theoretically available from either ^{1}H or ^{2}H NMR analyses. An early study of platinum catalysed exchange used fully deuterated substrates and proton NMR analysis of H_2O exchange products to give positional information of very high integrity.[7] This technique has not been widely used despite its obvious value, presumably because of the additional step of beginning with fully deuterated substrate. ^{2}H NMR spectroscopy has been used to analyse the products of homogeneous metal catalysed exchange reactions,[8] and the technique is very useful for small molecules. A comparison of deuterium and tritium NMR spectra of similarly labelled glucose products are shown in Figure 1, and the superior resolution and dispersion of tritium is obvious.

B. ^{3}H NMR Spectroscopy

After the initial high resolution ^{3}H NMR spectrum was described,[1] there was a long delay until the technique was reinvestigated[9] and many of the methodological questions resolved.[10,11] It should be noted that the development of tritium NMR spectroscopy coincided with advances in instrumentation that also allowed many other low abundance nuclei such as ^{13}C to be routinely observed. Once the methodological matters were clarified, and the power of the technique was obvious, the major point to be settled was the accuracy of orientations derived from ^{1}H-decoupled ^{3}H NMR studies. It is clear that nuclear Overhauser enhancements from ^{1}H will affect the reporting of ^{3}H intensities,[12] but in most cases differential enhancements are not observed.[13] In this, and more recent work,[14] it was shown that care should be exercised in analysis of very highly tritiated substrates or where tritium atoms do not have protons nearby.

A series of reviews have covered practical and theoretical aspects of ^{3}H NMR spectroscopy,[2,15-19] culminating in an excellent compilation of both techniques and results.[3] Table 1 gives a listing of a number of useful properties of tritium, especially in comparison to the other hydrogen isotopes. Containment of tritiated samples was originally a point of major concern,[1] but several simple and effective systems are currently being used.[3,20] Suffice it to say here that the handling and health physics of analysing tritiated samples is an important but trivially solved

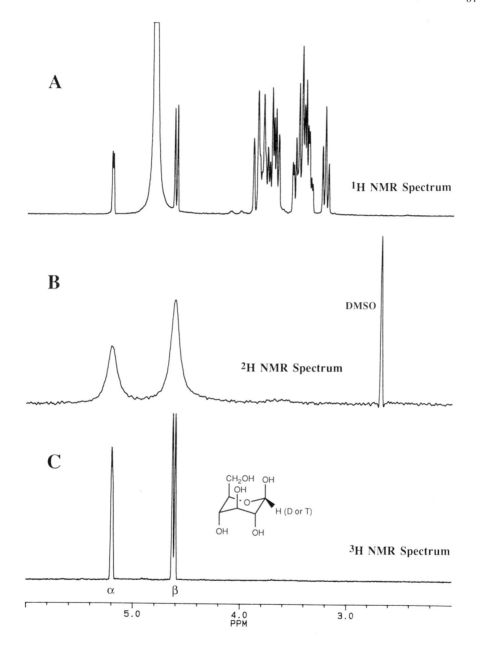

Figure 1. NMR spectra of D-glucose in water. (A). Proton spectrum of C-1 tritiated D-glucose; the large peak is due to HDO. (B). Proton decoupled deuterium NMR spectrum of C-1 labelled glucose; DMSO is an integration marker. (C). Proton coupled tritium spectrum of C-1 labelled glucose; labelled by the same catalytic technique as (B).

consideration, and that ^{3}H NMR spectroscopy can be readily executed in most laboratories having a pulsed Fourier transform NMR spectrometer.

Table 1

NMR Properties of [3]H

Nucleus	Natural Abundance	Spin	γ	Resonant [a] Frequency	Relative [b] Sensitivity
[1]H	99.984	1/2	26.7519	300.13	1.0
[2]H	0.0156	1	4.1064	46.07	9.65×10^{-3}
[3]H	**$<10^{-16}$**	**1/2**	**28.5336**	**320.13**	**1.21**
[13]C	1.11	1/2	6.7263	75.46	1.59×10^{-2}

δ range 0-20ppm $J_{HT}=J_{HH} \times \gamma_T/\gamma_H$ J range 0-20Hz
$$J_{TT}=J_{HH} \times (\gamma_T)^2 \times (1/\gamma_H)^2$$
Isotope effects 1°~ 9Hz (at 7T), 2°~ 4Hz (at 7T)
T_1 : 0-10sec T_2 : 0-10sec

Radiation properties
β(100%)   0.0186MeV   range: 4.5-6mm in air
Maximum specific activity: 28.76 Ci/mmole (1063 GBq/mmole)

a - 7.1 Tesla field. b - Sensitivity given for equal numbers of nuclei in the same field.

II.   ACID CATALYSIS

A.   Mineral Acids

Acids have been used to promote labelling of organic compounds since the 1930's.[21] The application of acids to alkane exchange was pursued by a number of groups including Ingold,[21-23] Burwell,[24-28] Beeck,[29-31] and Kursanov.[32] Since exchange only occurred when the subject hydrocarbon contained a tertiary carbon, and label was not incorporated into the tertiary carbon position, the proposed mechanism[31] was thought to involve three steps - initiation, exchange and propagation. The hydrogen on the tertiary carbon atom was removed to form a tertiary carbonium ion in the initiation step, and exchange of isotope into the carbonium ion proceeded by formation of an olefin and subsequent reprotonation by the acid catalyst. An ion was thought to terminate its exchange cycle by hydride (H⁻) transfer from the tertiary carbon of a neutral molecule, thereby also propagating the exchange.

Since alkanes not containing tertiary carbon atoms were not readily labelled, most acid-catalysed exchange studies have been concentrated on aromatic substrates. Labelling is generally thought to occur by electrophilic substitution, and there is a huge literature[33-36] based on nitration, chlorination and other such synthetic organic procedures which is immediately applicable.

Most fundamental research on acid-catalysed systems was complete before the advent of routine ^{3}H NMR spectroscopy. There are a number of published tritiation results[3,37] which rely on acids such as CF_3CO_2T, which were first explored in the 1960's.[38] Unfortunately, complex acids such as $T(F_3BOPO_3H_2)$[39] which have been reported as exceptionally powerful, have not been used in reactions where products were analysed by ^{3}H NMR spectroscopy.

General trends in electrophilic aromatic substitution reactions are summarized below, and will be illustrated in tritium NMR studies in this and following sections:

(1) groups such as -CH$_3$, -NH$_2$, -OCH$_3$, -NHCOCH$_3$ *etc* activate the aromatic nucleus and direct exchange to the ortho and para positions.

(2) halogens and -CH=CH-X are much less activating than the groups in (1), but are o/p directing.

(3) -N(CH$_3$)$_3$, -CHO, -NO$_2$, -CN and -CF$_3$ groups are deactivating and meta directing.

(4) no known groups are activating and meta directing.

A number of results from the literature are given in Table 2. The vinblastine sulfate study[37] was one of the first clear illustrations of the power of the ^{3}H NMR technique for resolution of catalytic labelling patterns. It also required care with reaction and workup of the compound (2 hours, room temperature), which is light sensitive. The structure of the compound is given in Figure 2(a), and the tritium and proton NMR spectra in Figures 2(b) and (c) respectively. It is easily seen that degradative assignment of the orientation of exchange would have been a daunting task. A very similar anti-cancer drug, vincristine, was similarly labelled and assigned.[2]

Specific labelling of a series of aliphatic acids was achieved[40] by the use of HTSO$_4$ and relatively forcing conditions. All the incorporated tritium was shown to be in the positions α to the carbonyl group. A similar position in glutamic acid (the α, or 2 position), which also bears the amino substituent, was specifically labelled by acid exchange.[3] In the same way, the 3 position of 8-methoxypsoralen (adjacent to the carbonyl) held 80% of the incorporated tritium.[3]

The labelling of dopamine[41] serves to illustrate several rules of electrophilic substitution reactions. The varying extent of labelling in the aromatic positions is a result of the substituent effects on the tritiation as outlined in the four rules above. Comparison of (1) and (2) clearly shows that the 5-position, which is meta to two substituents, is the least labelled of the three aromatic positions.

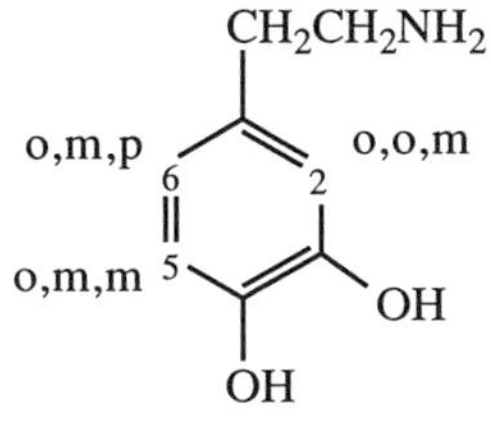

1 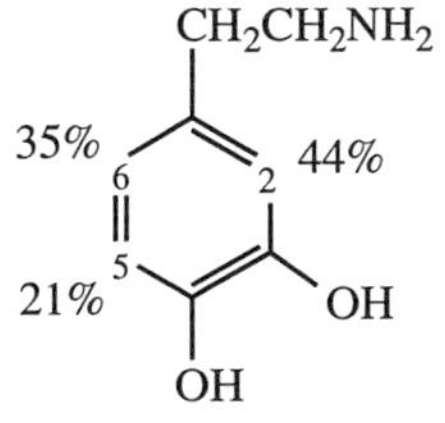   **2**

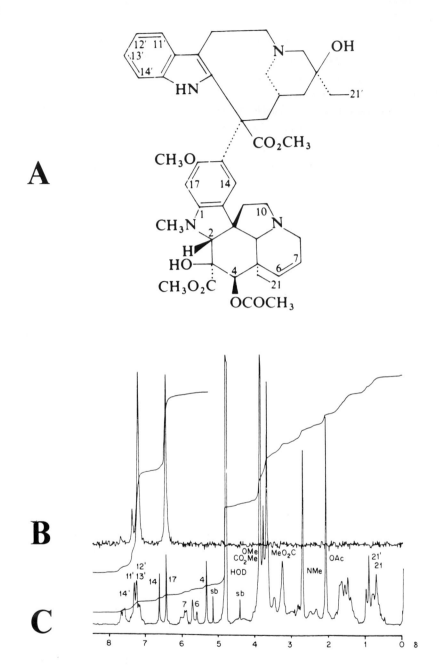

Figure 2. (A). Structure of Vinblastine Sulfate. (B). Proton decoupled tritium NMR spectrum. (C). Proton NMR spectrum. (Reproduced by permission, The Royal Society of Chemistry).

Table 2

Labelling by Acid-catalysed Exchange

Compound	Method	Orientation, %	Ref.
Vinblastine Sulfate	CF$_3$CO$_2$H/HTO/2 hrs	*11'-8; 12',13'-46; 14'-2; 17-44*	37
Vincristine Sulfate	CF$_3$CO$_2$H/HTO/2 hrs	*11'-11.4; 12',13'-76.1; 14'-5.7; 17-6.8*	2
Palmitic Acid	HTSO$_4$/64 hrs 100°C	*α-CH$_2$-100*	40
Myristic	HTSO$_4$/64 hrs 100°C	*α-CH$_2$-100*	40
Lauric	HTSO$_4$/64 hrs 100°C	*α-CH$_2$-100*	40
Stearic	HTSO$_4$/64 hrs 100°C	*α-CH$_2$-100*	40
Glutamic Acid	Acid/HTO	2-100	3
8-Methoxypsoralen	Acid/HTO	3-79.8; 5-8.9; 7-11.3	3
Dopamine HCl	HCl/HTO	2-44; 5-21; 6-35	41
Kainic Acid	Acid/HTO	(i) *CH$_3$-75; =CH$_2$-25* (ii) *CH$_3$-32; -CH$_2$-CO$_2$H-45; 5-CH$_2$-18; =CH$_2$-5*	41
Strychnine Sulfate	CF$_3$CO$_2$H/HTO	2-24; 4-23; *11α-27; 11β-26*	41
Benzo[a]pyrene	CF$_3$CO$_2$H/HTO	*1-9.6; 2-6.2; 3-9.6; 4-12.4; 5-6.7; 6-8.4; 7-7.3; 8-6.7; 9-12.4; 10-4.2; 11-4.2; 12-12.4*	3
7-Methylbenz[c]acridine	Acid/HTO	*CH$_3$-71.4; ring-6.3,8.7,9.5,3.2,0.8*	3
12-Methylbenz[a]acridine	Acid/HTO	*CH$_3$-24.3; ring-60.1,10.1,2.3,3.2*	3
Colchicine	Acid/HTO	4-100	3
Imipramine	CF$_3$CF$_2$CF$_2$CO$_2$H/HTO	4,6-45; *other aromatics-55*	42
Isoconessine	HTSO$_4$/0°C	*C1,C4,C6,C7,C10,C11, C14,C15,C19*-all labelled	43

Aromatic positions in compounds such as benzacridines, imipramine[42] and colchicine have also been exchanged, with extremely specific labelling in the case of the latter substrate (Table 2).[3] The HTSO$_4$-catalysed rearrangement and tritiation of isoconessine (**4**, from conessine, **3**) has been studied[43] and yielded highly labelled product, with nine resolved lines in the tritium NMR spectrum. Comparison with ^{2}H and ^{13}C NMR, and mass spectra of similarly deuterated substrates led to the conclusion that the product was tritiated on nine carbons (17 possible hydrogens), and that there were a considerable number of multiply tritiated molecules in the product mix. The carbons which were thought to bear tritium atoms include the 19-methyl group (near the 4 and 6 carbons) and the numbered positions on the structure below (**4**):

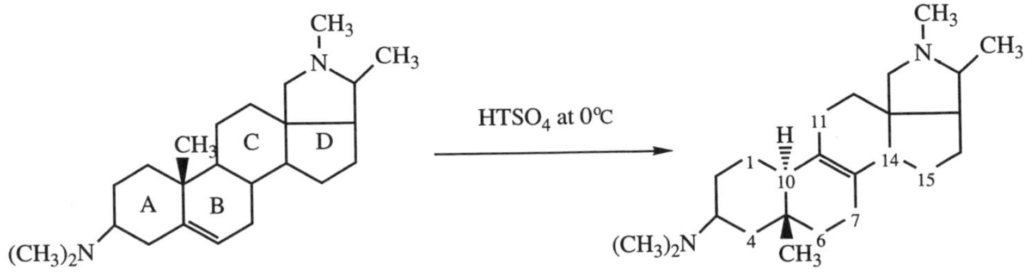

Acid catalysis is rapid and effective, but it should be kept in mind that tritium incorporated at low pH may also be lost under similar conditions. That is, positions labelled by acid (or base) catalysis may not be stable under biological conditions, where local pH excursions may be large.

B.   Lewis Acids

The possibility of using extremely reactive superacids and Lewis acids as isotope exchange catalysts has been recognized for many years. As with the mineral acid work, much of this research took place before the advent of routine ^{3}H NMR analyses, so positional information is scant.

Tritiation of alkanes[44] by HTO/EtAlCl$_2$ exchange was shown to occur preferentially at the CH positions, with considerably less labelling in CH$_2$ and CH$_3$ groups (Table 3). The aromatic labelling induced by EtAlCl$_2$ is rapid,[45] and is reported to be random, but until recently there was no published tritium NMR data to support this contention. Certainly, near-random tritium distributions were observed for the tritiation of 1,4-dimethylnaphthalene in the presence of both BBr$_3$ and EtAlCl$_2$[47] (see Table 3). Later work[48] with BBr$_3$ as the catalyst suggested that an o/p orientation in aromatic centres was obtained, supporting an electrophilic mechanism for this catalyst.

The most systematic study of Lewis acid catalysed exchange by ^{3}H NMR spectroscopy is that of Garnett, Long and Lukey.[49,50] A series of simple organic compounds were tritiated and the positions of labelling assessed (Table 3). The studies with BBr$_3$ as catalyst clearly show that the orientation of exchange in aromatic centres is o/p in every case where data was obtained.[49] Alkyl substituents were not labelled, and compounds such as α,α,α-trifluorotoluene, which are deactivated towards electrophilic attack, were not labelled. Important features of the BBr$_3$ system in contrast to EtAlCl$_2$ labelling, are:

(1)   that the aromatic labelling patterns are distinctly different (o/p vs random), and

(2)   that labelling may still be effected with BBr$_3$ even when it has previously been hydrolysed with HTO.

This latter point is very different from the EtAlCl$_2$ system, where the HTO appears to effect labelling by destruction of a catalyst-substrate complex, and the order of addition of reactants is critical.[45]

A wide variety of polycyclic aromatic hydrocarbons labelled with tritium are required for studies of the mechanisms of carcinogenesis. A range of compounds have been successfully labelled by application of the EtAlCl$_2$ technique with high specific activity HTO.[51,52] A number of these compounds have recently been analyzed by ^{3}H NMR spectroscopy,[53] and a typical spectrum is shown in Figure 3, acquired on 2mCi of purified substrate. The orientation in this and other (similar) substrates is almost random, and supports the widely accepted theory of the lack of specificity of aromatic labelling by EtAlCl$_2$.

Despite the wide use[54] of Lewis acids as tritiation catalysts, and their application to a number of classes of compounds,[52,55] very little orientational information exists in the literature.

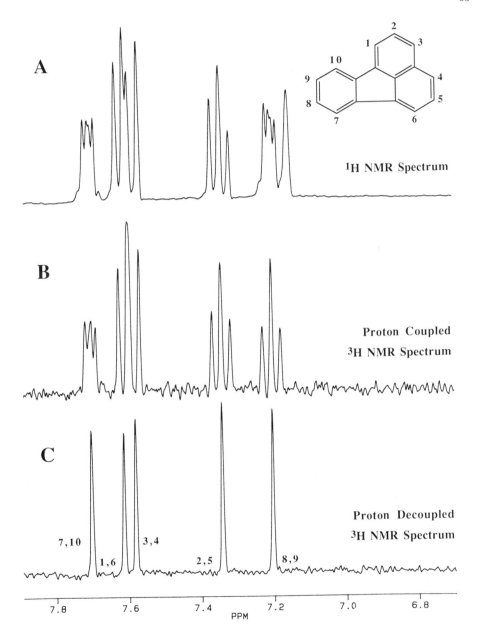

Figure 3. NMR spectra of Fluoranthene tritiated by Lewis acid catalysed exchange.

Table 3

Labelling by Lewis Acid Catalysts

Compound	Catalyst	Orientation, %	Ref.
2,3 Dimethylbutane	EtAlCl$_2$	*CH*-17; *CH$_3$*-83	44
3-Methylpentane	EtAlCl$_2$	*CH*-21; *CH$_2$*-29; *CH$_3$*-50	44
Methylcyclohexane	EtAlCl$_2$	*CH*-28; *CH$_2$*-32; *CH$_3$*-40	44
Bromobenzene	EtAlCl$_2$	*o*-38; *m*-36; *p*-25	48
Toluene	BBr$_3$	*o*-60; *m*<10; *p*-40; *CH$_3$*<3	48
iso-Propylbenzene	BBr$_3$	*o*-36; *m*-10; *p*-19; *CH*-16; *CH$_3$*-18	48
Bromobenzene	BBr$_3$	*o*-52; *m*<3; *p*-47	48
Naphthalene	BBr$_3$	*α*-76; *β*-24	48
Chlorobenzene	BBr$_3$	*o*-43; *m+p*-57	46
1,4 Dimethylnaphthalene	BBr$_3$	*2,3*-39; *5,8*-33; *6,7*-28; *CH$_3$*<1	47
1,3,5 Trimethylbenzene	BBr$_3$	*2,4,6*-100; *CH$_3$*<1	47
Chlorobenzene	EtAlCl$_2$	*o*-41; *m*-39; *p*-20	47
1,4 Dimethylnaphthalene	EtAlCl$_2$	*2,3*-32; *5,8*-33; *6,7*-35; *CH$_3$*<1	47
1,3,5 Trimethylbenzene	EtAlCl$_2$	*2,4,6*-100; *CH$_3$*<1	47
1-Chloronaphthalene	BBr$_3$	*2*-15; *3*-4; *4*-10; *5*-25; *6*-4; *7*-15; *8*-26	47
Toluene	EtAlCl$_2$	*o*-36.6; *m*-40; *p*-24.4; *CH$_3$*<0.1	50
Fluorobenzene	EtAlCl$_2$	*o*-42.4; *m*-35.2; *p*-22.3	50
Bromobenzene	EtAlCl$_2$	*o*-41.2; *m*-37.2; *p*-21.6	50
Naphthalene	EtAlCl$_2$	*α*-55.2; *β*-44.8	50
Toluene	BBr$_3$	*o*-63; *m*-10.2; *p*-26.9; *CH$_3$*<0.1	49
n-Propylbenzene	BBr$_3$	*o*-61; *m*-8.4; *p*-30.5; *alkyl*<0.1	49
iso-Propylbenzene	BBr$_3$	*o*-62; *m*-5.4; *p*-32.7; *alkyl*<0.1	49
iso-Butylbenzene	BBr$_3$	*o*-42.8; *m*<0.1; *p*-57.3; *alkyl*<0.1	49
t-Butylbenzene	BBr$_3$	*o*-60.4; *m*-8.6; *p*-31.0; *alkyl*<0.1	49
Cyclohexylbenzene	BBr$_3$	*o*-60.4; *m*-6.8; *p*-32.8; *alkyl*<0.1	49
Diphenylmethane	BBr$_3$	*o*-66.6; *m*<0.1 *p*-33.3; *alkyl*<0.1	49
Fluorobenzene	BBr$_3$	*o*-31.6; *m*<0.1; *p*-68.4	49
Chlorobenzene	BBr$_3$	*o*-54.6; *m*<0.1; *p*-45.5	49
Bromobenzene	BBr$_3$	*o*-52; *m*<0.1; *p*-48.0	49

C.   Zeolite Catalysis

Zeolites are crystalline aluminosilicates, and have long been known to have catalytic properties which allow them to act like very strong mineral acids. A detailed study of the hydrogen isotope exchange capabilities of these catalysts was undertaken with ^{3}H NMR spectroscopy as the major analytical tool.[56]

Preliminary results[57,58] for exchange of aromatic substrates with HTO over the large-pore zeolite, HNaY, showed very clearly that the orientation of labelling was that due to electrophilic aromatic substitution. As shown for toluene (**5**) aromatic exchange was confined almost exclusively to the ortho and para positions, and alkyl exchange was not observed. This pattern was generally true of straight-chain alkylbenzenes, but branched-alkyl aromatics gave alkyl exchange confined to the β-carbon atoms of molecules branched at the α-carbon (see **6, 7**), in addition to the ortho/para aromatic labelling. Such a substitution pattern is expected where exchange involves hydride transfer between the reactant molecule and an α-carbonium ion, as proposed for alkane exchange with strong mineral acids.[30,31] Incorporation of tritium into the carbonium ion may take place by deprotonation to an olefinic intermediate and reprotonation.[30,59]

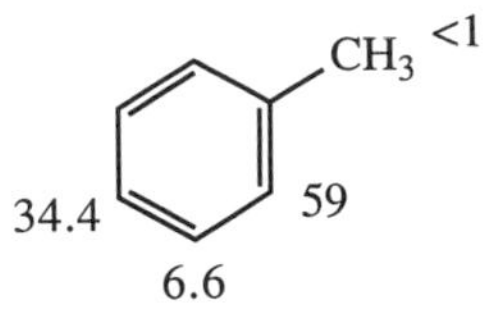

It was also observed that exchange between aromatic substrates was facile over HNaY and other zeolites, such as H-mordenite and HZSM-5.[58,60] Tritiated benzene and specifically tritiated toluene were used to characterize the mechanism of this aromatic-aromatic labelling.[61] As previously reported from deuteration studies,[62] exchange took place by transfer of isotope from one aromatic to the zeolite, followed by incorporation into the second aromatic centre. The extent of exchange observed in any given substrate depended on its adsorption and exchange in competition with the isotope source. Alkanes were not efficiently labelled with either an organic or water as the isotope source,[57,60] in contrast to metal-catalysed labelling systems.

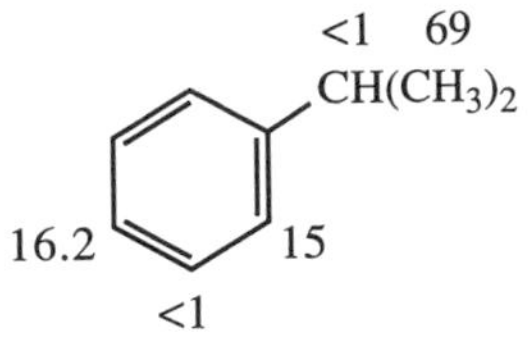

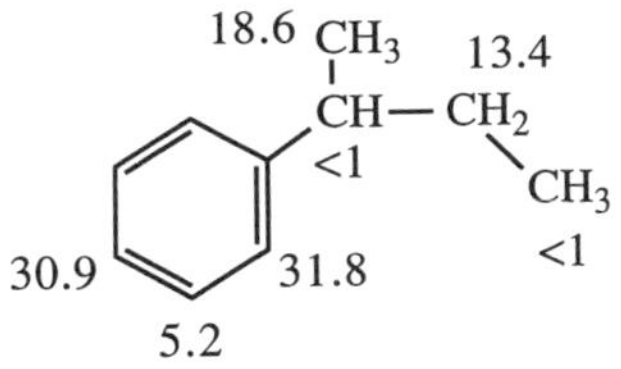

Although zeolites can activate tritium gas for exchange into organic substrates at modest temperatures,[63] the process is made much more efficient by the presence of a transition metal such as platinum or palladium.[58,61] The methyl position of toluene was efficiently labelled over PdY, and the orientations generally could be seen to be influenced both by guest metal and the zeolite. In addition to vastly increasing the uptake of tritium by a given substrate over a time period, the presence of metal caused positions and molecules not normally exchanged over zeolites to be labelled (compare **8** and **9** with **5**). Of particular interest was the efficient and relatively general labelling of alkanes (**10**).

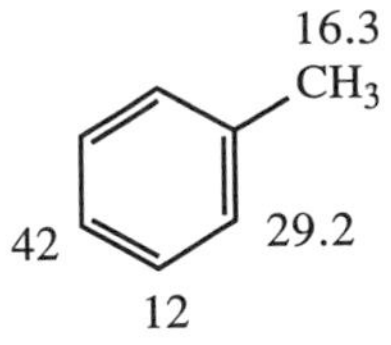

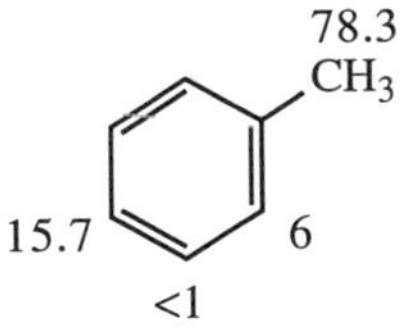

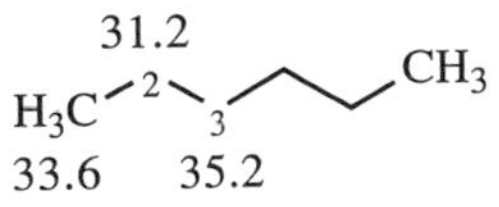

The appearance of "metal character" in labelling distributions was apparent with HTO as isotope source as well as with T_2, as indicated by extensive methyl exchange over Pd-Mordenite zeolite (**11**). However, as is well known, metal catalysis is poisoned by the presence of air, and methyl exchange (a metal catalysis feature) is quenched when air is included in the reaction tube (**12**).[64]

In summary of zeolite catalysis, a range of simple organic substrates may be readily labelled[56] using a variety of isotope sources. Alkanes and the straight-chain alkyl portions of alkylbenzenes require the presence of a metal with the zeolite to promote exchange. Pyridine also required metal presence, but other heterocyclic compounds such as thiophene and furan were readily exchanged. Exchange results were found to rely upon the following factors:

(1)   the form of zeolite: HNaY, H-Mordenite, HZSM-5 with Pt, Pd or Ni substitution

(2)   the isotope source - HTO, C_6D_6, C_6H_5T, T_2

(3)   the presence or absence of air in the reaction vessel.

Hence, the activation of tritium gas and its availability to the zeolite lattice is expedited by the presence of metal. In the case of HTO or organic isotope sources the metal and zeolite catalysts are in direct competition, but the metal contribution can be controlled by the admission or exclusion of air.

D.   Catalysis by Aluminophosphates

Direct exchange of tritium gas with organic substrates over aluminophosphates has been reported recently.[65,66] In comparison to the zeolite systems, where exchange with tritium gas was slow with the unmodified zeolite,[60,64] the AlPO-5 catalyst gave 5-20% incorporation of tritium over a few days at temperatures of 100-180°C.[65] The more remarkable facet of the exchange was the orientation of exchange in labelled substrates, as given in Table 4. Even though the acid properties of AlPO-5 have not been stressed in comparison to zeolites, the orientations show clearly that electrophilic aromatic substitution is likely to be the mechanism of labelling. However, the para position of toluene shows much more exchange than the ortho, and this is the first report of such a unique orientation.[65] This is also observed for labelling of bromo- and chlorobenzene, but not fluorobenzene.

The same orientation is observed when HTO is the isotope source,[66] but some of the specificity is lost as the temperature is raised in order to increase total incorporation. This was also a feature of zeolite labelling, and is expected from the kinetics of electrophilic aromatic substitution.

The use of the AlPO catalyst in a modified Wilzbach experiment with T_2[66] gave approx. 20% incorporation of tritium in toluene with excellent purity of the product (Table 4). The orientation of labelling shows almost random exchange in the aromatic ring, but a significant amount of tritium in the methyl positions (6%).

Table 4

Tritium Distribution in Compounds labelled over AlPO-5 Catalyst

Compound		Time (hr)	Temp. (°C)	Activity (mCi/mL)	Orientation, %	Ref.
Toluene	T_2	24	100	48	o-7.2; m<2; p-93; CH_3<1	65
Toluene	T_2	168	100	153	o-9.6; m-6.2; p-84; CH_3<1	65
Toluene	T_2	72	180	175	o-36; m-8.2; p-55; CH_3<1	65
m-Xylene	T_2	72	100	83	$C2$-25; $C4,6$-76; $C5$<1; CH_3<1	65
Naphthalene	T_2	72	100	31	α–100; β<1	65
Furan	T_2	72	100	36	α–100; β<1	65
2,3 Dimethylbutane	T_2	72	100	17	$Methine$-100; $Methyl$<1	65
Toluene	HTO	136	125	23	o<1; m<1; p-100; CH_3<1	66
Toluene	HTO	136	140	84	o-14.2; m<1; p-86; CH_3<1	66
Toluene	HTO	136	180	784	o-38; m-5; p-58; CH_3<1	66
Chlorobenzene	T_2	72	150	46	o<1; m<1; p-100	66
Bromobenzene	T_2	72	150	75	o<1; m<1; p-100	66
Fluorobenzene	T_2	72	150	35	o-58; m<1; p-42	66
Toluene[a]	T_2	120	RT	1410	o-36; m-36; p-21; CH_3-6	66

a - modified Wilzbach experiment with 2Ci of T_2 and 0.03mL of substrate

Once again, these results show that "solid acids" such as the crystalline aluminosilicates (zeolites) or aluminophosphates (AlPO's) may be used to give high levels of acid labelling in simple organic compounds. The methods have not been extended to larger molecules, and it remains to be seen whether efficient exchange will be observed in molecules excluded from the structural pores.

III.   BASE CATALYSIS

Most organic compounds can be regarded as carbon acids, even if very weak, and treatment with sufficiently strong base can lead to hydrogen isotope exchange.[67,68] Labelled compounds produced in this way may be used as tracers provided that the activity of the compound and the basicity of its solution medium are known. A series of studies by Jones and co-workers, recently reviewcd,[69] has explored the many factors affecting the detritiation of heterocyclic compounds in solution.

Table 5 contains the results of a large number of substrates labelled by base catalysis, and subsequently analysed by ^{3}H NMR spectroscopy. As is well known, the method provides a very specific procedure for introducing tritium adjacent to a carbonyl group, as illustrated in the labelling of acetophenone,[11] stearic acid,[40] diethyl malonate,[70] and a large series of substituted (2-acetyl) thiophenes.[71,72] In addition, a number of steroids (**13, 14**)[73] and 15,16-dihydrocyclopenta[a]-phenanthren-17-ones (**15**)[74] have been specifically labelled (Table 5) by use of the same characteristic, and (in some cases) subsequent chemical modification.

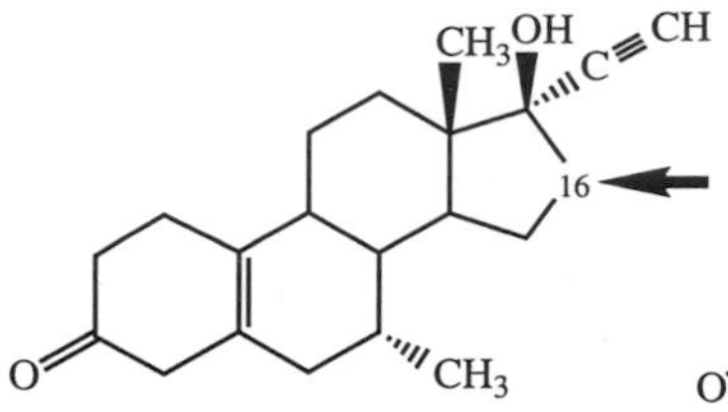

13 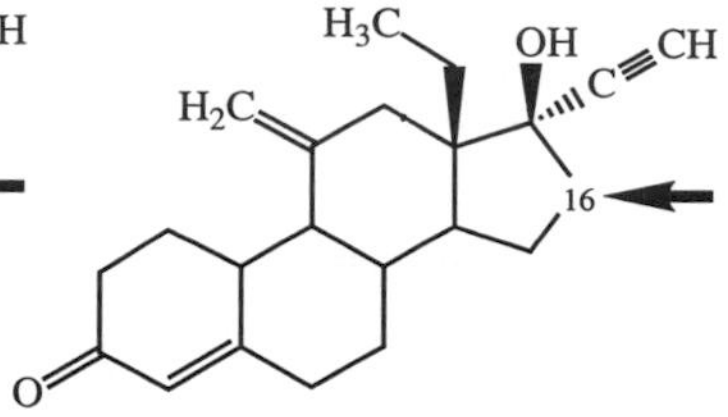   **14** 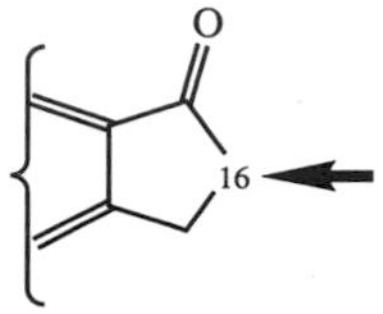   **15**

Similarly, compounds containing terminal triple bonds are specifically labelled on the terminal (methine) carbon,[10,75] and a series of nitriles have been tritiated on the 2 carbon.[10,11] Application of this technique leads to specific benzyl exchange in substrates such as phenylacetylene, phenylacetonitrile or benzoselenazole (Table 5). Aromatic tritiation has only been reported in a limited number of cases,[76-78] and the orientations are given below. The even distribution in orsellinic acid (**16**) probably just reflects equilibration of isotope, while the marked orientations in the other two substrates (**17, 18**) were used to propose canonical forms of intermediates,[78] and to provide evidence for carbenoid type delocalization in nitroaromatic systems. In addition, the results illustrate the particular usefulness of the method for labelling nitroaromatics, which are difficult to label by heterogeneous metal and other catalytic techniques.

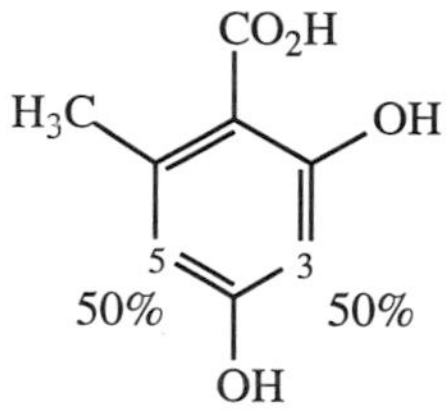

16 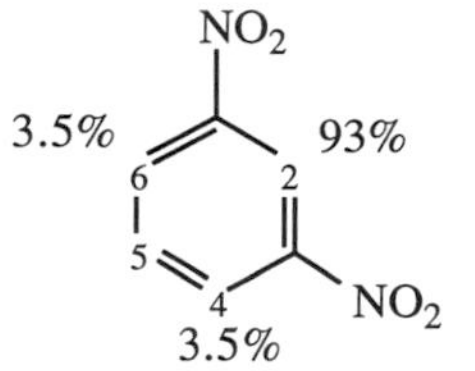   **17** 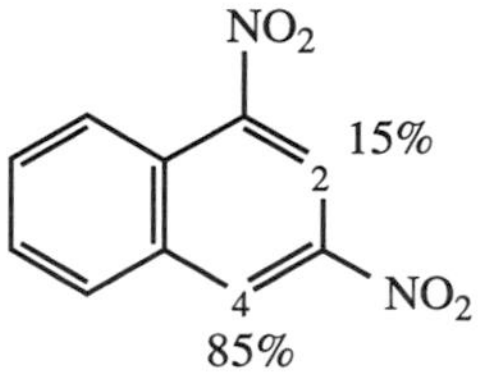   **18**

Base catalysed labelling has the advantages of providing rapid and specific exchange under conditions of mild temperature and pressure. The maximum specific activities attainable are controlled by the isotope source, and this can be an important disadvantage (as is also true of acid systems). However, since the results for base catalysed exchange can easily provide very specific labelling, and the parameters governing detritiation in a number of classes of compounds are well understood, it is surprising that the technique has not been more widely used and characterized by ^{3}H NMR spectroscopy. In particular, since the NMR technique allows the monitoring of multiple positions within one molecule, it is a little surprising that ^{3}H NMR studies of reactions in progress have not been conducted.

Table 5

Base Catalysis of Hydrogen Isotope Exchange

Compound	Method	Orientation, %	Ref.
Acetophenone	NaOH/20°C/18 hrs	CH_3-100	11
Stearic Acid	NaOH/160°C/16 hrs	α-CH_2-100	40
Diethyl Malonate	Na_2CO_3/RT/7 days	CH_2-100	76
Diethyl Malonate	Na_2CO_3/RT/2 days	CH_2-100	70
Sodium Acetate	NaOH/90°C/168 hrs	CH_3-100	11
Pentan-3-one	Na_2CO_3/60°C/3 days	2,4-100	75
α-Chloroacetophenone	Na_2CO_3/ /24 hrs	α-$ClCH_2$-100	75
α-Bromoacetophenone	Na_2CO_3/20°C/18 hrs	α-$BrCH_2$-100	75
Acetone	NaOH/20°C/18 hrs	CH_3-100	11
Sodium Malonate	NaOH/20°C/18 hrs	2-100	10
Isobutyric acid	NaOH/150°C/2 days	2-100	75
Sorbic Acid[a]	Pyridine/steam/2 hrs	α-36,γ-64	79
7αMethyl Norethinodrel	CH_3ONa/80°C/2 hrs	16 α/β-100	73
3-Oxo-desogestrol	CH_3ONa/80°C/2 hrs	16 α/β-100	73
15,16-Dihydrocyclopenta[a]-phenanthren-17-ones	NaOH/RT/2-3 days	16-100	74
2-Acetyl thiophene derivatives	NaOH/RT/48 hrs	Acetyl-100	71
3-Carboxy 2-acetyl thiophenes	NaOH/RT/48 hrs	Acetyl-100	72
Adenosine 3' monophosphate[b]	HTO/85°C/18 hrs	8-100	3
1-Methyl Inosine[b]	HTO/85°C/18 hrs	8-100	3
Phenylacetylene	NaOH/RT/36 hrs	Methine-100	75
Undec-10-yn-1-oic acid	NaOH/45°C/48 hrs	11-100	75
2-Methylbut-3-yn-2-ol	NaOH/45°C/48 hrs	4-100	75
Prop-2-yn-1-ol	NaOH/20°C/18 hrs	3-100	10
Prop-2-en-1-ol	NaOH/20°C/18 hrs	3-100	10
Phenylacetonitrile	Na_2CO_3/ /24 hrs	α-CHTCN-100	75
Acetonitrile	NaOH/20°C/18 hrs	CH_3 -100	11
Propionitrile	NaOH/20°C/18 hrs	2-100	10
Malononitrile	NaOH/20°C/18 hrs	2-100	10
Dimethylsulfoxide	NaOH/90°C/18 hrs	CH_3-100	11
Benzyl methyl sulfoxide	NaOH/RT/36 hrs	α-CH_2-100	75
Nitromethane	NaOH/20°C/18 hrs	CH_3-100	11
Chloroform	0.2N NaOH/20°C/1 hr	1-100	10
1,3 Dinitrobenzene	CH_3ONa/45°C/4 hrs	2-93, 4-7	77
1,3 Dinitronaphthalene	NaOH/RT/300 hrs	2-15, 4-85	78
Orsellinic Acid	NaOH/RT/4 days	3,5-50 each	76
2-Picoline	NaOH/20°C/18 hrs	1'-100	10
2 Methyl resorcinol	NaOH/20°C/18 hrs	4-100	10
Pyridine-1-oxide	NaOH/75°C/30 hrs	2-100	10
Quinoline-1-oxide	NaOH/75°C/20 hrs	2-100	10
Isoquinoline-1-oxide	NaOH/75°C/20 hrs	1,3-100	10
Benzoxazole	0.2N NaOH/20°C/1 hr	2-100	10
Benzothiazole	0.2N NaOH/20°C/1 hr	2-100	10
Benzoselenazole	0.2N NaOH/20°C/1 hr	2-100	10

a - Reaction of crotonaldehyde and malonic acid to give labelled product.

b - Simple heating with HTO is sufficient to label many purines.[69]

IV. GAS EXPOSURE TECHNIQUES

A. Wilzbach

The energy of the β-decay of tritium can sometimes be sufficient to activate an organic compound towards exchange with a tritium gas atmosphere.[80] Generally the percentage of the total tritium that is incorporated into the substrate is very small, and usually a large quantity and variety of other radioactive products are formed in addition to the desired product. As a consequence, a great number of variations on the technique have been tried,[81] and some of the more successful ones will be discussed in the next few sections.

In rare cases specific and efficient Wilzbach labelling is observed as illustrated in the following examples:

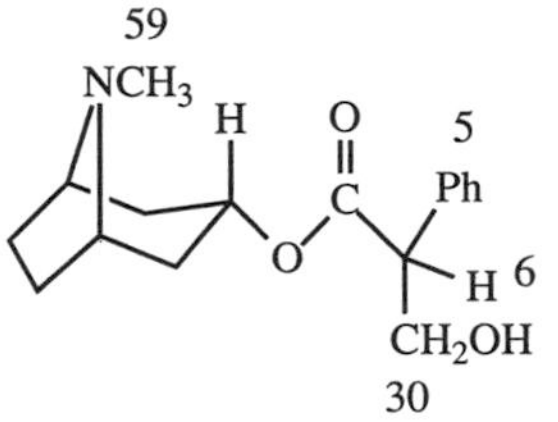

19 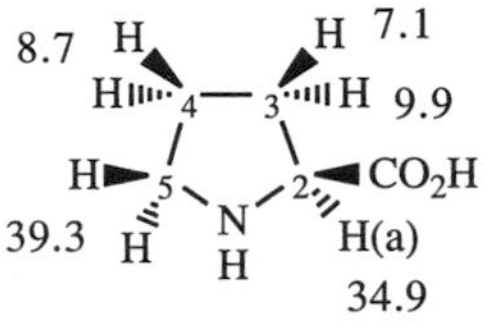   **20** 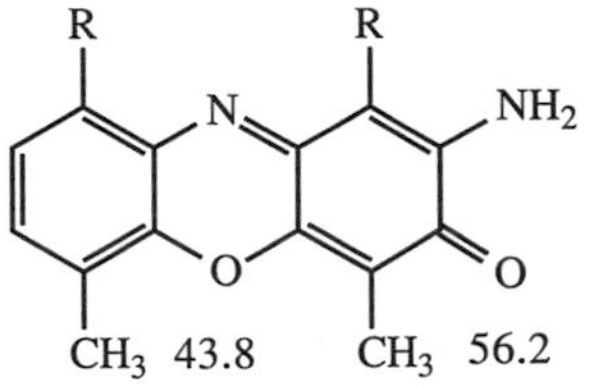   **21**

In the case of atropine (**19**) significant quantities of CTH_2, CT_2H, and CT_3 species were observed,[17] as well as the magnetic non-equivalence of the tritons of the -CH_2-OH group (adjacent to the chiral -CH-Ph group). The positions close to a nitrogen were once again most labelled in proline (**20**) with 75% of the incorporated tritium being on a carbon attached to nitrogen.[3] The Wilzbach exchange of Actinomycin D (where R represents a cyclic peptide) gave surprisingly specific labelling (**21**), and once again multiple tritiums were incorporated on each carbon atom, as revealed by the ^3H NMR spectra.[2]

B. Microwave Induced Exchange

Microwave discharge activation (MDA) was first published as a variation of gas exposure or Wilzbach labelling,[82] and has attracted a lot of attention as a potentially useful general labelling process.[81,83] Once again, very little ^3H NMR data has been published on the orientation of exchange induced by microwave techniques.[84,85] However, a number of studies on the best conditions for labelling have explored the critical parameters influencing yield and specific activities obtained.[84,86] The best reported results are those arising from the modification engineered by Peng and co-workers, whereby a support is used in the exchange reaction, and the microwave activated species are no longer acting directly on the exposed substrate.

In preliminary studies of peptide labelling using this system[84] it was reported that samples of Gly-Gly-Leu with specific activities of 1-5 Ci/mmole were obtained, and ^3H NMR evidence suggested most of the ^3H was in the CH_3 groups of the Leucine residue.

Studies of peptide labelling by Ehrenkaufer[83,87,88] have given several clues as to the mechanism of the exchange. It was noted[85] that the form of the substrate greatly affected the efficiency of labelling:[83] changing from the zwitterionic form of L-Valine to the neutral sodium salt gave a 45-fold increase in specific activity of product.[88] Similarly, experiments with the neutral form of L-Proline revealed that the orientation is also influenced by the overall charge on the molecule (zwitterion **22**, Na salt **23**)[85] where * denotes that the 3β resonance was insufficiently resolved from the 4 tritons in the NMR spectra for quantitation.

22   **23**

Studies of the labelling of steroids[89,90] have shown that the backbone of the molecule appears to provide protection against degradation during tritiation by microwave activated tritium. In addition, the presence and nature of supports was found to influence the orientation of labelling, as determined by ^{3}H NMR study.[90] Use of 5% Ru on silica-alumina pellets gave predominantly 2β labelled progesterone, whilst labelling without metal on the support yielded mainly the 2α tritiated product.

More recent work with the MDA system has shown that remarkably pure products with specific labelling may be obtained. Figure 4 shows the ^{3}H and ^{1}H NMR spectra of n-propylbenzene labelled by exposure to T_2 while supported on the same silica-alumina supported Ni catalyst as previously studied.[84] The pattern of labelling strongly favours ortho/para incorporation, which suggests attack by a T^+ or T_3^+ species, as proposed by Peng.[84] This is one of a series of simple organic substrates in which ^{3}H NMR analyses show similarly marked orientations.[91]

The method has also been valuable for labelling of a series of polycyclic aromatic hydrocarbons,[91] with similar high specific activities and excellent purity. An example is given in Figure 5 with the NMR analyses of phenanthrene, showing relatively even incorporation of tritium.

Although the parameters governing yield, purity and level of tritium incorporation by the MDA technique are still being pursued, the preceding results suggest that it may have great value as a simple, high level labelling process. In addition to the simple organic substrates, carcinogens[91] and steroids[89,90] labelled in this manner, benzodiazepines[92] have also been successfully tritiated.

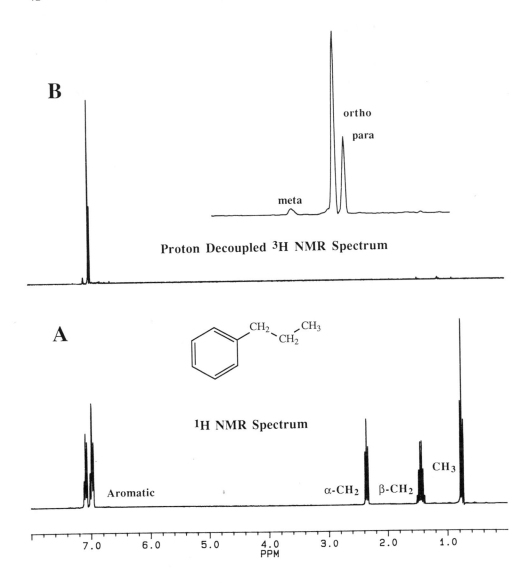

B

ortho

para

meta

Proton Decoupled ³H NMR Spectrum

A

CH₂　　CH₃
　　CH₂

¹H NMR Spectrum

CH₃

Aromatic

α-CH₂　　β-CH₂

7.0　　6.0　　5.0　　4.0　　3.0　　2.0　　1.0
PPM

Figure 4.　NMR spectra of n-Propylbenzene tritiated by the MDA labelling technique.

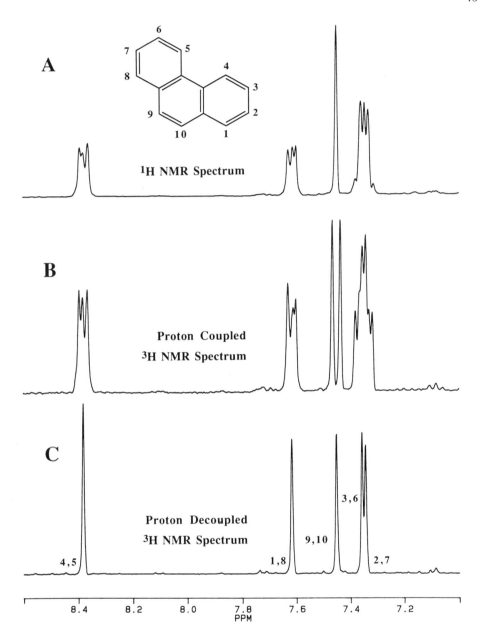

Figure 5. NMR spectra of Phenanthrene tritiated by the MDA labelling technique.

C.   Thermal Atom Labelling

Another variation on the gas exposure technique involves allowing tritium atoms produced by atomization of tritium gas on a hot tungsten wire to impinge on a substrate. These thermal tritons have been reported[93,94] to label molecules at high specific activity and with retention of biological activity; however, details on the purification and radiochemical identification of the products were sparse.

The technique has recently been reinvestigated[95,96] with the substantial benefit of ^{3}H NMR analytical capability. Initially benzene and m-xylene were used as model aromatic substrates for labelling studies under a variety of conditions, a tungsten wire was used for excitation, and volatile products were analysed by gas chromatography in addition to ^{3}H NMR spectroscopy. Reaction of benzene (frozen, 77K)[95] showed predominant saturation with 93% of the labelled products being cyclohexane (Figure 6(b)) - labelled benzene was present in trace amounts only. The standard mass markers for possible products are shown in Figure 6(a). In contrast, when the benzene was held at 213K, (and the surface of the substrate was considered mobile) the relative amount of saturation decreased, but cyclohexane was still the major product with appreciable amounts of hexane and polymers (Figure 6(c)). Labelled benzene accounted for 20% of the incorporated tritium; 1,4 cyclohexadiene was formed, but 1,3 cyclohexadiene was absent. NMR analysis of the reaction products (Figure 7) clearly supports the gas chromatographic data.

Under optimal conditions, the ratio of products could be modified to the point where the fraction of tritiated benzene was 60% of the product radioactivity.[96] Substituting platinum wire for tungsten further increased tritium exchange in benzene to 70%, with only small amounts of tritiated cyclohexane formed. Despite these promising improvements, specific radioactivity levels obtained under all conditions were several orders of magnitude below those previously reported.[93,94]

Table 6

Thermal Atom Labelling of Benzene and m-Xylene in the Presence of Various Metal Wires

Metal	% Incorporation into Benzene	mCi	% Incorporation into m-Xylene	mCi
Palladium	-	-	55	2.3
Nickel	-	-	55	1.0
Rhodium	51	1.3	45	0.1
Iridium	56	0.6	45	2.2
Tungsten	58	5.7	46	25.0
Titanium	-	-	32	5.0
Platinum	72	0.4	<5	0.5

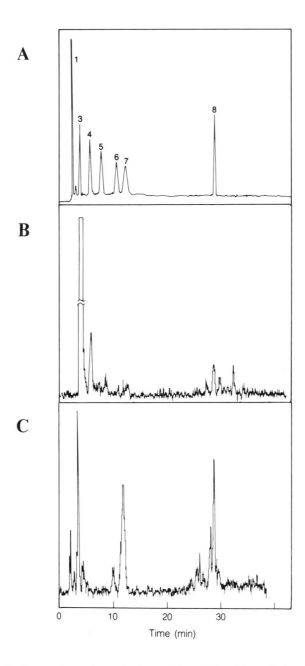

Figure 6. Radio-gas chromatography traces of products of thermal atom labelling of benzene: (A). Mass markers for possible products are (1) hexane, (2) contaminant, (3) cyclohexane, (4) cyclohexene, (5) 1,3 cyclohexadiene, (6) 1,4 cyclohexadiene, (7) benzene, and (8) bicyclohexyl. (B). Radioactivity trace, labelling at 77K. (C). Radioactivity trace, labelling at 213K. (Adapted and reproduced by permission, Elsevier Science Publishers, reference 96).

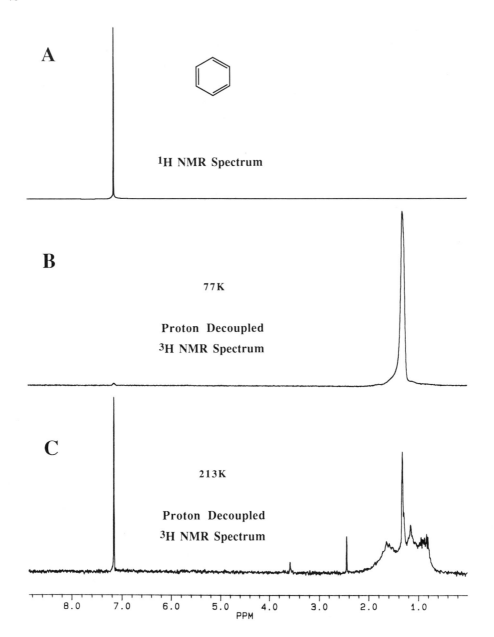

Figure 7. NMR spectra of the labelled products for which the gas chromatography analyses were shown in Figure 6. (Adapted and reproduced by permission, Elsevier Science Publishers, reference 96).

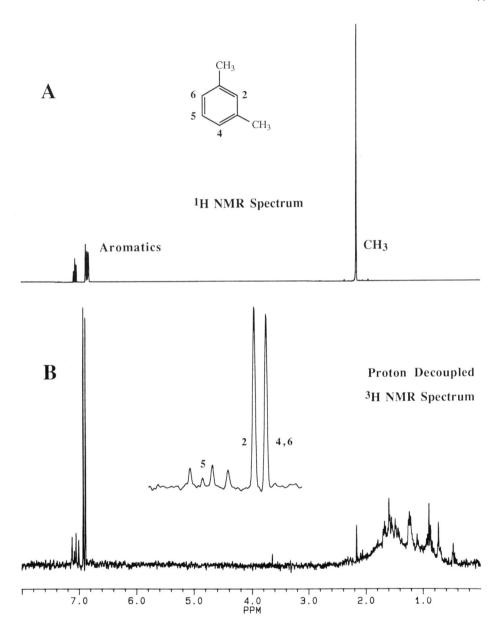

Figure 8. NMR spectra of m-Xylene tritiated by exposure to thermal atoms. (Adapted and reproduced by permission, Elsevier Science Publishers, reference 96).

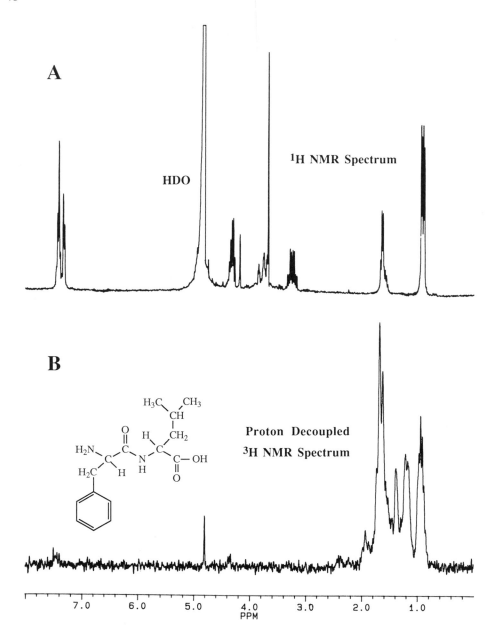

A

HDO

^{1}H NMR Spectrum

B

H$_3$C CH$_3$
   CH
   CH$_2$
O H
‖ | |
C C
H$_2$N   N   C—OH
   | | ‖
H$_2$C H H O

Proton Decoupled
^{3}H NMR Spectrum

7.0   6.0   5.0   4.0   3.0   2.0   1.0
            PPM

Figure 9.   NMR spectra of Phenylalanine-Leucine labelled by exposure to thermal atoms. (Adapted and reproduced by permission, Elsevier Science Publishers, reference 96).

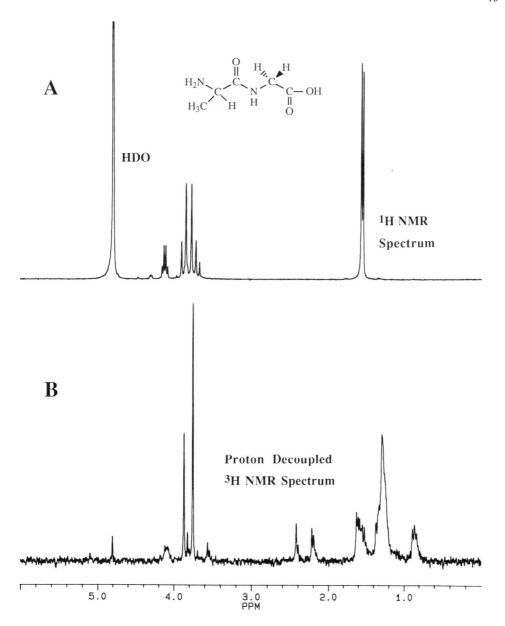

Figure 10. NMR spectra of Alanyl-Glycine tritiated by exposure to thermal atoms. (Adapted and reproduced by permission, Elsevier Science Publishers, reference 96).

Comparison of the labelling of benzene and m-xylene over a variety of wires (Table 6) showed very different reactivities for the two substrates. ^{3}H NMR analysis of the product formed with the tungsten wire and m-xylene as substrate is given in Figure 8, along with the ^{1}H NMR spectrum of the sample. Appreciable tritium is associated with m-xylene (some in the methyl position), and it appears there are also small amounts of other xylenes which are labelled in aromatic positions (6 peaks; 3-m-xylene; 1-p-xylene; 2-o-xylene). However, the majority of tritiated materials again appear to be saturation products (1-2ppm).

Since the main objective of radiation-induced or gas exposure techniques such as thermal atom labelling is to tritiate materials not readily accessible by other techniques, two dipeptides were studied.[95,96] The dipeptide solids were adsorbed onto filter papers, and exposed to the thermal tritons produced on a hot tungsten wire. Phenylalanyl leucine (Figure 9) showed a large amount of ring saturation (0.8-2ppm), but little exchange. Similarly, the NMR spectra of labelled alanyl-glycine in D_2O (Figure 10) suggested that only a little of the tritium was associated with the parent compound.

In summary, the major effect of thermal atom irradiation on solid aromatic centres is saturation. Even under conditions where the surface of the substrate is mobile and appreciable exchange is observed, the resultant mixture contains highly labelled but saturated products. Thus, when dealing with very large substrates, rigorous criteria of radiochemical purity and structural analysis are essential.

V.   METAL CATALYSIS

A.   Heterogeneous Metal Catalysed Hydrogen Isotope Exchange with HTO or Other Solvents

More substrates have been labelled by exchange over metal catalysts than by any other exchange technique. Consequently, the technique is also well represented in the ^{3}H NMR literature. The general technique is well reviewed,[97,98] and the purpose here is to show the detail available from the use of ^{3}H NMR spectroscopy.

The technique was first reported in the 1930's, and applied to a broad range of substrates following the seminal work of Garnett.[99] The mechanism of exchange with water as the isotope source has been thoroughly investigated,[97] and, although other metals have been studied, platinum appears to be the most active. The great majority of reported work is with reduced platinum oxide (Adams catalyst, or Platinum black). As with the other exchange techniques, this work was completed before the availability of the ^{3}H NMR technique. Recent investigations have served to confirm the majority of the earlier mechanistic proposals,[100] and to clarify a number of other points.[101,102] The mechanism of exchange in aromatic centres is thought to involve reversible dissociation of a π-complex to form a σ-bond, and desorption with incorporation of isotope (Figure 11). Sterically hindered positions will not form the complexes as readily as unhindered, and are therefore less labelled. This mechanism is supported by all the published ^{3}H NMR data, and the differences in steric hindrance are easily observed even amongst a series of halobenzenes,[100] as shown in Table 7. Protons adjacent to still larger substituents such as the ortho protons of t-butylbenzene, or in doubly hindered positions (as in m-xylene) show almost no exchange.

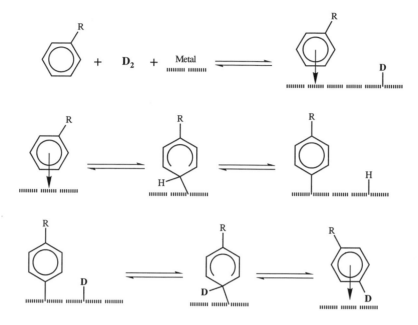

Figure 11. Dissociative π-complex mechanism of exchange of aromatic substrates.

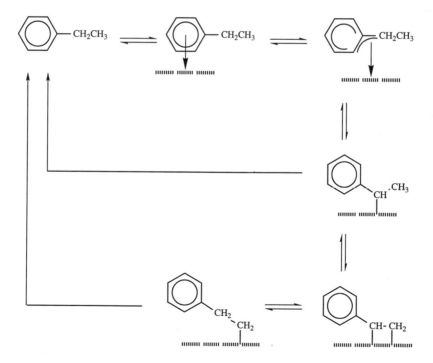

Figure 12. π-Allylic mechanism of exchange of alkylaromatic substrates.

Table 7

Heterogeneous Metal Catalysed Exchange with HTO

Compound	Catalyst	Orientation, %	Ref.
Benzene	Pt	-	104
Toluene	Pt	o-19.6; m-30; p-18; CH_3-33	100
m-Xylene	Pt	2-6; 4,6-16; 5-11; CH_3-66	47
Ethylbenzene	Raney Ni	CH_3-25; CH_2-75	75
n-Propylbenzene	X	α-CH_2-100	3
Fluorobenzene	Pt	o-16.6; m-58; p-25	100
Chlorobenzene	Pt	o-8; m-60; p-32	100
Naphthalene	X	α-49; β-51	3
1,4 Dimethylnaphthalene	Pt	2,3-14; 5,8-4; 6,7-39; CH_3-43	47
Phenanthrene	X	1,8-18; 2,7-21; 3,6-23; 4,5-19; 9,10-19	3
Triphenylene	X	1,4,5,8,9,12-59; 2,3,6,7,10,11-41	3
Pyrene	X	1,3,6,8-36; 2,7-29; 4,5,9,10-35	3
Anthracene	X	1,4,5,8-44; 2,3,6,7-38; 9,10-18	3
Benzo[e]pyrene	X	1,8-19; 2,7-14.5; 3,6-20; 4,5-14.5 9,12-18; 10,11-14	3
7,12 Dimethylbenz[a]-anthracene	PtO$_2$	7-CH_3-6; 12-CH_3-1.5; 9,10-27.1 5,3,2-43; 4-17.6; 8-4.9; 6,11,1<1	105
11-Methyl-15,16-dihydro-cyclopenta[a]phen-anthren-17-one	Pt	16-CH_2-19; 11-CH_3-23.1; 15-CH_2-14.8 2,3-17.4; 12-1.9; 6-8.7; 7-3.2 4-11.9; 1<1	106
5-Hydroxytryptamine Creatinine sulfate	X	2-19; 4-21; 6-22; 7-22; CH_2N-16	41
2'-Deoxyadenosine	X	2-15; 8-85	3
Propanolol hydrochloride	Pt	2-31; 3-10; 4-13; 6,7-46	42
iso-Quinoline	Pt	1-41.2; 3-45; 4-2.6; 5-3.2; 6-2.8 7<0.1; 8-5.3	101
Lutidine	Pt	CH_3-76; 3,5-7; 4-17	102
Phenanthridine	Pt	1,10-2; 2-14; 3,8-34; 4-11; 6-10; 7-13 9-16	102
2-Picoline	Pt	CH_3-47; 3-13; 4-13; 5-15; 6-12	102
Pyridine	Pt	2,6-43.6; 3,5-22.6; 4-33.8	101
Pyridoxine hydrochloride	X	CH_3-40; 4-CH_2OH-4.9; 6-55.1	3
Nicotine D-bitartrate	X	2-11.6; 4-8.3; 5-12; 6-11; 2'-12.4 3'a-5.1; 3'b-3.2; 4'a-6.5; 4'b-1.9 5'a-18.8; 5'b-9.2	3
Tryptamine hydrochloride	X	2-56.5; $ring$-18.5 and 24.9	3
L-Phenylalanine	Pt	β-CH_2-26; α-CH-2; o-27; m-29; p-16	107
L-Proline	X	2-29; 3α-6; 4α-5; 4β-5; 5α-28; 5β-26	108
D,L-Threonine	X	2-71.6; 3-28.4	3
L-Tryptophan	Pt	$sidechain$-8; 2-7; $ring$-85	103
L-Aspartic Acid	X	2-39.3; 3-25.2; 3-35.5	41
Shale oils	Raney Ni	General	109

X - Catalyst not given in publication, but most probably reduced PtO$_2$.

Alkanes exchange slowly relative to aromatic centres,[97] and the mechanism is thought to involve direct σ-bond formation. The alkyl groups of alkyl aromatics are also labelled by metal-catalysed exchange, and this exchange is both rapid and has a distinctive pattern. Initial adsorption is thought to occur by π-complexation, before formation of a π-allyl complex and an alkyl metal-carbon bond, as shown in Figure 12. Further exchange in the side-chain, beyond the α-position, could occur as shown in the Figure. Two results (**24**, **25**) illustrate this type of labelling:

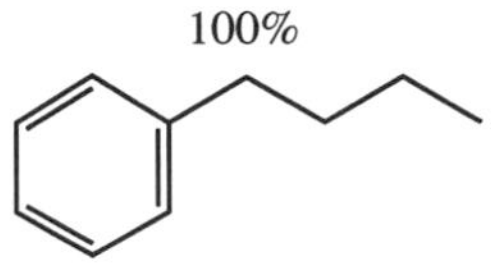

There is also considerable variation amongst the group VIII metals for aromatic vs alkyl labelling, as shown above for Raney nickel (**24**) and platinum (**25**) exchange of n-butylbenzene.

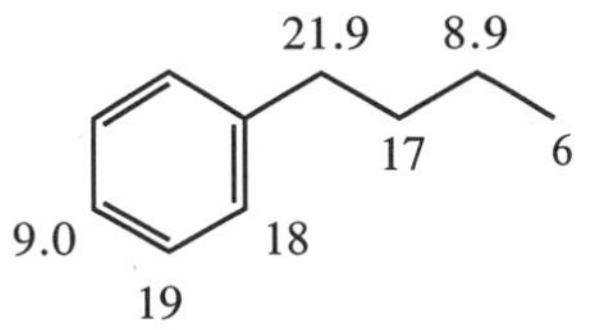

One of the great advantages of metal-catalysed exchange over other techniques is its applicability to a broad range of substrates. Notably these include compounds containing hetero-atoms, which are often poorly labelled by other methods. A selection of results for nitrogen-containing substrates, polycyclic aromatic hydrocarbons, amino acids and drugs are included in Table 7. Several other features of metal-catalysed exchange are borne out in the orientations derived from ^{3}H NMR study of the labelled products.

Comparison of the two results for 3-picoline (**26**, **27**) are very informative.[101] It is clear that primary adsorption involves the nitrogen atom, and most exchange is adjacent to this atom. Steric hindrance by the methyl group obviously affects exchange adjacent to that group. Even at higher temperatures (130 vs 60°C, **27** vs **26**) these effects are still apparent, although somewhat masked by the loss of specificity due to more vigorous exchange conditions. In general, orientational effects may be obscured after long reaction times or at high temperatures, but the sensitivity of the ^{3}H NMR technique allows analysis very early in the exchange cycle (*i.e.* ≤ 20% incorporation of tritium), when initial conditions exist and only low levels of isotope are incorporated.

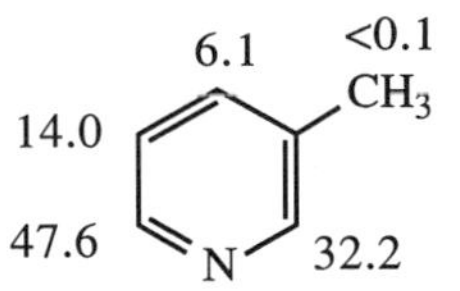

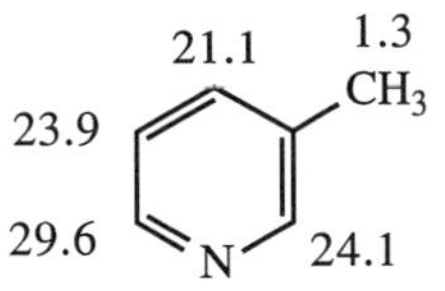

The analysis of 7-methylbenz(a)anthracene (**28**) is a good example of the variation in steric constraints over all the positions of one molecule. Amino acids and many drugs are readily labelled, and favour the least hindered, aromatic positions over aliphatic hydrogens for exchange (**29**, **30**).[42,103]

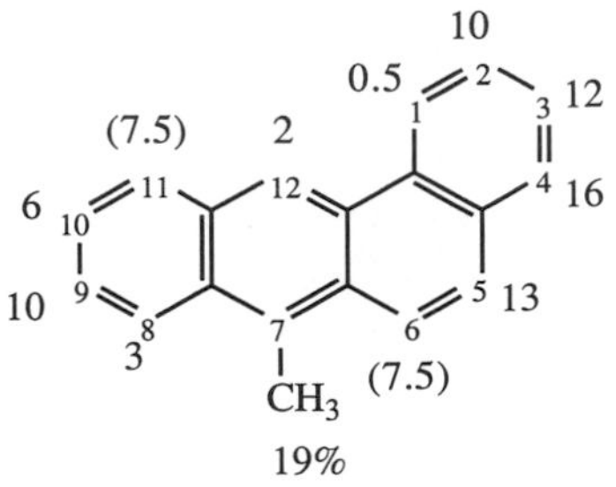

28

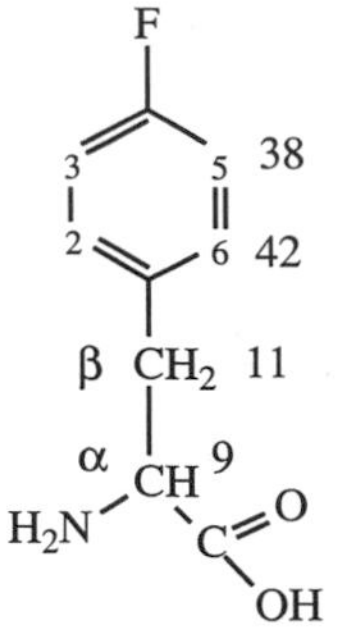

29

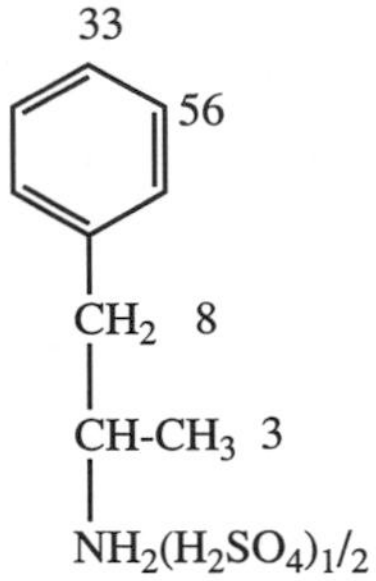

30

Limitations of the heterogeneous metal HTO labelling techniques include the inability to label nitro or iodo-containing substrates, racemisation of many optically active compounds and disproportionation of some reactants.[97]

B. Heterogeneous Metal Catalysed Exchange with T_2

This process was first published as a catalysed Wilzbach experiment.[110,111] The technique has several advantages, notably the possibility of introducing large amounts of tritium from a carrier free isotope source (T_2). However, this is also a disadvantage, since many substrates are catalytically hydrogenated under the reaction conditions. The fact that the use of a tritiated solvent as isotope source precludes these unwanted reactions has far outweighed the fact that extremely high specific activities are difficult to attain with the HTO method.

Evans *et al* [112] pioneered a variation of the tritium gas technique which yields high incorporation, specific labelling and very little degradation or hydrogenation. Briefly, the simple technique involves stirring a buffered solution (pH 7) of the substrate in the presence of an atmosphere of tritium gas and a supported metal catalyst for several hours. The method has been used to good effect in labelling a wide variety of substrates including purines, purine nucleosides and nucleotides, aromatic amines and amino acids, carbohydrates and steroids. A selection of results are given in Table 8, and the remarkable feature of the results is the specificity of the exchange, as illustrated by the following orientation data for glucose (**31**) and a blocked amino acid (**32**).[113]

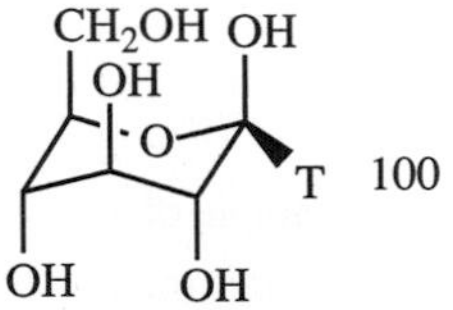

31

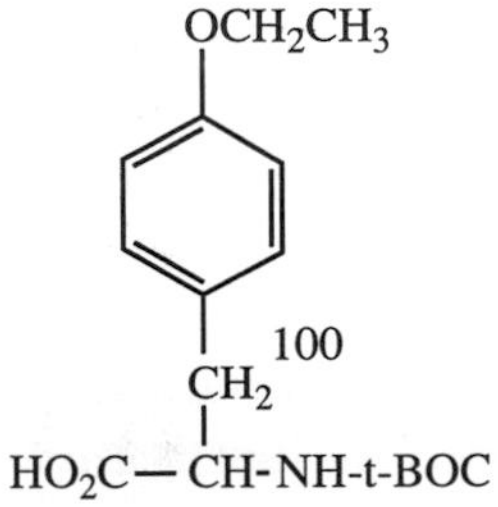

32

Table 8

Heterogeneous Metal Catalysed Exchange with T_2

Compound	Catalyst	Orientation, %	Ref.
Methotrexate	Pd/CaCO$_3$	*7-100*	117
Adenine β-D-arabinoside	PdO/BaSO$_4$	*2-48; 8-52*	3
Adenosine hydrochloride	PdO/BaSO$_4$	*5'-38,22; 2-4; 8-36*	41
Adenosine	PdO/BaSO$_4$	*2-22; 8-78*	3
Adenosine cyclic 5' monophosphate	PdO/BaSO$_4$	*8-100*	3
Adenosine 5' triphosphate	PdO/BaSO$_4$	*5'-62.2; 2-10.6; 8-27.2*	3
Caffeine	PdO/BaSO$_4$	*8-100*	41
β,γ-Methylene ATP	-	*2-73.8; 8-26.2*	3
Theophylline	PdO/BaSO$_4$	*8-100*	41
Estradiol 17-cyclopentyl ether	Pd/C	*6β-33; 7α-1; 9α-66*	73
Estriol	Pd/C	*6α-61; 9-39*	118
Estriol	Pd/C	*2-26.4; 4-19; 6α-31; 9-24*	118
Estrone β-D-glucuronide	Pd/C	*6α-59.2; 9-40.8*	3
Estrone sulfate	Pd/C	*6α-50; 9-50*	118
2-Hydroxyestradiol	-	*6α,β-58; 9-42*	3
2-Hydroxyestrone	-	*6-52.6; 9-42.7; 16-4.7*	3
2-Amino-6,7-dihydroxyl 1,2,3,4 tetrahydronaphthalene (ADTN)	PdO/BaSO$_4$	*1ax-26; 1eq-17; 3ax-6 3eq-4; 4-44; 5,8-3*	41
2,4 Diamino pteridine-β-carboxylic acid	Pd/CaCO$_3$	*7-100*	117
Folic acid	Pd/CaCO$_3$	*7-59; 9-CH$_2$-41*	117
Pyridoxine hydrochloride	-	*CH$_3$-63.4; 4-CH$_2$OH-22.4 5-CH$_2$OH-2; 6-12.2*	3
Tyramine hydrochloride	PdO/BaSO$_4$	*side-chain 2-100*	3
2',3'-Dideoxyadenosine	Pd/C	*8-100*	119
D-Fucose	PdO/BaSO$_4$	*1α-34; 1β-66*	3
L-Fucose	PdO/BaSO$_4$	*1α-34.4; 1β-65.6*	3
2-Deoxy-D-glucose	PdO/BaSO$_4$	*1α-50; 1β-50*	3
D-Galactose	PdO/BaSO$_4$	*1α-34.3; 1β-65.7*	3
D-Glucosamine	PdO/BaSO$_4$	*1α-63; 1β-37*	3
D-Glucose	PdO/BaSO$_4$	*1α-37.5; 1β-62.5*	120
D-Mannose	PdO/BaSO$_4$	*1α-61; 1β-39*	3
n-Hexylpropionate	Rh black	*CH$_3$'s most labelled*	5
n-Pentylpropionate	Rh black	*CH$_3$'s most labelled*	121
t-Boc-O-Ethyl-D-Tyrosine	Pd/BaSO$_4$	*Benzyl-100*	113
Toluene	Pd	*o<1; m-6.4; p-3; CH$_3$-90*	64
n-Hexylbenzene	Pt black	*o<1; m-20; p-10.1; α-CH$_2$-48 β-CH$_2$-15; γ-CH$_2$-3.2 δ,ε-CH$_2$'s-2.8; CH$_3$<1*	64
iso-Propylbenzene	Pt black	*o<1; m-40; p-19; CH-21 CH$_3$-20.4*	64
n-Hexane	Pt black	*CH$_3$-39.0; 2-CH$_2$-31.8; 3-CH$_2$-29.2*	64
n-propane	Raney Ni	*CH$_3$/CH$_2$ ratio = 2.4*	114
Dibenz[a,j]acridine	Pd/CaCO$_3$	H14	122
Dibenz[a,h]acridine	Pd/CaCO$_3$	H14	122
Dibenz[c,h]acridine	Pd/CaCO$_3$	H7	122
Thyroliberin (TRF)	Pd/Al$_2$O$_3$	C-2,C-5(1/6)	123
Triethylsilane	Raney Ni	*SiH-37; CH$_2$-38; CH$_3$-25*	115
Tetramethylsilane	Raney Ni	*CH$_3$-100*	115
Hexamethyldisiloxane	Raney Ni	*CH$_3$-100*	116
Chlorodimethylsilane	Raney Ni	*SiH-57; CH$_3$-43*	115

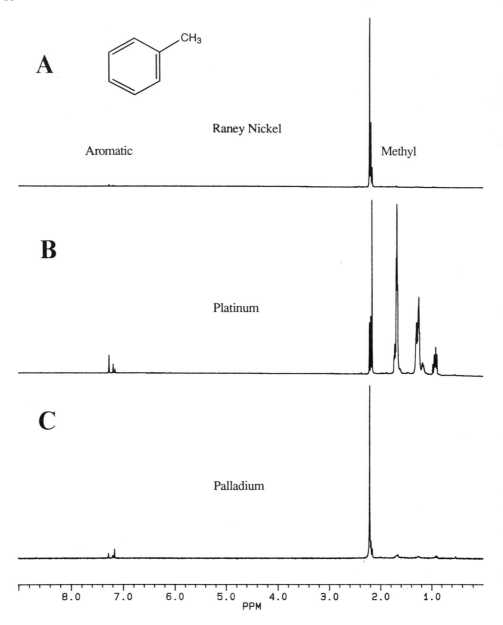

Figure 13. Proton decoupled tritium NMR spectra of Toluene labelled by exposure to tritium gas over reduced metal for five hours at room temperature. (Reproduced by permission, Elsevier Science Publishers, reference 124).

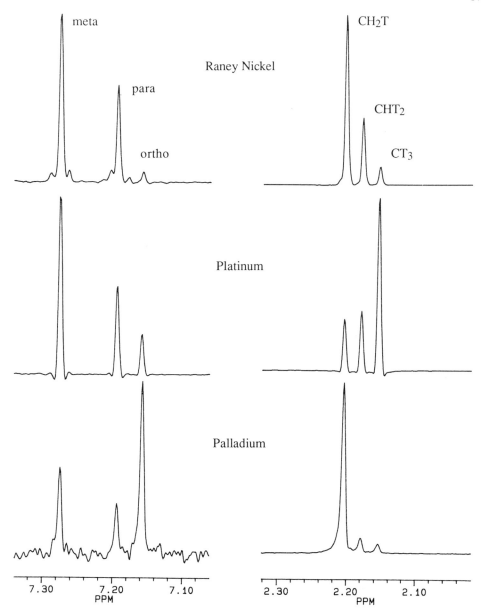

Figure 14. Expanded views of the aliphatic (2.02-2.32ppm) and aromatic (7.06-7.36ppm) regions of the spectra in Figure 13. (Reproduced by permission, Elsevier Science Publishers, reference 124).

In studies of unsupported metal catalysts Raney nickel has been demonstrated as an effective catalyst for the labelling of alkanes,[114] aromatic compounds[64] and silanes.[115,116] The variation of labelling patterns between metals can be very marked with tritium gas as isotope source, as previously noted with HTO. However, in general aromatic centres are not the predominant site of labelling as noted in Pt/HTO exchange, and most metals tend to the high α-CH incorporation previously observed with Raney nickel as the catalyst and HTO as isotope source.

The exquisite detail provided by careful ^{3}H NMR analysis of reaction products is illustrated by the data in Figure 13.[124] Toluene was labelled at room temperature over unsupported Pt, Pd and Ni, and the majority of the tritium was incorporated in the methyl groups in all three cases. Over platinum catalyst a large amount of hydrogenation also took place, yielding highly tritiated methyl cyclohexane (Figure 13B, 0.5-1.8ppm). The expanded methyl and aromatic regions of the spectra reveal additional information about the labelling processes under these reaction conditions (Figure 14). In particular, the multiply labelled methyl species would suggest that exchange is rapid over all catalysts relative to desorption. This is especially true in the case of platinum, where CT_2 and CT_3 labelled methyl groups are very abundant, and hydrogenation is also a major feature of the reaction. Analysis of the ring labelling patterns shows that slow exchange due to steric hindrance of the ortho position is obvious with platinum and nickel, but the toluene labelled over palladium has a large amount of ortho labelling. This effect was not observed when exchange was conducted at higher temperature and lower tritium pressure,[56,64] as given in Table 8.

These subtle differences, revealed only by ^{3}H NMR analyses, may suggest that there are several slightly different mechanisms of metal-catalysed exchange, and the predominant process (and therefore orientation) depends on exact conditions such as the particular metal, pH, isotope source and reaction temperature.

Recent work[125] has shown that nitrobenzene may be labelled in heterogeneous metal/T_2 systems, in contrast to the well-known lack of reactivity of this substrate in the comparable HTO experiment. Since the only obvious difference between the two cases is the presence of adsorbed OH on the metal surface in HTO exchange, it is not clear why the results differ. This phenomenon has been observed previously,[126] when the presence of water was found to influence the orientation of hexane exchange with D_2 over clean platinum surfaces.

In summary, metal/T_2 exchange procedures offer the opportunity for high specific activity products, and remarkable specificity in labelling, but have not been fully pursued.

C.   Homogeneous Metal Catalysed Exchange with HTO

The inability to label substrates such as nitrobenzene, naphthalene and acetophenone by heterogeneous exchange led to development of the homogeneous metal systems.[127,128] These catalytic systems were first proposed with tetrachloroplatinate as the metal salt, with acetic acid present to ensure a single phase for the aqueous isotope source, catalyst and organic substrate. Facile exchange was observed at 80°C, and the orientations of labelling were very similar to those from heterogeneous metal techniques.

Table 9

Homogeneous Tetrachloroplatinate Catalysed Exchange with HTO

Compound	Catalyst	Orientation, %	Ref.
1,3 Dinitronaphthalene	K_2PtCl_4	$C5$-10; $C6$-45; $C7$-45	78
Toluene	Na_2PtCl_4	o<0.1; m-44.8; p-23.0; CH_3-34.2	50
Ethylbenzene	Na_2PtCl_4	o<0.1; m-61.2; p-38.8; $alkyl$<0.1	50
n-Propylbenzene	Na_2PtCl_4	o<0.1; m-55.4; p-27.3; α-4.4; β-3.0; CH_3-10.2	50
n-Butylbenzene	Na_2PtCl_4	o<0.1; m-57.0; p-29.9; CH_2<0.1; CH_3-12.9	50
n-Hexylbenzene	Na_2PtCl_4	o<0.1; m-61.2; p-30.2; α–CH_2<0.1; $C2$-6<0.1; CH_3-8.7	50
n-Heptylbenzene	Na_2PtCl_4	o<0.1; m-57.0; p-27.1; α–CH_2-5.4; CH_2-1.8; CH_3-8.7	50
iso-Propylbenzene	Na_2PtCl_4	o<0.1; m-59.0; p-29.9; CH<0.1; CH_3-10.8	50
iso-Butylbenzene	Na_2PtCl_4	o<0.1; m-59.2; p-32.3; CH_2<0.1; CH-0.1 CH_3-8.4	50
sec-Butylbenzene	Na_2PtCl_4	o<0.1; m-55.2; p-30.7; CH<0.1; CH_2<0.1 α-CH_3-4.2; γ-CH_3-9.6	50
t-Butylbenzene	Na_2PtCl_4	o<0.1; m-62.2; p-37.9; CH_3<0.1	50

A selection of results using tetrachloroplatinate catalyst are given in Table 9. There are several features of the orientations which should be stressed, and are shown in the Figures below:[50]

| **33** | **34** | **35** |

Firstly (**33**), ortho deactivation appears more pronounced than in the case of heterogeneous catalysis; it would seem more difficult for the substrate to complex with the multiply substituted metal atom in homogeneous catalysis, than with a metal surface as in the heterogeneous catalysis case. Secondly, bulky alkyl groups cause some deactivation of meta positions (as well as ortho, of course) in comparison to para labelling (**34**). The labelling of long-chain alkylbenzenes shows a curious pattern of alkyl labelling, where the terminal (CH_3) positions appear to be as well tritiated as the α-CH_2 hydrogens (**35**). To the everlasting credit of the authors[129] this was first detected by difference ^{1}H NMR spectroscopy. It is a small effect, but is clearly shown by ^{3}H NMR spectroscopic studies, and close analysis of the alkylbenzene results in Table 9 suggests that it may be observed for all chain-lengths greater than 3-carbons, or with a reasonable amount of chain branching. The phenomenon was explained[130] in terms of a "terminal abstraction π-complex" (TAPC) mechanism which requires initial complexation through the aromatic centre and curling around of the chain for exchange in remote positions. Since alkanes also exchange under the reaction conditions,[131] it's entirely possible that the result is caused by direct competition of aromatic and alkyl complexation. One of the major advantages of homogeneous catalysis over heterogeneous is illustrated by the labelling of 1,3 dinitronaphthalene.[78]

Other metal salts have been shown to catalyse exchange under homogeneous conditions, and iridium[132] and rhodium[133] were used to label both alkylbenzenes and alkanes. Generally, salts other than platinum require more stringent conditions, especially for alkane labelling. These conditions can often cause metal precipitation, at which point the exchange observed may be heterogeneously catalysed, or be due to the acid present. A selection of results for Ir^{3+} as catalyst are given in Table 10.[50] Several features are immediately clear: comparison of the two toluene results reveals that orientational differences may be obscured after long exchange times. Survey of the results shows that alkyl exchange was never observed, in contrast to the results with tetrachloroplatinate as catalyst. As an extension of this, one might expect that alkane exchange would be slow, and this is the case.[50] As reported for heterogeneous platinum catalysis,[100] the labelling of halobenzenes shows the effect of steric hindrance of the halogen eg. the larger the halogen, the less ortho labelling observed.

Table 11 lists a number of homogeneous labelling results with $RhCl_3$[47,134] and $(Ph_3P)_3RuCl_2$[135] as catalysts. The orientation of exchange in aromatic rings with $RhCl_3$ as catalyst is markedly different from other homogeneous metal systems,[47,134] and a mechanism has been proposed to explain this difference.[136,137] In any case, advantage has been taken of the phenomenon to yield very specifically labelled benzamides and benzoic acids,[134] as shown below (36-38):

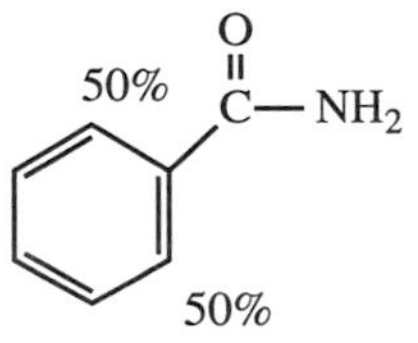

36 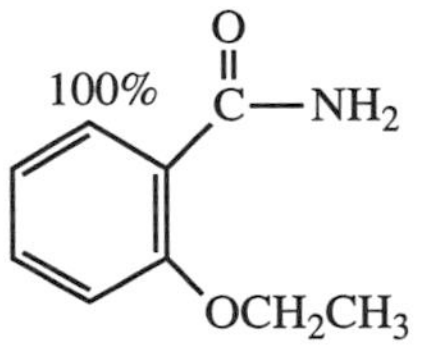   **37** 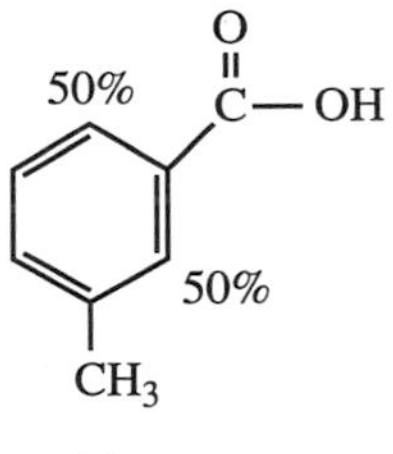   **38**

A series of primary and secondary alcohols has been specifically tritiated by the application of $(Ph_3P)_3RuCl_2$ as catalyst.[135] The specificity of the labelling is excellent for primary alcohols, but some specificity loss occurs for secondary alcohol substrates.

Homogeneous metal catalysed exchange systems offer several advantages over the related heterogeneous technique. Nitro and other substrates which poison heterogeneous systems are readily tritiated. Most substrates are soluble in the solvent systems used for homogeneous catalysts, and so there is no fear of extra solvents poisoning exchange or competing with the substrate. Disadvantages include the difficulty of retrieving labelled products, and the mid-to-low specific activities attainable, as a result of the proton rich solvent systems used in homogeneous exchange.

Table 10

Homogeneous Iridium Catalysed Exchange with HTO

Compound	Catalyst	Orientation, %	Ref.
Toluene[a]	Na_3IrCl_6	o-18.8; m-43.0; p-38.3; $alkyl$<0.1	50
Toluene[b]	Na_3IrCl_6	o-23.8; m-47.4; p-28.9; $alkyl$<0.1	50
iso-Propylbenzene	Na_3IrCl_6	o-24.0; m-35.0; p-41.0; $alkyl$<0.1	50
iso-Butylbenzene	Na_3IrCl_6	o-18.6; m-48.8; p-32.6; $alkyl$<0.1	50
s-Butylbenzene	Na_3IrCl_6	o-10.0; m-54.4; p-35.7; $alkyl$<0.1	50
t-Butylbenzene	Na_3IrCl_6	o-18.0; m-52.6; p-29.3; $alkyl$<0.1	50
Cyclohexylbenzene	Na_3IrCl_6	o-7.8; m-56.4; p-35.7; $alkyl$<0.1	50
o-Xylene	Na_3IrCl_6	$3,6$-41.2; $4,5$-58.8; CH_3<0.1	50
m-Xylene	Na_3IrCl_6	2-21.4; $4,6$-50.4; 5-28.2; CH_3<0.1	50
1,2,4 Trimethylbenzene	Na_3IrCl_6	3-84.6; 5-8.9; 6-6.5; CH_3<0.1	50
Fluorobenzene	Na_3IrCl_6	o-20.4%; m-39.0; p-40.7	50
Chlorobenzene	Na_3IrCl_6	o-13.8%; m-41.4; p-44.9	50
Bromobenzene	Na_3IrCl_6	o-6.6%; m-63.0; p-30.4	50
Naphthalene	Na_3IrCl_6	α-20.8%; β-79.2	50
Biphenyl	Na_3IrCl_6	o-18.4; m-56.8; p-24.8	50

a. 8 hours - 16.1% Approach to Eqm.
b. 264 hours - 54.2% Approach to Eqm.

Table 11

Homogeneous Rhodium or Ruthenium Catalysed Exchange with HTO

Compound	Catalyst	Orientation, %	Ref.
iso-Propylbenzene	$RhCl_3$	$2,6$-47; 4-42; CH_3-10	47
1,3,5 Trimethylbenzene	$RhCl_3$	$2,4,6$-100; CH_3<1	47
Benzamide	$RhCl_3$	$2,6$-97	134
Benzoic acid	$RhCl_3$	$2,6$-99	134
2-Ethoxybenzamide	$RhCl_3$	6-96	134
2-Hydroxybenzamide	$RhCl_3$	6-66; $3,5$-34	134
2-Hydroxybenzoic acid	$RhCl_3$	6-9; $3,5$-91	134
4-Methoxybenzoic acid	$RhCl_3$	$2,6$-99	134
2-Methoxybenzoic acid	$RhCl_3$	6-98	134
3-Methylbenzoic acid	$RhCl_3$	$2,6$-98	134
4-Methylbenzoic acid	$RhCl_3$	$2,6$-99	134
4-Oxo-4H-chromene-2-carboxylic acid	$RhCl_3$	3-97	134
Ethanol	$(Ph_3P)_3RuCl_2$	α-100	135
1-Heptanol	$(Ph_3P)_3RuCl_2$	α-86; β-14	135
3-Phenyl-1-propanol	$(Ph_3P)_3RuCl_2$	α-96; β-4	135
1-Octadecanol	$(Ph_3P)_3RuCl_2$	α-88; β-4; $other$-8	135
Benzyl alcohol	$(Ph_3P)_3RuCl_2$	α-100	135
2-Pentanol	$(Ph_3P)_3RuCl_2$	α-CH-14; 1-CH_3-48; 3-CH_2-38	135
2-Decanol	$(Ph_3P)_3RuCl_2$	α-CH-18; 1-CH_3-49; 3-CH_2-34	135
2-Hexadecanol	$(Ph_3P)_3RuCl_2$	α-CH-20; 1-CH_3-47; 3-CH_2-33	135

D.   Homogeneous Metal Catalysed Exchange with T_2

     There are no reported examples of this type of exchange, although the first report of homogeneous alkane exchange[131] with D_2O also discussed attempts to exchange alkanes with D_2. The over-riding reason for the lack of activity in this area is probably due to reports that metal complexes will be reduced to the heterogeneous metal by the presence of hydrogen (tritium) gas.[98] This may be the case for some systems, but the success of Wilkinson's catalyst for homogeneous hydrogenation reactions, and the recent reports of ^3H[138] and ^2H NMR studies[139] of hydrogen complexed by metals in homogeneous organometallic systems suggests that T_2 exchange should be possible. If realized, this eventuality would quickly remove the specific activity limitations imposed by the use of HTO as isotope source. It is also clear that development of a homogeneous tritium gas exchange technique will have to take account of such factors as the exchange of solvents, *etc.*

VI.   SUMMARY

     Hydrogen isotope exchange techniques are many and varied, and can be exceptionally powerful. Extremely specific labelling may be obtained, as in the case of base systems or $Pd/BaSO_4$ catalysis. High incorporation may also be observed (up to 10's of Ci/mmole), but not always in the same case as specific labelling. The availability of extremely high specific activity HTO (max. 2650 Ci/g) would eliminate all the specific activity limitations of the catalytic systems discussed here. However, this reagent should be treated with caution since a lethal dose (LD_{50}) of HTO is reported to be approximately 1Ci/kg in man,[140] and 70Ci of T_2O is only 27µL (70kg=157lb man).

     Exchange techniques offer the ability to label a substrate without any prior or subsequent chemical synthesis, and with complicated biological molecules this is often essential. In a similar way, ^3H NMR spectroscopy[3] offers the opportunity to non-destructively assay the results of labelling experiments on ever-smaller quantities of radiochemical. Although the sensitivity of the NMR technique will never rival that of liquid scintillation counting, modern high field spectrometers allow the observation of µCi quantities of tritium.

ACKNOWLEDGEMENTS

     Thank you to Mervyn Long, Hiromi Morimoto, Chin-Tzu Peng, Alexander Susan and James Wiley for permission to use data prior to publication. Special thank you to Vangie Peterson for patient and painstaking preparation of the manuscript and diagrams. PGW is supported by the U.S. National Institutes of Health, Biotechnology Resources Program, Division of Research Resources, under Grant P41 RR01237, and the U.S. Department of Energy under Contract DE-AC03-76SF00098.

REFERENCES

1   G.V.D. Tiers, C.A. Brown, R.A. Jackson and T.N. Lahr, J. Am. Chem. Soc., **86** (1964) 2526-2527.

2   J.A. Elvidge, "Tritium Nuclear Magnetic Resonance Spectroscopy", in: The Multinuclear Approach to NMR Spectroscopy, J.B. Lambert and F.G. Riddell (Eds.), Reidel: London (1983) 169-206.

3   E.A. Evans, D.C Warrell, J.A. Elvidge and J.R. Jones, "Handbook of Tritium NMR Spectroscopy and Applications", Wiley and Sons: Chichester (1985).

4   E.A. Evans, D.C. Warrell, J.A. Elvidge and J.R. Jones, J. Radioanal. Chem., **64** (1981) 41-45.

5   J.L. Garnett, M.A. Long and A.L. Odell, Chem. Aust., **47** (1980) 215-220.

6   J.R. Jones, Synthesis and Applications of Isotopically Labeled Compounds, Proc. Int. Symp., W.P. Duncan and A.B. Susan (Eds.), Elsevier: Amsterdam (1983) 303-308.

7   R.R. Fraser and R.N. Renaud, J. Am. Chem. Soc., **88** (1966) 4365-4370.

8   P.A. Colfer, T.A. Foglia and P.E. Pfeffer, J. Org. Chem., **44** (1979) 2573-2575.

9   J. Bloxsidge, J.A. Elvidge, J.R. Jones and E.A. Evans, Org. Magn. Reson., **3** (1971) 127-138.

10 J.M.A. Al-Rawi, J.P. Bloxsidge, C. O'Brien, D.E. Caddy, J.A. Elvidge, J.R. Jones and E.A. Evans, J. Chem. Soc., Perkin Trans. 2, (1974) 1635-1638.

11 J.M.A. Al-Rawi, J.A. Elvidge, J.R. Jones and E.A. Evans, J. Chem. Soc., Perkin Trans. 2, (1975) 449-452.

12 L.J. Altman and N. Silberman, Steroids, **29** (1977) 557-565.

13 J.P. Bloxsidge, J.A. Elvidge, J.R. Jones, R.B. Mane and E.A. Evans, J. Chem. Res. (S), (1977) 258-259.

14 F.M. Kaspersen, C.W. Funke, E.M.G. Sperling and G.N. Wagenaars, J. Labelled Comp. Radiopharm., **24** (1987) 219-225.

15 V.M.A. Chambers, E.A. Evans, J.A. Elvidge and J.R. Jones, "Tritium Nuclear Magnetic Resonance (tnmr) Spectroscopy", Review 19, Radiochemical Centre: Amersham (1978).

16 J.A. Elvidge, J.R. Jones, V.M.A. Chambers and E.A. Evans, "Tritium Nuclear Magnetic Resonance Spectroscopy", in: Isotopes in Organic Chemistry, Vol. 4, Tritium in Organic Chemistry, E. Buncel and C.C. Lee (Eds.) Elsevier: Amsterdam (1978) 1-49.

17 J.A. Elvidge, "Deuterium and Tritium Nuclear Magnetic Resonance Spectroscopy", in: Isotopes: Essential Chemistry and Applications, Spec. Publ. No. 35, Chemical Society: London (1980) 123-194.

18 J.P. Bloxsidge and J.A. Elvidge, Prog. Nucl. Magn. Reson. Spectrosc., **16** (1983) 99-114.

19 A.L. Odell, "Tritium NMR", in: NMR of Newly Accessible Nuclei, Vol. 2, P. Laszlo (Ed.) Academic: New York (1983) 27-48.

20 P.G. Williams, Fusion Technology, **14** (1988) 840-844.

21 C.K. Ingold, C.G. Raisin and C.L. Wilson, Nature, **134** (1934) 734.

22 C.K. Ingold, C.G. Raisin and C.L. Wilson, J. Chem. Soc., (1936) 1643-1645.

23 S.K. Hsu, C.K. Ingold, C.G. Raisin, E. de Salas and C.L. Wilson, J. Chim. Phys., **45** (1948) 232-236.

24   R.L. Burwell and G.S. Gordon, J. Am. Chem. Soc., **70** (1948) 3128-3132.

25   G.S. Gordon and R.L. Burwell, J. Am. Chem. Soc., **71** (1949) 2355-2359.

26   R.L. Burwell, R.B. Scott, L.G. Maury and A.S. Hussey, J. Am. Chem. Soc., **76** (1954) 5822-5827.

27   R.L. Burwell, L.G. Maury and R.B. Scott, J. Am. Chem. Soc., **76** (1954) 5828-5831.

28   R.L. Burwell and A.D. Shields, J. Am. Chem. Soc., **77** (1955) 2766-2771.

29   O. Beeck, J.W. Otvos, D.P. Stevenson and C.D. Wagner, J. Chem. Phys., **16** (1948) 255-256.

30   D.P. Stevenson, C.D. Wagner, O. Beeck and J.W. Otvos, J. Am. Chem. Soc.,**74** (1952) 3269-3282.

31   J.W. Otvos, D.P. Stevenson, C.D. Wagner and O. Beeck, J. Am. Chem. Soc., **73** (1951) 5741-5746.

32   D.N. Kursanov, V.N. Setkina and A. Mescheryakov, Dokl. Chem. (Engl. Transl.), **105** (1965) 279.

33   P.B.D. de la Mare, "Aromatic Substitution", Butterworths: London (1959).

34   G.A. Olah, "Freidel Crafts and Related Reactions", Interscience: New York (1963).

35   L.M. Stock, "Aromatic Substitution Reactions", Prentice-Hall: New Jersey (1968).

36   R.O.C. Norman and R. Taylor, "Electrophilic Substitution in Benzenoid Compounds", Elsevier: Amsterdam (1965) 25-30.

37   J.P. Bloxsidge, J.A. Elvidge, J.R. Jones, R.B. Mane, V.M.A. Chambers, E.A. Evans and D. Greenslade, J. Chem. Res. (S), (1977) 42-43.

38   B. Aliprandi and F. Cacace, Ann. Chim. (Rome), **51** (1961) 397-401.

39   P.M. Yavorsky and E. Gorin, J. Am. Chem. Soc., **84** (1962) 1071-1072.

40   P.C. Crossley, R.W. Martin, J.B. Mawson and A.L. Odell, J. Labelled Comp. Radiopharm., **17** (1980) 779-784.

41   J.P. Bloxsidge, J.A. Elvidge, M. Gower, J.R. Jones, E.A. Evans, J.P. Kitcher and D.C. Warrell, J. Labelled Comp. Radiopharm., **18** (1981) 1141-1165.

42   J.A. Elvidge, J.R. Jones and M. Saljoughian, J. Pharm. Pharmacol., **31** (1979) 508-511.

43   F. Frappier, M. Audinot, J.P. Beaucourt, L. Sergent and G. Lukacs, J. Org. Chem., **47** (1982) 3783-3785.

44   J.A. Elvidge, J.R. Jones, M.A. Long and R.B. Mane, Tetrahedron Lett., **49** (1977) 4349-4350.

45   M.A. Long, J.L. Garnett and R.F.W. Vining, J. Chem. Soc., Perkin Trans. 2, (1975) 1298-1303.

46   J.A. Elvidge and J.R. Jones, personal communication.

47   J.M.A. Al-Rawi, J.A. Elvidge, J.R. Jones, R.B. Mane and M. Saieed, J. Chem. Res. (S), (1980) 298-299.

48   M.A. Long, J.L. Garnett and J.C. West, Tetrahedron Lett., **43** (1978) 4171-4174.

49   J.L. Garnett, M.A. Long and C.A. Lukey, J. Labelled Comp. Radiopharm., **22** (1985) 641-647.

50   C.A. Lukey, Ph.D. Thesis, University of New South Wales (1983).

51   W.P. Duncan, E.J. Ogilvie and J.F. Engel, J. Labelled Comp., **11** (1975) 461-463.

52   A.B. Susan and J.C. Wiley, Polynuclear Aromatic Hydrocarbons: Formation, Metabolism, and Measurement, Proc. Seventh Int. Symp., M. Cooke and A.J. Dennis (Eds.), Battelle Press: Columbus, Ohio, (1983) 1153-1159.

53   A.B. Susan, J.C. Wiley and P.G. Williams, to be published.

54   E.A. Evans, Synthesis and Applications of Isotopically Labeled Compounds, Proc. Int. Symp., W.P. Duncan and A.B. Susan (Eds.), Elsevier: Amsterdam (1982) 1-13.

55   R.A. Brooks, M.A. Long and J.L. Garnett, J. Labelled Comp. Radiopharm., **19** (1982) 659-667.

56   P.G. Williams, Ph.D. Thesis, University of New South Wales (1984).

57   M.A. Long, J.L. Garnett, P.G. Williams and T. Mole, J. Am. Chem. Soc., **103** (1981) 1571-1572.

58   J.L. Garnett and M.A. Long, Synthesis and Applications of Isotopically Labeled Compounds, Proc. Int. Symp., W.P. Duncan and A.B. Susan (Eds.), Elsevier: Amsterdam (1983) 415-420.

59   W.H. Calkins and T.D. Stewart, J. Am. Chem. Soc., **71** (1949) 4144-4145.

60   M.A. Long, J.L. Garnett and P.G. Williams, J. Chem. Soc., Perkin Trans. 2, (1984) 2105-2109.

61   J.L. Garnett, M.A. Long and P.G. Williams, Synthesis and Applications of Isotopically Labeled Compounds, Proc. Second Int. Symp., R.R. Muccino (Ed.), Elsevier: Amsterdam (1986) 395-400.

62   P.B. Venuto and E.L. Wu, J. Catal., **15** (1969) 205-208.

63   M.A. Long, J.L. Garnett and P.G. Williams, Aust. J. Chem., **35** (1982) 1057-1059.

64   M.A. Long, J.L. Garnett and P.G. Williams, Synthesis and Applications of Isotopically Labeled Compounds, Proc. Int. Symp., W.P. Duncan and A. B. Susan (Eds.), Elsevier: Amsterdam (1983) 315-320.

65   J.L. Garnett, E.M. Kennedy, M.A. Long, C. Than and A.J. Watson, J. Chem. Soc., Chem. Commun., (1988) 763-764.

66   M.A. Long, J.L. Garnett and C. Than, Synthesis and Applications of Isotopically Labelled Compounds, Proc. Third Int. Symp., T.A. Baillie and J.R. Jones (Eds.), Elsevier: Amsterdam (1989) 111-116.

67   J.R. Jones, "The Ionisation of Carbon Acids", Academic Press: London (1973).

68   E. Buncel, "Carbanions. Mechanistic and Isotopic Aspects", Elsevier: Amsterdam (1975).

69   J.R. Jones and S.E. Taylor, Chem. Soc. Rev., (1981) 329-344.

70   J.A. Elvidge, D.K. Jaiswal, J.R. Jones and R. Thomas, J. Chem. Soc., Perkin Trans. 2, (1976) 353-356.

71   J.R. Jones, G.M. Pearson, D. Spinelli, G. Consiglio and C. Arnone, J. Chem. Soc., Perkin Trans. 2, (1985) 557-558.

72   E. Buncel, J.R. Jones, K. Sowdani, D. Spinelli, G. Consiglio and C. Arnone, J. Chem. Soc., Perkin Trans. 2, (1985) 559-561.

73   C.W. Funke, F.M. Kaspersen, J. Wallaart and G.N. Wagenaars, J. Labelled Comp. Radiopharm., **28** (1983) 843-853.

74   J.A. Elvidge, J.R. Jones, J.C. Russell, A. Wiseman and M.M. Coombs, J. Chem. Soc., Perkin Trans. 2, (1985) 563-565.

75   J.P. Bloxsidge, J.A. Elvidge, J.R. Jones, R.B. Mane and M. Saljoughian, Org. Magn. Reson., **12** (1979) 574-578.

76   J.A. Elvidge, D.K. Jaiswal, J.R. Jones and R. Thomas, J. Chem. Soc., Perkin Trans. 1, (1977) 1080-1083.

77   E. Buncel, J.A. Elvidge, J.R. Jones and K.T. Walkin, J. Chem. Res. (S), (1980) 272-273.

78   E. Buncel, A.R. Norris, J.A. Elvidge, J.R. Jones and K.T. Walkin, J. Chem. Res. (S), (1980) 326-327.

79   J.A. Elvidge, J.R. Jones, R.B. Mane and M. Saljoughian, J. Chem. Soc., Perkin Trans. 1, (1978) 1191-1194.

80   K.E. Wilzbach, J. Am. Chem. Soc., **79** (1957) 1013.

81   C.T. Peng, "Radiation Induced Methods of Labelling", in: Isotopes in the Physical and Biomedical Sciences, Labelled Compounds (Part A), Vol. 1, E. Buncel and J.R. Jones (Eds.), Elsevier: Amsterdam (1987) 6-51.

82   N.A. Ghanem and T. Westermark, J. Am. Chem. Soc., **82** (1960) 4432-4433.

83   W.C. Hembree, R.E. Ehrenkaufer, S. Lieberman and A.P. Wolf, J. Biol. Chem., **248** (1973) 5532-5540.

84   R. Hua and C.T. Peng, J. Labelled Comp. Radiopharm., **24** (1987) 1095-1106.

85   R.L.E. Ehrenkaufer, W.C. Hembree and A.P. Wolf, Synthesis and Applications of Isotopically Labelled Compounds, Proc. Third Int. Symp., T.A. Baillie and J.R. Jones (Eds.), Elsevier: Amsterdam (1989) 105-110.

86   R.E. Ehrenkaufer, W.C. Hembree and A.P. Wolf, J. Labelled Comp. Radiopharm., **22** (1985) 819-831.

87   R.L.E. Ehrenkaufer, W.C. Hembree, S. Lieberman and A.P. Wolf, J. Am. Chem. Soc., **99** (1977) 5005-5009.

88   R.L.E. Ehrenkaufer, A.P. Wolf, W.C. Hembree and S. Lieberman, J. Labelled Comp. Radiopharm., **13** (1977) 359-365.

89   G.Z. Tang and C.T. Peng, J. Labelled Comp. Radiopharm., **25** (1988) 585-601.

90   C.T. Peng, Synthesis and Applications of Isotopically Labelled Compounds, Proc. Third Int. Symp., T.A. Baillie and J.R. Jones (Eds.), Elsevier: Amsterdam (1989) 525-528.

91   C.T. Peng, to be published.

92   J. Hiltunen and C.T. Peng, to be published.

93   A.V. Shishkov, L.A. Neiman and V.S. Smolyakov, Russ. Chem. Rev. (Engl. Transl.), **53** (1984) 656-671.

94   L.A. Neiman, L.P. Antropova, M.A. Zalesskaya and E.I. Budovskii, Bioorg. Khim., **12** (1986) 1070-1072.

95   H. Morimoto, P.G. Williams and B.E. Gordon, Trans. Am. Nucl. Soc., **55** (1987) 47-48.

96   H. Morimoto, P.G. Williams and M. Saljoughian, Synthesis and Applications of Isotopically Labelled Compounds, Proc. Third Int. Symp., T.A. Baillie and J.R. Jones (Eds.), Elsevier: Amsterdam (1989) 123-128.

97   J.L. Garnett, Catal. Rev., **5** (1971) 229-268.

98   J.L. Garnett and M.A. Long, "Catalytic Exchange Methods of Hydrogen Isotope Labelling", in: Isotopes in the Physical and Biomedical Sciences, Labelled Compounds (Part A), Vol. 1, E. Buncel and J.R. Jones (Eds.), Elsevier: Amsterdam (1987) 86-121.

99   W.G. Brown and J.L. Garnett, J. Am. Chem. Soc., **80** (1958) 5272-5274.

100   J.L Garnett, M.A. Long and C.A. Lukey, J. Chem. Soc., Chem. Commun., (1979) 634-635.

101   J.L. Garnett, M.A. Long, C.A. Lukey and P.G. Williams, J. Chem. Soc., Perkin Trans. 2, (1982) 287-289.

102   J.A. Elvidge, J.R. Jones, R.B. Mane and J.M.A. Al-Rawi, J. Chem. Soc., Perkin Trans. 2, (1979) 386-388.

103   J.M.A. Al-Rawi, J.A. Elvidge, J.R. Jones, V.M.A. Chambers and E.A. Evans, J. Labelled Comp. Radiopharm., **12** (1976) 265-273.

104   M.A. Long and C.A. Lukey, Org. Magn. Reson., **12** (1979) 440-441.

105   J.M.A. Al-Rawi, J.P. Bloxsidge, J.A. Elvidge, J.R. Jones, V.M.A. Chambers and E.A. Evans, J. Labelled Comp. Radiopharm., **12** (1976) 293-307.

106   M.M. Coombs, J. Labelled Comp. Radiopharm., **17** (1980) 147-152.

107   M.C. Clifford, E.A. Evans, A.E. Kilner and D.C. Warrell, J. Labelled Comp., **11** (1975) 435-443.

108   L.J. Altman and N. Silberman, Anal. Biochem., **79** (1977) 302-309.

109   D.S. Farrier, J.R. Jones, J.P. Bloxsidge, L. Carroll, J.A. Elvidge and M. Saieed, J. Labelled Comp. Radiopharm., **19** (1982) 213-227.

110   T. Meshi and T. Takahashi, Bull. Chem. Soc. Jpn., **35** (1962) 1510-1514.

111   T. Meshi and Y. Sato, Bull. Chem. Soc. Jpn., **37** (1964) 683-687.

112   E.A. Evans, H.C. Sheppard, J.C. Turner and D.C. Warrell, J. Labelled Comp., **10** (1974) 569-587.

113   S.W. Landvatter, J.R. Heys and S.G. Senderoff, J. Labelled Comp. Radiopharm., **24** (1987) 389-396.

114   R. Cipollini and M. Schuller, J. Labelled Comp. Radiopharm., **15** (1978) 703-713.

115   M.A. Long, J.L. Garnett, C.A. Lukey and P.G. Williams, Aust. J. Chem., **33** (1980) 1393-1395.

116   M.A. Long, J.L. Garnett and C.A. Lukey, Org. Magn. Reson., **12** (1979) 551-552.

117   E.A. Evans, J.P. Kitcher and D.C. Warrell, J. Labelled Comp. Radiopharm., **16** (1979) 697-710.

98

118   J.M.A. Al-Rawi, J.P. Bloxsidge, J.A. Elvidge and J.R. Jones, Steroids, **28** (1976) 359-375.

119   G.F. Taylor and J.A. Kepler, J. Labelled Comp. Radiopharm., **27** (1989) 683-690.

120   J.A. Elvidge, J.R. Jones and R.B. Mane, J. Labelled Comp. Radiopharm., **15** (1978) 141-151.

121   D. Calvert, A. Kazakevics, W. Martin and A.L. Odell, JEOL News, **14A** (1977) 5-7.

122   C.A. Rosario, C.C. Duke, J.H. Gill, M. Dawson, G.M. Holder, T. Ghazy and M.A. Long, J. Labelled Comp. Radiopharm., **24** (1987) 23-29.

123   H. Levine-Pinto, P. Pradelles, J.L. Morgat and P. Fromageot, J. Labelled Comp. Radiopharm., **17** (1980) 231-246.

124   P.G. Williams, H. Morimoto and M. Saljoughian, Synthesis and Applications of Isotopically Labelled Compounds, Proc. Third Int. Symp., T.A. Baillie and J.R. Jones (Eds.), Elsevier: Amsterdam (1989) 55-60.

125   J.L. Garnett, M.A. Long, C. Than and P.G. Williams, to be published.

126   M.A. Long, R.B. Moyes, P.B. Wells and J.L. Garnett, J. Catal., **52** (1978) 206-217.

127   J.L. Garnett and R.J. Hodges, J. Am. Chem.Soc., **89** (1967) 4546-4547.

128   J.L. Garnett and R.J. Hodges, J. Chem. Soc., Chem. Commun., (1967) 1001-1003.

129   J.L. Garnett and R.S. Kenyon, J. Chem. Soc., Chem. Commun., (1971) 1227-1228.

130   J.L. Garnett and R.S. Kenyon, Aust. J. Chem., **27** (1974) 1033-1045.

131   N.F. Gol'dshleger, M.B. Tyabin, A.E. Shilov and A.A. Shteinman, Russ. J. Phys. Chem., **43** (1969) 1222-1223.

132   J.L. Garnett, M.A. Long, A.B. McLaren and K.B. Peterson, J. Chem. Soc., Chem. Commun., (1973) 749-750.

133   M.R. Blake, J.L. Garnett, I.K. Gregor, W. Hannan, K. Hoa and M.A. Long, J. Chem. Soc., Chem. Commun., (1975) 930-932.

134   L. Carroll, J.R. Jones and W.J.S. Lockley, Synthesis and Applications of Isotopically Labeled Compounds, Proc. Second Int. Symp., R.R. Muccino (Ed.), Elsevier: Amsterdam (1986) 389-393.

135   J.A. Elvidge, E.A. Evans, J.R. Jones and L.M. Zhang, Synthesis and Applications of Isotopically Labeled Compounds, Proc. Second Int. Symp., R.R. Muccino (Ed.), Elsevier: Amsterdam (1986) 401-408.

136   W.J.S. Lockley, Tetrahedron Lett., **23** (1982) 3819-3822.

137   W.J.S. Lockley, J. Labelled Comp. Radiopharm., **21** (1984) 45-57.

138   K.W. Zilm, D.M. Heinekey, J.M. Millar, N.G. Payne and P. Demou, J. Am. Chem. Soc., **111** (1989) 3088-3089.

139   E. Rosenberg, Polyhedron, **8** (1989) 383-405.

140   E.A. Evans, "Tritium and Its Compounds", Butterworths: London (1974) 210-237.

Isotopes in the Physical and Biomedical Science, Vol. 2, edited by E. Buncel and J.R. Jones
© 1991 Elsevier Science Publishers B.V., Amsterdam

Chapter **3**

Deuterium Nuclear Magnetic Resonance Spectroscopy in Partially Ordered Systems

J. H. Davis
Physics Department, University of Guelph, Guelph, Ontario, Canada, N1G 2W1

Contents

I Introduction

Deuterium has a nucleus with spin angular momentum $I = 1$ and its magnetic moment is a factor of 6.5 smaller than that of the spin $I = \frac{1}{2}$ nucleus of hydrogen, for which it is often substituted. The utility of deuterium NMR is a result of its electric quadrupole moment which is small enough that the quadrupolar interaction can be treated as a first order perturbation on the Zeeman interaction in the magnetic fields in common use (1 - 12 Tesla). The quadrupolar moment is also small enough that the pulse techniques of solid state NMR can be applied. The maximum quadrupolar splitting for ^{2}H in a C-^{2}H bond is about 252 kHz (it is somewhat larger in an O-^{2}H or an N-^{2}H bond), which is much less than its Larmor frequency of 55 MHz in a magnetic field of 8.5 T. For comparison, ^{2}H-^{2}H dipolar splittings for a methylene group are at most a few kHz, ^{2}H-^{1}H dipolar splittings for a similar geometry are of the order of 10 kHz, and the chemical shift dispersion is only about 1 kHz. Thus, it is often possible to neglect the chemical shift dispersion and the dipolar interactions with other nuclei and treat deuterium as an isolated spin 1 nucleus, using first order perturbation theory to describe the quadrupolar interaction. The theoretical analysis of such a system is relatively simple.

Molecular solids, liquid crystals, as well as biological and model membranes, are examples of what can be called partially ordered systems. In this class of systems, molecular motions are generally anisotropic, resulting in the incomplete motional averaging of the orientation dependent (second rank tensor) interactions. These include the anisotropic chemical shift, the nuclear dipole-dipole and quadrupolar interactions. In contrast, the motions of isotropic liquids completely remove these interactions, leaving only the isotropic part of the chemical shift and the scalar J-coupling. In both ordered systems and isotropic liquids the tensor interactions can dominate relaxation processes, however.

The techniques of solid state NMR enable us to measure the residual tensor interactions, i.e., that part of the interaction which has not been averaged away by motion. This provides information on the nature of the system under study, e.g., on the symmetry of the system and of the molecular motions, on the amplitudes and time scales of the motions, and on the occurrence of phase transitions of the system.

There have been a large number of applications of ^{2}H NMR to all types of ordered systems. One of the most active areas of application has been biological and model membranes, which are examples of lyotropic liquid crystals. The theory, experimental techniques and methods of analysis developed for membrane studies are applicable to all other areas of ^{2}H NMR.

This chapter begins with the theoretical description of isolated spin 1 nuclei using first order perturbation theory. The form of the quadrupolar Hamiltonian is developed first, introducing the quadrupolar splitting, orientational order parameters, spherical tensors and the transformations between principal axis coordinates and both the crystal and laboratory reference frames. Then, spin 1 spectroscopy is described by the density matrix using an operator space formalism. The quadrupolar echo serves as the first example of the use of this formalism to calculate the effects of NMR pulse sequences on the spin system. In addition to the three components of magnetization, the spin 1 density operator can describe all the other observables of the system. These include quadrupolar polarization and double quantum coherence. These quantities are described and the methods of measuring and manipulating them using appropriate pulse sequences are presented.

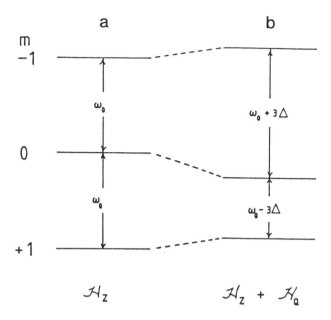

Figure 1: Energy levels of a spin 1 nucleus: a) splittings due to the Zeeman interaction; b) first order quadrupolar shifts of the Zeeman levels.

Next, the applications of ^{2}H NMR are illustrated by studies of molecular structure, conformation and orientational order. These include studies of the influence of surface charge on phospholipid head group conformation or orientation in lipid bilayers, cholesterol orientation in membranes, and peptide backbone structure of gramicidin A in a lyotropic nematic liquid crystal. Then, the use of ^{2}H NMR to study the phase equilibria of two component mixtures is illustrated first by cholesterol/ 1,2-dipalmitoyl-sn-glycero3-phosphocholine (DPPC) bilayers and then with a model membrane system composed of synthetic bilayer spanning peptides in DPPC. The technique of ^{2}H NMR difference spectroscopy is used to determine the phase boundaries of two-phase coexistence regions in both of these studies.

II Theory

A The Quadrupolar Interaction

In an external magnetic field, in the absence of any electric field gradient at the nucleus, the Zeeman interaction splits the ground state nuclear energy levels of a spin 1 nucleus as shown in Figure 1a. The eigenvalues of the Zeeman Hamiltonian,

$$\mathcal{H}_z = -\vec{\mu} \cdot \vec{H}_0 = -\gamma \hbar \omega_0 I_z \tag{1}$$

where $\omega_0 = \gamma H_0$ and $\vec{H}_0$ defines the axis of quantization of the nuclear angular momentum (the z-axis), are

$$E_m = -m\hbar\omega_0 \tag{2}$$

For a system in thermal equilibrium with the lattice at a temperature T, these energy levels are populated according to the Boltzmann factor so that the relative populations are, in the high temperature approximation (i.e., for temperatures greater than or of the order of 1 K),

$$
\begin{aligned}
p_1 &= \frac{1}{3}\left(1 + \frac{\hbar\omega_0}{kT}\right) \\
p_0 &= \frac{1}{3} \\
p_{-1} &= \frac{1}{3}\left(1 - \frac{\hbar\omega_0}{kT}\right)
\end{aligned}
\tag{3}
$$

For a nucleus with an electric quadrupole moment, eQ, (the nuclear spin I must be ≥ 1), an electric field gradient, eq, at the nucleus (due primarily to the electronic charge distribution of the atom or molecule containing the nucleus) results in a shift of these nuclear Zeeman levels, as shown in Figure 1b. The energy of the nuclear charge distribution $\rho(\vec{r})$ in an electrical potential $V(\vec{r})$ can be written as

$$E = \int \rho(\vec{r})V(\vec{r})d\tau \tag{4}$$

where the integral is over the nuclear volume. Expanding $V(\vec{r})$ in a Taylor's series about $\vec{r} = 0$,

$$V(\vec{r}) = V(0) + \sum_\alpha x_\alpha \frac{\partial V}{\partial x_\alpha}\bigg|_{\vec{r}=0} + \frac{1}{2!}\sum_{\alpha,\beta} x_\alpha x_\beta \frac{\partial^2 V}{\partial x_\alpha \partial x_\beta}\bigg|_{\vec{r}=0} + \cdots \tag{5}$$

we find the expression for the energy[1,2]

$$E = V(0)\int \rho d\tau + \sum_\alpha V_\alpha \int x_\alpha \rho d\tau + \frac{1}{2}\sum_{\alpha,\beta} V_{\alpha\beta} \int x_\alpha x_\beta \rho d\tau + \cdots \tag{6}$$

In this expansion, the first term, $E^{(0)}$, is the energy of a point charge in the electrical potential $V(0)$, the integrals which involve odd powers of x_α all vanish for nuclear states with definite parity, so the second term is zero, and the third term, $E^{(2)}$, is due to the electric quadrupole interaction. The quantity

$$V_{\alpha\beta} = \frac{\partial^2 V}{\partial x_\alpha \partial x_\beta}\bigg|_{\vec{r}=0} \tag{7}$$

is the electric field gradient tensor at the nucleus ($\vec{r} = 0$). This tensor is symmetric with respect to an interchange of α and β, $V_{\alpha\beta} = V_{\beta\alpha}$. Further, if the electron density at the nucleus is zero, then Laplace's equation must hold, so that

$$\nabla^2 V = \sum_\alpha V_{\alpha\alpha} = 0 \tag{8}$$

i.e., the electric field gradient tensor has zero trace ($Tr\{V_{\alpha\beta}\} = \sum_\alpha V_{\alpha\alpha} = 0$). Only s-electron orbitals have a non-zero density at $\vec{r} = 0$, but because they are spherically symmetric then do not contribute to the electric field gradient, so we ignore them.

A traceless, symmetric, second rank tensor tensor has only five independent components

$$V_{\alpha\beta} = \begin{pmatrix} V_{11} & V_{12} & V_{13} \\ V_{12} & V_{22} & V_{23} \\ V_{13} & V_{23} & V_{33} \end{pmatrix} \tag{9}$$

There always exists some coordinate transformation R which will diagonalize a symmetric second rank tensor. There will be then, some principal axis system such that

$$V^P = RVR^{-1} = \begin{pmatrix} V_{11}^P & 0 & 0 \\ 0 & V_{22}^P & 0 \\ 0 & 0 & V_{33}^P \end{pmatrix} \tag{10}$$

Now the diagonal electric field gradient tensor, whose trace is still 0, has only two independent components, but we must still define the coordinate transformation[3,4] $R(\alpha, \beta, \gamma)$, which brings us from the laboratory coordinate system into the principal axis system of the electric field gradient tensor. This requires, in the general case, rotations through the three Euler angles[3,4] $(\alpha\beta\gamma)$.

The nuclear quadrupole moment is defined by

$$Q_{\alpha\beta}^N = \int (3x_\alpha x_\beta - \delta_{\alpha\beta} r^2)\rho d\tau \tag{11}$$

where, as before, the integral is over the nuclear volume. The electric quadrupole interaction energy is now

$$E^{(2)} = \frac{1}{2} \sum_{\alpha,\beta} V_{\alpha\beta} Q_{\alpha\beta}^N \tag{12}$$

where we have made use of Laplace's equation ($\nabla^2 V = 0$) once again. With the help of the Wigner-Eckart theorem[2,3] we can write the electric quadrupolar Hamiltonian as

$$\mathcal{H}_Q = \frac{eQ}{6I(2I-1)} \sum_{\alpha,\beta} V_{\alpha\beta} \left[\frac{3}{2}(I_\alpha I_\beta + I_\beta I_\alpha) - \delta_{\alpha\beta} I^2 \right] \tag{13}$$

In the principal axis system of the electric field gradient tensor, $V_{\alpha\beta} = V_{\alpha\alpha}\delta_{\alpha\alpha}$, so that

$$\mathcal{H}_Q = \frac{eQ}{6I(2I-1)} \left[V_{x'x'}(3I_{x'}^2 - I^2) + V_{y'y'}(3I_{y'}^2 - I^2) + V_{z'z'}(3I_{z'}^2 - I^2) \right] \tag{14}$$

Introducing the principal value of the electric field gradient,

$$eq = V_{z'z'}, \tag{15}$$

and the asymmetry parameter

$$\eta = \left| \frac{\mathcal{V}_{z'z'} - \mathcal{V}_{y'y'}}{\mathcal{V}_{z'z'}} \right|, \tag{16}$$

we obtain

$$\mathcal{H}_Q = \frac{e^2 qQ}{4I(2I-1)} \left[(3I_{z'}^2 - I^2) + \eta(I_{x'}^2 - I_{y'}^2) \right] \tag{17}$$

We have used primes on our coordinates (x', y', z') since the principal axis system of the electric field gradient (efg) tensor need not coincide with our laboratory system defined by the direction of the magnetic field $\vec{H}_0$. To proceed any further, we must make the transformation to the laboratory reference frame. This transformation can be carried out in Cartesian coordinates as the inverse of the diagonalization procedure of equation 10, using the Euler angles $(\alpha\beta\gamma)$ and the rotation matrix $R(\alpha\beta\gamma)$[3,4]. It is more convenient, however, to switch to a spherical coordinate system and use the formalism of the irreducible spherical tensors[3,4,5].

In spherical coordinates, the electric field gradient tensor in its principal axis system is,

$$\mathcal{F}_{2,0}^P = \frac{3}{2} eq$$
$$\mathcal{F}_{2,\pm 1}^P = 0 \tag{18}$$
$$\mathcal{F}_{2,\pm 2}^P = \frac{1}{2} eq\eta$$

where η and eq are defined by equations 15 and 16. The second rank spherical tensors representing the spin variables are[5,6,7]

$$T_{2,0} = \frac{1}{\sqrt{6}} \left[3I_z^2 - I(I+1) \right]$$
$$T_{2,\pm 1} = \mp \frac{1}{2} [I_z I_\pm + I_\pm I_z] \tag{19}$$
$$T_{2,\pm 2} = \frac{1}{2} I_\pm^2 \tag{20}$$

These spherical tensors transform in the same fashion as the spherical harmonics[3,4], namely, the electric field gradient tensor in the laboratory system is given by

$$\mathcal{F}_{2,m}^L = \sum_{m'=-2}^{2} \mathcal{F}_{2,m'}^P \mathcal{D}_{m'm}^{(2)}(\alpha\beta\gamma) \tag{20}$$

where $(\alpha\beta\gamma)$ are the Euler angles defining the transformation, and $\mathcal{D}_{m'm}^{(2)}$ is the Wigner rotation matrix[3,4].

The electric quadrupolar Hamiltonian is the product of the spin and spatial tensors so that, in the laboratory frame,

$$\mathcal{H}_Q = \frac{eQ}{2} \sum_{m'=-2}^{2} (-1)^m T_{2,m} \mathcal{F}_{2,-m}^L$$

(21)

$$= \frac{eQ}{2} \sum_{m=-2}^{2} (-1)^m T_{2,m} \sum_{m'=-2}^{2} \mathcal{F}_{2,m'}^P \mathcal{D}_{m'-m}^{(2)}(\alpha\beta\gamma)$$

We now make two observations: firstly, since the magnetic field defining the laboratory coordinate system is axially symmetric, we can choose our laboratory frame so that the directions z and z' lie in a plane perpendicular to the (laboratory) y-axis. Then, the first Euler angle, γ, will be zero (see ref. 3, pp. 77-80); secondly, since $\mathcal{H}_Q$ is much smaller than $\mathcal{H}_z$ for deuterium NMR at high fields (e.g., at 8.5 T), we can calculate the spectrum using first order perturbation theory[8], so we need only keep that part of $\mathcal{H}_Q$ which commutes with the Zeeman Hamiltonian, $\mathcal{H}_Z$. Thus, we need only keep the terms in $T_{2,0}$ so that

$$\mathcal{H}_Q = \frac{e^2 qQ}{2} T_{2,0} \left\{ \frac{3}{2} \mathcal{D}_{00}^{(2)}(\alpha\beta0) + \frac{1}{2}\eta \left[\mathcal{D}_{20}^{(2)}(\alpha\beta0) + \mathcal{D}_{-20}^{(2)}(\alpha\beta0) \right] \right\}$$

(22)

Substituting for $\mathcal{D}_{m0}^{(2)}(\alpha\beta0)$ we find[3,4,9]

$$\mathcal{H}_Q = \frac{e^2 qQ}{8} \left[3I_z^2 - I(I+1) \right] \left[(3\cos^2\beta - 1) + \eta \sin^2\beta \cos 2\alpha \right]$$

(23)

so that to first order in $\mathcal{H}_Q$, the nuclear Zeeman energy levels are shifted by the amount

$$E_m = \frac{e^2 qQ}{8} (3m^2 - 2) \left[(3\cos^2\beta - 1) + \eta \sin^2\beta \cos 2\alpha \right]$$

(24)

As shown in Figure 1b, the levels with $m = \pm 1$ are shifted upwards by an amount

$$\Delta = \frac{e^2 qQ}{8} \left[(3\cos^2\beta - 1) + \eta \sin^2\beta \cos 2\alpha \right]$$

(25)

while the $m = 0$ level is shifted downwards by the amount 2Δ. Instead of observing a singlet at frequency ω_0, we now observe a doublet which is symmetrically displaced about ω_0 with a quadrupolar splitting of

$$\omega_Q = 6\Delta = \frac{3e^2 qQ}{4\hbar} \left[(3\cos^2\beta - 1) + \eta \sin^2\beta \cos 2\alpha \right]$$

(26)

This quadrupolar splitting is dependent on the orientation of the electric field gradient tensor principal axis system relative to the magnetic field through the angles $(\alpha\beta)$.

In the presence of molecular motion, it is instructive to introduce another intermediate coordinate system and make the transformation from the molecule fixed principal axis system to the intermediate system and then finally to the laboratory system. A particular example of this is used in phospholipid bilayers where the bilayer normal is an axis of symmetry for the motion. In other situations, certain crystallographic directions may be axes of symmetry or may allow a convenient model for molecular reorientations. The transformation now proceeds in two steps, first from the efg principal axis to the intermediate or crystal frame

$$\mathcal{F}^D_{2,m'} = \sum_{m''} \mathcal{F}^P_{2,m''} \mathcal{D}^{(2)}_{m''m'}(\alpha\beta\gamma) \tag{27}$$

where $\mathcal{F}^D$ is the efg tensor in this new crystallographic coordinate system, and the Euler angles $(\alpha\beta\gamma)$ now define the transformation from the efg principal axis system to this new frame. Next, we transform from the crystal to the laboratory frame,

$$\mathcal{F}^L_{2,m} = \sum_{m'} \mathcal{F}^D_{2,m'} \mathcal{D}^{(2)}_{m'm}(\alpha'\beta'\gamma') \tag{28}$$

where $(\alpha'\beta'\gamma')$ define the transformation from the crystal frame to the laboratory frame. If the crystal frame z'' axis is an axis of at least three fold symmetry with respect to molecular motion, then, by the same argument used earlier with respect to the axial symmetry of the laboratory frame, we may write

$$\mathcal{H}_Q = \frac{e^2qQ}{2} T_{20} \sum_{m''} \mathcal{F}^P_{2,m''} \mathcal{D}^{(2)}_{m''0}(\alpha\beta0)\mathcal{D}^{(2)}_{00}(0\beta'0) \tag{29}$$

where we have kept only the secular part (that part of $\mathcal{H}_Q$ which commutes with the Zeeman Hamiltonian), as before, and the symmetry of the laboratory and crystal frames has allowed us to choose α', γ' and γ all equal to zero. In more general cases with non-axially symmetric motion, we will not be able to set α' and γ equal to zero. Substituting expressions for the $\mathcal{D}_{m'm}$ and T_{20} into equation 29, we obtain

$$\mathcal{H}_Q = \frac{e^2qQ}{8} \left(3\cos^2\beta' - 1\right) \left[\frac{1}{2}(3\cos^2\beta - 1) + \frac{1}{2}\eta\sin^2\beta\cos 2\alpha\right] \left[3I_z^2 - 2\right] \tag{30}$$

The angle β' may often be taken as fixed while molecular reorientation may cause α and β to be time dependent. In this case, it may be convenient to introduce the orientational order parameters, S_{ii}, defined by[5,9]

$$\langle\omega_Q\rangle = \frac{3e^2qQ}{4\hbar} \left(3\cos^2\beta' - 1\right) \frac{1}{2}\langle(3\cos^2\beta - 1) + \eta\sin^2\beta\cos 2\alpha\rangle$$

$$\tag{31}$$

$$= \frac{3e^2qQ}{4\hbar} \left(3\cos^2\beta' - 1\right) \left[S_{zz} + \frac{1}{3}\eta(S_{xx} - S_{yy})\right]$$

where the angular brackets denote an average over the molecular motions which are fast relative to the inverse of the width of the spectrum (i.e., relative to $1/\omega_Q$).

The above analysis assumes that, in the NMR sample being studied, the electric field gradient tensor has the same orientation for all deuterium nuclei, or that the crystal axis system has the same orientation relative to the magnetic field for all nuclei. Often the

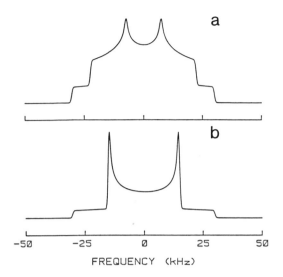

FREQUENCY (kHz)

Figure 2: First order spin 1 powder pattern line shapes: a) non-axially symmetric case with $\eta \neq 0$; b) axially symmetric case, $\eta = 0$. (Reproduced from ref. 5.)

sample consists of a "powder" where the crystal axes of the individual crystallites are randomly oriented. In this case, the angle β' takes on all values and we observe quadrupolar splittings ranging all the way from 0 to $\omega_Q^0 = 3e^2qQ/4\hbar$. The resulting superposition of quadrupolar split doublets leads to the characteristic spin 1 powder pattern lineshape. In the absence of molecular motion we can observe spectra like those of Figure 2. For a system with a non- zero asymmetry parameter we expect a spectrum similar to that of Figure 2a, the sharp edges and the peaks can be used to determine the values of eq and η (though not their orientation within the crystal or molecule)[5]. If $\eta = 0$ we observe the axially symmetric lineshape of Figure 2b. The two sharp peaks in this spectrum correspond to crystallites whose efg principal z'-axis is aligned perpendicular to $\vec{H}_0$. The small shoulders correspond to the orientations parallel to $\vec{H}_0$. The line shape in the axially symmetric case, neglecting broadening due to relaxation, is[5]

$$f(x) = \frac{1}{\sqrt{6}} \left[\frac{1}{(\frac{1}{2} + x)^{\frac{1}{2}}} + \frac{1}{(\frac{1}{2} - x)^{\frac{1}{2}}} \right], 0 \leq x \leq \frac{1}{2}$$

(32)

$$f(x) = \frac{1}{\sqrt{6}} \frac{1}{(\frac{1}{2} + x)^{\frac{1}{2}}}, \frac{1}{2} < x \leq 1$$

where $x = \omega/\omega_Q^0$. The expression for negative values of x is obtained replacing x by -x in equation 32 (the spectrum is symmetric about x=0). The case where $\eta \neq 0$ is more complex and must be evaluated numerically[5].

B Spin 1 Spectroscopy

1 Density Matrix Formalism

The theoretical description of a system of isolated (non-interacting) spin 1 nuclei is most easily accomplished using the density operator formalism[5,7,10,11,12]. In the following development we neglect the interactions between nuclear spins and the small differences in chemical shift (of the order of 20 ppm) among deuterium nuclei. In addition, we will assume that the rf pulses are pure rotations, i.e., that the rf field is much stronger than the quadrupolar interaction and that there is no relaxation during the pulses. In actual fact, the rf pulses are usually *not* much larger than the quadrupolar interaction. However, the calculation of the spin system's response to rf excitation is greatly simplified if we can neglect the quadrupolar interaction during the pulse. This is the only justification for making this assumption. The exact calculations, including finite rf pulse effects, have been carried out previously[12]. In a realistic situation, where the rf pulse strength, ω_1, is comparable to the quadrupolar splitting, ω_Q, there will be a considerable distortion of the signal from even the simplest experiment and even a single pulse can transform the equilibrium Zeeman polarization along the z-axis into a combination of the other observable coherences discussed below.[11,12]

If we have a quantum mechanical system which can be described by a state vector Ψ, then the expectation value of an operator $\mathcal{M}_x$ is given by

$$\langle \mathcal{M}_x \rangle = (\Psi, \mathcal{M}_x \Psi) \tag{33}$$

where (Ψ, Ψ') is defined as the scalar product between two state functions Ψ and Ψ',

$$(\Psi, \Psi') = \int dq_1 dq_2 \cdots dq_N \Psi^*(q_1 q_2 \cdots q_N) \Psi'(q_1 q_2 \cdots q_N) \tag{34}$$

where the integral is over all of configuration space $(q_1 q_2 \cdots q_N)$. We will require that all state functions be normalized, i.e., that $(\Psi, \Psi) = 1$ for all Ψ. The configuration space of the system is a linear N-dimensional vector space so that it is possible to define a complete orthonormal set of N basis functions, $\{|n\rangle\}$, with $\langle n|n'\rangle = \delta_{nn'}$, where $\delta_{nn'}$ is the Kronecker delta which has the value 1 if $n = n'$ and is 0 otherwise.

Since this set of basis functions spans all of configuration space, the state vector Ψ can be expanded in terms of this complete set as

$$\Psi = \sum_n c_n |n\rangle \tag{35}$$

The basis vectors, $|n\rangle$, are fixed in vector space, i.e., are independent of time. If Ψ is time dependent, then the time dependence is contained entirely within the coefficients $c_n(t)$. The expectation value of $\mathcal{M}_x$ is now

$$\langle \mathcal{M}_x \rangle = (\Psi, \mathcal{M}_x \Psi) = \sum_m \sum_n c_m^* c_n \langle m|\mathcal{M}_x|n\rangle \tag{36}$$

With this expression, we should make the following observations:

(i) the matrix elements of the operator $\mathcal{M}_x$, $\langle n|\mathcal{M}_x|m\rangle$ are time independent and are the same for all Ψ;

(ii) the $c_m^* c_n$ may be time dependent, they will be different for different state vectors Ψ, but they are not dependent on which of the operators or observables we may be evaluating.

The $c_m^* c_n$ may be considered to be the matrix elements of an operator $\mathcal{P}$ defined by $\mathcal{P}_{kl} = |l\rangle\langle k|$

$$
\begin{aligned}
P_{kl} &= (\Psi, \mathcal{P}\Psi) \\
&= \sum_{n,m} c_m^* c_n \langle m|\mathcal{P}|n\rangle \\
&= \sum_{n,m} c_m^* c_n \langle m|l\rangle\langle k|n\rangle \\
&= c_k c_l^* \\
&= \langle k|\mathcal{P}|l\rangle
\end{aligned}
\tag{37}
$$

Now the equation for the expectation value of $\mathcal{M}_x$ becomes, in terms of this new operator $\mathcal{P}$

$$
\begin{aligned}
\langle \mathcal{M}_x \rangle &= \sum_{m,n} c_m^* c_n \langle m|\mathcal{M}_x|n\rangle \\
&= \sum_{m,n} \langle n|\mathcal{P}|m\rangle\langle m|\mathcal{M}_x|n\rangle \\
&= \sum_n \langle n|\mathcal{P}\mathcal{M}_x|n\rangle \\
&= \text{Tr}\,\{\mathcal{P}\mathcal{M}_x\}
\end{aligned}
\tag{38}
$$

since, by closure, $\sum_m |m\rangle\langle m| = 1$. We should note that

$$
\begin{aligned}
\langle n|\mathcal{P}|m\rangle^* &= (c_n c_m^*)^* \\
&= c_n^* c_m \\
&= \langle m|\mathcal{P}|n\rangle
\end{aligned}
\tag{39}
$$

so that the operator $\mathcal{P}$ is Hermitian, $\mathcal{P} = \mathcal{P}^\dagger$.

In a statistical ensemble of systems the state vectors, Ψ, may vary from system to system within the ensemble, but the matrix elements, $\langle m|\mathcal{M}_x|n\rangle$ will always be the same. Denoting the ensemble average as a $\overline{bar}$,

$$
\overline{\langle \mathcal{M}_x \rangle} = \sum_{n,m} \overline{c_n c_m^*} \langle m|\mathcal{M}_x|n\rangle
\tag{40}
$$

We now have a definition of the density operator, ρ, in terms of its matrix elements

$$\rho_{nm} = \langle n|\rho|m \rangle = \overline{c_n c_m^*} \tag{41}$$

so that the ensemble averaged expectation value of $\mathcal{M}_x$ is

$$
\begin{aligned}
\overline{\langle \mathcal{M}_x \rangle} &= \sum_{n,m} \langle n|\rho|m \rangle \langle m|\mathcal{M}_x|n \rangle \\
&= \sum_{n} \langle n|\rho\mathcal{M}_x|n \rangle \\
&= Tr\{\rho\mathcal{M}_x\}
\end{aligned}
\tag{42}
$$

A knowledge of the density operator, ρ, is sufficient to determine the expectation values of any and all observables.

When the state vector, Ψ, is a function of time,

$$\Psi(t) = \sum_{n} c_n(t)|n \rangle \tag{43}$$

then we can write the time dependent Schrödinger equation

$$-\frac{\hbar}{i}\frac{\partial \Psi}{\partial t} = \mathcal{H}\Psi \tag{44}$$

as

$$-\frac{\hbar}{i}\sum_{n}\frac{dc_n(t)}{dt}|n \rangle = \sum_{n} c_n(t)\mathcal{H}|n \rangle \tag{45}$$

where $\mathcal{H}$ is the Hamiltonian of the system. Multiplying from the left by $\langle k|$, we obtain

$$-\frac{\hbar}{i}\sum_{n}\frac{dc_n(t)}{dt}\langle k|n \rangle = \sum_{n} c_n(t)\langle k|\mathcal{H}|n \rangle \tag{46}$$

or, using the orthonormality of the basis vectors $\{|n \rangle\}$,

$$-\frac{\hbar}{i}\frac{dc_k(t)}{dt} = \sum_{n} c_n(t)\langle k|\mathcal{H}|n \rangle \tag{47}$$

for each k between k=1 and k=N, where N is the number of dimensions of the vector space spanned by the basis set $\{|n \rangle\}$.

The equation of motion of the density operator is obtained by taking the time derivative of the matrix elements of the operator $\mathcal{P}$.

$$
\begin{aligned}
\frac{d}{dt}\langle k|\mathcal{P}|m \rangle &= \frac{d}{dt}(c_k c_m^*) \\
&= c_k \frac{dc_m^*}{dt} + \frac{dc_k}{dt} c_m^* \\
&= c_k \left(\frac{i}{\hbar}\sum_{n} c_n^*(t)\langle n|\mathcal{H}|m \rangle \right) - \left(\frac{i}{\hbar}\sum_{n} c_n(t)\langle k|\mathcal{H}|n \rangle \right) c_m^*
\end{aligned}
\tag{48}
$$

$$= \frac{i}{\hbar} \sum_n c_k c_n^* \langle n|\mathcal{H}|m\rangle - \frac{i}{\hbar} \sum_n c_n \langle k|\mathcal{H}|n\rangle c_m^*$$

On taking the ensemble average we obtain the equation for ρ,

$$
\begin{aligned}
\frac{d\rho}{dt} &= \frac{d}{dt}\overline{(c_k c_m^*)} \\
&= \frac{i}{\hbar} \sum_n \{\langle k|\rho|n\rangle\langle n|\mathcal{H}|m\rangle - \langle k|\mathcal{H}|n\rangle\langle n|\rho|m\rangle\} \quad (49) \\
&= \frac{i}{\hbar} \{\rho\mathcal{H} - \mathcal{H}\rho\}
\end{aligned}
$$

or, using commutator notation, we obtain the Liouville equation

$$\frac{d\rho}{dt} = \frac{i}{\hbar} [\rho, \mathcal{H}] \quad (50)$$

If the Hamiltonian is not explicitly time dependent, then

$$\rho(t) = \exp\left(-\frac{i}{\hbar}\mathcal{H}t\right)\rho(0)\exp\left(\frac{i}{\hbar}\mathcal{H}t\right) \quad (51)$$

is a formal solution to the Liouville equation.

If we choose the basis states $|n\rangle$ to be the eigenfunctions of this time independent Hamiltonian, with eigenvalues E_n, then

$$
\begin{aligned}
\langle k|\rho(t)|m\rangle &= \langle k|e^{-\frac{i}{\hbar}\mathcal{H}t}\rho(0)e^{\frac{i}{\hbar}\mathcal{H}t}|m\rangle \\
&= e^{\frac{i}{\hbar}(E_m-E_k)t}\langle k|\rho(0)|m\rangle \quad (52)
\end{aligned}
$$

We notice, from this equation, that if a matrix element of the density operator, $\rho_{km}(0)$, is equal to 0 at t=0, then $\rho_{km}(t) = 0$ for all t. Under the influence of this static Hamiltonian, non-zero diagonal terms (i.e., those with m=k) will remain constant, while non-zero off-diagonal terms will oscillate with an angular frequency $\omega = (E_m - E_k)/\hbar$.

To find an explicit form for the equilibrium density operator, we consider a spin system in equilibrium with the lattice at a temperature T. We use as a basis set, $\{|n\rangle\}$ the eigenfunctions of the time independent Hamiltonian $\mathcal{H}_0$. As before, any state vector of the system can be written

$$\Psi = \sum_n c_n(t)|n\rangle \quad (53)$$

Then, if Ψ is normalized,

$$(\Psi, \Psi) = \sum_{n,m} c_n c_m^* \langle m|n\rangle = \sum_n c_n c_n^* = 1 \quad (54)$$

Taking the ensemble average, we find that

$$\rho_{nn} = \overline{c_n c_n^*} = \frac{1}{\mathcal{Z}} e^{-E_n/kT} \tag{55}$$

where

$$\mathcal{Z} = \sum_m e^{-E_m/kT}$$

The n'th diagonal element of the density operator is the fractional population of the n'th energy level, with energy E_n, and these energy levels are populated according to the Boltzmann factor at the lattice temperature T.

The off diagonal terms have the form

$$\overline{c_n c_m^*} = \overline{|c_n||c_m| e^{i(\alpha_n - \alpha_m)}} \tag{56}$$

where α_n and α_m are phase factors. We will assume that the phases in different members of the ensemble are random (or uncorrelated). Then, on performing the ensemble average indicated in equation 56, all off-diagonal elements of the equilibrium density operator will be zero. Of course, we can still prepare the system in a non-equilibrium state where off-diagonal elements are not zero.

In operator form we can write the equilibrium density operator as

$$\rho_0 = \frac{1}{\mathcal{Z}} e^{-\mathcal{H}_0/kT} \tag{57}$$

$$\mathcal{Z} = Tr\left\{ e^{-\mathcal{H}_0/kT} \right\}$$

The density operator contains the complete statistical description of the spin system, expectation values of all observables can be calculated from it using equation 42, and the time evolution of the density operator is given by the Liouville equation (equation 50). A pulsed NMR experiment typically involves allowing the spin system to come to equilibrium in a strong magnetic field, H_0, so that our choice of basis states are the eigenfunctions of the static Zeeman Hamiltonian

$$\mathcal{H}_Z = -\gamma \hbar \vec{I} \cdot \vec{H}_0 \tag{58}$$

where γ is the nuclear gyromagnetic ratio, which for ^2H has a value $\gamma = 2\pi \times 6.536 \times 10^6$ Hz/T. Once equilibrium has been achieved, the Hamiltonian is changed, for example, by turning on an intense rf magnetic field rotating at or near the nuclear Larmor frequency, ω_0. The evolution of the spin system under this new Hamiltonian is described by the Liouville equation. The result of a complex sequence of rf pulses can be calculated using the approach outlined above. However, the procedure can be considerably simplified as described below.

2   Operator Space

In analogy to the vector space discussed above, which was spanned by the complete set of basis vectors $\{|n\rangle\}$, we can define an operator space spanned by a complete set of operators, $\{||Q_k\rangle\rangle\}$, where the "double ket" notation is used to distinguish these from normal kets ($|n\rangle$). We will choose our set of operators to be orthonormal with respect to the trace, i.e.,

$$\langle\langle Q_k||Q_l\rangle\rangle = \mathit{Tr}\left\{Q_k^\dagger Q_l\right\} = \delta_{kl} \tag{59}$$

One of the operators in this basis set will be proportional to the unit operator, $\mathcal{E}$. Then, for any $Q_i \neq \mathcal{E}$,

$$\mathit{Tr}\left\{\mathcal{E}Q_i\right\} = 0 \tag{60}$$

so that all operators except $\mathcal{E}$ have zero trace (are traceless) and

$$\mathit{Tr}\left\{\mathcal{E}\mathcal{E}\right\} = 1 \tag{61}$$

For a spin $I = 1$ system, operator space has $(2I + 1)^2 = 9$ dimensions, so our basis set $\{||Q_k\rangle\rangle\}$ consists of nine independent orthonormal operators. It is convenient to represent the operators in this nine dimensional space in terms of 3×3 matrices. For example, the identity operator can be represented as

$$\mathcal{E} = \frac{1}{\sqrt{3}}\begin{pmatrix} 1 & 0 & 0 \\ 0 & 1 & 0 \\ 0 & 0 & 1 \end{pmatrix} \tag{62}$$

where the normalization factor $\frac{1}{\sqrt{3}}$ is chosen to satisfy equation 61.

Just as we expanded the state vector Ψ in terms of the vector basis set $\{|n\rangle\}$, we can expand the density operator in terms of our new complete set of basis operators

$$||\rho\rangle\rangle = \sum_q c_q(t)||Q_q\rangle\rangle \tag{63}$$

Furthermore, any operator corresponding to any physical observable (on this spin space) can be expanded in the same fashion.

The expectation value of an observable $\mathcal{O}$ is given by

$$\begin{aligned} \langle \mathcal{O}\rangle &= \mathit{Tr}\left\{\mathcal{O}\rho\right\} \\ &= \mathit{Tr}\left\{\mathcal{O}\sum_q c_q(t)Q_q\right\} \\ &= \sum_q c_q(t)\mathit{Tr}\left\{\mathcal{O}Q_q\right\} \end{aligned} \tag{64}$$

If, for example, the operator $\mathcal{O}$ is given by

$$||\mathcal{O}\rangle\rangle = a||\mathcal{Q}_q\rangle\rangle + b||\mathcal{Q}_b\rangle\rangle \tag{65}$$

then, its expectation value is

$$\langle \mathcal{O} \rangle = a^* c_a(t) + b^* c_b(t) \tag{66}$$

From the cyclic property of the trace,

$$Tr\{AB\} = Tr\{BA\}$$

and also since both ρ and $\mathcal{O}$ are Hermitian ($\mathcal{O} = \mathcal{O}^\dagger$),

$$\langle\langle \mathcal{O}||\rho\rangle\rangle = \langle\langle \rho||\mathcal{O}\rangle\rangle$$

we find that $a = a^*$ and $b = b^*$ (they are real) so that

$$\langle \mathcal{O} \rangle = a c_a(t) + b c_b(t) \tag{66'}$$

From this relation, it is clear that a knowledge of all the $c_q(t)$ will be sufficient to describe the time dependence of all physical observables. In the following we will drop the double ket notation and refer to operators only through the use of script characters (such as $\mathcal{Q}$).

The equations of motion of the $c_q(t)$ are given by the Liouville equation (cf., equation 50), with $\rho = \sum_q c_q(t)\mathcal{Q}_q$, so that

$$
\begin{aligned}
\frac{d\rho}{dt} &= \frac{i}{\hbar}[\rho, \mathcal{H}] \\
&= \sum_q \frac{dc_q}{dt}\mathcal{Q}_q \\
&= \frac{i}{\hbar}\left[\sum_q c_q(t)\mathcal{Q}_q, \mathcal{H}\right] \\
&= \frac{i}{\hbar}\sum_q c_q(t)[\mathcal{Q}_q, \mathcal{H}]
\end{aligned}
\tag{67}
$$

If we multiply both sides of equation 67 from the left by $\mathcal{Q}_p$ and take the trace, we obtain

$$\frac{dc_p(t)}{dt} = \frac{i}{\hbar}\sum_q c_q(t)\,Tr\{\mathcal{Q}_p[\mathcal{Q}_q, \mathcal{H}]\} \tag{68}$$

For a spin $I = 1$ system we have, in the most general case, a system of nine coupled linear first order differential equations for the $c_p(t)$. The equation corresponding to $\mathcal{Q}_p = \mathcal{E}$ is trivial, since the identity commutes with the Hamiltonian and all the other operators. We are then left with at most eight coupled equations.

To find explicit expressions for these equations for a given Hamiltonian, we need to evaluate the commutator of $\mathcal{H}$ with each of the basis operators. Since the Hamiltonian can also be expanded in terms of this basis set, we need to evaluate, for spin I, at most

$[(2I + 1)^2 - 1]^2 - [(2I + 1)^2 - 1]$ commutators (for spin 1 this requires a table of 56 commutators). Since $[Q_p, Q_q] = -[Q_q, Q_p]$ the number that need to be evaluated is reduced by a factor of two (to 28 for spin 1)[7]. In some cases, such as when we deal with several coupled spins, it may be more practical to rearrange equation 68 using the cyclic property of the trace

$$
\begin{aligned}
Tr\,\{Q_p[Q_q, \mathcal{H}]\} &= Tr\,\{Q_p Q_q \mathcal{H} - Q_p \mathcal{H} Q_q\} \\
&= Tr\,\{\mathcal{H} Q_p Q_q - Q_p \mathcal{H} Q_q\} \\
&= Tr\,\{[\mathcal{H}, Q_p]\,Q_q\}
\end{aligned}
\tag{69}
$$

to obtain

$$
\frac{dc_p(t)}{dt} = \frac{i}{\hbar} \sum_q c_q(t)\, Tr\,\{Q_q\,[\mathcal{H}, Q_p]\}
\tag{70}
$$

Now, in order to obtain the equation for $c_p(t)$ we need only evaluate $[\mathcal{H}, Q_p]$. A given Hamiltonian couples a specific $c_p(t)$ to the other coefficients $c_q(t)$ through the commutator $[\mathcal{H}, Q_p]$. Before proceeding with an example we must select a suitable basis set of operators $\{Q_q\}$.

The best choice of a basis set depends on the Hamiltonian. In pulsed NMR experiments, the Hamiltonian is different during different time intervals (due to application of rf pulses, field gradients, etc.). The basis set which provides the simplest sets of coupled equations under the conditions of the experiment is, in a practical sense, the best choice. We will be concerned with experiments involving rf pulses of strength w_i applied along either the x or y directions in the reference frame rotating at the Larmor frequency w_0 (i.e., on resonance). In addition, the ^2H quadrupolar interaction, of strength w_Q, will be neglected during the rf pulses. This is often not a good assumption but the effects of finite pulse width have been treated elsewhere[12].

During an rf pulse, of strength w_1, along the x-axis, the Hamiltonian is

$$
\mathcal{H}_1 = -\hbar w_1 I_x
\tag{71}
$$

when the rf pulses are off, the effective Hamiltonian in the rotating frame is simply that due to the quadrupolar interaction

$$
\mathcal{H}_Q = \hbar \frac{w_Q}{3} \left(3I_z^2 - 2\right)
\tag{72}
$$

With these Hamiltonians, a good set of spin 1 basis operators is:

$$
Q_1 = \frac{1}{\sqrt{2}} I_x = \frac{1}{2}
\begin{pmatrix}
0 & 1 & 0 \\
1 & 0 & 1 \\
0 & 1 & 0
\end{pmatrix}
\tag{73a}
$$

$$Q_2 = \frac{1}{\sqrt{2}} I_y = \frac{1}{2} \begin{pmatrix} 0 & -i & 0 \\ i & 0 & -i \\ 0 & i & 0 \end{pmatrix} \tag{73b}$$

$$Q_3 = \frac{1}{\sqrt{2}} I_z = \frac{1}{\sqrt{2}} \begin{pmatrix} 1 & 0 & 0 \\ 0 & 0 & 0 \\ 0 & 0 & -1 \end{pmatrix} \tag{73c}$$

$$Q_4 = \frac{1}{\sqrt{6}} \left(3I_z^2 - 2 \right) = \frac{1}{\sqrt{6}} \begin{pmatrix} 1 & 0 & 0 \\ 0 & -2 & 0 \\ 0 & 0 & 1 \end{pmatrix} \tag{73d}$$

$$Q_5 = \frac{1}{\sqrt{2}} \left(I_x I_z + I_z I_x \right) = \frac{1}{2} \begin{pmatrix} 0 & 1 & 0 \\ 1 & 0 & -1 \\ 0 & -1 & 0 \end{pmatrix} \tag{73e}$$

$$Q_6 = \frac{1}{\sqrt{2}} \left(I_y I_z + I_z I_y \right) = \frac{1}{2} \begin{pmatrix} 0 & -i & 0 \\ i & 0 & i \\ 0 & -i & 0 \end{pmatrix} \tag{73f}$$

$$Q_7 = \frac{1}{\sqrt{2}} \left(I_x^2 - I_y^2 \right) = \frac{1}{\sqrt{2}} \begin{pmatrix} 0 & 0 & 1 \\ 0 & 0 & 0 \\ 1 & 0 & 0 \end{pmatrix} \tag{73g}$$

$$Q_8 = \frac{1}{\sqrt{2}} \left(I_x I_y + I_y I_x \right) = \frac{1}{\sqrt{2}} \begin{pmatrix} 0 & 0 & -i \\ 0 & 0 & 0 \\ i & 0 & 0 \end{pmatrix} \tag{73h}$$

$$Q_9 = \mathcal{E} = \frac{1}{\sqrt{3}} \begin{pmatrix} 1 & 0 & 0 \\ 0 & 1 & 0 \\ 0 & 0 & 1 \end{pmatrix} \tag{73i}$$

with these definitions, the Hamiltonian during an rf pulse along the x-axis is

$$\mathcal{H}_1 = -\sqrt{2} \hbar \omega_1 Q_1 \tag{71'}$$

and the quadrupolar Hamiltonian becomes

$$\mathcal{H}_Q = \left(\frac{2}{3} \right)^{\frac{1}{2}} \hbar \omega_Q Q_4 \tag{72'}$$

The total Hamiltonian, with the rf off, is

$$\begin{aligned} \mathcal{H}_0 &= \mathcal{H}_Z + \mathcal{H}_Q \\ &= -\sqrt{2} \hbar \omega_0 Q_3 + \left(\frac{2}{3} \right)^{\frac{1}{2}} \hbar \omega_Q Q_4 \end{aligned} \tag{74}$$

For temperatures above 1 K, the exponential in the expression for the equilibrium density matrix, equation 57, can be expanded in a Taylor's series and we need keep only the linear term

$$\rho_0 \approx [1 - \mathcal{H}_0/kT] / Tr\{1 - \mathcal{H}_0/kT\} \tag{75}$$

But, $Tr\{\mathcal{H}_0\} = 0$, and the zeroth order term in the expansion, proportional to the unit matrix has a trace of three, so that, neglecting the term in the numerator which is proportional to the unit matrix (since it contributes nothing in the usual calculation of quantities of interest to NMR), we obtain

$$\rho_0 \approx -\frac{\mathcal{H}_0}{3kT} \tag{76}$$

At equilibrium, there are two non-zero quantities of interest, the Zeeman polarization, proportional to $\mathcal{I}_z$ and the quadrupolar order, proportional to $\langle \frac{1}{2}(3\mathcal{I}_z^2 - 2)\rangle$. The equilibrium expectation values of these quantities are readily calculated using equations 42 and 76,

$$\langle \mathcal{I}_z \rangle = Tr\{\mathcal{I}_z\rho_0\} \tag{77}$$

Substituting expressions for $\mathcal{H}_0$ and $\mathcal{I}_z$ from equations 73 and 74, we obtain

$$\langle \mathcal{I}_z \rangle = \frac{1}{3kT} Tr\left\{ (\sqrt{2}\mathcal{Q}_3)\left(\sqrt{2}\hbar\omega_0\mathcal{Q}_3 - \left(\frac{2}{3}\right)^{\frac{1}{2}}\hbar\omega_Q\mathcal{Q}_4 \right) \right\} \tag{78}$$

Using the orthonormality of the $\{\mathcal{Q}_q\}$ we obtain

$$I_0 = \langle \mathcal{I}_z \rangle = \frac{2\hbar\omega_0}{3kT} \tag{79}$$

Similarly, the expectation value for $\frac{1}{2}(3\mathcal{I}_z^2 - 2)$ is

$$\begin{aligned} Q_0 &= \langle\frac{1}{2}(3\mathcal{I}_z^2 - 2)\rangle \\ &= \langle\left(\frac{3}{2}\right)^{\frac{1}{2}}\mathcal{Q}_4\rangle \\ &= -\frac{\hbar\omega_Q}{3kT} \end{aligned} \tag{80}$$

For a spin system, such as deuterium, where the Larmor frequency, ω_0, is much larger than the quadrupolar splitting, ω_Q, the equilibrium value of Q_0 is much smaller than I_0 and may often be neglected. It is possible, however, to convert the large Zeeman polarization, I_0, into quadrupolar order, Q_0, using pulse techniques. In these non-equilibrium situations it may be useful to replace the lattice temperature, T, with the Zeeman temperature, T_Z, for I_0 and with the quadrupolar temperature, T_Q, for Q_0. Then, transferring the large equilibrium Zeeman polarization (at room temperature) into quadrupolar order corresponds to a very low quadrupolar temperature [7,12].

3   The Quadrupolar Echo

The usual single pulse free induction decay experiment used in high resolution NMR is not appropriate for NMR of solids because the finite receiver dead time, often as long as tens of microseconds, is comparable to the inverse of the spectral width. In such a situation, the free induction decay signal decays appreciably during this dead time. For example, deuterium in the rigid lattice case (no motional averaging of the quadrupolar interaction) has a maximum quadrupolar splitting of 252 kHz, for C-^{2}H bonds. An fid with such a broad splitting will decay appreciably in 1 microsecond. A typical receiver dead time for deuterium is about 20 microseconds, so that much of the signal will be lost before the receiver is ready to detect it. The quadrupolar echo technique was invented to avoid the problem of receiver dead time. This simple two pulse sequence, illustrated in Figure 3a, consists of a 90° pulse (rotation) with a 0° phase shift relative to the receiver reference phase ($90_{0°}$) applied at time $t = 0$, followed by a delay, τ, which is greater than the receiver dead time. After this delay, a second 90° pulse with a phase shift of 90°, ($90_{90°}$) is applied. Then, at a time τ following this second pulse (at time $t = 2\tau$), the nuclear magnetization refocusses to form the quadrupolar echo. The spectrum is obtained from this echo by taking the Fourier transform of the time domain signal, defining the top of the echo (occurring at $t = 2\tau$) as $t' = 0$. Insofar as the dephasing of the magnetization is due only to the quadrupolar interaction, this refocussing is complete and the dead time problem is circumvented[5,12,13]. The calculation of the quadrupolar echo provides a simple demonstration of the application of the density operator formalism described above.

At $t = 0$ we assume the spin system is in equilibrium with the lattice so that the initial density operator has only two non-zero elements, proportional to I_0 and Q_0, as given by equations 79 and 80. For deuterium in high fields, the quadrupolar interaction is a small perturbation on the Zeeman inteaction so that Q_0 is small compared to I_0. Consequently, we will neglect the small equilibrium quadrupolar order term Q_0 in the following, and only concern ourselves with the evolution and transformation of the Zeeman polarization, I_0, during the experiment. Initially, then, the density matrix has the form

$$\rho(0) = \sum_q c_q(0) Q_q = c_3(0) Q_3 \tag{81}$$

where

$$c_3(0) = \frac{\sqrt{2}\hbar\omega_0}{3kT} \tag{82}$$

At $t = 0$ we apply an rf pulse of strength ω_1 for a time t_{w_1} along the x-axis in the rotating frame, as in Figure 3a. The rotating frame Hamiltonian during this time interval is

$$\mathcal{H}_1 = -\sqrt{2}\hbar\omega_1 Q_1 \tag{83}$$

The evolution of the density operator under the influence of $\mathcal{H}_1$ is given by the Liouville

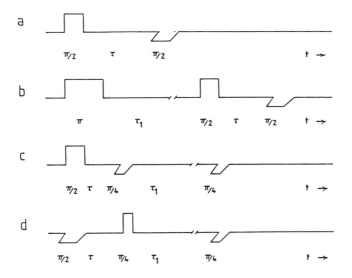

Figure 3: Pulse sequences used in spin 1 echo spectroscopy and relaxation experiments: a) the quadrupolar echo sequence; b) the inversion recovery sequence; c) the Jeener-Broekaert sequence; and d) the modified Jeener- Broekaert sequence.

equation, in particular, by equation 70.

$$\frac{dc_3(t)}{dt} = \frac{i}{\hbar} \sum_q c_q(t)\, Tr\,\{Q_q\,[\mathcal{H}_1, Q_3]\} \tag{84}$$

The commutator,

$$[\mathcal{H}_1, Q_3] = -\sqrt{2}\hbar\omega_1\,[Q_1, Q_3] = i\hbar\omega_1 Q_2 \tag{85}$$

is easily evaluated using the matrix representation of the operators given in equation 73. Thus, during a pulse along the x-axis, the z component of magnetization, Q_3, is coupled to the y component, Q_2, as expected from a vector model. Clearly we need also the equation for $c_2(t)$

$$\frac{dc_2(t)}{dt} = \frac{i}{\hbar} \sum_q c_q(t)\, Tr\,\{Q_q\,[\mathcal{H}_1, Q_2]\} \tag{86}$$

Again, the commutator is readily evaluated to give

$$[\mathcal{H}_1, Q_2] = -\sqrt{2}\hbar\omega_1\,[Q_1, Q_2] = -i\hbar\omega_1 Q_3 \tag{87}$$

The resulting pair of coupled equations is

$$\frac{dc_2(t)}{dt} = \omega_1 c_3(t) \tag{88a}$$

$$\frac{dc_3(t)}{dt} = -\omega_1 c_2(t) \tag{88b}$$

All of the other coefficients $(c_q(t))$ remain zero during this first pulse. Already we see the convenience of this approach. During a particular time interval, we need only evaluate the coefficients of those elements of the density operator which were non-zero at the beginning of that interval and the coefficients to which they are coupled by the Hamiltonian appropriate to that interval. For simple pulse experiments we may need to evaluate only a few commutators.

Differentiating the first of these two coupled equations with respect to t, and substituting for dc_3/dt from the second equation we obtain the second order differential equation for $c_2(t)$

$$\frac{d^2 c_2(t)}{dt^2} = -\omega_1^2 c_2(t) \tag{89}$$

which has the solution

$$c_2(t) = A\cos(\omega_1 t) + B\sin(\omega_1 t) \tag{90}$$

at $t = 0$, $c_2(0) = 0$ so that $A = 0$. Taking the derivative of the expression for $c_2(t)$ gives the solution for $c_3(t)$, from equation 90. The initial condition, that at $t = 0$, $c_3(0)$ is given by equation 82, leads to the pair of solutions, with $t = t_{w_1}$,

$$c_2(t_{w_1}) = \frac{\sqrt{2}\hbar\omega_0}{3kT}\sin(\omega_1 t_{w_1})$$

$$\tag{91}$$

$$c_3(t_{w_1}) = \frac{\sqrt{2}\hbar\omega_0}{3kT}\cos(\omega_1 t_{w_1})$$

During the rf pulse one can think of the vector magnetization precessing about the direction of the rf magnetic field in the rotating frame. The precession frequency is ω_1, the strength of the rf field. If the pulse amplitude and duration are such that $\omega_1 t_{w_1} = 90°$, then at the end of the pulse the equilibrium z-component of magnetization will have been rotated into the positive y-axis. In operator space we can think of this evolution as a precession in the $(\mathcal{Q}_2, \mathcal{Q}_3)$ plane.

Following this 90° rotation the only non-zero element of the density operator is $\mathcal{Q}_2$ with the coefficient

$$c_2(t_{w_1}) = \frac{\sqrt{2}\hbar\omega_0}{3kT} \tag{92}$$

The spin system is next allowed to evolve for a time τ under the influence of the quadrupolar Hamiltonian, equation 72'. The equation of motion for $c_2(t)$ during this time period is

$$\frac{dc_2(t)}{dt} = \frac{i}{\hbar}\sum_q c_q(t) \, Tr\,\{\mathcal{Q}_q\,[\mathcal{H}_Q, \mathcal{Q}_2]\} \tag{93}$$

(where t starts at t_{w_1}). The commutator here is

$$[\mathcal{H}_Q, Q_2] = \left(\frac{2}{3}\right)^{\frac{1}{2}} \hbar\omega_Q [Q_4, Q_2] = -i\hbar\omega_Q Q_5 \qquad (94)$$

so we see that the quadrupolar Hamiltonian couples Q_2 to Q_5. We will also need the corresponding equation for Q_5 and the commutator of $\mathcal{H}_Q$ with Q_5

$$[\mathcal{H}_Q, Q_5] = \left(\frac{2}{3}\right)^{\frac{1}{2}} \hbar\omega_Q [Q_4, Q_5] = i\hbar\omega_Q Q_2 \qquad (95)$$

so again we have a set of two coupled linear differential equations during the time following the first pulse

$$\frac{dc_2(t)}{dt} = \omega_Q c_5(t)$$

$$\qquad (96)$$

$$\frac{dc_5(t)}{dt} = -\omega_Q c_2(t)$$

Solving these equations in the same manner as for equation 88, and using the initial conditions that at $t = t_{w_1}$, c_2 is given by equation 92, and $c_5 = 0$, we obtain, with $t = \tau$,

$$c_2(\tau) = \frac{\sqrt{2}\hbar\omega_0}{3kT} \cos(\omega_Q\tau)$$

$$\qquad (97)$$

$$c_5(\tau) = -\frac{\sqrt{2}\hbar\omega_0}{3kT} \sin(\omega_Q\tau)$$

Now the precession, under the influence of the quadrupolar Hamiltonian, is in the (Q_2, Q_5) plane of operator space.

The usual detection scheme for NMR uses a receiving coil which is transverse to the static magnetic field, H_0. This coil is only capable of detecting transverse magnetization, M_x and M_y. The signal following a 90° pulse applied along the x-axis (in the rotating frame) is $\langle M_y \rangle$ which is proportional to $\langle I_y \rangle$

$$\langle I_y(t) \rangle = \sqrt{2} Tr \{Q_2 \, \rho(t)\} = \sqrt{2} c_2(t) = \frac{2\hbar\omega_0}{3kT} \cos(\omega_Q\tau) \qquad (98)$$

This signal is the normal free induction decay following a 90° pulse. The x component of magnetization, proportional to I_x is zero. The other non-zero element of the density matrix, proportional to Q_5 does not give rise to a directly detectable free induction signal. However, $c_5(t)$ does represent a real coherence and it can be converted into observable signal (by varying τ) or into quadrupolar order or double quantum coherence with suitable pulse sequences. At a time $\tau = \pi/2\omega_Q$ the y component of magnetization is zero, and $c_5(\tau)$ is maximal. The coherence represented by c_5, which can be called quadrupolar single quantum coherence, has the same form as the "antiphase" coherence referred to in discussions of systems of two coupled spin 1/2 nuclei[14]. Thus, following a 90° rotation about the x-axis, the spin system oscillates between y-magnetization, proportional to $c_2(t)$ and an "antiphase" coherence proportional to $c_5(t)$ at a frequency ω_Q under the influence of the quadrupolar Hamiltonian. Later we will return to see how to convert Q_5 to quadrupolar order and double quantum coherence.

A second pulse, this time a rotation about the y-axis, is used to form the quadrupolar echo, as in Figure 3a. At time $t = \tau + t_{w_1}$ the only two non-zero elements of the density operator have coefficients $c_2(t)$ and $c_5(t)$. Thus, we need to evaluate the evolution of these two coefficients during the rf pulse. The Hamiltonian in this region is

$$\mathcal{H}_2 = -\hbar\omega_2 I_y = -\sqrt{2}\hbar\omega_2 Q_2 \tag{99}$$

where ω_2 is the strength of the pulse. The equation of motion for $c_2(t)$ is (with t starting at $t_{w_1} + \tau$),

$$\frac{dc_2(t)}{dt} = \frac{i}{\hbar} \sum_q c_q(t) \, Tr \, \{Q_q \, [\mathcal{H}_2, Q_2]\} \tag{100}$$

but, since the Hamiltonian commutes with Q_2,

$$[\mathcal{H}_2, Q_2] = -\sqrt{2}\hbar\omega_2 \, [Q_2, Q_2] = 0 \tag{101}$$

the coefficient c_2 is independent of time during this interval. The coefficient $c_5(t)$ does not commute with $\mathcal{H}_2$ so we must solve the equation for its evolution

$$\frac{dc_5(t)}{dt} = \frac{i}{\hbar} \sum_q c_q(t) \, Tr \, \{Q_q \, [\mathcal{H}_2, Q_5]\} \tag{102}$$

The commutator is again easily evaluated using the matrix representation given in equation 73

$$[\mathcal{H}_2, Q_5] = -\sqrt{2}\hbar\omega_2 \, [Q_2, Q_5] = 2i\hbar\omega_2 \frac{1}{2} \left\{ \sqrt{3}Q_4 - Q_7 \right\} \tag{103}$$

so that equation 102 becomes

$$\frac{dc_5(t)}{dt} = -2\omega_2 \left[\frac{1}{2} \left\{ \sqrt{3}c_4(t) - c_7(t) \right\} \right] \tag{104}$$

In an analogous fashion, we can write

$$\frac{d}{dt} \left[\frac{1}{2} \left\{ \sqrt{3}c_4 - c_7 \right\} \right] = \frac{i}{\hbar} \sum_q c_q(t) \, Tr \, \left\{ Q_q \left[\mathcal{H}_2, \frac{1}{2} \left\{ \sqrt{3}Q_4 - Q_7 \right\} \right] \right\} \tag{105}$$

which, on evaluating the commutator, reduces to

$$\frac{d}{dt} \left[\frac{1}{2} \left\{ \sqrt{3}c_4(t) - c_7(t) \right\} \right] = 2\omega_2 c_5(t) \tag{106}$$

and we again have two coupled linear differential equations.

The solutions are found just as for equation 96. The initial conditions are that at $t = \tau + t_{w_1}$, c_5 is given by equation 97, and $\frac{1}{2}\left(\sqrt{3}c_4 - c_7\right) = 0$. Then, the solutions to equations 104 and 106 are, for pulse length $t = t_{w_2}$,

$$c_5(t_{w_2}) = -\frac{\sqrt{2}\hbar\omega_0}{3kT}\sin\left(\omega_Q\tau\right)\cos\left(2\omega_2 t_{w_2}\right)$$

(107)

$$\frac{1}{2}\left\{\sqrt{3}c_4(t_{w_2}) - c_7(t_{w_2})\right\} = -\frac{\sqrt{2}\hbar\omega_0}{3kT}\sin\left(\omega_Q\tau\right)\sin\left(2\omega_2 t_{w_2}\right)$$

If the pulse length is such that $\omega_2 t_{w_2} = 90°$, or $t_{w_2} = \pi/2\omega_2$, then, at the end of the pulse we have

$$c_5(\pi/2\omega_2) = \frac{\sqrt{2}\hbar\omega_0}{3kT}\sin\left(\omega_Q\tau\right)$$

(108)

$$\frac{1}{2}\left\{\sqrt{3}c_4(\pi/2\omega_2) - c_7(t)\right\} = 0$$

On the other hand, if the pulse length is such that $\omega_2 t_{w_2} = 45°$, or $t_{w_2} = \pi/4\omega_2$, then

$$c_5(\pi/4\omega_2) = 0$$

(109)

$$\frac{1}{2}\left\{\sqrt{3}c_4(\pi/4\omega_2) - c_7(\pi/4\omega_2)\right\} = -\frac{\sqrt{2}\hbar\omega_0}{3kT}\sin\left(\omega_Q\tau\right)$$

The first case, equation 108, corresponds to the refocussing pulse of the quadrupolar echo experiment. The second case, equation 109, is used to prepare the spin system in a state of quadrupolar or double quantum polarization. We will return to this latter case in the next section.

Starting from equation 108, at the end of the second pulse, we allow the spin system to evolve under the quadrupolar Hamiltonian, equation 72'. Since both $c_5(t)$ and $c_2(t)$ are non-zero immediately following the pulse, we need to examine the time dependence of both of these coefficients. The commutators we require in equation 70 are $[\mathcal{H}_Q, \mathcal{Q}_5]$, which is given by equation 95, and $[\mathcal{H}_Q, \mathcal{Q}_2]$, which is given by equation 94. Just as during the time interval between pulses, in the region following the second pulse the quadrupolar Hamiltonian couples $c_5(t)$ and $c_2(t)$. The two equations are identical to those in equation 96, but the initial conditions are different. At the end of the second pulse, we have c_5 from equation 108, and c_2 from equation 97 (we neglect the evolution during the second pulse because of our assumption that $\omega_Q \ll \omega_2$). The solutions at a time t' following the end of the second pulse are

$$c_2(t') = \frac{\sqrt{2}\hbar\omega_0}{3kT}\left\{\sin\left(\omega_Q\tau\right)\sin\left(\omega_Q t'\right) + \cos\left(\omega_Q\tau\right)\cos\left(\omega_Q t'\right)\right\} \qquad (110a)$$

$$c_5(t') = \frac{\sqrt{2}\hbar\omega_0}{3kT} \left\{ \sin\left(\omega_Q\tau\right)\cos\left(\omega_Q t'\right) - \cos\left(\omega_Q\tau\right)\sin\left(\omega_Q t'\right) \right\} \tag{110b}$$

In terms of the real time $t = t' + \tau$, as defined in Figure 3a, these solutions can be written (neglecting t_{w_1} and t_{w_2})

$$c_2(t) = \frac{\sqrt{2}\hbar\omega_0}{3kT}\cos\left(\omega_Q\left[t - 2\tau\right]\right)$$

$$\tag{111}$$

$$c_5(t) = -\frac{\sqrt{2}\hbar\omega_0}{3kT}\sin\left(\omega_Q\left[t - 2\tau\right]\right)$$

The expression for $c_2(t)$ gives us the quadrupolar echo signal. As we did in the case of the fid following the first pulse, equation 98, we evaluate the trace of $\mathcal{I}_y(t)$

$$\langle \mathcal{I}_y(t)\rangle = \sqrt{2}\, Tr\left\{Q_2\rho(t)\right\} = \sqrt{2}\,c_2(t) = \frac{2\hbar\omega_0}{3kT}\cos\left(\omega_q\left[t - 2\tau\right]\right) \tag{112}$$

this signal has a maximum at $t = 2\tau$ and is symmetric about that point. The amplitude at $t = 2\tau$ is equal (neglecting relaxation effects) to the amplitude of the fid at $t = 0$. Thus, insofar as we can neglect the finite width and other pulse imperfections, the refocussing is complete and independent of the value of ω_Q. For this reason quadrupolar echo spectroscopy has solved the problem of receiver dead time in ^{2}H NMR. Figure 4a shows a quadrupolar echo from a $[2,2-^2H_2]$-potassium laurate/water unoriented sample. The spectrum obtained by Fourier transforming the time domain signal starting at the top of the echo is shown in Figure 4b.

As we have already mentioned, the assumption that ω_1 and ω_2 are much greater than ω_Q is often not satisfied. For example, ω_Q in C-^{2}H bonds can be as large as $2\pi \times 252$ kHz $= 1.583 \times 10^6$ s^{-1}. A 2 μs 90° pulse has an $\omega_1 = 0.785 \times 10^6$ s^{-1}. Thus, even a 2 μs pulse, which is beyond the power of many commercial spectrometers, actually has an ω_1 which is less than this maximum value of ω_Q. Finite pulse width effects are very important in such cases[12]. For quadrupolar splittings much less than this maximum value, however, the simple analysis above is adequate. In any case, this analysis illustrates the procedure used. The other major concern is the relaxation which may occur between the two pulses used to form the quadrupolar echo. If the echo decay time, T_{2e} is comparable to the pulse separation (typically about 20 to 30 μs), then the spectrum obtained from the quadrupolar echo may have a different shape than the $\tau = 0$ spectrum would have had. This is especially true for powder spectra in situations where the relaxation is orientation dependent. The changes in the shape of the spectrum with τ can be quite dramatic and can be analyzed in terms of models for the motions responsible for the relaxation[15].

If we had applied our first rf pulse along the negative x-direction, a $(90_{180°})$ pulse, the Hamiltonian in equation 83 would have been positive. Then, the commutators in equation 85 and 87 would have changed sign as would the right hand sides of the two coupled

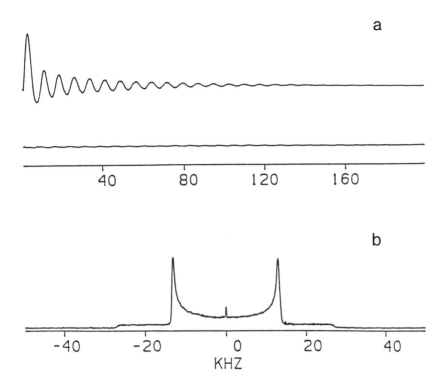

Figure 4: a)The quadrupolar echo for [2, 2−²H₂]-potassium laurate/water. b) The spectrum obtained by Fourier transforming the echo in a) starting at the top of the echo.

equations 88. The solution for $c_2(t)$ would change sign while that for c_3 would remain the same, i.e., the magnetization would have been rotated to the negative y-direction. Keeping the second pulse the same as before, a $(90_{90°})$ pulse, all of the other equations remain the same, but their initial conditions will have changed so that the final result would be a quadrupolar echo forming along the negative y-direction. Thus, alternating the phase of the first pulse between 0° and 180° results in the alternation of the sign of the echo. On the other hand, the alternation of the phase of the second pulse between 90° and 270° has no effect on the sign of the echo. Pulse sequences which cycle the phases of the two pulses between these values are used to eliminate artifacts caused by residual receiver recovery transients and pulse imperfections. In addition, if the first pulse is applied along the y-axis (instead of the x-axis) and the second pulse applied along the x-axis, we obtain a quadrupolar echo signal in the receiver channel corresponding to the x-direction (instead of the y-direction). This can be used to average over the differences between the two receiver channels.

If we apply a 180° pulse at $t = 0$, then, from equation 91, following the pulse we would have the nuclear magnetization aligned along the negative z-direction. The density operator would be proportional to $-\mathcal{I}_z$, and the populations would be inverted compared to their equilibrium values. After this 180° pulse, the negative z-magnetization is invariant under $\mathcal{H}_Q$ (which commutes with $-\mathcal{I}_z$) so the only evolution following the inverting pulse would be the recovery of the magnetization to its equilibrium value along the positive z-direction due to spin-lattice relaxation[5,16], with a characteristic time T_{1z}. We can monitor the recovery to equilibrium by applying the full quadrupolar echo sequence at a time τ_1 following the inverting pulse, as in Figure 3b. A plot of the echo amplitude, or the spectrum area, vs τ_1 gives the recovery curve and allows us to determine T_{1z}. In a similar fashion, measuring the echo amplitude or spectrum area as a function of τ in the normal quadrupolar echo experiment, Figure 3a, provides a measure of the transverse relaxation time[5,16] T_{2e}.

4 Quadrupolar Order and Double Quantum Coherence

We now return to the situation following the application of a single pulse where the system is described by equation 97. We continue the development from there, showing how we can create quadrupolar order and double quantum coherence, and later, how, by changing the phases of the first two pulses, we can separate these two quantities.

The quadrupolar interaction, involving the nuclear quadrupolar moment and the electronic electric field gradient at the nucleus, results in a shift of the three spin 1 nuclear Zeeman energy levels. The $m = \pm 1$ levels are shifted upwards by an amount Δ while the $m = 0$ level is shifted down by an amount 2Δ. This is the origin of the quadrupolar splitting, $\omega_Q = 6\Delta$. At equilibrium, the deuterium spin 1 system possesses non-zero quadrupolar order, as described by equation 80, corresponding to the shift in the populations of the three Zeeman energy levels due to the quadrupolar interaction. The value $Q_0 = -\{\hbar\omega_Q/3kT\}$ corresponds to the decrease in the populations of the $m = \pm 1$ energy levels and an increase in the population of the $m = 0$ level. From equations 74 and 75 we find the populations

$$
\begin{aligned}
p_1 &= \frac{1}{3}\left[1 + \frac{\hbar\omega_0}{kT} - \frac{\hbar\omega_Q}{3kT}\right] \\
p_0 &= \frac{1}{3}\left[1 + \frac{2\hbar\omega_Q}{3kT}\right] \\
p_{-1} &= \frac{1}{3}\left[1 - \frac{\hbar\omega_0}{kT} - \frac{\hbar\omega_Q}{3kT}\right]
\end{aligned}
\tag{113}
$$

In the discussion of the quadrupolar echo sequence we found that, if the second pulse was $(45_{90°})$ instead of the usual $(90_{90°})$, the coefficients $c_4(t)$ and $c_7(t)$ are optimized. These correspond to quadrupolar polarization and double quantum coherence. In order to evaluate these independently we need to solve the equations for c_4 and c_7 separately. The real time at this point is $t_{w_1} + \tau$, i.e., we are considering the second pulse whose width is

$t = t_{w_2}$, and whose strength is w_2. The equation for $c_5(t)$ is the same as before and the other two equations require the commutators $[Q_2, Q_4]$ and $[Q_2, Q_7]$.

$$\frac{dc_4(t)}{dt} = \sqrt{3}w_2 c_5(t)$$

$$\frac{dc_5(t)}{dt} = -2w_2 \left[\frac{1}{2}\left\{\sqrt{3}c_4(t) - c_7(t)\right\}\right] \tag{114}$$

$$\frac{dc_7(t)}{dt} = -w_2 c_5(t)$$

The solutions, for $w_2 t_{w_2} = 45°$ are

$$c_4(t_{w_2}) = -\frac{1}{2}\left(\frac{2}{3}\right)^{\frac{1}{2}} \frac{\hbar w_0}{kT} \sin w_Q \tau$$

$$c_5(t_{w_2}) = 0 \tag{115}$$

$$c_7(t_{w_2}) = \frac{\sqrt{2}}{2} \frac{\hbar w_0}{3kT} \sin w_Q \tau$$

so that immediately following the pulse, the density operator is

$$\rho(t_{w_2}) = \frac{1}{3}\left\{\sqrt{3}\mathcal{E} - \left(\frac{3}{2}\right)^{\frac{1}{2}} \frac{\hbar w_0}{kT} \sin w_Q \tau \, Q_4 + \frac{1}{\sqrt{2}} \frac{\hbar w_0}{kT} \sin w_Q \tau \, Q_7\right\} \tag{116}$$

where we have kept the term in the density operator which is proportional to the unit operator ($\sqrt{3}\mathcal{E}$) so that we can calculate the populations. Our measure of quadrupolar order is

$$Q_0 = \langle\frac{1}{2}\left(3I_z^2 - 1\right)\rangle = \langle\left(\frac{3}{2}\right)^{\frac{1}{2}} Q_4\rangle = -\frac{\hbar w_0}{2kT} \sin w_Q \tau \tag{117}$$

which is larger than the equilibrium value by the factor $\{3w_0/2w_Q\} \sin w_Q \tau$, and is comparable to the Zeeman polarization. Clearly this is a maximum if $w_Q \tau = 90°$, which is easily managed if we have an oriented sample with a single quadrupolar splitting. Thus, it is easy to prepare the system so that it has a large quadrupolar order. However, using this pulse sequence mixes in a significant amount of double quantum coherence since the term in c_7 in equation 115 is of approximately the same magnitude as that for c_4. We will use the quantity $D_Q = \langle\sqrt{2}Q_7\rangle$ as a measure of double quantum coherence. Then, following the $(90_{0°}) - \tau - (45_{90°})$ sequence,

$$D_Q = \langle\sqrt{2}Q_7\rangle = \frac{\hbar w_0}{3kT} \sin w_Q \tau \tag{118}$$

Neither quadrupolar polarization nor double quantum coherence are directly observable in the usual NMR apparatus. These two quantities must first be converted back to magnetization (single quantum coherence) in order to be measured.

The operators Q_7 and Q_8, which correspond to double quantum coherence, have matrix elements which connect Zeeman energy levels with $m = \pm 1$. Under the Hamiltonian

$$\mathcal{H} = -\sqrt{2}\hbar\delta\omega_0 Q_3 + \left(\frac{2}{3}\right)^{\frac{1}{2}} \hbar\omega_Q Q_4 \tag{119}$$

(i.e., when we are off resonance by the amount $\delta\omega_0$), we find that

$$[\mathcal{H}, Q_7] = -\sqrt{2}\hbar\delta\omega_0[Q_3, Q_7] = -2i\hbar\delta\omega_0 Q_8 \tag{120}$$

since $[Q_4, Q_7] = 0$, and

$$[\mathcal{H}, Q_8] = 2i\hbar\delta\omega_0 Q_7 \tag{121}$$

The two coupled equations for c_7 and c_8 have solutions

$$c_7(t) = \frac{1}{\sqrt{2}}\frac{\hbar\omega_0}{3kT} \sin\omega_Q\tau \cos 2\delta\omega_0 t$$

$$c_8(t) = -\frac{1}{\sqrt{2}}\frac{\hbar\omega_0}{3kT} \sin\omega_Q\tau \sin 2\delta\omega_0 t \tag{122}$$

where we have taken our initial condition from equation 115. Thus, double quantum coherence precesses at the twice the frequency of single quantum coherence. The range of frequency offsets due, for example, to different isotropic chemical shifts, is multiplied by a factor of two. In higher spin systems, or systems of coupled spins, similar behaviour of n-quantum coherences can be observed[17].

If the Hamiltonian in the interval following the creation of quadrupolar polarization and double quantum coherence is simply that due to the quadrupolar interaction, then, since $[\mathcal{H}_Q, Q_4] = [\mathcal{H}_Q, Q_7] = 0$, these two quantities remain invariant except for relaxation back towards their equilibrium values due to fluctuations in the quadrupolar Hamiltonian, caused, for example, by molecular reorientations. The two quantities are sensitive to these fluctuations in different ways so that their recovery to equilibrium will be described by different characteristic relaxation times T_{1Q}, for quadrupolar polarization, and T_{DQ}, for double quantum coherence[5,16]. During this time interval the density matrix can be written as

$$\rho(t') = -\frac{1}{\sqrt{2}}\frac{\hbar\omega_0}{3kT} \sin\omega_Q\tau \left[\sqrt{3}\exp\{-t'/T_{1z}\}Q_4 - \exp\{-t'/T_{DQ}\}Q_7\right] \tag{123}$$

where $t' = 0$ at real time $t_{w_1} + \tau + t_{w_2}$. Here we have again neglected the small non-zero equilibrium value of Q_0 and allow both terms to relax back to 0 with their characteristic decay times. After a time $t' = \tau_1$ we apply a third pulse, namely a $(45)_{90°}$ rotation, of strength ω_3. During this pulse the equations of motion are

$$\frac{dc_5(t)}{dt} = -2\omega_3 \left[\frac{1}{2}\left\{\sqrt{3}c_4(t) - c_7(t)\right\}\right] \tag{124}$$

$$\frac{d}{dt}\left[\frac{1}{2}\left\{\sqrt{3}c_4 - c_7\right\}\right] = 2\omega_3 c_5(t)$$

with solutions, when $\omega_3 t_{w_3} = 45°$,

$$c_5(t_{w_3}) = \frac{1}{2\sqrt{2}} \frac{\hbar\omega_0}{3kT} \sin\omega_Q\tau \, [3\exp\{-\tau_1/T_{1Q}\} + \exp\{-\tau_1/T_{DQ}\}]$$

(125)

$$\frac{1}{2}\{c_4(t_{w_3}) - c_7(t_{w_3})\} = 0$$

During the final evolutionary period under $\mathcal{H}_Q$, c_5 and c_2 are again coupled as in equations 94 to 96. The solutions in this case are (with t' starting at $t_{w_1} + \tau + t_{w_2} + \tau_1 + t_{w_3}$)

$$c_2(t') = \frac{1}{2\sqrt{2}} \frac{\hbar\omega_0}{3kT} \sin\omega_Q\tau \sin\omega_Q t' \, [3\exp\{-\tau_1/T_{1Q}\} + \exp\{-\tau_1/T_{DQ}\}]$$

(126)

and

$$c_5(t') = \frac{1}{2\sqrt{2}} \frac{\hbar\omega_0}{3kT} \sin\omega_Q\tau \cos\omega_Q t' \, [3\exp\{-\tau_1/T_{1Q}\} + \exp\{-\tau_1/T_{DQ}\}]$$

(127)

These equations give the response of the spin system to the Jeener-Broekaert[18] pulse sequence $(90)_{0°} - \tau - (45)_{90°} - \tau_1 - (45)_{90°} - t'$ at time $t = t' + \tau_1 + \tau$, neglecting the small pulse widths t_{w_1}, t_{w_2}, and t_{w_3}.

We can separate the contributions due to quadrupolar order and double quantum coherence by changing the phases of the first two pulses, generating a modified Jeener-Broekaert sequence. If we had prepared the system with the sequence $(90)_{90°} - \tau - (45)_{0°}$, then the density matrix at the end of this preparation period would have been

$$\rho(\tau_1) = \frac{1}{3}\left\{\sqrt{3}\mathcal{E} + \left(\frac{3}{2}\right)^{\frac{1}{2}} \frac{\hbar\omega_0}{kT} \sin\omega_Q\tau\, Q_4 + \frac{1}{\sqrt{2}} \frac{\hbar\omega_0}{kT} \sin\omega_Q\tau\, Q_7\right\}$$

(128)

The quadrupolar order, analogous to the result of equation 117, corresponding to this preparation sequence is

$$Q_0 = \langle\left(\frac{3}{2}\right)^{\frac{1}{2}} Q_4\rangle = +\frac{\hbar\omega_0}{2kT} \sin\omega_Q\tau$$

(129)

which has the same magnitude as previously but the opposite sign. The measure of double quantum coherence, equation 118, is the same for both pulse sequences. Allowing the system to evolve under $\mathcal{H}_Q$ for a time τ_1, as before, gives

$$\rho(\tau_1) = \frac{1}{\sqrt{2}} \frac{\hbar\omega_0}{3kT} \sin\omega_Q\tau \, [3\exp\{-\tau_1/T_{1Q}\}Q_4 + \exp\{-\tau_1/T_{DQ}\}Q_7]$$

(130)

instead of the result given by equation 123.

Applying the same $(45)_{90°}$ pulse as before, we obtain again equations 114 and 87, but with initial conditions given by equation 130 instead of by equation 117. The solutions in this new case are, for $\omega_3 t_{w_3} = 45°$,

$$c_5(t_{w_3}) = -\frac{1}{\sqrt{2}}\frac{\hbar\omega_0}{3kT} \sin\omega_Q\tau \, [3\exp\{-\tau_1/T_{1Q}\} - \exp\{-\tau_1/T_{DQ}\}]$$

$$(131)$$

$$\frac{1}{2}\left\{\sqrt{3}c_4(t_{w_3}) - c_7(t_{w_3})\right\} = 0$$

Again, allowing an evolutionary period t' under the influence of $\mathcal{H}_Q$, c_5 and c_2 are coupled according to equations 94 to 96, with solutions in this case of

$$c_2(t') = -\frac{1}{2\sqrt{2}}\frac{\hbar\omega_0}{3kT} \sin\omega_Q\tau \sin\omega_Q t' \, [3\exp\{-\tau_1/T_{1Q}\} - \exp\{-\tau_1/T_{DQ}\}] \qquad (132)$$

and

$$c_5(t') = -\frac{1}{2\sqrt{2}}\frac{\hbar\omega_0}{3kT} \sin\omega_Q\tau \cos\omega_Q t' \, [3\exp\{-\tau_1/T_{1Q}\} - \exp\{-\tau_1/T_{DQ}\}] \qquad (133)$$

Thus, it is possible to prepare the spin system in states of double quantum coherence and

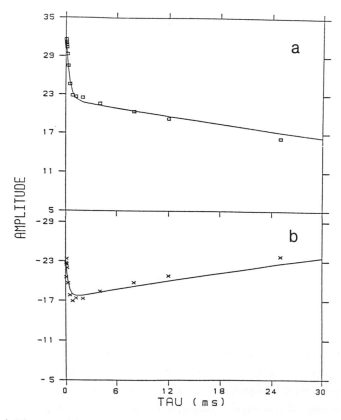

Figure 5: a) The signal following the Jeener-Broekaert sequence of Figure 3c; and b) the signal following the modified Jeener-Broekaert sequence of Figure 3d.

either positive or negative quadrupolar order. Allowing these two coherences to evolve under the quadrupolar Hamiltonian in the interval τ_1 we can then convert them to observable magnetization and measure their recovery to equilibrium.

We combine these two pulse sequences in a single experiment, in one part of computer memory we add the signal following the two sequences, and in another part of memory we subtract the signal from the second sequence from that of the first, and obtain the two distinct signals

$$
\begin{aligned}
\langle D_Q(\tau_1, t') \rangle &= \sqrt{2}\left[c_2^{(1)}(t') + c_2^{(2)}(t')\right] \\
&= \frac{\hbar\omega_0}{3kT}\sin\omega_Q\tau\,\sin\omega_Q t'\,\exp\{-\tau_1/T_{DQ}\} \quad (134) \\
\langle Q_0(\tau_1, t') \rangle &= \sqrt{2}\left[c_2^{(1)}(t') - c_2^{(2)}(t')\right] \\
&= \frac{\hbar\omega_0}{kT}\sin\omega_Q\tau\,\sin\omega_Q t'\,\exp\{-\tau_1/T_{1Q}\} \quad (135)
\end{aligned}
$$

where we have used the fact that the observed signal is proportional to $\langle \mathcal{M}_y \rangle = Tr\{\rho\mathcal{I}_y\} = \sqrt{2}c_2$. These two pulse sequences are illustrated in Figures 3c and 3d. The τ_1 dependence of signals corresponding to equations 134 and 135 for a sample of $[2,2\text{-}^2H_2]$-potassium laurate are shown in Figure 5. The two-component character is displayed nicely in both cases, the signal corresponding to double quantum coherence decreasing with τ_1 in both cases and the signal corresponding to quadrupolar order being positive in Figure 5a, and negative in Figure 5b. The relaxation times for the decay of double quantum coherence and quadrupolar order depend on molecular motion in different ways. The very short values of T_{DQ} in this particular case are quite dramatic.

III   Applications

In partially ordered systems, such as solids, liquid crystals and model or biological membranes, the residual quadrupolar splittings of ^2H labelled molecules provide information on the degree of orientational order and on molecular structure and conformation. Several review articles have been written on the use of ^2H NMR spectra to study molecular conformation and orientational order[5,9,19-21], so we will present here only a few recent examples.

Nematic liquid crystals often spontaneously orient in a large magnetic field with the director (the local axis along which the molecules are preferentially aligned) either parallel or perpendicular to the field. Molecules dissolved in such a liquid crystal will also be oriented so that ^2H NMR spectra will appear as doublets rather than powder patterns. The doublet splitting will depend on the degree of orientational order and on the average orientation of the bond containing the deuterium atom with respect to the director. Flexible

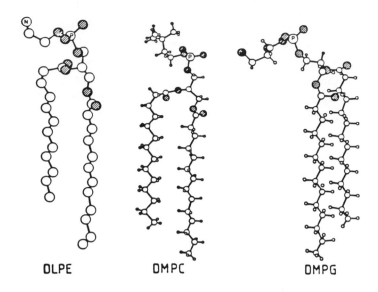

OLPE   DMPC   DMPG

Figure 6: Single crystal structures of three phospholipids. The lipids are: 1,2-dilauroyl-
sn-glycero-3-phosphoethanolamine (DLPE); 1,2-dimyristoyl-*sn*-glycero-3-phosphocholine
(DMPC); and 1,2-dimyristoyl-*sn*-glycero-3-phosphoglycerol (DMPG). (Reproduced
from ref. 28.)

solute molecules will experience a uniaxial ordering potential due to the oriented liquid
crystalline "solvent". The quadrupolar splittings of deuterium labelled fatty acids, soaps,
phospholipids and n-alkanes[22] reflect the nature of this ordering potential and the degree
of orientational averaging due to rotational isomerization (*trans-gauche* isomers) about the
C-C bonds. The behaviour of flexible molecules in liquid crystalline solutes is very similar
to that in lipid bilayers and biological membranes. It appears that the local environment
(i.e., the uniaxial ordering potential) is similar in these systems and that the degrees of
molecular order, and even the dynamics, are also very similar.

   Other liquid crystalline phases, such as smectic, discotic and cholesteric phases give
interesting and sometimes unique ^{2}H NMR spectra[21,23]. In such systems, ^{2}H NMR is a
convenient and powerful way of studying the properties of each phase and of the transitions
between phases.

A   Molecular Structure, Conformation and Orientational Order

Phospholipid bilayers are examples of lyotropic liquid crystals where the phase behaviour
is a function not only of temperature but also of solvent concentration, the solvent in this
case being water. Lipid bilayers are also excellent models of biological membranes so that
a large variety of these systems has been studied by ^{2}H NMR. Figure 6 shows the x-ray

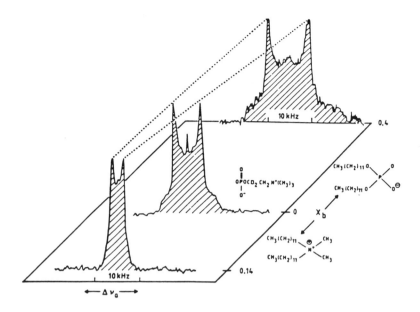

Figure 7: Charged amphiphiles in POPC bilayers. The figure shows three deuterium NMR spectra of head group α-deuterated POPC, namely, those of pure POPC membranes ($X_b = 0$), POPC with cationic amphiphile ($X_b = 0.14$), and POPC with anionic amphiphile ($X_b = 0.14$). Positive charges decrease the quadrupolar splitting of the α-segment; negative charges increase it. (Reproduced from ref. 28.)

crystal structures of several phospholipids [24,25,26].

To a large degree this structure is similar to that found in fully hydrated phospholipid bilayers. The major differences are that, in the commonly observed liquid crystalline or "fluid" bilayer phase, the hydrocarbon chains are highly disordered (by *trans-gauche* isomerization), the molecules reorient rapidly about their long axis which is aligned roughly along the bilayer normal, and they diffuse rapidly within the plane of the bilayer.

Phospholipid molecules, as seen in Figure 6, are amphiphilic. The long hydrocarbon chains are hydrophobic (or nonpolar) while the "head group" region is polar, containing the negatively charged phosphate group and, for example a choline (for PC), ethanolamine (PE), serine (PS), or glycerol (PG) group, among others. The structures, or phases, which are formed by phospholipid/water mixtures have hydrophobic regions, where water is largely excluded (such as the bilayer interior), and polar surfaces which form an interface with the water[27].

134

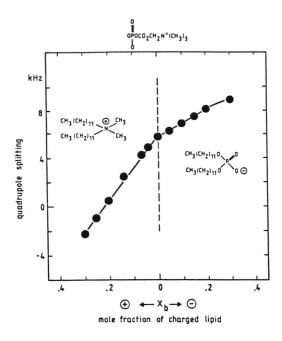

Figure 8: Influence of electric charge on phosphatidylcholine head group. (Reproduced from ref. 28.)

1 Influence of Surface Charge on Head Group Orientation

The polar head groups of PC's and PE's are electrically neutral but have an electric dipole moment formed by the negatively charged phosphate group, PO_4^-, and the positively charged choline, $N^+(CH_3)_3$, or ethanolamine, N^+H_3. Phosphatidylglycerol (PG) and phosphatidylserine (PS) both have a net negative charge. The orientation of the head group relative to the bilayer normal is sensitive to the local distribution of electric charge due to electrostatic interactions[28-30]. By specifically ^2H labelling the choline α- and β-segments of the 1-palmitoyl-2-oleoyl-sn-glycero-3-phosphocholine (POPC) head

$$\text{OPOCD}_2\text{CH}_2\text{N}^+(\text{CH}_3)_3$$
$$\alpha \quad \beta$$

group, Seelig, et al.[28] have demonstrated the influence of positively and negatively charged amphiphiles on the head group orientation.

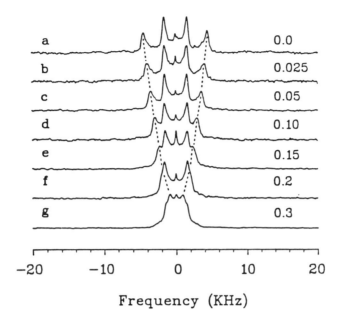

Figure 9: ^{2}H NMR spectra at 46 MHz of DMPS α-C^{2}H$_2$ recorded from DMPC/DMPS (5:1 M/M) bilayers in the presence of amphiphilic peptide K$_2$GL$_{20}$K$_2$A, with molar lipid to peptide ratios a) 6:0, b) 6:0.025, c) 6:0.05, d) 6:0.1, e) 6:0.15, f) 6:0.2, and g) 6:0.3. Measuring temperature was 34° C. (Reproduced from ref. 30.)

Figure 7 shows the ^{2}H NMR spectra of α-choline deuterated POPC in the presence of these amphiphiles. The positively charged amphiphiles give the bilayer surface a positive charge density and results in a significant concentration dependent decrease in the quadrupolar splitting. The negative surface charge resulting when bilayers composed of mixtures of POPC with the negatively charged amphiphiles leads to an increase in the quadrupolar splitting. The dependence of the splitting on amphiphile concentration is shown in Figure 8, the effect of positive amphiphiles on the left, that of the negative amphiphiles on the right. The variation in quadrupolar splitting can be explained simply in terms of a change in the orientation of the head group's electric dipole[28] (i.e., by a change in head group conformation). Similar effects of charged lipids on POPC head group orientation have also been reported[31].

A similar study using analogously head group labelled 1,2-dimyristoyl- *sn*-glycero-3-phosphoserine (DMPS) and DMPC

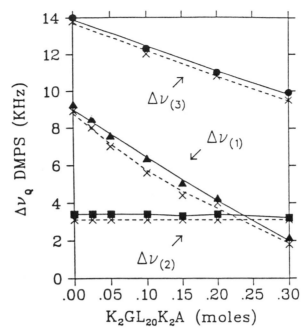

Figure 10: ^{2}H de-Paked quadrupolar splittings measured for the α-C^{2}H$_2$ group $\delta\nu_{(1)}$(▲) and $\delta\nu_{(2)}$(■) and the β-C^{2}H group $\delta\nu_{(3)}$(•) of the DMPS head group in DMPC/DMPS (5:1 M/M) membranes, as a function of the molar amount of amphiphilic peptide K$_2$GL$_{20}$K$_2$A incorporated in the lipid bilayers. Note the systematically smaller values measured from original and non-de-Paked spectra (dotted line (×)). (Reproduced from ref. 30.)

$$\text{DMPS:}\quad \overset{\alpha}{}\overset{\beta}{}$$
$$\text{DMPS:}\quad ^-\text{O}_3\text{POC}\overset{\alpha}{\text{H}}_2\overset{\beta}{\text{C}}\text{H}(\text{NH}_3{}^+)\text{COO}^-$$

in the presence of positively charged amphiphilic peptides (amino acid sequence K$_2$GL$_{20}$K$_2$A-amide) has been performed by Roux, et al[30]. As in the previous study, the positive surface charge provided by the bilayer spanning peptides leads to a reduction in the splitting of the α-C^{2}H$_2$ labelled choline. There is a concommitant increase in the splitting for the β-C^{2}H$_2$ labelled choline. The influence of the positively charged peptide on the spectra of α-C^{2}H$_2$ labelled serine head group is shown in Figure 9. Here the two deuterons in the α-position are inequivalent. The splitting of one of these deuterons is unaffected by the peptide while that of the other decreases with increasing peptide concentration. The splitting of the single β-position deuteron of PS also decreases as peptide concentration is increased. These results are summarized in Figure 10 which shows the peptide concentration dependence of all of the three quadrupolar splittings. The data are interpreted in terms of a model which treats the peptide charge as a "continuous uniform membrane charge density" located at a specific height z_k in the membrane, defined in Figure 11. The serine head group has three charges q_j whose positions are indicated by vectors $\vec{r}_j$ in the figure. The model considers the head group to be a quasi-rigid rod hinged at the glycerol backbone of the

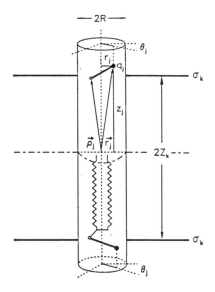

Figure 11: Schematic representation of phospholipid molecules in a bilayer. (Reproduced from ref. 30.)

phospholipid. The continuous charge distribution exerts a torque on the head group dipole moment which results in a rotation of the head group. The fact that the splitting of one of the α- C^2H_2 deuterons is independent of peptide concentration indicates that the head group rotates about this bond direction. In order to explain the changes in the other two splittings the continuum surface charge model requires that the effective average vertical distance from the surface charge layer to the head group charges lie in the range of 0.17 to 1.7 $\mathring{A}$ (i.e., the surface charge layer lies 0.17 to 1.7 $\mathring{A}$ below the head group). A study of PS head group orientation in the presence of penta-lysine, a positively charged hydrophilic peptide, shows the opposite behaviour. In this case, one α-C^2H_2 position has a splitting which does not change when penta-lysine is added, while the other α–position and the β-position splittings both increase. This is consistent with the continuum model if now the positive surface charge layer lies above the head group.

2 Cholesterol Orientation in Lipid Bilayers

The cytoplasmic membanes of eukaryotic cells are characteristically rich in cholesterol. The cholesterol molecule has a rigid core ring structure with a hydroxyl group at one end and a short flexible hydrocarbon tail at the other. It has an amphiphilic character but does not form a bilayer structure when mixed with water. In mixtures with phospholipids or in natural membranes it can be incorporated into a bilayer structure at concentrations greater than 50 mole %. In the lipid bilayer it adopts the position and orientation in-

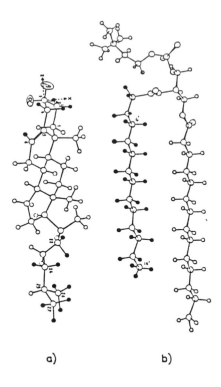

a)   b)

Figure 12: Structures of cholesterol (a) and DMPC (b): α =H, β =OH, β-cholesterol; α =OH, β =H, α -cholesterol. The filled circles represent the available deuterium labels. The sterol fixed axis system (x,y,z) has its origin at C-3. (Reproduced from ref. 35.)

dicated in Figure 12, with the hydroxyl group located at the lipid/water interface, and the hydrocarbon tail at the center of the bilayer. By deuterium labelling the rigid moiety of cholesterol at several inequivalent positions it is possible to determine the molecule's orientation relative to the bilayer normal from the ^{2}H NMR spectra[31-35]. In the liquid crystalline, or fluid phase, cholesterol reorients rapidly about an axis which in turn may wobble relative to the local bilayer normal.

A ^{2}H spectrum of β-[2,2,3,4,4,6-^{2}H$_6$]-cholesterol in DMPC bilayers[35] (at 30 mole % cholesterol) is shown in Figure 13b. For axially symmetric ^{2}H powder patterns it is possible to extract the "oriented sample" spectrum by a procedure called "de- Pakeing"[36]. Figure 13a shows the "de-Paked" or oriented sample spectrum obtained from the powder pattern of Figure 13b. The different doublets have been assigned to different label positions in the figure.

If we know the molecular structure of cholesterol (the atomic coordinates and bond angles), then since the ring structure is rigid and the C-^{2}H bonds at the different labelled positions have different orientations with respect to the axis of motion, we can determine

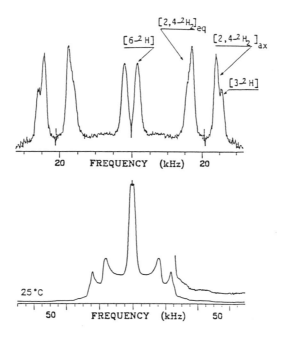

Figure 13: ^{2}H NMR spectrum (bottom) and de-Paked spectrum (top) of β-[2,2,3,4,4,6-^{2}H$_6$]cholesterol in DMPC (3:7 ratio), at 25° C. (Reproduced from ref. 35.)

the orientations of the bonds relative to that axis from the ^{2}H spectrum. If α_i is the angle between the i'th C-^{2}H bond and the motional axis, then the quadrupolar splitting is

$$\delta\omega_{Qi} = \frac{3e^2qQ}{4\hbar}\left(\frac{1}{2}(3\cos^2\alpha_i - 1)\right)\left(\frac{1}{2}(3\cos^2\gamma - 1)\right) \tag{136}$$

where γ is the angle between the axis of reorientation of the molecule and the bilayer normal, and the angular brackets indicate an average over the wobbling motion of that axis and we have used the axial symmetry of the C-^{2}H bond. By taking ratios of splittings for the different positions labelled we can eliminate the common factor depending on the orientation of the motional axis and the bilayer normal[32]. This allows us to determine the orientation of the motional axis within the molecular coordinate system. For example, if we consider first one particular splitting then there will be either one or two values of α which satisfy equation 136, one if $(3\cos^2\alpha - 1)/2 > 0.5$, and two if it is ≤ 0.5). The motional axis must lie on one of the two cones formed by the locus of all lines making either of these angles with the bond vector. For a second splitting, equation 136 again has two solutions defining two new cones. The possible directions are at the intersections between the two sets of cones (there can be as many as 8 possible directions if we have only two splittings). With a third splitting, we can eliminate most of these possibilities provided that the splitting is significantly different from the first two and that the errors in

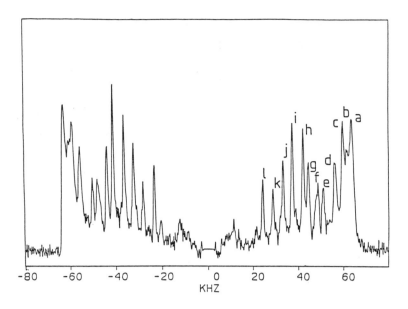

Figure 14: ^{2}H NMR spectrum of 12 mg of exchange ^{2}H labelled gramicidin in the oriented potassium laurate/decanol/KCl/buffer phase at 46° C. The twelve major doublets are labelled a-l in the figure. (Reproduced from ref. 43.)

measuring the splittings are small. In the case of cholesterol we can argue that the long axis of the molecule must be aligned roughly along the bilayer normal, as in Figure 12. With this constraint, three splittings is sufficient to identify the rotation axis of cholesterol[32,35].

In oriented liquid crystals, either nematics or lyotropic nematics, the ^{2}H NMR spectra of β-[2,2,3,4,4,6-^{2}H$_6$]-cholesterol are similar[37]. In particular, the ratios of the different splittings are nearly the same, indicating that the axis of reorientation is similar. In addition, studies of the dynamics of cholesterol in bilayers and in liquid crystals[32,35,37] by deuterium spin lattice relaxation find that in all cases the correlation time for cholesterol reorientation is of the order of 3×10^{-9} s with a T_{1z} minimum near 25° C.

3 Gramicidin in a Lyotropic Nematic Liquid Crystal

The pentadecapeptides, gramicidins A, B, and C, from *Bacillus brevis*, serve as simple analogues of integral membrane proteins. The natural mixture, referred to as gramicidin D, contains approximately 85% gramicidin A, and the remainder consists of gramicidins B and C, which differ from A only in the replacement of tryptophan[11] by either phenylalanine or tyrosine. Two molecules of gramicidin can form a trans-membrane channel capable of conducting an ionic current. The structure of this channel has not been determined to high resolution, but various structures have been proposed[38-41].

It is possible to exchange the amide hydrogens of each of the peptide's 15 amino acids for deuterium by dissolving gramicidin in deuterated methanol. In addition, the 4 tryptophan residues each have an exchangeable hydrogen on their indole group. The exchange labelled gramicidin can then either be intercalated into phospholipid bilayers, such as with DMPC[42], or into a lyotropic nematic liquid crystalline phase (K-laurate:decanol: KCl:H_2O, at weight ratios of 33:7.1:2.1:57.8) which spontaneously aligns with its director perpendicular to the magnetic field[43]. The spectrum of exchange labelled gramicidin D in an oriented lyotropic nematic at 46° C is shown in Figure 14. In this spectrum there are twelve resolved doublets. The quadrupolar splittings of these doublets depend on the orientation of the N-^{2}H bonds relative to the helix axis (about which the molecule reorients rapidly at temperatures above about 30°C) and on the orientation of the helix axis with respect to the magnetic field. In addition, the quadrupolar coupling constant depends on the length of the associated hydrogen bond to the opposing carbonyl oxygen in the helical channel structure. Given the peptide structure it is possible to predict the spectrum; however, a precise structure is still unavailable. The model structures, for example Urry's left handed $\beta_{LD}^{6.3}$ helix[38,39] can be used for comparison and the splittings observed are not inconsistent with this model structure. Slight distortions of the model structure, angular displacements of only a few degrees, could explain the experimental spectrum[43]. However, without an assignment of the different doublets to specific amino acid residues we cannot reject any of the proposed structures. Further work with specifically deuterated gramicidin A in oriented liquid crystals and in lipid bilayers should provide a stringent test of the model structures.

From the spectrum we can conclude that the helix axis is aligned perpendicular to the magnetic field (i.e., is parallel to the director of the liquid crystal). The temperature dependence of the quadrupolar splittings, summarized in Figure 15, indicates that there is little internal flexibility in the peptide backbone. The data of Figure 15 are the splittings normalized to their value at 30° C. Only two of the doublets have splittings which change by more than 7% over the temperature range of from 30 to 70° C. The two doublets with the stronger temperature dependence have chemical shifts which are about 3 ppm larger than the other doublets[43], suggesting that they are from the tryptophan indole groups which may move more freely than the peptide backbone.

Spin lattice relaxation measurements[43], made using a modification of the inversion recovery experiment of Figure 3b, are shown in Figure 16. Values of $1/T_{1z}$ at 50° C are plotted vs $(1 - S^2)$, where

$$S = \frac{\delta\omega_Q}{3e^2qQ/4\hbar} = \frac{1}{2}(3\cos^2\theta - 1) \tag{137}$$

where θ is the angle between the N-^{2}H bond and the helix axis. For a rigid molecule undergoing rapid axial diffusion about an axis perpendicular to the magnetic field, the spin lattice relaxation rate is[5,43]

142

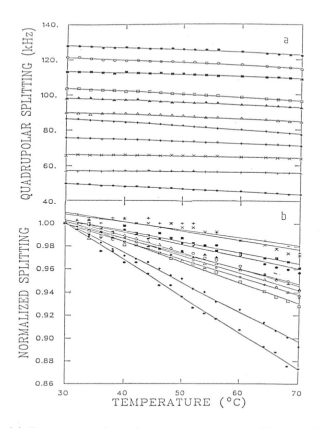

Figure 15: (a) Temperature dependence of quadrupolar splittings of exchange labelled gramicidin. (b) Ratio of the splittings at temperature T to their value at 30° C. The different symbols refer to the different doublets labelled in Figure 14 as follows (symbol, peak): •, a; o, c; ■, d; □, e; ▲, f; △, g; ◆, h; ◇, i; ×, j; +, k; and *, l. (Reproduced from ref. 43.)

$$\frac{1}{T_{1z}} = \frac{9}{256} \left[\sin^4\theta + 4\cos^2\theta\sin^2\theta \right] \left(\frac{2\tau_c}{1 + \omega_0^2\tau_c^2} + \frac{2\tau_c}{1 + 4\omega_0^2\tau_c^2} \right) \tag{138}$$

where $\omega_0 = 2\pi \times 55.264$ MHz and τ_c is the correlation time for diffusion through one radian about the axis. In terms of $S = (3\cos^2\theta - 1)/2$, this equation becomes

$$\frac{1}{T_{1z}} = \frac{3}{32} \left(\frac{e^2 qQ}{\hbar} \right)^2 \left(1 - S^2 \right) f \tag{139}$$

where

$$f = \frac{2\tau_c}{1 + \omega_0^2\tau_c^2} + \frac{2\tau_c}{1 + 4\omega_0^2\tau_c^2}$$

The solid line in Figure 16 is the least squares fit to $1/T_{1z} = A + B(1 - S^2)$. The linear

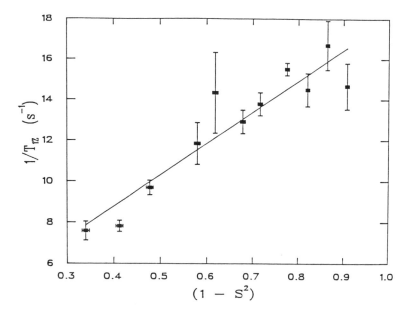

Figure 16: Variation of spin-lattice relaxation rate, $1/T_{1z}$, with the deuterium orientational order parameter, S, at 50° C. The vertical error bars express the uncertainty in the values of $1/T_{1z}$ determined by least-squares fit to the peak areas in an inversion-recovery T_{1z} experiment. The horizontal error bars are due to the uncertainty in measuring the splittings, assumed to have a constant error of ±500 Hz. The solid straight line is the least-squares fit to the function $1/T_{1z} = A + B(1 - S^2)$, giving $A = 2.64$ s^{-1} and $B = 15.32$ s^{-1}. (Reproduced from ref. 43.)

variation with $(1 - S^2)$ suggests that the model of a rigid peptide backbone undergoing rapid axial diffusion is appropriate. These results, together with T_{2e} measurements at 50° and a similar set of data taken at 30° C have been used to determine the value of the correlation time for axial diffusion[43], namely, $\tau_c \approx 10^{-7}$ s.

B   Phase Equilibria of Binary Mixtures

^{2}H NMR spectra are sensitive to the amplitudes, rates and symmetry of molecular reorientation as well as to the static (or average) orientation of the electric field gradient tensor relative to $\vec{H}_0$. At a first order phase transition of a pure one component system, we may observe a discontinuous change in the ^{2}H NMR spectrum. Figure 17a shows a series of spectra as a function of temperature in the vicinity of the chain melting transition of the chain perdeuterated phospholipid DPPC (1,2-dipalmitoyl-*sn*-glycero-3-phosphocholine) in excess water. At 38° C the lipid chains are disordered, i.e., undergo rapid *trans-gauche* isomerization, and the phospholipids reorient rapidly about the phospholipid bilayer normal[5,9]. At 37.5° C the chains are much more rigid, being nearly all-*trans*, the axial reorientation rate

144

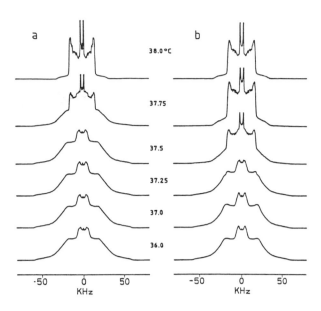

Figure 17: ^{2}H NMR spectra of chain perdeuterated DPPC vs temperature near the chain melting transition: a) 0% cholesterol, and b) 2.5 mole % cholesterol. All spectra were taken at a frequency of 41 MHz. (Reproduced from ref. 48.)

is much slower[44], and the diffusion within the bilayer plane is greatly reduced[45,46]. The ^{2}H spectrum in this lower temperature gel phase are much broader than the spectra of the higher temperature liquid crystalline phase. In addition, the higher temperature spectra are characteristic of rapid axially symmetric reorientation while at lower temperatures the rates of motion are so slow that, on the ^{2}H NMR time scale (defined by the inverse of the spectral width), they are not axially symmetric. At 37.75° C, the spectrum consists of a superposition of two components, one characteristic of liquid crystalline phase domains, the other of gel phase domains. The area of each component in the spectrum is a measure of the amount of lipid in each type of domain. The use of excess water guarantees, by Gibb's phase rule, that there will be only one temperature where two phospholipid phases co-exist. Thus, we can treat this two component mixture (DPPC plus water) as a pseudo one component system.

1 Cholesterol/Phospholipid Mixtures

The plasma membranes of eukaryotic cells are characteristically rich in cholesterol. Its abundance in eukaryotic membranes and absence from the membranes of prokaryotes suggest that it plays an important role in the function of the membranes of higher organisms. Its precise function in the membrane is still unclear, but its effect on the physical prop-

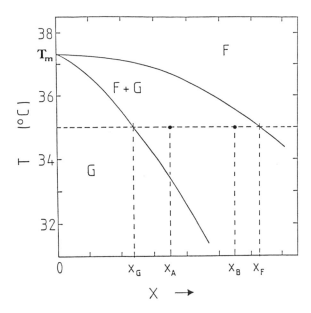

Figure 18: A hypothetical fluid plus gel two-phase coexistence region for a two component mixture. (Reproduced from Davis, J.H., Adv. Magn. Reson. 13 (1989), in press.)

erties of membranes is becoming clearer. A wide range of experimental techniques have been applied to the study of the effects of cholesterol on membranes. These have recently been briefly reviewed by Vist[47], Vist and Davis[48] and by Ipsen, et al.[49].

The addition of 2.50 mole % cholesterol to the DPPC/water mixture results in a small downward shift and a broadening of the phase transition region, as illustrated by the spectra in Figure 17b. By Gibb's phase rule, at constant pressure a two component mixture may exhibit a continuous range of temperatures over which two phases may coexist in equilibrium. The ^{2}H NMR spectra, such as those in Figure 17b, can be used to map the phase boundaries by identifying the range of temperatures over which the spectra consist of two components.

Figure 18 illustrates a hypothetical two-phase region with coexisting liquid crystalline (L_α) and gel phases. A sample of composition x_i at a temperature T_0 will contain domains of liquid crystalline and gel phases. The amount of each type of domain is given by the lever rule

$$\alpha_i(T_0) = \frac{[x_i - x_g(T_0)]}{[x_f(T_0) - x_g(T_0)]} \tag{140}$$

where α_i is the fraction of the sample in liquid crystalline or fluid domains, x_f is the cholesterol concentration of the fluid domains (determined by the intersection of the tie-line or isotherm at T_0 and the fluidus or upper phase boundary), and x_g is the concentration of the gel phase domains (at the intersection of the tie-line with the solidus or lower phase boundary). The gel phase fraction is just $1 - \alpha_i$. The amount of phospholipid in the fluid domains is determined by the relation

$$1 - x_i = \alpha_i(1 - x_f) + (1 - \alpha_i)(1 - x_g) \tag{141}$$

where $1 - x_i$ is the phospholipid concentration of the sample. Then, the fraction of phospholipid in the fluid domains is

$$f_i = \alpha_i \frac{(1 - x_f)}{(1 - x_i)} \tag{142}$$

and the fraction of phospholipid in the gel phase is $1 - f_i$. The distinction between α_i and f_i is important because the NMR experiment measures f_i, not α_i. The relative areas of the two components in the ^{2}H NMR spectrum is given by the relative amounts of phospholipid in the two phases. It should be emphasized that this analysis assumes that the rate of exchange of phospholipid molecules between the two types of domains is slow on the spectroscopic time scale. In this case, the spectrum can be written as

$$S(x_i, T_0) = f_i S_f(x_f, T_0) + (1 - f_i)S_g(x_g, T_0) \tag{143}$$

where $S_f(x_f, T_0)$ is the fluid phase end point spectrum at T_0 and $S_g(x_g, T_0)$ is the gel phase end point spectrum. For two or more samples of different composition at a temperature such that both samples are within the two-phase region, we can use the experimental spectra to invert this equation to determine both $S_f(x_f, T_0)$ and $S_g(x_g, T_0)$

$$S_g(x_g, T) = \frac{1}{(f_B - f_A)} [f_B S(x_A, T) - f_A S(x_B, T)]$$

$$\tag{144}$$

$$S_f(x_f, T) = \frac{1}{(f_A - f_B)} [(1 - f_B)S(x_A, T) - (1 - f_A)S(x_B, T)]$$

Clearly we can generate the end point spectra by subtracting a fraction of one of the normalized experimental spectra from the other.

If we let x_B correspond to the concentration of the sample which has the higher fluid phase content at T, then it will be possible to subtract a fraction K of $S(x_B, T)$ from $S(x_A, T)$ such that no fluid phase component remains in the difference spectrum. This occurs when

$$K = \frac{\alpha_A(1 - x_B)}{\alpha_B(1 - x_A)} \tag{145}$$

Conversely, the fraction K' of $S(x_A, T)$ which must be subtracted from $S(x_B, T)$ to yield a difference spectrum with no gel phase component is,

$$K' = \frac{(1 - \alpha_B)(1 - x_A)}{(1 - \alpha_A)(1 - x_B)} \tag{146}$$

From equations 140, 142, 145 and 146 we can solve for the end point concentrations

$$x_g = \frac{[(1 - x_B)x_A - K(1 - x_A)x_B]}{[(1 - x_B) - K(1 - x_A)]}$$

$$\tag{147}$$

$$x_f = \frac{[(1 - x_A)x_B - K'(1 - x_B)x_A]}{[(1 - x_A) - K'(1 - x_B)]}$$

This procedure will be valid for any two-phase coexistence region (not only phospholipid fluid and gel phases), provided that the basic assumption of slow exchange of molecules between domains holds, and that the end point spectra characteristic of the two phases are sufficiently different that one can carry out the subtraction procedure.

The fluid/gel two-phase coexistence region illustrated by the spectra of Figure 17b is too narrow for us to fruitfully use the difference spectroscopy technique. Instead, we simply monitor the spectra in this region for appearance or disappearance of one of the components. At higher cholesterol concentrations, greater than 7.5 mole %, and at temperatures below the pure phospholipid's chain melting transition (at 37.75° C), we find another two-phase region, one where difference spectroscopy is applicable. Figure 19 shows spectra for two samples, 10 mole % (Figure 19a), and 15 mole % cholesterol (Figure 19b) at 31° C. Figures 19c and 19d show the end point difference spectra (renormalized to unit area) obtained by the above procedure. While neither of the experimental spectra in Figures 19a and 19b appear to be obvious superpositions of distinct end point spectra, subtraction of 59 % of the 10 mole % spectrum of Figure 19a from that of the 15 mole % sample, in Figure 19b, yields the axially symmetric difference spectrum shown in Figure 19d. The value $K' = 0.59$ gives the upper end point concentration $x_\beta = 0.214$. This difference spectrum is compared with the normalized experimental spectrum obtained from a sample with 22.5 mole % cholesterol, also taken at 31° C, in Figure 20. The spectrum of the 22.5 mole % sample (Figure 20b) is subtracted from the end point spectrum (Figure 20a) and the result is shown in Figure 20c. The agreement between the end point spectrum and the 22.5 mole % cholesterol spectrum is striking. A comparison of the end point spectrum with the spectra of samples with either 20 mole% or 25 mole % cholesterol shows large differences[47].

Subtraction of 34 % of the 15 mole % cholesterol spectrum (Figure 19b) from the 10 mole % spectrum (Figure 19a) gives the end point spectrum of Figure 19c. The value $K = 0.34$ gives the end point concentration $x_g = 0.072$. Comparison of this end point difference spectrum, Figure 20d, with the spectrum of a sample with 7.5 mole % cholesterol, Figure 20e, again shows remarkable agreement, the difference is shown in Figure 20f. Comparison with the spectrum of a sample containing 6.25 mole % cholesterol again shows large differences[47].

148

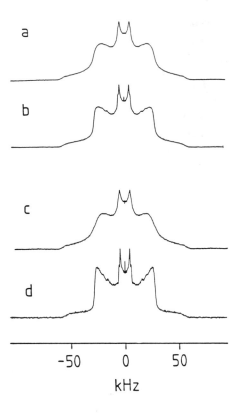

Figure 19: ^{2}H NMR difference spectroscopy for cholesterol/DPPC mixtures at 31° C. a) The spectrum of a sample with 10 mole % cholesterol; b) that of a 15 mole % cholesterol sample; c) the gel phase end point spectrum, given by the spectrum in 'a' minus 0.34 times the spectrum in 'b'; d) the β-phase end point spectrum given by the spectrum in 'b' minus 0.59 times the spectrum in 'a'. (Reproduced from ref. 48.)

In this fashion, we have mapped the phase boundaries of this two-phase region from about 30 to 37° C. Using other features of the ^{2}H NMR spectra and the results of DSC (differential scanning calorimetry) experiments on the same samples, we generate the partial phase diagram shown in Figure 21. We have identified three separate two-phase regions: i) a narrow region of coexistence of L_α and gel phases getween 0 and 7.5 mole % cholesterol; ii) a gel phase/β-phase coexistence region below 37° C, between about 7 and 22 mole % cholesterol; and iii) an L_α/β-phase region above 37° C, starting at a eutectic point somewhere between 7.5 and 10 mole %. The L_α/gel phase coexistence region, because it is so narrow (about 1° C), has not been previously discussed in the literature even though its existence is expected from fundamental principles. The second two-phase region, that of gel/β-phase coexistence, has been seen in neutron scattering experiments on DMPC/cholesterol mixtures where phase boundaries of 8 and 24 mole % were reported[50] and in EPR (electron paramagnetic resonance) studies of that same system[51]. The third two-phase region,

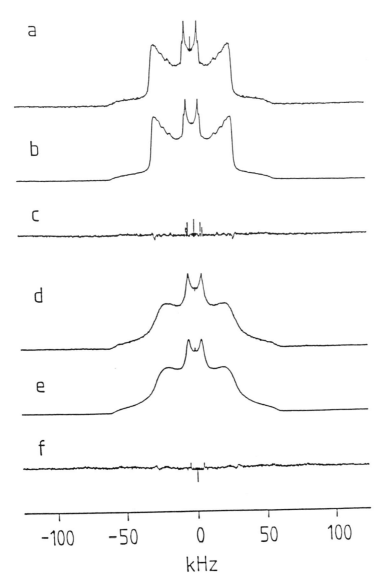

Figure 20: Comparison of normalized end point difference spectra with the normalized experimental spectra of samples with compositions near the end point compositions determined by the lever rule and the difference spectra of Figure 19: a) the end point spectrum of Figure 19d, which corresponds to an end point concentration of 21.4 mole % cholesterol; b) the spectrum of a sample with 22.5 mole % cholesterol at 31° C; c) the difference between the spectra in 'a' and 'b'; d) the end point spectrum of Figure 19c, which corresponds to an end point concentration of 7.2 mole % cholesterol; e) the spectrum of a sample with 7.5 mole % cholesterol at 31° C; and f) the difference between the spectra in 'd' and 'e'. (Reproduced from ref. 48.)

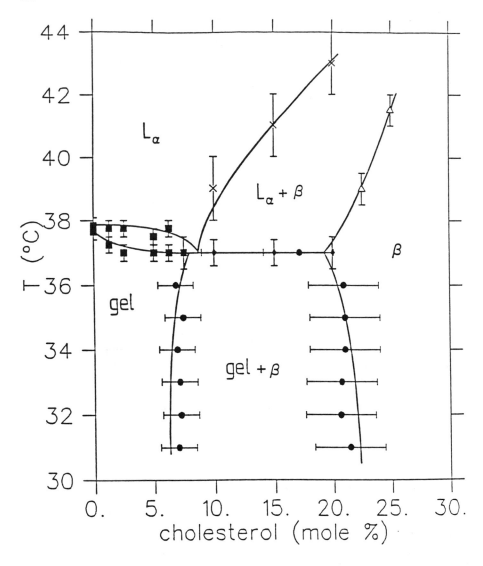

Figure 21: The partial phase diagram for cholesterol/DPPC mixtures. The phases are labelled as follows: L_α, the fluid or liquid crystalline phase; G, the gel phase; and β, the high cholesterol phase characteristic of biological membranes. The squares, ■, are determined by inspection of spectra such as those of Figure 17; the circles, •, are from difference spectroscopy; the diamonds, ♦, show the three phase line determined by the temperature of the sharp peak in the DSC traces; the crosses, ×, are from the upper limit of the broad component in the DSC traces; and the triangles, △, are from the sharpening of the resonanes at high cholesterol concentrations. The solid lines simply delineate the phase boundaries as described above. (Reproduced from ref. 48.)

L_α/β-phase coexistence, is suggested by DSC[47], fluorescence[52] and EPR[51,53] data. While ^{2}H NMR difference spectroscopy allowed us to determine the phase boundaries of the gel/β-phase region, the high rate of lateral diffusion and/or the small domain size in the L_α/β-phase region prevent us from using that method there.

2   Amphiphilic Peptide/Phospholipid Mixtures

How protein-lipid interactions influence membrane function is a question of fundamental interest. Thes interactions may provide a means of modulating protein function through changes in membrane lipid properties or they may provide a means of controlling lipid bilayer physical properties through changes in the membrane protein. At the molecular level, protein-lipid interactions may manifest themselves as local or site-specific perturbations. Accordingly, many investigations of protein-lipid interactions have been directed towards finding examples of a lipid specificity for protein function, while others have concentrated on studying the changes in lipid orientational order and dynamics induced by the presence of protein.

The local perturbations in molecular motion and orientational order are only one manifestation of protein-lipid interactions. A potentially important effect is the modification of membrane phase equilibria caused by this interaction. Changes in membrane phase equilibria reflect the changes in the total free energy of the system. The measurement of phase boundaries in binary protein-lipid mixtures provides valuable information about the molecular interactions between the components. To determine the form and relative importance of the different contributions to the free energy in a thermodynamic model of membrane phase equilibria, one needs to be able to change the characteristic variables of the system. These include not only the temperature and composition (or concentration) but also the quantities which determine the strength of the various interactions. In the case of the hydrophobic mismatch between lipids and proteins, for example, it is useful to be able to vary the amount of this mismatch. This can be accomplished by changing the length of the lipid hydrocarbon chains and by varying the hydrophobic length of the peptide. Systematically modifying the chemical structure of both protein and lipid will facilitate the determination of the character of protein-lipid interactions.

By using synthetic amphiphilic peptides as protein analogues, one can avoid many of the problems of large scale protein preparation, purification and characterization associated with the use of natural proteins in reconsitituted membranes. In addition, it allows the flexibility of being able to modify the peptide's sequence as well as its length, hydrophobicity and polarity.

Two peptides, Lys_2-Gly-Leu_N-Lys_2-Ala-amide, differing only in N, the number of leucines forming the hydrophobic core, were studied in bilayers of chain perdeuterated DPPC as a function of temperature and peptide concentration[54,55,56]. Figure 22 shows a series of spectra of the peptide with N=24, at a lipid to peptide mole ratio of 66:1, as a

152

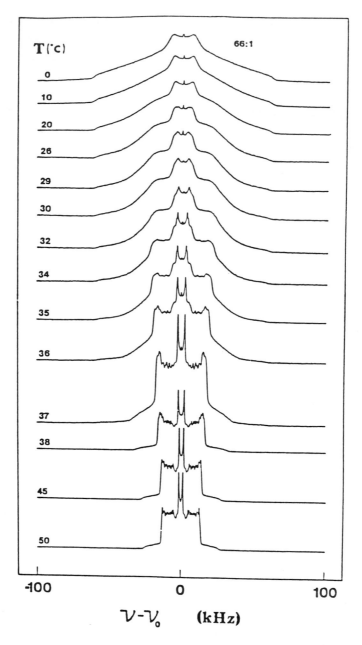

Figure 22: Temperature dependence of the ^{2}H NMR spectrum of the peptide 24-DPPC-d_{62} sample at molar peptide concentration $x_p = 0.0149$ (mole ratio 66:1). The temperature for each spectrum is displayed on the left. (Reproduced from ref. 54.)

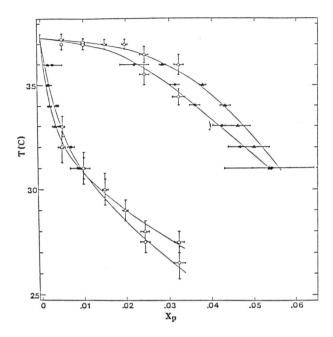

Figure 23: Temperature-composition plots of the two peptide-lipid systems. The triangles refer to peptide 24 and the circles to peptide 16. The open symbols were obtained by visual inspection of series of spectra as in Figure 22. In these cases the horizontal error bars represent the uncertainties in sample concentration and the vertical error bars the uncertainty in the temperature where two components in the spectra can be identified. The solid symbols were obtained by spectral subtraction and represent the average values of the end-point concentrations. The horizontal error bars represent the uncertainties in these values. (Reproduced from ref. 54.)

function of temperature. As in the cholesterol/DPPC mixtures discussed above, we can identify a region of two-phase coexistence between about 34 and 37° C for this sample. From such temperature scans, using a series of samples with different peptide concentrations, it is possible to map the phase boundaries in a vertical fashion. In addition, the technique of ^{2}H NMR difference spectroscopy can be used to determine the phase boundaries horizontally, as described earlier. In the present study there were 7 different peptide compositions between 0 and 3.2 mole % for the peptide with N=24, and 6 different compositions covering the same range for the peptide with N=16. In both cases we found a two-phase coexistence region below the melting point of the pure lipid. The phase diagrams determined by these experiments are shown in Figure 23. Here, for a number of temperatures, there were several samples with compositions placing them inside the two phase region so that we could form a large number of pair-wise spectral subtractions in determining the phase boundaries. The points on the boundaries with horizontal error bars were obtained by difference spectroscopy. The points with vertical error bars were obtained from temperature scans like that shown in Figure 22.

Having determined the phase boundaries from NMR, we were able to develop a non-ideal solution theory expression for the free energy of this two component system which could explain these phase boundaries[55]. This model was capable of generating the phase boundaries of both peptide/lipid mixtures and also predicted the slopes of the linear enthalpy of melting plots (ΔH, vs peptide concentration, x_p) determined by DSC. This linear behaviour of ΔH vs x_p is frequently observed in protein/lipid mixtures and is a consequence of the non-ideal mixing.

IV   Acknowledgements

I would like to thank M. R. Morrow and S. Prosser for their critical comments on the manuscript and the authors who have allowed me to reproduce the figures describing their work. I would also like to thank the Natural Sciences and Engineering Research Council of Canada for financial support.

V   References

[1] A. Abragam, *The Principles of Nuclear Magnetism* (1961), Oxford University Press, Oxford.

[2] C. P. Slichter, *Principles of Magnetic Resonance*, 2nd. Edition (1980), Springer-Verlag, Berlin.

[3] R. Zare, *Angular Momentum* (1988) John Wiley and Sons, New York.

[4] M. E. Rose, *Elementary Theory of Angular Momentum* (1967), John Wiley and Sons, New York.

[5] J. H. Davis, Biochim. Biophys. Acta 737 (1983) 117-171.

[6] A. Abragam and M. Goldman, *Nuclear Magnetism: Order and Disorder* (1982) Oxford University Press, Oxford.

[7] M. Bloom, in *Physics of NMR Spectroscopy in Biology and Medicine*, Proceedings of the International School of Physics "Enrico Fermi" (1988) North-Holland, Amsterdam. pp. 121-152.

[8] C. Cohen-Tannoudji, B. Diu, and F. Laloe, *Quantum Mechanics*, vol. 2 (1977) Hermann, Paris.

[9] J. Seelig, Quart. Revs. Biophys. 10 (1977) 353-418.

[10] E. Sternin, *Some Mechanisms of Transverse Nuclear Magnetic Relaxation in Model Membranes* (1988) PhD. Thesis, University of British Columbia, Vancouver, B.C.

[11] S. Vega and A. Pines, J. Chem. Phys. 66 (1977) 5624-5644.

[12] M. Bloom, J. H. Davis and M. I. Valic, Can. J. Phys. 58 (1980) 1510-1517.

[13] J. H. Davis, K. R. Jeffrey, M. Bloom, M. I. Valic, and T. P. Higgs, Chem. Phys. Lett. 42 (1976) 390-394.

[14] R. R. Ernst, G. Bodenhausen and A. Wokaun, *Principles of Nuclear Magnetic Resonance in One and Two Dimensions* (1987) Oxford University Press, Oxford.

[15] R. J. Wittebort, E. T. Olejniczak and R. G. Griffin, J. Chem. Phys. 86 (1987) 5411-5420.

[16] K. R. Jeffrey, Bull. Magn. Res. 3 (1981) 69-82.

[17] A. Pines, in *Physics of NMR Spectroscopy in Biology and Medicine*, Proceedings of the International School of Physics "Enrico Fermi" (1988) North-Holland, Amsterdam. pp. 43-115.

[18] J. Jeener and P. Broekaert, Phys. Rev. 157 (1967) 232-240.

[19] J. Seelig and A. Seelig, Quart. Revs. Biophys. 13 (1980) 19-61.

[20] H. W. Spiess, Adv. Polymer Sci. 66 (1985) 23-58.

[21] J. W. Emsley, ed., *Nuclear Magnetic Resonance of Liquid Crystals* (1985) D. Reidel Publishing Co., Dordrecht.

[22] B. Janik, E. T. Samulski and H. Toriumi, J. Phys. Chem. 91 (1987) 1842-1850.

[23] B.-G. Wu and J. W. Doane, J. Magn. Reson. 75 (1987) 39-49.

[24] P. G. Hitchcock, R. Mason, K. M. Thomas and G. G. Shipley, Proc. Natl. Acad. Sci. (USA) 71 (1974) 3036-3040.

[25] R. H. Pearson and I. Pascher, Nature (London) 281 (1979) 499-501.

[26] I. Pascher, S. Sundell, K. Harlos and H. Eibl, Biochim. Biophys. Acta. 896 (1987) 77-88.

[27] G. Lindblom and L. Rilfors, Biochim. Biophys. Acta 988 (1989) 221-256.

[28] J. Seelig, P. M. Macdonald and P. G. Scherer, Biochemistry 26 (1987) 7535-7541.

[29] M. Roux, J.-M. Neumann, M. Bloom and P. F. Devaux, Eur. Biophys. J. 16 (1988) 267-273.

[30] M. Roux, J.-M. Neumann, R. S. Hodges, P. F. Devaux and M. Bloom, Biochemistry 28 (1989) 2313-2321.

[31] P. G. Scherer and J. Seelig, The EMBO Journal 6 (1987) 2915-2922.

[32] M. G. Taylor, T. Akiyama and I. C. P. Smith, Chem. Phys. Lipids 29 (1981) 327-339.

[33] M. G. Taylor, T. Akiyama, H. Saito and I. C. P. Smith, Chem. Phys. Lipids 31 (1982) 359-379.

[34] E. C. Kelusky, E. J. Dufourc and I. C. P. Smith, Biochim. Biophys. Acta 735 (1983) 302-304.

[35] E. J. Dufourc, E. J. Parish, S. Chitrakorn and I. C. P. Smith, Biochemistry 23 (1984) 6062-6071.

[36] M. Bloom, J. H. Davis and A. L. MacKay, Chem. Phys. Lett. 80 (1981) 198-202.

[37] J. H. Davis, in *Physics of NMR Spectroscopy in Biology and Medicine*, Proceedings of the International School of Physics "Enrico Fermi" (1988) North-Holland, Amsterdam. pp. 302-312.

[38] C. M Venkatachalam and D. W. Urry, J. Comput. Chem. 4 (1983) 461-469.

[39] C. M Venkatachalam and D. W. Urry, J. Comput. Chem. 5 (1984) 64-71.

[40] B. A. Wallace, Biopolymers 22 (1983) 397-402.

[41] A. S. Arseniev, I. L. Barsukov, V. F. Bystrov, A. L. Lomize, and Yu. A. Ovchinnikov, FEBS Lett. 186 (1985) 168-174.

[42] K. P. Datema, K. P. Pauls and M. Bloom, Biochemistry 25 (1986) 3796-3803.

[43] J. H. Davis, Biochemistry 27 (1988) 428-436.

[44] P. Meier, E. Ohmes and G. Kothe, J. Chem. Phys. 85 (1986) 3598-3614.

[45] A.-L. Kuo and C. G. Wade, Biochemistry 18 (1979) 2300-2308.

[46] W. L. C. Vaz, Z. I. Derzko and K. A. Jacobson, Cell Surf. Rev. 8 (1982) 83-136.

[47] M. R. Vist, *Partial Phase Behavior of Perdeuteriated Dipalmitoylphosphatidylcholine-Cholesterol Model Membranes*, (1984) MSc. Thesis, University of Guelph, Guelph, Ontario.

[48] M. R. Vist and J. H. Davis, Biochemistry (1989), in press.

[49] J. H. Ipsen, G. Karlstrom, O. G. Mouritsen, H. Wennerstrom and M. H. Zuckermann, Biochim. Biophys. Acta 905 (1987) 162-172.

[50] W. Knoll, G. Schildt, K. Ibel and E. Sackmann, Biochemistry 24 (1985) 5240-5246.

[51] D. J. Recktenwald and H. M. McConnell, Biochemistry 20 (1981) 4505-4510.

[52] B. R. Lentz, D. A. Barrow and M. Hoechli, Biochemistry 19 (1980) 1943-1954.

[53] E. J. Shimshick and H. M. McConnell, Biochem. Biophys. Res. Comm. 53 (1973) 446-451.

[54] J. C. Huschilt, R. S. Hodges and J. H. Davis, Biochemistry 24 (1985) 1377-1386.

Isotopes in the Physical and Biomedical Science, Vol. 2, edited by E. Buncel and J.R. Jones 159
© 1991 Elsevier Science Publishers B.V., Amsterdam

Chapter 4

Alkali Metal NMR Studies of Synthetic and Natural Ionophore Complexes

Christian Detellier, Helen P. Graves and Kathleen M. Brière
The Ottawa-Carleton Chemistry Institute, University of Ottawa Campus,
Ottawa, Ont. K1N 9B4 Canada

CONTENTS:

SYMBOLS AND ABBREVIATIONS

K_f	formation constant
K_1, K_c	equilibrium constant of $(M^+\text{-}L)$
K_2	equilibrium constant of $(L\text{-}M^+\text{-}L)$
L, (C)	ligand, (crown ether)
$[L]_{tot}$	total concentration of ligand
M^+	metal cation
$[M^+]_{tot}$	total concentration of metal cation
$(M^+)_{sol}$	solvated cation
$(M\text{-}L)^+$	1:1 complex
R	gas constant
k_B	Boltzmann's constant

h	Planck's constant
$\hbar$	Planck's constant/2π
γ	gyromagnetic ratio
γ_H	gyromagnetic ratio of proton
γ_C	gyromagnetic ratio of carbon
μ_o	permeability of the vacuum
χ	quadrupolar coupling constant
I	spin quantum number
B_o	external magnetic field
q	electric field gradient at the nucleus
Q	quadrupole moment
ρ	$[L]_{tot}/[M^+]_{tot}$
$\nu_{1/2}$	linewidth at half-height
δ	chemical shift (ppm)
δ_A	chemical shift of site A; $(M^+)_{sol}$
δ_B	chemical shift of site B; $(M-L)^+$
P_A	population of site A
P_B	population of site B
ν_A	resonance frequency of site A (Hz)
ν_B	resonance frequency of site B (Hz)
k_A	rate constant for formation of complex (equation 18)
k_B	rate constant for dissociation of complex (equation 18)
k_1	rate constant for dissociative exchange (formation of complex; equation 23)
k_{-1}	rate constant for dissociative exchange (dissociation of complex; equation 23)
k_2	rate constant for associative exchange (equation 24)
T_1	longitudinal (spin-lattice) relaxation time
T_2	transverse (spin-spin) relaxation time
T_1^{-1}	longitudinal relaxation rate
T_2^{-1}	transverse relaxation rate
T_q^{-1}	quadrupolar relaxation rate
$T_{2,q}^{-1}$	quadrupolar transverse relaxation rate
$T_{1,obs}$	observed longitudinal relaxation time
$T_{2,obs}^{-1}$	observed transverse relaxation rate
$T_{2,inh}^{-1}$	inhomegeneity contribution to the transverse relaxation rate.
$T_{2,ex}^{-1}$	exchange contribution to the transverse relaxation rate.
$T_{1,f}^{-1}$	longitudinal relaxation rate of "free" metal
$T_{1,c}^{-1}$	longitudinal relaxation rate of the complexed cation

T_1^{DD}	dipolar longitudinal relaxation time
nOe	nuclear Overhauser enhancement
N_H	number of attached hydrogens (equation 7)
r_{CH}	C-H bond length
τ_{eff}	effective correlation time
τ_q	correlation time associated with quadrupolar relaxation (equation 10)
E_a	energy of activation
ΔH^o	standard enthalpy of formation
ΔS^o	standard entropy of formation
ΔG^o	standard Gibb's free energy of formation
ΔH^*	enthalpy of activation
ΔS^*	entropy of activation
ΔG^*	free energy of activation
OM16C4	octamethyl-16-crown-4
15C5	15-crown-5
B15C5	benzo-15-crown-5
18C6	18-crown-6
DB18C8	dibenzo-18-crown-6
DC18C6	dicyclohexano-18-crown-6
DA18C6	1,10-diaza-18-crown-6
DNDB18C6	dinitrodibenzo-18-crown-6
AMDB18C6	diaminodibenzo-18-crown-6
DB21C7	dibenzo-21-crown-7
DB24C8	dibenzo-24-crown-8
DB27C9	dibenzo-27-crown-9
DB30C10	dibenzo-30-crown-10
$C21C_5$	cryptand-C21(C_5H_{10})
C211	cryptand-211
C221	cryptand-221
C222	cryptand-222
C222B	benzo-cryptand-222
C222D	dilactam of cryptand-222
C322	cryptand-322
DN	Gutmann donor number
AC	acetone
AN	acetonitrile
EA	ethylamine

EDA	ethylenediamine
FOR	formamide
DMF	dimethyl formamide
DME	dimethoxyethane
DMSO	dimethyl sulfoxide
1,3DO	1,3 dioxalane
1,4DO	1,4 dioxane
MA	methylamine
NM	nitromethane
PC	propylene carbonate
PY	pyridine
THF	tetrahydrofuran
TMG	tetramethylguanadine

I. INTRODUCTION

Phenomena involving the complexation of natural ionophores by alkali metal cations have been the focus of scientists for two decades. These include important biological processes such as drug action, enzyme catalysis and ion transfer through lipophilic membranes. The presence of ionophores facilitates the transport of sodium and potassium cations across membranes by either serving as a carrier or forming a channel. Neutral ionophores, such as valinomycin (Figure 1a) and enniatin (Figure 1b), form a charged species when complexed with an alkali metal cation.

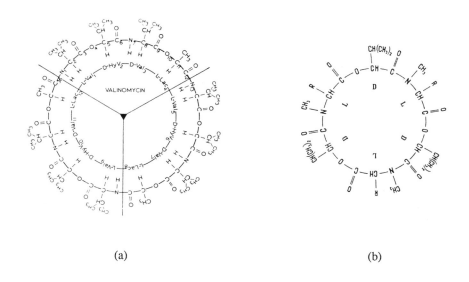

(a)   (b)

163

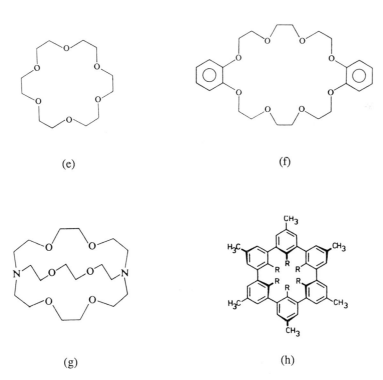

(c)

(d)

(e)   (f)

(g)   (h)

Figure 1. (a) valinomycin, (b) enniatin A, R = CH(CH₃)C₂H₅; enniatin B, R = CH(CH₃)₂, (c) monensin, (d) lasalocid, (e) 18-crown-6, (f) dibenzo-24-crown-8, (g) cryptand-222 and (h) an example of a spherand.

The anionic ionophores, such as monensin (Figure 1c) and lasalocid (Figure 1d), are often characterized by hydroxyl groups at one end which may hydrogen bond with a carboxylic acid group at the other to surround the metal cation, forming a neutral complex.[1]

Structurally simpler synthetic analogs of naturally occurring ionophores were first reported in 1967 by Charles J. Pedersen.[2] These cyclic polyethers were found to have an unusual affinity for alkali metal cations. 1,4,7,10,13,16-hexaoxacyclooctadecane (Figure 1e) and 6,7,9,10,12,13,20,21,23,24,26,27-dodecahydrodibenzo[b,n] [1,4,7,10,13,16,19,22] octaoxacyclotetracosin (Figure 1f) are two common examples of crown ethers which are understandably referred to as 18-crown-6 (18C6) and dibenzo-24-crown-8 (DB24C8) respectively. Pedersen's classical work with crown ethers, which are part of the *coronands*, launched a whole new field of synthesis. One year later, in 1968, Jean-Marie Lehn and co-workers synthesized the first bicyclic ionophore, cryptand-222 (Figure 1g). This new class of ligands, called the *cryptands*, also has a high affinity for alkali metal cations and their three dimensional spheroidal cavities result in the formation of complexes more stable than those with the crown ethers.[3a,b] Another class of synthetic ionophores known as the *spherands* (Figure 1h), was first reported in by Donald J. Cram in 1973.[4] Since then, several other types of synthetic ionophores have been developed, including acyclic hosts known as *podands*. The pioneering work done by Pedersen, Lehn and Cram earned them the Nobel Prize in Chemistry in 1987.

Macrocyclic complexes of alkali metal cations in solution continue to be studied extensively by several experimental techniques.[3c] *Electrochemical methods*, such as potentiometry and electrical conductivity, and to a lesser extent polarography and voltammetry, have been used successfully in the investigation of complexation equilibria. Thermodynamic parameters have been determined for several alkali metal ionophore complexes in aqueous and nonaqueous media. Other methods, including *extraction studies* and *calorimetry*[5], have also been successful in yielding thermodynamic information. *Relaxation techniques*, such as temperature jump, ultrasonic absorption and stopped flow methods have been used to investigate the kinetics and dynamics of complex formation and dissociation.

Of all the *spectroscopic techniques* used, including electronic and vibrational spectroscopies, electron spin resonance (ESR) and nuclear quadrupole resonance (NQR), it is nuclear magnetic resonance (NMR) which has been the most popular. ^{1}H and ^{13}C NMR have been used to obtain structural, thermodynamic and kinetic information about the complexes in solution. From measurements of the ^{13}C spin lattice relaxation time (T_1) information about molecular motions of some macrocyclic ligands and their metal complexes has been obtained. A great wealth of information can also be obtained with alkali metal NMR[6] which has been one of the most useful probes concerning the nature of these complexes.

A list of the common alkali metal NMR active nuclei along with their nuclear properties is found in Table 1. The combined influence of their high gyromagnetic ratios and natural abundances make the ^{7}Li, ^{23}Na, ^{87}Rb and ^{133}Cs nuclei easily observable as compared to ^{13}C.

TABLE 1

Nuclear Properties of the Alkali Metals.[7]

Isotope	Natural abundance (%)	Spin	Gyromagnetic ratio γ (10^7 rad T^{-1} s^{-1})	Quadrupole moment Q (10^{-28} m^{-2})	Receptivity referred to ^{13}C	Resonance frequency ^{1}H at 300 MHz
^{6}Li	7.42	1	3.9366	-8×10^{-4}	3.58	44.146
^{7}Li	92.58	3/2	10.3964	-4.5×10^{-2}	1540	116.571
^{23}Na	100	3/2	7.0761	0.12	525	79.346
^{39}K	93.1	3/2	1.2483	5.5×10^{-2}	2.69	13.997
^{41}K	6.88	3/2	0.6851	6.7×10^{-2}	0.0328	7.684
^{85}Rb	72.15	5/2	2.5828	0.25	43	28.965
^{87}Rb	27.85	3/2	8.7532	0.12	277	98.164
^{133}Cs	100	7/2	3.5087	-3×10^{-3}	269	39.342

It is from three useful NMR parameters that the characterization of these complexes takes place. The origin of the *chemical shift* in alkali metal NMR is not as well understood as the ^{1}H chemical shift. It is, however, governed by the environment of the alkali metal cation which can include the solvent and/or the ligand. In cases where the chemical shift of the solvated cation can be differentiated from that of the complexed cation, one may follow the complexation and it is usually possible to obtain structural and thermodynamic data. The measurement of *relaxation times* (T_1 and T_2) along with *spectral lineshape analyses* can yield information about chemical exchange including mechanistic modelling.

This field of host-guest chemistry is now well established. Whereas earlier studies focussed mostly on complex structures and the determination of thermodynamic parameters, it was not until a few years later that the kinetics and dynamics of alkali metal cations and ionophore interactions were studied. Through a review of the literature, the development of alkali metal NMR, as a probe for the direct observation of the environment of alkali metal cation complexed by ionophores, will be followed and its applications will be illustrated.

II. COMPLEX STRUCTURES

Alkali metal NMR provides a probe whereby structural elucidation of a complex may be achieved as viewed through the cation. In this context, structural elucidation in solution involves characterizing the environment of a cation, which can include the ligand, solvent and counteranion (or combination thereof). This is in contrast with the direct observation of protons and carbons by ^{1}H and ^{13}C NMR which are traditionally used for the purpose of identification. As solids, alkali metal-ionophore complexes are often characterized by X-ray crystallography[8], giving conformational information, but in solution this task is not so readily achieved. With ^{1}H and ^{13}C NMR, it is possible to observe the differences in the solution conformations of complexed and uncomplexed crown ethers.[9] In alkali metal NMR the environment of the cation is characterized by two main parameters: the *chemical shift* is affected by shielding of the cation by the surrounding electrons and atoms whereas the *linewidth* is controlled by coupling between the nuclear quadrupole and the electric field gradients and on the correlation times for re-orientation of these electric field gradients about the nucleus (see equation 10 section III). Each cation species, be it solvated, complexed or otherwise, will have its own characteristic chemical shift and linewidth. A sample containing both solvated and complexed cation will display a spectrum which depends on the rate of chemical exchange between the two species (see section V). It is common practice, in alkali metal NMR, to use dilute aqueous alkali salt solutions as references and without corrections for the differences in solvent bulk magnetic susceptibilities comparisons of results obtained in different experiments are impossible. Mainly due to limitations imposed by the equipment, early data neglected these corrections which depend upon both the sample solvent and the orientation of the magnetic field with respect to the sample.[10]

The first alkali metal NMR paper with relevance to ionophore complexation was puplished in 1970.[11] In this solvation study by Popov and co-workers, an anion effect of the ^{23}Na NMR chemical shift was seen and a linear relationship between the chemical shift of solvated sodium and Gutmann donicity number[12] (DN) of the solvent was observed (see Figure 2). An interaction between sodium cations and DC18C6 was reported and this was the first observation of its kind by alkali metal NMR. The addition of this ionophore to solutions of sodium resulted in a broadening of the signal to such an extent that it was no longer visible with the instrument used. Another early solvation study in 1971, by Grotens et al., focussed on the linewidth of the ^{23}Na NMR signal.[13] An increase in linewidth was observed when cyclic and acyclic polyethers were added to THF solutions of sodium. This confirmed that the THF solvation shell was replaced by the complexing agent.

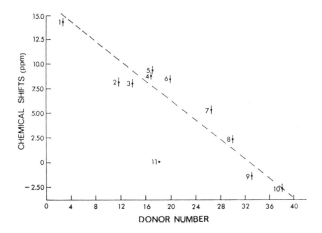

Figure 2. Plot of [23]Na chemical shifts versus the donor numbers of the solvents (1) nitromethane; (2) benzonitrile; (3) acetonitrile; (4) acetone; (5) ethyl acetate; (6) tetrahydrofuran; (7) dimethylformamide; (8) dimethyl sulfoxide; (9) pyridine; (10) hexamethyl phosphoramide; (11) water. (With permission from reference 11.)

The complexation of sodium by natural ionophores was studied by [23]Na as early as 1971 by Haynes et al..[14] The addition of ionophores to the sodium solutions showed that the observed chemical shifts (δ_{obs}) and linewidths (related to the observed transverse relaxation rate, $T_{2,obs}^{-1}$) were linear functions of the fractions of sodium in the bound form. That is, the observed signal was the weighted average of all the contributions.

$$\delta_{obs} = \Sigma \, \delta_i P_i \qquad (1)$$

$$T_{2,obs}^{-1} = \Sigma \, T_{2,i}^{-1} P_i \qquad (2)$$

In the above equations P refers to the population and i refers to the species (for example, i=1: solvated sodium and i=2: complexed sodium). It was also noted that both the ionic strength and the concentration a sodium salt, in the range 50-800 mM, did not affect the NMR signal suggesting no influence from ion pairing in 80% MeOH-20% $CHCl_3$(v/v). Recent studies have since then shown the influence of ion pairing, which alters both the chemical shift and linewidth.[15,16] The quadrupolar coupling constants (χ) of the ionophore complexes, another important parameter in alkali metal NMR (see section III), were found to vary between 0.47 and 1.64 MHz and increased in the order monactin < enniatin B < valinomycin < monensin < DC18C6 < nigericin. This constant characterizes the symmetry of the cation environment and these relatively low values for χ indicate a high degree of symmetry about the sodium cation and a very small amount of covalent interaction between the cation and ligand oxygens. Illustrations of a few of these solid complex structures are shown in Figure 3.

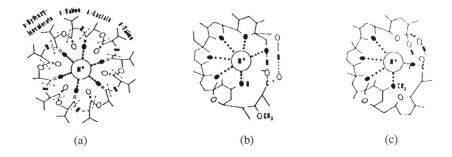

(a)   (b)   (c)

Figure 3. Structures illustrating complexed (a) valinomycin, (b) monensin and (c) nigericin. The dark circles denote the oxygen atoms which have been shown to be within van der Waals contact with the cation. (With permission from reference 14.)

Popov and his group have used several alkali metal NMR nuclei to characterize a variety of ionophore complexes.[17-22] A strong sodium-monensin interaction in MeOH was evidenced by the broad [23]Na linewidth at half-height (~700 Hz!) which precluded the determination of the [23]Na chemical shift.[17] As well, evidence of a weak lithium-monensin interaction was concluded from the slight increase in the [7]Li linewidth. These differences in behavior show that the ionic size of the cation is an important factor in governing metal cation-ionophore complex selectivity. This factor may be more important when the ligand has a fixed cavity size (crown ethers and cryptands). In ^{7}Li NMR studies of lithium cryptates[18,19], a competition between the solvent and the ligand has been shown. C222 and C221, having cavities larger than the lithium cation, could not effectively compete with the strongly solvating DMSO (DN=29.8). In NM (DN=2.7), however, a change in the [7]Li chemical shift, upon the addition of the ligands, signalled complex formation. The chemical shift of Li$^+$-C211 cryptates (C211 has a cavity size comparable to the lithium cation), was essentially independent of solvent indicating that there is no solvent in the coordination sphere of lithium in the complex.[18,19]

Thus, the nature of a cryptate in solution may be either exclusive or inclusive. In an inclusive complex, the cation sits directly in the middle of the cavity unexposed to the solvent and in an exclusive complex (see Figure 4) the solvent has access to the first coordination sphere of the cation. In certain situations there may be an equilibrium between both types of complexes. This was observed in [133]Cs studies of cesium-cryptates.[20,21] C322 completely encloses the cesium cation and only the inclusive complex was shown to exist between 178 and 298 K. This is not surprising due to the relatively large cavity size which can comfortably accommodate this cation. Only the exclusive complex was formed with the smaller and more rigid C222B. With C222, which is smaller than C322 but more flexible than C222B, the equilibrium given in equation 3 was shown to exist.

$$[Cs^+ \cap C222] \rightleftharpoons [Cs^+ \subset C222] \qquad (3)$$
$$\text{exclusive} \qquad \qquad \text{inclusive}$$

In the polar solvents MeOH and nitropropane, only the inclusive complex was observed but in the non-polar THF, cation-anion interactions hindered the formation of the inclusive complex.[20]

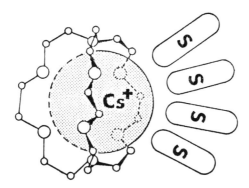

Figure 4. Schematic illustration of the configuration of an "exclusive" complex. (With permission from reference 21.)

Whereas ^{7}Li,^{23}Na and [133]Cs NMR studies of ionophore complexes have been relatively numerous, those involving Rb cations have been rather sparse. The [87]Rb nucleus has desirable NMR properties (see Table 1) but due to its large quadrupole moment, the resulting spectral linewidths are typically very broad. In one [87]Rb NMR study, however, rubidium cations were shown to form complexes with 18C6 and C222, as seen by an increase in the linewidth upon the addition of these ligands to rubidium solutions.[22]

Several studies departing from the traditional alkali metal cation-simple crown ether/cryptand interactions, show the diversity of systems which are looked at by alkali metal NMR. As shown in Figure 5, Li$^+$-ethyl acetoacetate enolate triple ion associates with Li$^+$-C211 in solution, which has been characterized by ^{7}Li NMR where 2 peaks for the two lithium sites are seen.[23] A similar association with sodium as the cation and 15C5 and 18C6 as the ionophore was characterized by ^{23}Na NMR.[24] Sodium cations have also been shown to form complexes with substituted crown ethers such as monopiperidine and dipiperidine.[25]

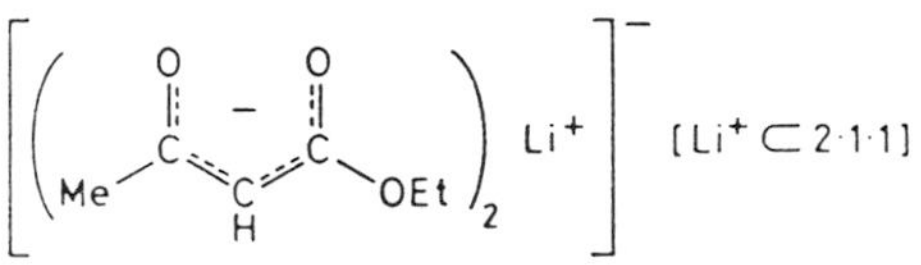

Figure 5. A Li$^+$ - ethyl acetoacetate enolate triple ion - Li$^+$ cryptate complex; [E-Li-E]$^-$ [Li$^+$-C11]. (With permission from reference 23.)

Due to the importance of polymers, it is not surprising that studies involving crown ethers supported by a polymer backbone have surfaced.[26] In one such study, DA18C6 is attached to krytofix (2,2), as shown in Figure 6, in a water soluble or gel polyacrylamide. Usual ^{23}Na experiments have shown that there is complex formation at the DA18C6 moiety.[26(a)] In a ^{133}Cs study of DB18C6 in a polyamide chain in DMSO, it was shown that after the formation of a 1:1 complex a subsequent formation of a 2:1 "sandwich" complex was reported.[26(b)] The cesium cation is larger than the crown ether which cannot wrap itself completely about the cation resulting in the coordination of a second ligand.

Another NMR parameter which can give information about the nature of complexes is the spin-lattice relaxation time, T_1. Each alkali site has its own characteristic T_1 and since the T_1 variations usually parallel the chemical shift variations, it could be substituted for δ in equation 1. The interaction of alkali cations with ionophores is expected to affect T_1 through the alteration of both the electric field gradient and the correlation time for reorientation. For example, the complexation of a cation is accompanied by a decrease in motion and therefore the T_1 decreases. This parameter was used in a ^{133}Cs NMR study of 18C6 in MeOH[27] and a 2:1 complex between 18C6 and Cs$^+$ is suggested by the results shown in Figure 7.

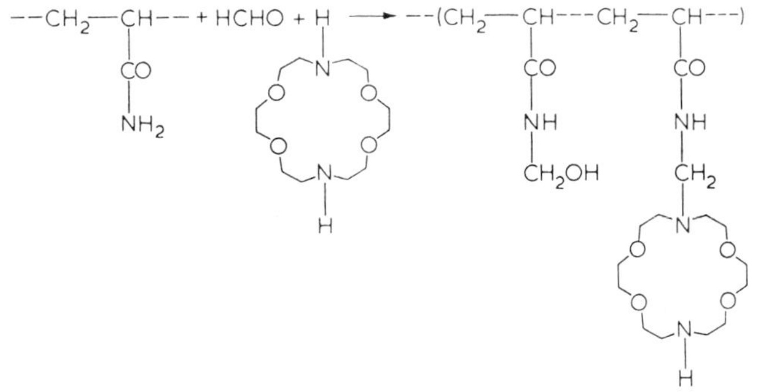

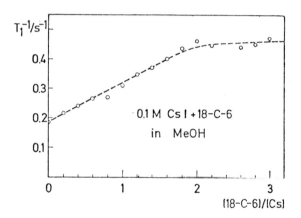

Figure 6. 1,10-diaza-18-crown-6 attached to krytofix (2,2). (With permission from reference 26(a).)

T_1^{-1}/s^{-1}

0.1 M Cs I + 18-C-6

in MeOH

$[18\text{-}C\text{-}6]/[Cs]$

Figure 7. ^{133}Cs spin-lattice relaxation rates, T_1^{-1}, as a function of the 18C6 to Cs^+ molar ratio at 301 K. (With permission from reference 27.)

The variation of T_1^{-1} as a function of ρ clearly indicates the 2:1 stoichiometry of the complex.

Alkali metal NMR studies on ionophore complexes have not been restricted to alkali metal cations. In solution in the presence of crown ethers and cryptands, alkali metal anion (alkalides) species have been detected.[28] Often the NMR spectra of these samples are composed of two peaks, corresponding to the alkali metal cation in the complexed site and to the free alkali metal anion. This is illustrated in Figure 8, where the broad peak represents K^+ in a sandwich complex with the small 15C5 and the high spherical symmetry of the potasside is attested by the

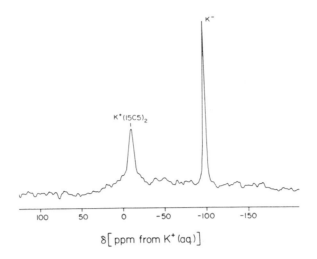

Figure 8. ^{39}K NMR spectrum of (K(15C5)$_2$)$^+$,K$^-$) in dimethyl ether (0.06M) at 220K; 20400 acquisitions with 10 Hz exponential broadening. (With permission from reference 29.)

narrowness of its peak.[29] Other similar studies with sodium[28,30], rubidium[31], cesium[32] as well as potassium[33] by their respective NMR have shown the existence of the alkali metal anion which actually is a stable entity having only weak interactions with its surroundings.[33] That is, the lifetime of the anion must not be shortened by electron and cation exchange processes if it is to be observed. The two equilibria shown in equations 4 and 5 (exchange and dissociation equilibria) must be comparatively inefficient or the NMR line will be broadened beyond detection.

$$K^- + {}^*K_s^+ \rightleftharpoons K_s^+ + {}^*K^- \tag{4}$$
$$K^- \rightleftharpoons K_s^+ + 2e_s^- \tag{5}$$

In a ^{133}Cs NMR study by Ellaboudy et al. in 1988, three cesium species were observed in same solution.[32] This was confirmed by a NMR spectrum which contained three peaks. In the presence of 12C4 and cesium metal in THF, the three cesium sites were identified as Cs$^+$(12C4)$_2$e$^-$, Cs$^+$(12C4)$_2$ and Cs$^-$.

III. MICRODYNAMICS OF COMPLEXATION

^{13}C spin-lattice relaxation times (T$_1$) of carbons provide a probe of the molecular motion of organic entities in solution.[34,35] The observed T$_1$ originates from several relaxation mechanisms and is often a composite of these mechanisms. They include mainly contributions from the dipole-dipole (T_1^{DD}), spin-rotation, chemical shift anisotropy, scalar and electron-nuclear spin relaxation mechanisms.

In the case of a protonated carbon of an organic molecule in a deoxygenated and deuterated solvent, the contribution of the ^{13}C -^1H intramolecular dipole-dipole relaxation mechanism is significant and usually dominates the relaxation. T_1^{DD} can be obtained from the observed T_1 (T_1^{obs}) by the measurement of the {^{13}C - ^1H} nuclear Overhauser enhancement (nOe), through equation 6, which is valid under extreme narrowing conditions.

$$T_1^{DD} = T_{1,obs} \times (1.988/nOe) \tag{6}$$

T_1^{DD} is described by equation 7 in which μ_o is the permeability of the vacuum, N_H the number of directly attached hydrogens, r_{CH} the C-H bond length, and τ_{eff} an effective correlation time.[30,36] Intermolecular contributions may usually be neglected since T_1^{DD} depends on the inverse 6th power of r_{CH}. Quite reasonably, one has to consider only the intramolecular ^1H -^{13}C dipole-dipole contributions from the C-H bond, which is assumed to have a constant length of 0.1085 nm.

$$(T_1^{DD})^{-1} = (\mu_o/4\pi) N_H \gamma_H^2 \gamma_C^2 \hbar^2 r_{CH}^{-6} \tau_{eff} \tag{7}$$

Equation 7 can be reduced to equation 8, in which the constant term has units of s^{-2}.

$$(T_1^{DD})^{-1} = (2.2086 \times 10^{10}) N_H \tau_{eff} \tag{8}$$

From equations 6 and 8, one can obtain equation 9 where it is shown that τ_{eff} is directly obtained from two measurements which can be routinely done on modern NMR spectrometers, with some necessary precautions in the preparation of the sample and in the error analysis. The factors affecting accuracy in ^{13}C spin-lattice measurements have been reviewed and analyzed.[37]

$$\tau_{eff} = (22.78 \times 10^{-12}) \times nOe / (N_H \times T_{1,obs}) \tag{9}$$

Since the constant term in equation 9 has s^2 units, a $T_{1,obs}$ of 1s for a -CH_2- fragment (fully dipolar; nOe = 1.988), would correspond to a τ_{eff} of 23 ps.

Dipole-dipole relaxation originates from time fluctuations of local magnetic fields associated with magnetic nuclei (^{13}C and ^1H in this case) whose internal orientation relative to B_o is fluctuating. The relaxation is thus dependent on molecular motions and the relative orientation of the two nuclei with respect to the applied magnetic field B_o is constantly changing.[34] The reorientational motion of a molecule in a liquid is very complex; it depends upon the geometry of the molecule and can not be described by a single time constant.[35] One considers, however, an effective correlation time, τ_{eff}, which is assumed to represent the average time for a molecule to progress through one radian, in the case of random molecular tumbling.[35,36] It includes not only the rotational diffusion constants but also the internal segmental motion of an aliphatic chain for example.

So, this effective correlation time, obtained from equation 9, has no real physical significance, but it could be qualitatively considered as a parameter reflecting the "fluidity" or the "stiffening" of a molecule in a liquid. For example, the segmental methylenic motions of a crown ether should be lost, or at least diminished during the complexation process. This should result in a longer effective correlation time and in a decrease of the dipole-dipole ^{13}C longitudinal relaxation time. This reasoning can be exemplified with the following two cases, chosen from a study by Echegoyen et al.[38] of lariat crown ether dynamics.

The side chain of the lariat crown ether (1) was shown to be involved in the complexation process of Na^+, since the C-3 T_1 of (1) decreased from 2.61s for uncomplexed (1) to 1.62s upon complexation. This 38% decrease was close, in value, to the 42% decrease observed for C-α and the 55% decrease for the CH_2 carbons of the crown ether.

(1)

Another particularly interesting example of the use of ^{13}C T_1's for probing the participation of side-arm in a complexation process can be found in the comparison of ^{13}C T_1 of the methyl groups in (2)and (3), which is decreased by 50% upon the complexation of Na^+ with (2) and is increased (!) by 7% upon the Na^+ complexation of (3).

This simple T_1 measurement stresses the importance of the correct geometric disposition of the sidearm heteroatoms, which is accordingly reflected in the values of the equilibrium constants of complex formation: 297 and 364 respectively for (2) and (3).

Generally, for similar reasons, the spin-lattice relaxation time of carbon atoms from a crown ether ring is smaller after the complexation of an alkali metal cation. This has been reported in a number of studies.[9,38-40] For example, the ^{13}C T_1 of 15C5 in NM was reported to be 3.98s , while it was 1.86s in the case of its K^+ complex.[39] A notable exception is 18C6 in NM. 18C6 forms a weak complex with one or two molecules of NM, which restricts the ring flexibility of the crown in a way similar to the case of the complexation by a cation.[39]

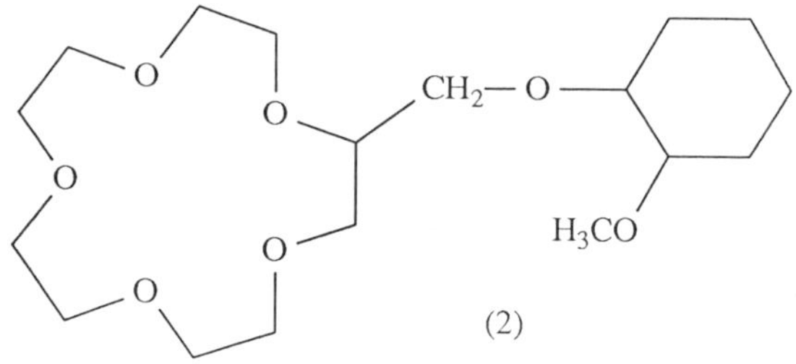

(2)

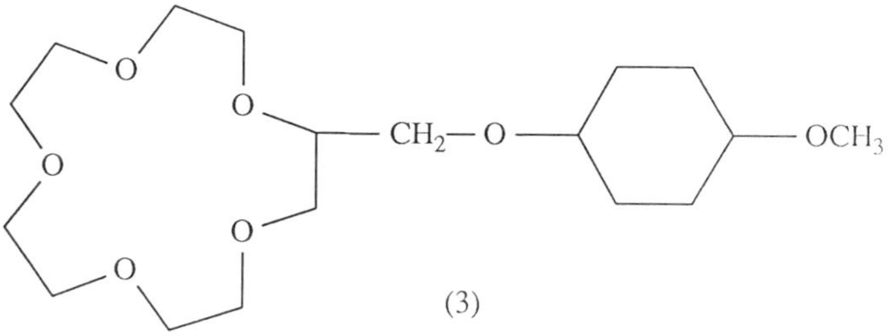

(3)

The effective correlation time, τ_{eff}, obtained from T_1^{DD}, reflects the movement in the solution of the C-H bond responsible for the dipole-dipole relaxation. Since, upon complexation, the ring of the crown ether is rigidified in such a way that the segmental motion does no longer contribute to the relaxation process and that, on the time scale of 10^{-11}s, the complex appears rigid[9], it is reasonable to make a further assumption: in the complex, the reorientational movement of the C-H vector is highly correlated with the reorientation of the O-M$^+$ vector, where M$^+$ is an alkali metal cation and O is an oxygen in the crown ether ring. This is in agreement with the conclusion of a paper by Stoddart et al.[41] stating that: "It seems not only reasonable but logical that constitution, configuration and conformation must define the structures of non covalently bonded species in much the same way as they define the structures of covalently bonds species."

The time fluctuation of the reorientation of the O-M$^+$ vector is responsible for the quadrupolar relaxation of ^{23}Na$^+$ or ^{39}K$^+$, which is described by equation 10, in the case of a spin 3/2 nucleus:[42]

$$T_q^{-1} = \pi\, v_{1/2} = (2\pi^2/5)\, (e^2qQ/h)^2\, \tau_q \tag{10}$$

where $v_{1/2}$ is the linewidth at half height of the Lorentzian signal, (e^2qQ/h) is the quadrupolar coupling constant (χ), q is the electric field gradient at the quadrupolar nucleus site, Q is the quadrupole moment and τ_q is the correlation time associated to the quadrupolar relaxation. In analogy with τ_{eff}, τ_q can be assumed to represent an average time for complex reorientation through one radian in the solution. τ_q is associated with the O-M$^+$ vector, τ_{eff} with the C-H vector, but both pertain to the reorientation of the same rigid complex in the solution. So, τ_{eff} can be reasonably identified with the ^{23}Na or the ^{39}K quadrupolar correlation time, τ_q. The measurement of $T_{2,q}^{-1}$ (from the linewidth at half-height of the ^{23}Na or ^{39}K Lorentzian signal; equation 10) coupled with the determination of τ_{eff} permits an evaluation of the quadrupolar coupling constant of the cation in the complex. This approach has been first described by Kintzinger and Lehn[43] who determined χ (^{23}Na) for four cryptands. They found a linear relationship between δ (^{23}Na) and χ (^{23}Na), in agreement with a previously proposed linear relation between χ (^{23}Na) and the paramagnetic shielding term of δ (^{23}Na).[44,45] A similar relation has also been demonstrated for Na$^+$-amine complexes.[46] The ^{23}Na quadrupolar coupling constant and the ^{23}Na chemical shift for the complex (Na-DB24C8)$^+$, in which the crown ether is wrapped around the cation, fit the χ - δ relationship of the cryptands.[47] A value of χ around 1 MHz is typical of an oxygen symmetrical environment surrounding the complexed sodium cation.[39,43,47] For example χ (^{23}Na) = 1.0 MHz for (Na-DB24C8)$^+$ [47] and (Na-C222)$^+$ [43], while it reaches 2.2 MHz in the case of the unsymmetrical (Na-C211)$^+$ [43] complex. The same type of observations could be extracted from ^{39}K NMR results. While χ (^{39}K) is equal to 1.2 MHz for (K-Valinomycin)$^+$ [48] and to 1.4 MHz for (K-DB30C10)$^+$ [49], in which cases K$^+$ is surrounded by a symmetrical shell of oxygen atoms, χ (^{39}K) reaches 2.4 MHz for 18C6 in MeOH[48] and 3.4 MHz in NM[39], in which cases the first coordination sphere of K$^+$ involves not only oxygen atoms, but also some solvent molecules. The similarity of the coordination spheres of ^{23}Na$^+$ and ^{39}K$^+$ in, respectively, (Na-DB24C8)$^+$ and (K-DB30C10)$^+$ is illustrated in Figure 9.

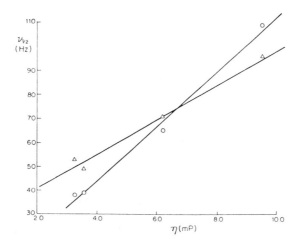

Figure 9. Linewidths at half-height of the ^{39}K NMR signals ($\triangle$; reference 49) and of the ^{23}Na NMR signals (O ; reference 47) for the complexes (K-BD30C10)$^+$ and (Na-DB24C8)$^+$ respectively, as a function of the solvent viscosity increasing in the order AC, AN, NM and PY. (With permission.)

The slopes of the linear relationships between the linewidths of ^{23}Na and ^{39}K signals in four solvents as a function of the solvent viscosity are in the same order of magnitude.[49] From known values of Q, the ratio of the two electric field gradients in the above complexes could be estimated: $q(^{39}K)/q(^{23}Na) = 1.0$. Quite remarkably, this value attests of the similarity of the wrapping process in the complexation of Na$^+$ and K$^+$ by DB24C8 and DB30C10 respectively. In the case of (Na-DB30C10)$^+$ the large crown ether cannot adopt a conformation fitting as perfectly with the size of the cation, and some solvent molecules are kept on the first coordination sphere of Na$^+$.[51]

IV. THERMODYNAMICS OF COMPLEXATION

The relationships between chemical shifts, relaxation rates, and mole ratios of complexed to free cation (ρ) can aid in the determination of complex stoichiometry and formation constants for alkali metal-ionophore complexes. If the formation of the complex is quantitative, ($K_f > 10^5$), the variation of chemical shift is linear with respect to the mole ratio until it reaches 1.0, after which it remains constant. This is the case for (Na-18C6)$^+$ shown in Figure 10, in several non-aqueous solvents, namely those of medium and low donicity.[52] If the formation of the 1:1 complex is not quantitative because of its weak nature, the chemical shift increases but does not reach a plateau at ρ = 1. This is shown in Figure 11 for (Na-15C5)$^+$ in DMF, DMSO and H$_2$O.[52] This is most common in high donicity solvents where the solvent can effectively compete with the ligand for the cation. K_f can be determined from the resulting curve.

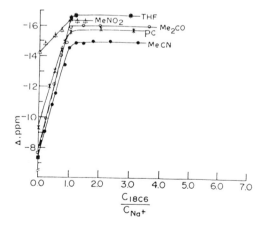

Figure 10. The variation of the ^{23}Na chemical shift as a function of ρ in five non-aqueous solvents. (With permission from reference 52.)

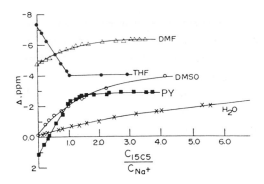

Figure 11. The variation of the ^{23}Na chemical shift as a function of 15C5/Na$^+$ mole ratio in a variety of solvents. (With permission from reference 52.)

A third possibility when observing changes in chemical shift, is a linear increase of δ until ρ=1, followed by a change in slope, with no plateau, upon further addition of ligand. This is seen in Figures 12 and 13 for (Na-15C5)$^+$ and (Na-B15C5)$^+$ in NM, and suggests the formation of a sandwich (Na-L$_2$)$^+$ complex.[52]

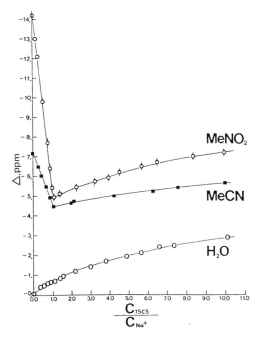

Figure 12. The variation of the ^{23}Na chemical shift as a function of 15C5/Na$^+$ mole ratio. (With permission from reference 52.)

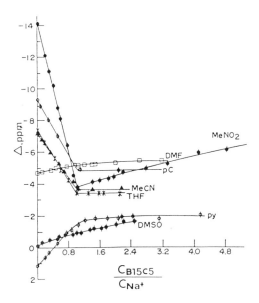

Figure 13. The variation of the ^{23}Na chemical shift as a function of B15C5/Na$^+$ mole ratio. (With permission from reference 52.)

^{133}Cs NMR has been a popular choice for thermodynamic studies of alkali metal-ionophore complexes over the years. Popov's group has been particularly active in this field[21,53-64]. The stabilities of cesium cryptates in various non-aqueous solvents, vary in the order (Cs-C221)$^+$ ≥ (Cs-C222)$^+$ > (Cs-C222B)$^+$ >> (Cs-C211)$^+$ [21,53]. As described in section II, two types of 1:1 complexes may form between Cs$^+$ and cryptands. At low temperatures the inclusion complex of (Cs-C222)$^+$ is favoured whereas at higher ones the exclusive complex predominates. The rigidity of C222B precludes the formation of inclusive complexes, while exclusive complexes definitely result with C211 (1.6 Å) and C221 (2.2 Å) which have cavities smaller than the cesium cation (3.3 Å).[21] At 233 K, Cs$^+$ is completely surrounded by C222 so its chemical shift is independent of solvent. As the temperature increases the concentration of exclusive complex increases, as shown by an increased dependence on solvent.[53] The thermodynamic parameters for the formation of the exclusive complex, followed by its conversion to the inclusive complex, are given for (Cs-C222)$^+$ in Table 2. Further studies with (Cs-C222)$^+$ revealed the influence of solvent through the use of binary solvent mixtures.[54] In the neat solvents the formation constants increased in the order DMSO < DMF < AC < PC < AN which opposes the trend in the Gutmann donicity numbers. From this data and the amounts of each solvent component in the binary mixtures, the donating ability of a particular solvent mixture can be qualitatively predicted.[54]

TABLE 2

Thermodynamic parameters for (1) the formation of the exclusive complex (Cs-C222)$^+$ and (2) the conversion to the inclusive complex. (With permission from reference 53.)

Solvent	ΔH_1° kJ.mol^{-1}	ΔS_1° J.mol^{-1}.deg^{-1}	ΔH_2° kJ.mol^{-1}	ΔS_2° J.mol^{-1}.deg^{-1}
AC	-54	-113	-10.5	-23.4
PC	-36.0 ± 1.7	-57.3 ± 6	-12.1	-29.3
DMF	-23.8	-46.9	-10.9	-31.8

Alkali metal cations form complexes with cryptands which typically have higher stabilities than with crown ethers. This is easily explained by the presence of the "extra" bridge in the macrocycle which allows the cation to often be shielded from the outside environment when it is complexed. However, crown ether complexes are not necessarily weaker than those with cryptands. The 1:1 cesium complex with 18C6[21] is stronger than that with C222B[57] in several non-aqueous solvents. The strength of the former is seen in Figure 14 by the linearity of the relationship between the chemical shift versus ρ, until $\rho = 1$, in PY, AC and PC.[56] The 18C6 is also capable of forming a "sandwich" complex where two crown ether molecules surround the cesium cation.[56,57] This can be seen in Figure 14 in the three same solvents after $\rho = 1$.

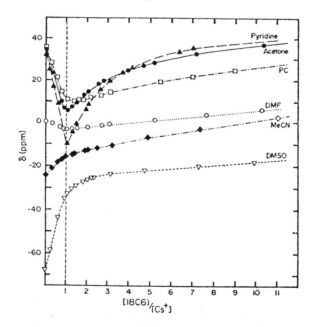

Figure 14. ^{133}Cs NMR chemical shifts versus $\rho = [C]/[Cs^+]_t$ in various solvents. The solutions contained 0.01M CsBPh$_4$. (With permission from reference 57.)

Another consideration should be introduced however. There is the possibility of ion-pairing between the cesium cation and the counteranion, which may influence formation constants of the complexes. For example the following equilibria must be considered for the the the formation of $(Cs-18C6)^+$ and $(Cs-(18C6)_2)^+$ in PY:[56]

$$Cs^+ + X^- \rightleftharpoons (Cs^+,X^-) \tag{11}$$
$$Cs^+ + C \rightleftharpoons (Cs-C)^+ \tag{12}$$
$$(Cs-C)^+ + C \rightleftharpoons (Cs-C_2)^+ \tag{13}$$
$$(Cs-C)^+ + X^- \rightleftharpoons ((Cs-C)^+,X^-) \tag{14}$$
$$(Cs-C_2)^+ + X^- \rightleftharpoons ((Cs-C_2)^+,X^-) \tag{15}$$

where $X^- = BPh_4^-$ and $C = 18C6$. Since it is unlikely that the sandwich compound would form ion-pairs equation 15 can be omitted. The chemical shift of the 1:1 complex was shown to be independent of sodium salt concentration so equation 14 may also be omitted, thus reducing the reaction scheme to three equations.[56] The formation of the "sandwich" complex is characterized by the following thermodynamic parameters: $\Delta H_2° = -24.3 \pm 0.8$ kJ.mol^{-1}, $\Delta S_2° = -44.8 \pm 2.5$ J.mol^{-1}.K^{-1} and $(\Delta G_2°)_{298} = -10.79 \pm 0.08$ kJ.mol^{-1}.[56]

While the formation constant (K_2) for $(Cs-(18C6)_2)^+$ can be measured in many non-aqueous solvents, Dye et al. found that there was also evidence for 2:1 complexes between cesium cations and either DB18C6 or DC18C6 but only K_2 for DB18C6 in PY could be measured.[57] In certain cases even K_1 was too large to be measured precisely but lower limits for it were obtained. Values obtained for the formation constants are summarized for these three crown ethers in Table 3.

TABLE 3
Formation constants of 1:1 and 1:2 $(Cs-C)^+$ complexes at 298 K. (With permission from reference 57.)

Solvent		18C6	DB18C6	DC18C6
PY	K_1	$> 5 \times 10^5$	$(7 \pm 2) \times 10^3$	$> 10^5$
	K_2	74 ± 2	$(2.3 \pm 0.2) \times 10^2$	yes
PC	K_1	$(1.5 \pm 0.6) \times 10^4$	$\sim 10^3$	~ 10
	K_2	11 ± 4	yes	yes
AC	K_1	$> 2 \times 10^5$	$> 10^3$	$> 10^4$
	K_2	34.0 ± 0.5	yes	yes
DMF	K_1	$(9 \pm 3) \times 10^3$	30 ± 3	$(2.8 \pm 0.9) \times 10^3$
	K_2	2.44 ± 0.05	-	-
DMSO	K_1	$(1.1 \pm 0.1) \times 10^3$	22 ± 3	$(1.6 \pm 0.1) \times 10^2$
	K_2	1.0 ± 0.4	-	-
AN	K_1	$> 10^4$	35 ± 2	$> 10^4$
	K_2	3.7 ± 0.6	-	yes

More detailed studies as to the role of counteranions, and the influence of ion-pair formation, focused on the $(Cs-(18C6)_n)^+$ complex (where n=1,2) in MA and liquid ammonia, where the competition between ion-pair formation and complex formation was shown.[58,59] For 1:1 complexes, cationic and anionic triple ions are considered as well as equations 11, 12, 14 and 16.

$$(Cs^+,X^-) + C \rightleftharpoons ((Cs-C)^+,X^-) \tag{16}$$

where $X^- = SCN^-$, I^-, BPh_4^- and C = 18C6. Table 4 gives the different equilibrium constants associatied with equations 11, 12, 14 and 16 with the counter-anions which were studied. Relative formation constants may also be determined by looking at the competition of two cations for one ligand. This was done with ^{133}Cs NMR as a probe with Cs^+, Na^+ and 18C6 in AN.[60,61]

TABLE 4

Formation constants determined for various counteranions. (With permission from reference 59.)

	CsI	CsBPh$_4$	CsSCN
K_{14}	$(1.51 \pm 0.06) \times 10^5$	$(1.2 \pm 0.4) \times 10^4$	-
K_{16}	$(6.3 \pm 0.4) \times 10^3$	$(8 \pm 1) \times 10^3$	$(4.9 \pm 0.5) \times 10^3$
K_{12}	$(1.1 \pm 0.2) \times 10^4$	$(1.1 \pm 0.2) \times 10^4$	$(1.1 \pm 0.2) \times 10^4$
K_{11}	$(2.5 \pm 0.3) \times 10^5$	$(1.4 \pm 0.3) \times 10^4$	-

When cesium cations complex with larger crown ethers the character of the complex is still very dependent on the macrocycle cavity size even though this has little effect on the stabilities.[62] DB21C7 forms a stable two-dimensional complex in which all of the crown ether oxygens are electrostatically bound to the cation. DB27C9 and DB30C10 form three-dimensional 'wrap-around' complexes, where all of the crown ether oxygens are also bound. DB24C8 is intermediate in size so can only form a two-dimensional complex with only part of the donor sites bound to the cation, yielding a less stable entity. Table 5 is a summary of the thermodynamic parameters for these four crown ethers.[62]

Rounaghi and Popov furthered the study of DB24C8 and DB27C9 to include mixed solvent systems[63] just as they had done for $(Cs-C222)^+$, as mentioned earlier. In contrast with $(Cs-DB24C8)^+$, the thermodynamic parameters for $(Cs-DB27C9)^+$ do not increase monotonically with increased amounts of AN in the AN-DMF mixture. The two complexes are enthalpy stabilized but entropy destabilized, which is also the case for $(Cs-DB30C10)^+$.[64]

TABLE 5

Free energies, enthalpies and entropies of complexation for Cs^+ complexes in various solvents.
(With permission from reference 62; the data were converted from cal to J.)

	$\Delta G°_{303K}$ (kJ.mol^{-1})	$\Delta H°$ (kJ.mol^{-1})	$\Delta S°$ (J.mol^{-1}.K^{-1})
DB21C7			
NM	-24.08 ± 0.42	-31.87 ± 1.05	-26.17 ± 3.27
AN	-23.00 ± 0.21	-34.51 ± 0.84	-38.48 ± 2.76
AC	-22.86 ± 0.34	-46.61 ± 1.72	-78.06 ± 5.65
MeOH	-23.03 ± 0.34	-27.68 ± 0.59	-14.82 ± 1.84
PY	-24.83 ± 0.42	-30.23 ± 0.75	-18.34 ± 2.35
DMF	-16.12 ± 0.29	-43.05 ± 3.85	-88.61 ± 11.89
PC	-22.07 ± 0.29	-46.65 ± 4.77	-81.03 ± 15.28
DB24C8			
NM	-23.91 ± 0.46	-26.17 ± 0.42	-7.50 ± 0.42
AN	-22.91 ± 0.42	-34.00 ± 0.67	-36.26 ± 2.01
AC	-21.57 ± 0.54	-46.90 ± 2.18	-84.09 ± 7.16
MeOH	-21.19 ± 0.29	-41.33 ± 1.93	-67.42 ± 6.24
PY	-23.24 ± 0.21	-25.00 ± 0.67	-5.82 ± 1.93
DMF	-12.48 ± 0.17	-258.63 ± 0.50	-43.09 ± 1.76
PC	-18.89 ± 0.25	-33.96 ± 3.10	-47.70 ± 9.63
DB27C9			
NM	-24.92 ± 0.25	-31.11 ± 3.81	-18.97 ± 11.64
AN	-22.57 ± 0.17	-33.21 ± 1.76	-33.29 ± 5.86
AC	-24.62 ± 0.88	-60.72 ± 7.50	-120.60 ± 23.45
MeOH	-20.44 ± 0.50	-21.31 ± 2.26	-22.40 ± 7.83
PY	-24.08 ± 0.29	-37.86 ± 2.18	-29.73 ± 6.41
DMF	-12.77 ± 0.04	-30.15 ± 1.55	-57.41 ± 4.98
PC	-21.15 ± 0.17	-46.94 ± 2.35	-85.72 ± 7.50
DB30C10			
NM	-25.00 ± 0.29	-33.29 ± 1.63	-27.89 ± 0.23
AN	-19.72 ± 0.54	-21.48 ± 1.17	-6.41 ± 3.73
AC	-23.03 ± 0.42	-56.45 ± 2.14	-109.67 ± 0.08
MeOH	-24.33 ± 0.42	-53.27 ± 1.42	-95.56 ± 4.65
PY	-25.67 ± 0.59	-33.25 ± 1.51	-24.83 ± 4.73

When sodium and potassium cations are complexed with DB30C10 or DB24C8, aggregates may form.[50,51,64-66] In NM and AN, using ^{23}Na NMR, inflection points on the chemical shift versus ρ plot at $\rho = 0.5, 0.7, 1.0$ indicate the presence of three species: $(Na-C)^+$, $(Na_2-C)^{++}$ and $(Na_3-C)^{+++}$.[50,64] In NM, using ^{39}K NMR, 1:1 and 2:1 complexes with DB30C10 are formed[65] but the presence of thiocyanate anions must also be taken into account. The model needed to fit the data must consider the ion-paired potassium cation, the favoured 'wrapped around' complex and the 2:1 $(K_2-DB30C10)^{++}$ complex. The data thus obtained is probably a reflection of the weighted average of the equilibria shown in equation 17. A similar aggregation model is needed to account for the ^{23}Na NMR transverse relaxation rates for $0 < \rho < 1$ for the Na-DB24C8 system in NM.[56]

$$(K-C-K)^{++} + 2SCN^- \rightleftharpoons (SCN^-,(K-C-K)^{++}) + SCN^- \rightleftharpoons (SCN^-,(K-C-K)^{++},SCN^-) \qquad (17)$$

In section III the similarity of the coordination spheres of ^{23}Na$^+$ and ^{39}K$^+$ in $(Na-DB24C8)^+$ and $(K-DB30C10)^+$ have been discussed but the $(Na-DB30C10)^+$ is different in that it has been shown that for this complex the sodium cation is not isolated from the solvent.[51]

The principles which have been used to study systems involving cesium cations can also be used for systems involving other alkali metal cations and ionophores.[67-70] A 1:1 $(Li-12C4)^+$ complex forms quantitatively in NM, PC, AN, PY, AC; with a further $(Li-C_2)^+$ complex forming in NM and PC. The strongly solvating MeOH, DMSO, TMG and H_2O preclude total complexation with both 12C4 and 15C5. No $(Li-(15C5)_2)^+$ complexes were seen in the other solvents where the Li$^+$ fits comfortably inside the crown cavity.[67] Hourdakis and Popov's study of the dilactam of C222 with ^{7}Li, ^{23}Na and ^{133}Cs NMR shows that this ligand, shown in Figure 15, forms a weaker complex than C222 in most non-aqueous solvents, a circumstance attributed to the fact that the dilactam does not completely isolate the cation from the solvent.[68]

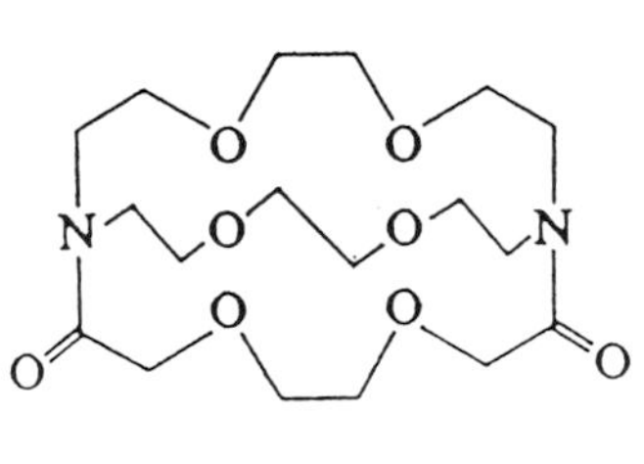

Figure 15. The structure of the dilactam of C222.

A study involving derivatives of other previously mentioned compounds is that by Laszlo and co-workers on spiro-bis-crown ethers, where ring sizes and proximities yield a variety of 1:1, 1:2,

2:1 complexes.[69] The thermodynamic studies discussed in this section have tended to be those of crown ether and cryptand complexes but there have been studies with other ligands, a final example of which is one on podands[70] which are compounds that undergo conformational changes to wrap around that cation since they have no preformed cavity. Once again solvent molecules may or may not participate in the co-ordination sphere, depending on the number of co-ordination sites available on the ionophore.

V. KINETICS OF COMPLEXATION

The use of alkali metal NMR to study the kinetics and mechanisms of the associations and dissociations of alkali metal cations to ionophores is uncommon, but its application has been well understood since the 1970's. The values of the rate constants involved in the kinetic processes of many of these systems, fall in the range where they can be measured by alkali metal NMR (typically 10^2 - 10^5 s^{-1}) making this a choice technique. To this date, several review articles have been published concerning interactions between alkali metal cations and ionophores[3c,71-74], but because of its relatively recent application, the use of alkali metal NMR for kinetic investigations has received very little attention.

The kinetic interactions of these systems have also been probed by other methods. The first such study was performed by Chock[75] in 1972 where he found, using the temperature jump method, that the complexations between DB30C10 and Na$^+$, K$^+$, Rb$^+$, and Cs$^+$ were diffusion-controlled. Ultrasonic absorption techniques[76-78], stopped flow methods[79,80] and ^{13}C NMR[81,82] have been used to study alkali metal cation-ionophore complexes. Stopped flow techniques were also used by Ishihara et al. to look at the dissociation of the (Na-C221)$^+$ complex in various non-aqueous solvents. This study, with a high pressure apparatus, was the first to report volumes of activation for the dissociation of cation-ionophore complexes.[83] This shows the diversity of techniques currently used in order to elucidate the kinetics and mechanisms of cation-macrocycle interactions in solution. The use of alkali metal NMR as a viable kinetic and mechanistic tool, will now be shown by discussing the specific information which may obtained when using technique. First, a brief overview of the background involved in dynamic NMR spectroscopy and exchange kinetics studied by NMR is presented.[84-87]

It has been found that the complexation reactions in question are usually diffusion controlled[69,75] and the respective high rate constants are not accesible by NMR. Rather, it is from the much slower dissociation processes, that relevant kinetic and mechanistic information may be obtained. The rate constant for the complexation (forward reaction in equation 18), k_A, may still be obtained if both the formation constant of the complex, K_f, and the dissociation rate constant, k_B, are known ($K_f = k_A/k_B$).

For these systems, two metal cation sites for the exchange are generally considered:

$$M^+ + L \overset{k_A}{\underset{k_B}{\rightleftharpoons}} (M\text{-}L)^+ \qquad (18)$$

where M^+ (site A) is the solvated alkali metal cation, L is the ligand, $(M\text{-}L)^+$ (site B) is the metal cation complexed to the ionophore, and k_A and k_B are the rate constants for forward and reverse reactions respectively. Systems which undergo chemical exchange where the rates can be measured by NMR display spectra which are quite different from those obtained in cases where the exchange rates are too fast on the NMR chemical shift timescale. In the latter case, one peak having a Lorentzian lineshape is observed, characteristic of the population average of the two sites, and its chemical shift is given by:

$$\delta_{obs} = P_A \delta_A + P_B \delta_B \qquad (19)$$

where P_A and P_B are the populations of the free and complexed sites while δ_A and δ_B are the corresponding chemical shifts. In equation 19 one could substitute the chemical shift δ, for the transverse relaxation rate T_2^{-1} (this is related to the linewidth as shown below). When the exchange rates are measurable on the NMR chemical shift timescale, three types of spectra, which depend on the magnitude of the exchange rates, may be obtained. In the fast exchange limit (case 1), one peak is observed, still with Lorentzian lineshape, but broadened due to the chemical exchange. Equation 19 still applies for the chemical shift and the linewidth of the peak is related to the transverse relaxation time as follows:

$$T_2^{-1} = \pi \, \nu_{1/2} \qquad (20)$$

where T_2^{-1} is the transverse relaxation rate and $\nu_{1/2}$ is the width of the peak at half-height. An increase in the transverse relaxation rate (corresponding to a decrease in the relaxation time) will therefore result in a broadening of the peak. The longitudinal relaxation time , T_1, is not affected by the exchange and an exchange contribution, $T_{2,ex}^{-1}$, to the transverse relaxation rate may be obtained:

$$T_{2,ex}^{-1} = T_{2,obs}^{-1} - T_{2,q}^{-1} - T_{2,inh}^{-1} \qquad (21)$$

where $T_{2,obs}^{-1}$ is the observed transverse relaxation rate, $T_{2,q}^{-1}$ is the quadrupolar contribution to the relaxation and $T_{2,inh}^{-1}$ is the inhomogeneity contribution. In the extreme narrowing limit, $T_{2,q}^{-1}$ and $T_{1,q}^{-1}$, the quadrupolar transverse and longitudinal relaxation rates, are equivalent. Essentially then, the exchange contribution to the transverse relaxation rate is experimentally obtained from the differences in the measured transverse and longitudinal relaxation rates. A graphical representation of this is shown in Figure 16.

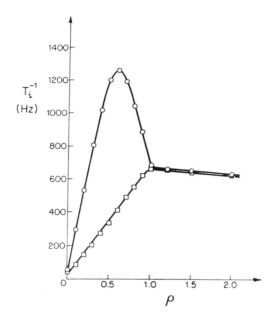

Figure 16. ^{23}Na transverse ($i=2$;o) and longitudinal ($i=1$;□) relaxation rates as a function of the ratio ρ = [B15C5]/[NaBPh$_4$] nitromethane. v_o = 79.35 MHz; T=301.5 ± 0.5 K; [NaBPh$_4$] = 8.00 × 10^{-2} M. (With permission from reference 88.)

In this figure both the transverse and longitudinal relaxation rates are plotted as a function of ρ, [B15C5]/[NaBPh$_4$], and it can be seen that in the region where the sodium cation can occupy the two different sites ($0 < \rho < 1$), the linewidth of the ^{23}Na NMR spectrum has been broadened because of chemical exchange, as denoted by the transverse relaxation rate. It can be shown[87] that in these conditions, the lifetime of the cation in the two sites A and B, $\tau = P_A/k_B = P_B/k_A = (k_A + k_B)^{-1}$, is related to the $T_{2,ex}^{-1}$ as shown:

$$T_{2,ex}^{-1} = 4\, P_A P_B\, \pi^2\, (v_A - v_B)^2\, (k_A + k_B)^{-1} \tag{22}$$

where v_A and v_B are the chemical shifts expressed in Hz.

As the exchange rates slow down on the NMR chemical shift timescale, there are spectacular changes in the lineshape as it ceases to be Lorentzian (case 2). In the slow exchange limit (case 3), two separate peaks, corresponding to the two metal cation sites, are observable. If there is no chemical exchange, the chemical shifts and the linewidths of these two peaks will correspond to the those of the two separate sites. In cases 2 and 3, full lineshape analyses[86], must be performed in order to obtain quantitative results for the rates of exchange. Figure 17 shows a series of spectra illustrating the changes from case 1 to case 3 as the rate of exchange of sodium between its solvated and complexed to DB24C8 sites decreases.

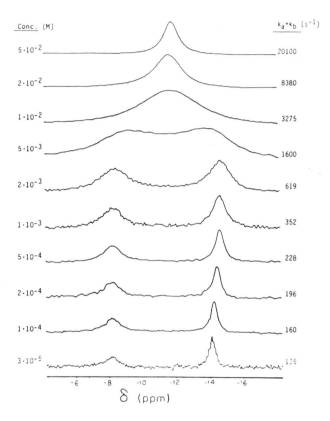

Figure 17. ^{23}Na NMR spectra (79.35 MHz) of sodium hexafluorophosphate solutions in nitromethane in the presence of DB24C8 ([DB24C8]/[NaPF$_6$] = 0.50) at different concentrations (295K). (With permission from reference 89.)

The above discussion gives background in the quantitative determination of rate constants for chemical exchange, but how may one obtain mechanistic information from alkali metal NMR? Typically there are two possible mechanisms, originally postulated by Shchori et al. in 1971[90], considered for the exchange between solvated and complexed alkali metal cations. The first has been referred to as unimolecular dissociation or dissociative exchange and is shown to be:

$$(M\text{-}L)^+ \underset{k_1}{\overset{k_{-1}}{\rightleftharpoons}} M^+ + L \tag{23}$$

Here, the complex simply dissociates to give the free (solvated) cation and ligand. The chemical exchange may also proceed via a second mechanism referred to as bimolecular cation interchange or associative exchange and is shown to be:

$$M^{*+} + (M\text{-}L)^+ \overset{k_2}{\rightleftharpoons} M^+ + (M^*\text{-}L)^+ \tag{24}$$

In this mechanism one can visualize the simultaneous binding of two cations by the ligand in the transition state before effecting the exchange. Taking these two possible mechanisms into account (equations 23 and 24) for the two site exchange (equation 18), the following expression is derived:

$$(k_A + k_B) = k_{-1}(1\text{-}\rho)^{-1} + k_2[M^+]_{tot} \tag{25}$$

where $[M^+]_{tot}$ is the total metal ion concentration and $\rho = [L]/[M^+]_{tot}$. It will be seen later how from this relationship, that experimental results can give information on the mechanistic aspects of the chemical exchange. These two possible mechanisms have been very successful in accounting for experimental data. One must keep in mind however, that this is only a model and does not preclude the possibility of other mechanisms which could be implicated in the chemical exchange, such as the possible participation of the salt counteranion. As well, in the case of the solvated crown ether, C, coexisting in solution with the complex (M-C)$^+$, an associative mechanism for the ligand exchange should be considered:[91]

$$C^* + (M\text{-}C)^+ \rightleftharpoons C + (M\text{-}C^*)^+ \tag{26}$$

Once the exchange mechanism and the corresponding rate constants have been determined (this will be shown later), the free energies of activation, ΔG^*, can be calculated:

$$\Delta G^* = RT \ln(kh/k_B T) \tag{27}$$

R is the gas constant, T is the temperature in Kelvin, k is the rate constant for the chemical exchange, h is Planck's constant and k_B is Boltzmann's constant. Furthermore, the enthalpies, ΔH^*, and entropies, ΔS^*, of activation may also be determined with a linear Eyring plot.

At this time, approximately 30 published papers are kinetic studies by alkali metal NMR involving alkali metal-ionophore systems. Most of these have been published since the early 1980's indicating of the relatively recent application of NMR for this purpose. The number of these studies is increasing steadily and with the development of higher field instruments, higher rate constants become accessible by NMR. Table 6 at the end of this section shows a summary of kinetic and mechanistic results, including dissociation rate constants and activation parameters, which have been obtained to date for these systems.

The selectivity of cation-ionophore interactions are very dependant on the ratio of the diameter of the alkali metal cation to that of the cavity size of the macrocycle. With the open-chain ionophores, this factor is less important since these ionophores simply wrap themselves around the cation. Lithium is the smallest of the alkali metal cations and is the most sensitive of the alkali metal NMR nuclei (see Table 1) but despite this, very few NMR studies with this cation have been published. Popov and co-workers[92] were the first to use this nucleus in 1975 as a kinetic probe to investigate the exchange kinetics for $(Li-C211)^+$ in various solvents and $(Li-C221)^+$ in PY. They found that the mechanisms for the exchange were dissociative (equation 23) in all cases, and that the energies of activation (E_a) for the release of the lithium cation from the cryptates increased with increasing donicity of the solvent as expressed by the Gutmann[12] donor number. This implies that the solvent is directly involved in the transition state. This phenomenon is often observed when the exchange mechanism proceeds via a dissociative pathway.[84,93-95] Another study revealed that the dissociation of $(Li-C221)^+$ in AN and PC proceeds via the associative mechanism (equation 24) and for this mechansim the barrier for the activation appears to be independent of the nature of the solvent.[93] Two factors contributed to this finding: (1) C221 is a relatively rigid cryptand and can conformationally satisfy the associative mechanism better than the more flexible C222 which can wrap itself around the Li^+ cation in such a way that the only possible pathway for the exchange of the cation between the solvated and the complexed sites would be the dissociative mechanism and (2) the high dielectric constants and donicities of these solvents could more easily reduce the charge-charge repulsion of the Li^+ ions in the transition state. Again, for the cases where the dissociative mechanism predominates, $(Li-C222)^+$ in AN, PC and AC, there is a correlation between the donicity of the solvent and the activation parameters for the reaction. A ^7Li study by Aalmo and Krane[82] revealed that the exchange between the solvated lithium and lithium complexed to OM16C4 in NM proceeded via the associative mechanism and that the free energy of activation (79 kJ.mol^{-1}) far exceeded any conformational barriers in the ligand. Very recently, Detellier and co-workers have used ^7Li NMR to look at the exchange kinetics for the $(Li-15C5)^+$ and $(Li-B15C5)^+$ systems.[96] In the low donicity solvent NM, it was shown that the preferred exchange mechanism was via the associative pathway.

^{23}Na is the most popular nucleus with these NMR studies, and since the size of the sodium cation is well matched with 18C6[73], this ligand and its derivatives are also the most popular. Shchori et al.[90,97] were the first to obtain kinetic results from ^{23}Na NMR studies of sodium cations with 18C6 derivatives. Having proposed the two mechanisms shown in equations 23 and 24, they determined that for sodium complexed to various 18C6 derivatives in DMF, MeOH and DME, the dissociative mechanism predominated. The exchange kinetics of 18C6 with sodium has been studied by a number of groups.[95,98-101] Popov and coworkers[101] found that the exchange mechanism for the dissociation of $(Na-18C6)^+$ in THF depends on the counteranion. With the highly coordinating SCN⁻ counteranion, contact ion pairs are formed which offset the charge-charge repulsion of the two sodium cations in the transition state thus allowing the associative mechanism to predominate for the exchange. When the BPh_4^- anion is used, this repulsion in the transition state is not offset and the predominant mechanism is the dissociative one. The influence of the counteranion has also been shown by Stöver and Detellier[16] where, for the associative exchange of (Na-DB24C8)⁺ in NM, the rate consant k_2 increased in the order $BPh_4^- < PF_6^- < I^- < SCN^-$. In another study by Strasser and Popov[95], it was found that the solvent also played an important role in determining the exchange mechanism. For the system (Na-18C6)⁺ it was found that the preferred exchange mechanism was dissociative in the solvents MeOH, THF-MeOH mixture and 80-20% THF-PC mixture while in neat PC and 40-60% THF-PC the associative mechanism predominated. As pointed out earlier, PC has a high dielectric constant and is able to reduce the charge-charge repulsion in the transition state allowing the associative mechanism to be in effect. Again it was found that in the cases where the dissociatve mechanism is favoured, there is a correlation between the donicity of the solvent and the activation parameters. Lincoln and coworkers[99] found the dissociative mechanism to be predominant for the exchange of (Na-18C6)⁺ in AC, MeOH and PY with SCN⁻ as the counteranion. A recent exchange study of (Na-18C6)⁺ by Graves and Detellier[98] with tetraphenylborate as the counter-anion showed the competition between the two exchange mechanisms in PC, AN and PY. The mechanism was purely dissociative in AC and in PC it was possible, for the first time, to cleanly separate the contributions of both mechanisms and to determine the activation parameters in both cases. Figure 18 shows this separation where the linear plots of $(k_A + k_B)$ as a function of $(1-\rho)^{-1}$ indicate the presence of the two competing mechanisms, as shown by equation 25, from which the rate constants for the dissociative, k_{-1}, and associative, k_2, mechanisms can be obtained from the respective values of the slopes and intercepts.

In section II, the presence of the alkalides has been discussed. When 15C5 and 18C6 are complexed with Na⁺ in MA, the species (Na⁺-C,Na⁻) is formed which implicates the sodium in a site different than the two which have been considered thus far.[30] The ²³Na spectra displays two peaks at 246 K, showing sodium in Na⁺-C and Na⁻, which coalesces upon increasing the temperature. The exchange mechanisms were modified to account for the data obtained.

So far we have seen the effect of the solvent and the counteranion on the operative exchange mechanism. In the dissociation of (Na-DA18C6)⁺ in THF, it was found that the exchange mechanism was dependent on the temperature: above 253 K the exchange is dominated by the dissociative pathway whereas below 233K the mechanism changes to the associative one.[102] A study of the exchange kinetics of (Na-DB24C8)⁺ in NM, showed that at low sodium concentrations, <10⁻³ M, the exchange mechanism is dissociative whereas at higher sodium concentrations the associative exchange predominates.[84,89] For (Na-DB18C6)⁺ in AN it was found[89] that the exchange mechanism was mainly dissociative whereas in NM there is competition between the two mechanisms. Recent studies by Brière and Detellier[88,103] again

192

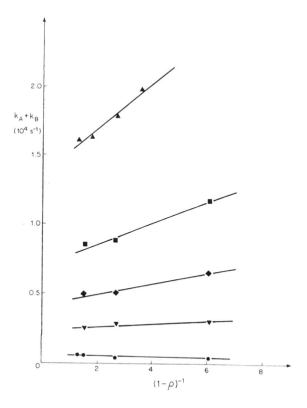

Figure 18. $(k_A + k_B)$ as a function of $(1-\rho)^{-1}$ in proplylene carbonate $(\rho = [18C6]_{tot} / [Na^+]_{tot})$ at these temperatures: ($\bullet$), 249.4; ($\blacktriangledown$), 268.8; ($\blacklozenge$), 279.2; ($\blacksquare$), 288.2; ($\blacktriangle$), 301.5 K. (With permission from reference 98.)

showed this solvent dependence of the mechanism. The dissociation of $(Na-15C5)^+$ and $(Na-B15C5)^+$ proceeds via the associative pathway in NM while in AN the dissociative mechanism for the exchange is preferred. Figures 19 and 20 again show plots of $(k_A + k_B)$ as a function of $(1-\rho)^{-1}$ where in AN (Figure 19) the absence of any residual associative mechanism is confirmed by an intercept equal to zero within experimental uncertainties, and in NM (Figure 20) the perfectly horizontal lines indicate the absence of any residual dissociative exchange. These two solvents have many similar properties but differ in donicity numbers (nitromethane = 2.7 and acetonitrile = 14.1).[12,52] It would thus seem that the dissociative mechanism is favoured in higher donicity solvents.

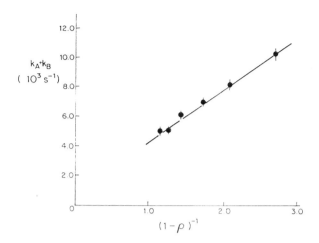

Figure 19. $(k_A + k_B)$ as a function of $(1-\rho)^{-1}$ in acetonitrile. $\rho = [B15C5]/[NaBPh_4]$. T = (258.0 ± 0.5) K. $[NaBPh_4] = 1.00 \times 10^{-2}$ M. The result of the linear regression is $(k_A + k_B) = (3.5 \pm 0.1) \times 10^3 (1-\rho)^{-1} + (9 \pm 3) \times 10^2$ (see equation 25). (With permission from reference 103.)

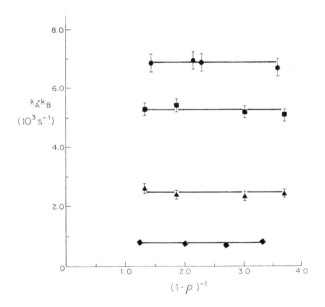

Figure 20. $(k_A + k_B)$ as a function of $(1-\rho)^{-1}$ in nitromethane. $\rho = [15C5]/[NaBPh_4]$. T = (301.5 ± 0.5) K. $[NaBPh_4] = 5.01 \times 10^{-2}$ (●); 4.00×10^{-2} (■); 2.00×10^{-2} (▲); 8.00×10^{-3} M (◆). (With permission from reference 103.)

TABLE 6

Kinetic Parameters for the dissociation of $(M-L)^+$ complexes.

	Ligand	Alkali Nucleus	Counter-anion	Solvent	Mech.[a]	Rate constant[b]
1.	15C5	^{7}Li	ClO_4^-	NM	II	$(1.9 \pm 0.1) \times 10^6$
2.	B15C5	^{7}Li	ClO_4^-	NM	II	$(1.88 \pm 0.08) \times 10^5$
3.	OM16C4	^{7}Li	ClO_4^-	NM	II	67
4.	C211	^{7}Li	I^-	H_2O	I	$(4.9 \pm 2.0) \times 10^3$
5.	C211	^{7}Li	ClO_4^-	PÝ	I	$(0.12 \pm 0.24) \times 10^3$
6.	C211	^{7}Li	ClO_4^-	DMSO	I	$(2.32 \pm 0.54) \times 10^4$
7.	C211	^{7}Li	ClO_4^-	DMF	I	$(1.30 \pm 0.33) \times 10^4$
8.	C211	^{7}Li	ClO_4^-	FOR	I	$(7.4 \pm 2.9) \times 10^3$
9.	C221	^{7}Li	ClO_4^-	PY	I	$(1.23 \pm 0.20) \times 10^6$
10.	C221	^{7}Li	ClO_4^-	AN	II	680 ± 42
11.	C221	^{7}Li	ClO_4^-	PC	II	892 ± 50
12.	C222	^{7}Li	ClO_4^-	AN	I	420 ± 14
13.	C222	^{7}Li	ClO_4^-	PC	I	507 ± 31
14.	C222	^{7}Li	ClO_4^-	AC	I	794 ± 42
15.	15C5	^{23}Na	BPh_4^-	NM	II	$(1.45 \pm 0.05) \times 10^5$
16.	B15C5	^{23}Na	BPh_4^-	NM	II	$(1.09 \pm 0.04) \times 10^5$
17.	B15C5	^{23}Na	BPh_4^-	NM	I	$(1.1 \pm 0.1) \times 10^2$
18.	B15C5	^{23}Na	BPh_4^-	AN	I	$(3.5 \pm 0.1) \times 10^3$
19.	B15C5	^{23}Na	BPh_4^-	AN	II	$(0.9 \pm 0.3) \times 10^5$
20.	18C6	^{23}Na	Br^-	MA	I	4.3×10^5
21.	18C6	^{23}Na	BPh_4^-	THF	I	53 ± 6
22.	18C6	^{23}Na	SCN^-	THF	II	$(9.16 \pm 0.12) \times 10^4$
23.	18C6	^{23}Na	SCN^-	MeOH	I	$(3.65 \pm 0.07) \times 10^4$
24.	18C6	^{23}Na	BPh_4^-	PC	II	$(1.30 \pm 0.07) \times 10^5$
25.	18C6	^{23}Na	BPh_4^-	THF/MeOH (0.6/0.4)	I	$(3.56 \pm 0.09) \times 10^3$
26.	18C6	^{23}Na	BPh_4^-	THF/PC (0.8/0.2)	I	$(9.15 \pm 0.13) \times 10^2$
27.	18C6	^{23}Na	BPh_4^-	THF/PC (0.4/0.6)	II	$(5.00 \pm 0.23) \times 10^4$
28.	18C6	^{23}Na	SCN^-	AC	I	4.34×10^5
29.	18C6	^{23}Na	SCN^-	MeOH	I	7.2×10^4
30.	18C6	^{23}Na	SCN^-	PY	I	1.03×10^4
31.	18C6	^{23}Na	BPh_4^-	AN	I	$(3.8 \pm 0.5) \times 10^3$
32.	18C6	^{23}Na	BPh_4^-	AN	II	$(2.6 \pm 0.4) \times 10^5$
33.	18C6	^{23}Na	BPh_4^-	PC	I	$(1.6 \pm 0.2) \times 10^3$
34.	18C6	^{23}Na	BPh_4^-	PC	II	$(6.9 \pm 0.3) \times 10^5$
35.	18C6	^{23}Na	BPh_4^-	AC	I	$(1.48 \pm 0.02) \times 10^4$
36.	18C6	^{23}Na	BPh_4^-	PY	I	$(5.3 \pm 0.2) \times 10^2$
37.	18C6	^{23}Na	BPh_4^-	PY	II	$(2.0 \pm 0.3) \times 10^4$
38.	DB18C6	^{23}Na	BF_4^-	AN	I	4×10^3
39.	DB18C6	^{23}Na	BPh_4^-	AN	I	4×10^3
40.	DB18C6	^{23}Na	BPh_4^-	NM	I	2×10^2

	Temp. K	ΔG^* kJ.mol^{-1}	ΔH^* kJ.mol^{-1}	ΔS^* J.mol^{-1}.K^{-1}	Reference
1.	300	37.4 ± 0.2	21 ± 1	-54 ± 3	96
2.	300	43.2 ± 0.2	32.2 ± 0.4	-36 ± 1	96
3.	373	79	-	-	82
4.	298	86.3 ± 0.8	86 ± 5	2 ± 12	92
5.	298	95 ± 5	80 ± 14	-52 ± 39	92
6.	298	82.4 ± 0.4	65 ± 2	-58 ± 6	92
7.	298	83.7 ± 0.4	65 ± 2	-65 ± 6	92
8.	298	87.0 ± 0.8	57 ± 3	-95 ± 8	92
9.	298	74.9 ± 0.4	54 ± 2	-62 ± 4	92
10.	298	56.90 ± 0.04	28.7 ± 0.6	-101 ± 2	93
11.	298	56.8 ± 0.4	26.2 ± 0.7	-109 ± 2	93
12.	298	58.28 ± 0.04	11.6 ± 0.5	-157 ± 2	93
13.	298	57.53 ± 0.08	15.0 ± 0.4	-143 ± 2	93
14.	298	56.32 ± 0.08	20.2 ± 0.8	-121 ± 3	93
15.	301.5	44.1 ± 0.1	21 ± 1	-76 ± 2	103
16.	301.5	44.8 ± 0.1	28 ± 3	-57 ± 3	88
17.	301.5	62.1 ± 0.3	-	-	88
18.	258	45.4 ± 0.4	-	-	103
19.	258	38.4 ± 0.8	-	-	103
20.	298	41	25	-54	100
21.	298	63.1 ± 0.4	47 ± 1	-52 ± 5	101
22.	298	44.66 ± 0.04	12 ± 2	-113 ± 4	101
23.	298	46.96 ± 0.08	38 ± 1	-30 ± 4	95
24.	298	43.82 ± 0.17	17 ± 2	-91 ± 7	95
25.	298	52.7 ± 0.1	39 ± 1	-47 ± 4	95
26.	298	56.08 ± 0.08	40.4 ± 0.4	-53 ± 2	95
27.	298	16.9 ± 0.1	36 ± 2	-35 ± 6	95
28.	298.2	40.8	36.7	-14.0	99
29.	298.2	45.3	53.6	27.8	99
30.	298.2	50.1	50.5	1.4	99
31.	301.5	53.2 ± 0.4	32 ± 2	-65 ± 8	98
32.	301.5	42.6 ± 0.4	-	-	98
33.	301.5	55.4 ± 0.3	54 ± 6	-5 ± 20	98
34.	301.5	40.2 ± 0.1	35 ± 1	-16 ± 3	98
35.	301.5	49.8 ± 0.1	44 ± 1	-21 ± 2	98
36.	301.5	58.1 ± 0.1	49 ± 4	-30 ± 10	98
37.	301.5	49.0 ± 0.4	-	-	98
38.	300	53	40 ± 2	-44 ± 8	89
39.	300	53	40 ± 2	-44 ± 8	89
40.	300	60	37 ± 3	-78 ± 8	89

	Ligand	Alkali Nucleus	Counter-anion	Solvent	Mech.[a]	Rate constant[b]
41.	DB18C6	^{23}Na	BPh_4^-	NM	II	3×10^4
42.	DB18C6	^{23}Na	PF_6^-	NM	I	2×10^2
43.	DB18C6	^{23}Na	PF_6^-	NM	II	4×10^4
44.	DB18C6	^{23}Na	SCN^-	DMF	I	1×10^5
45.	DB18C6	^{23}Na	SCN^-	DME	I	1.5×10^4
46.	DB18C6	^{23}Na	SCN^-	MeOH	I	1.4×10^4
47.	DC18C6 (cis-anti-cis)	^{23}Na	SCN^-	THF	II	$(4.0 \pm 0.2) \times 10^5$
48.	DC18C6 (cis-syn-cis)	^{23}Na	SCN^-	THF	II	$(2.2 \pm 0.2) \times 10^6$
49.	DC18C6 (cis-anti-cis)	^{23}Na	SCN^-	MeOH	I	5.2×10^4
50.	DC18C6 (cis-trans)	^{23}Na	SCN^-	MeOH	I	$> 9.6 \times 10^2$
51.	DA18C6	^{23}Na	SCN^-	THF	I	$(2.34 \pm 0.09) \times 10^3$
52.	DA18C6	^{23}Na	SCN^-	THF	I	$(1.74 \pm 0.05) \times 10^3$
53.	DA18C6	^{23}Na	SCN^-	THF	II	$(9.8 \pm 0.4) \times 10^4$
54.	DAMDB18C6	^{23}Na	SCN^-	DMF	I	1.9×10^5
55.	DNDB18C6	^{23}Na	SCN^-	DMF	I	2.0×10^5
56.	DB24C8	^{23}Na	BPh_4^-	NM($>10^{-3}$M)	II	$(2.8 \pm 0.2) \times 10^5$
57.	DB24C8	^{23}Na	BPh_4^-	NM($<10^{-3}$M)	I	19
58.	DB24C8	^{23}Na	PF_6^-	NM($>10^{-3}$M)	II	$(3.5 \pm 0.2) \times 10^5$
59.	DB24C8	^{23}Na	PF_6^-	NM($<10^{-3}$M)	I	67
60.	DB24C8	^{23}Na	I^-	NM	II	$(4.5 \pm 1.0) \times 10^5$
61.	DB24C8	^{23}Na	SCN^-	NM	II	$(1.3 \pm 2) \times 10^6$
62.	C211	^{23}Na	ClO_4^-	H_2O	I	47.6 ± 0.5
63.	C211	^{23}Na	ClO_4^-	DMF	I	12.1 ± 0.2
64.	C211	^{23}Na	ClO_4^-	DMSO	I	34.0 ± 0.7
65.	C221	^{23}Na	Br^-	MeOH/H_2O (25/75)	I	36.4
66.	C222	^{23}Na	Br^-	EDA	I	165 ± 5
67.	C222	^{23}Na	Br^-	PY	I	1.14 ± 0.09
68.	C222	^{23}Na	Br^-	THF	I	8.03 ± 0.27
69.	C222	^{23}Na	Br^-	H_2O	I	147.4 ± 2.6
70.	C21C$_5$	^{23}Na	ClO_4^-	AN	I	84.8 ± 1.6
71.	C21C$_5$	^{23}Na	ClO_4^-	PC	I	19.4 ± 0.5
72.	C21C$_5$	^{23}Na	ClO_4^-	AC	I	878 ± 6
73.	C21C$_5$	^{23}Na	ClO_4^-	MeOH	I	1800 ± 50
74.	C21C$_5$	^{23}Na	ClO_4^-	DMF	I	$(2.88 \pm 0.03) \times 10^4$
75.	C21C$_5$	^{23}Na	ClO_4^-	PY	I	93.5 ± 0.5
76.	Enniatin B	^{23}Na	SCN^-	MeOH	I	$> 3.1 \times 10^2$
77.	Valinomycin	^{23}Na	SCN^-	MeOH	I	$> 1.2 \times 10^3$
78.	Nonactin	^{23}Na	SCN^-	MeOH	I	$> 1.0 \times 10^3$
79.	Monensin	^{23}Na	Mon^-	MeOH	I	63
80.	Monensin	^{23}Na	Mon^-	MeOH	I	83 ± 4

	Temp. K	ΔG^* kJ.mol^{-1}	ΔH^* kJ.mol^{-1}	ΔS^* J.mol^{-1}.K^{-1}	Reference
41.	300	48	-	-	89
42.	300	60	-	-	89
43.	300	47	-	-	89
44.	298	45	51	21	90
45.	298	50	53	12	97
46.	298	50	47	-10	97
47.	298	41.06 ± 0.04	8.8 ± 0.8	-109 ± 3	102
48.	298	36.88 ± 0.04	8 ± 2	-98 ± 5	102
49.	298	46.0	32	46	97
50.	294	< 55			14
51.	298	53.74 ± 0.13	5.4 ± 0.8	-162 ± 2	102
52.	273	49.72 ± 0.08	5.9 ± 0.8	-161 ± 2	102
53.	233	33.70 ± 0.04	37 ± 2	10 ± 4	102
54.	298	43	-	-	97
55.	298	43	-	-	97
56.	295	41 ± 3	31 ± 3	-32 ± 10	84
57.	295	$\cong 65$	-	-	84
58.	300	40 ± 2	30 ± 2	-37 ± 10	89
59.	300	$\cong 63$	-	-	89
60.	294	40.7	-	-	16
61.	294	38.5	-	-	16
62.	298.2	63.4 ± 0.1	67.2 ± 0.3	12.6 ± 0.7	106
63.	298.2	66.8 ± 0.1	83.5 ± 0.5	55.8 ± 1.2	106
64.	298.2	64.2 ± 0.1	69.5 ± 0.4	17.4 ± 1.2	106
65.	298	64.1	66.0 ± 1.2	6.5 ± 3.8	108
66.	298	60.50 ± 0.08	51.5 ± 0.8	-32 ± 3	105
67.	298	72.65 ± 0.02	56.9 ± 0.8	-53 ± 3	105
68.	298	67.96 ± 0.08	57.7 ± 0.8	-34 ± 3	105
69.	298	60.60 ± 0.04	67.4 ± 0.8	22 ± 3	105
70.	298.2	62.0 ± 0.1	57.9 ± 0.7	-13.8 ± 2.1	107
71.	298.2	65.7 ± 0.1	70.3 ± 0.5	15.3 ± 1.4	107
72.	298.2	56.2 ± 0.1	54.4 ± 0.4	-6.1 ± 1.2	107
73.	298.2	54.4 ± 0.1	44.9 ± 0.1	-31.9 ± 0.4	107
74.	298.2	46.6 ± 0.1	40.0 ± 0.1	-25.3 ± 0.5	107
75.	298.2	61.8 ± 0.1	62.8 ± 0.2	3.3 ± 0.5	107
76.	294	< 58	-	-	14
77.	294	< 55	-	-	14
78.	294	< 55	-	-	14
79.	298	63	43.0	-66.2	109
80.	298	62.0 ± 0.2	46.8 ± 0.7	-51.1 ± 2.5	80

	Ligand	Alkali Nucleus	Counter-anion	Solvent	Mech.[a]	Rate constant[b]
81.	18C6	^{39}K	AsF_6^-	AC	II	$(4.1 \pm 0.9) \times 10^5$
82.	18C6	^{39}K	AsF_6^-	AC/1,4DO (80/20)	II	$(5.7 \pm 1.1) \times 10^5$
83.	18C6	^{39}K	AsF_6^-	1,3DO	II	$(1.65 \pm 0.04) \times 10^4$
84.	18C6	^{39}K	I$^-$	MeOH	II	$(6.8 \pm 2.7) \times 10^5$
85.	18C6	^{39}K	I$^-$	MeOH	I	$(6.8 \pm 2.7) \times 10^4$
86.	DB18C6	^{39}K	I$^-$	MeOH	I	610
87.	DB18C6	^{87}Rb	SCN$^-$	MeOH	I	$> 10^4$
88.	18C6	^{133}Cs	BPh_4^-	PY	I	$(9.5 \pm 0.2) \times 10^3$
89.	DC18C6 (cis-syn-cis)	^{133}Cs	BPh_4^-	PC	I	$(1.1 \pm 0.1) \times 10^4$
90.	DA18C6	^{133}Cs	SCN$^-$	NM	II	$(2.31 \pm 0.23) \times 10^7$
91.	DB21C7	^{133}Cs	SCN$^-$	AC	II	$(9.4 \pm 3) \times 10^3$
92.	DB21C7	^{133}Cs	SCN$^-$	MeOH	II	$(2.7 \pm 0.4) \times 10^4$
93.	DB24C8	^{133}Cs	SCN$^-$	AC	II	$(7.4 \pm 1.0) \times 10^4$
94.	DB24C8	^{133}Cs	SCN$^-$	MeOH	II	$(5.6 \pm 2.0) \times 10^4$
95.	DB30C10	^{133}Cs	SCN$^-$	NM	I	$(1.7 \pm 0.1) \times 10^5$
96.	DB30C10	^{133}Cs	SCN$^-$	AN	II	$(4.2 \pm 0.3) \times 10^7$
97.	DB30C10	^{133}Cs	SCN$^-$	PC	II	$(5.4 \pm 0.2) \times 10^6$
98.	DB30C10	^{133}Cs	SCN$^-$	MeOH	II	$(7.6 \pm 0.2) \times 10^6$
99.	C221	^{133}Cs	SCN$^-$	NM	I	$(3.90 \pm 0.06) \times 10^2$
100.	C221	^{133}Cs	SCN$^-$	AN	I	$(2.17 \pm 0.05) \times 10^3$
101.	C221	^{133}Cs	SCN$^-$	MeOH	I	$(3.29 \pm 0.11) \times 10^3$
102.	C221	^{133}Cs	SCN$^-$	DMF	I	$(4.49 \pm 0.17) \times 10^4$
103.	C222	^{133}Cs	BPh_4^-	DMF	I	$(9.0 \pm 0.9) \times 10^6$
104.	C222B	^{133}Cs	BPh_4^-	PC	I	344 ± 38

(a) Mechanism I: dissociative exchange (equation 23); Mechanism II; associative exchange (equation 24). (b) In mechanism I units for k: s^{-1}; in mechanism II units for k: $M^{-1}s^{-1}$.

	Temp. K	ΔG^* kJ.mol^{-1}	ΔH^* kJ.mol^{-1}	ΔS^* J.mol^{-1}.K^{-1}	Reference
81.	298	41.0 ± 0.4	36 ± 2	-17 ± 8	111
82.	298	41.1 ± 0.4	55 ± 2	50 ± 8	111
83.	298	49.02 ± 0.04	68 ± 1	63 ± 4	111
84.	298	40.0 ± 0.8	36 ± 3	-13 ± 12	111
85.	298	45.6 ± 0.8		-33 ± 3	111
86.	239	45	-	-	110
87.	223	< 37	-	-	110
88.	298	50.3 ± 0.1	33 ± 2	-60 ± 2	56
89.	298	49.9 ± 0.2	33 ± 2	-59 ± 8	57
90.	283	27.3 ± 0.1	21.3 ± 0.4	-19 ± 3	94
91.	220	36 ± 1	34 ± 2	-11 ± 10	114
92.	220	34.7 ± 0.4	25 ± 1	-44 ± 6	114
93.	220	32.6 ± 0.4	28 ± 1	-21 ± 6	114
94.	220	34 ± 1	12 ± 2	-97 ± 10	114
95.	253	36.3 ± 0.1	13.8 ± 0.4	-90 ± 2	113
96.	253	27.8 ± 0.2	34 ± 1	26 ± 6	113
97.	253	29.0 ± 0.1	46 ± 2	69 ± 6	113
98.	253	28.4 ± 0.1	38 ± 1	36 ± 3	113
99.	298	58.2 ± 0.1	-	-	94
100.	298	53.82 ± 0.04	31 ± 1	-75 ± 4	94
101.	298	52.4 ± 0.1	25 ± 1	-93 ± 3	94
102.	298	46.50 ± 0.04	21.7 ± 0.4	-83 ± 1	94
103.	298	33.3 ± 0.3	54 ± 1	69 ± 4	53
104.	298	41.4 ± 0.3	60 ± 3	63 ± 8	57

Sodium cations also interact with cryptands and natural ionophores. Using ^{23}Na NMR, Ceraso and Dye[104] in 1973 were the first to observe two separate peaks for the two sodium sites in a mixture of solvated sodium and sodium complexed to C222 in EDA. In a subsequent study[105], this system was studied in other solvents where it was shown that the exchange proceeded via the dissociative mechanism. For the systems (Na-C211)$^{+}$ [106], (Na-C21C$_5$)$^{+}$ [107] and (Na-C222)$^{+}$ [104,105] in several nonaqueous solvents, the exchange between solvated and complexed cation all proceed via the dissociative mechanism. Studies of natural ionophores with sodium have yielded relatively little quantitative kinetic information.[14,80,108,109] The exchange of Na^{+} with the natural ionophore monensin was found to proceed via the dissociative mechanism.[109] The dissociative rate constant, k_{-1}, and the activation parameters for this dissociation was found in MeOH.[80] Relevant information is listed in Table 6.

Despite the great importance of potassium in biological systems, very little use has been made of ^{39}K NMR because of its low gyromagnetic ratio and hence its low NMR sensitivity (see Table 1). Nevertheless, ^{39}K NMR has yielded information about the exchange kinetics between solvated potassium and potassium complexed by DB18C6 in MeOH and a measurable dissociative exchange rate, k_{-1}= 610 s^{-1} at 239 K, was reported.[110] In this same study, the rate of exchange of rubidium in the solvated and complexed (DB18C6) states in MeOH was too fast to be measured by ^{87} Rb NMR, even at 223 K. It has also been possible to obtain mechanistic information with ^{39}K NMR for potassium in the solvated and complexed to 18C6 sites. In the solvents MeOH, AC, 1,3 dioxalane and a AC/1,4 dioxane mixture they found that the associative mechanism for the exchange predominated.[111] In MeOH[95,99] and AC[98,99] it has been shown that the predominant mechansim for the exchange is the dissociative one when sodium is the interacting cation. It is known, from solid crystal structures[8] and molecular mechanics studies[112], that the positions of these two cations in the cavity of 18C6 differ when they are complexed; K^{+} lies in the center of the crown ring whereas complexation with the smaller Na^{+} results in slight distortions in the symmetry of the complex. This could explain the discrepancy in the exchange mechanisms.

Popov and co-workers[53,57,94,113,114] have used ^{133}Cs NMR to study the exchange kinetics with the cesium cation. Since cesium is the largest of the stable alkali metal cations, ligands such as DB21C7, DB24C8, DB30C10, C221 and C222, all having cavities large enough to accommodate this cation, have been used. With the crown ethers as ligands[113,114] the mechanism for the exchange between the two cesium sites is generally found to be associative and the free energy barrier for the exchange process appears to be independent of the solvent. The large crown ether DB30C10 wraps itself around the cesium cation permitting minimal contact with the solvent once complexed. In the associative exchange, two cations would be attached to the ligand in the transition state which would also involve a more "open" type structure permitting more solvent interactions. In this case, the transition state would be further stabilized by solvents with higher donicities. This explains why Shamsipur and Popov[113] found the associative mechanism to be predominant for (Na-DB30C10)$^{+}$ in AN, PC and MeOH but in NM the dissociative mechanism predominated. It was noted[114] that the predominant exchange mechanism varies from that of either the dissociative or associative process for the Na^+ cation to primarily the associative process for the Cs^+ cation as one goes through the series Na^+, K^+, Cs^+. This trend may be due to the decrease in charge density as one goes to the larger cation which would minimize the charge-charge repulsion in the transition state of the associative exchange. The dissociation of $(Na-C221)^+$ in NM, AN, MeOH and DMF proceeded via the dissociative exchange mechanism.[99] Again there was a correlation between the donicity of the solvent and values for the activation parameters.

As more data is becoming available, the task of understanding the underlying nature of kinetic activity is becoming clearer. It is evident that this is not a simple endeavour and many factors such as the ratio of the cation diameter to the crown ether, the nature of the solvent and the possible participation of counteranions may be implicated.

VI. TRANSPORT STUDIES

This subject has been extensively reviewed very recently by Grandjean and Laszlo.[115] Essentially, there are two strategies: (a) the direct observation of the cation or anion moving from one compartment to the other, and (b) The observation of a nucleus from the membrane itself, whose NMR characteristics will depend on the presence of the cation under study. This review will focus on the first strategy which, for example, involves ^7Li, ^{23}Na, ^{39}K or ^{35}Cl. The presence of liposomes or of biological membranes in water permits a distinction between two regions of the space, the inner and the outer compartments. However, the chemical shifts of cations occupying one or the other compartments are usually identical, and this would not permit their differentiation. Some anionic paramagnetic species have been introduced as shift reagents for alkali metal cations[116-117], so that well separated signals could be obtained for cations in the inner and outer compartments. This strategy has been successfully chosen by Fernandez et al. who have followed the Na^+ and K^+ fluxes mediated by Lasalocid A across erythrocyte membranes.[119] While quantitative results could be obtained from ^{23}Na or ^{35}Cl chemical shifts, peak areas (^{23}Na) or signal intensities (^{35}Cl), only qualitative results could be obtained in the case of ^{39}K, due to the lack of resolution of the intra- and extracellular $^{39}K^+$ signals.[119] Pettegrew et al.[120] have followed by ^7Li NMR the uptake of lithium cations in normal human erythrocytes, in the absence of any ionophore. The intra- and extra-cellular lithium cation signals were differentiated by the presence of a dysprosium tripolyphosphate shift reagent in the extra-cellular compartment.

The observation of a longitudinal relaxation time much longer than the transverse relaxation time for intracellular $^7Li^+$ ($T_1 \approx 5$ s; $T_2 \approx 0.15$ s) is particularly interesting and has been interpreted as being either due to some fixation of Li^+ on the membrane-associated cytoskeleton, or due to slow diffusion through heterogeneous electrostatic field gradients.[120] Another study of lithium transport in human erythrocytes[121] has appeared almost simultaneously with the study of

Pettegrew et al. in which a shift reagent, 3 mM in concentration, was sufficient to achieve chemical shift separation between the $^7Li^+$ signals of the intra- and extra-cellular cations. This is shown in Figure 21.

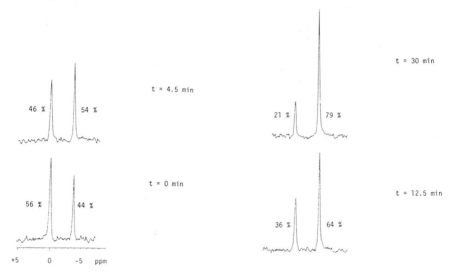

Figure 21. Time dependence of the $^7Li^+$ NMR signal of a suspension of red blood cells that had been incubated with LiCl. The suspension medium contained K^+ as the extra-cellular cation, and a dyprosium shift reagent, inter alia. The time indicated is the time elapsed after monensin was added to the cell suspension. (With permission from reference 121.)

This experiment was done on red blood cells which had been previously incubated with LiCl, and contained 56% of the total Li^+ concentration. The extra-cellular $^7Li^+$ signal was shifted to lower frequencies ($\delta \approx -4$ ppm) by the presence of dysprosium triphosphate. After incubation, the cells were immersed in a medium containing only K^+, in the presence of the ionophore monensin. A passive transport, completed in about 75 min, was observed (see Figure 21).

The potential transport properties of a novel organocobalt ionophore[122] have been studied by following the transport rates of $^{23}Na^+$ and $^7Li^+$ across artificial phospholipid vesicles.[123] Again, a dysprosium shift reagent was used and two well-separated $^7Li^+$ signals were obtained.

A ^7Li magnetization transfer method has also been applied to the study of lithium transport through phospholipid bilayers. Since the system consisted only of two sites (Li^+ *in* and *out*), Riddell et al.[124] used a simple three pulse sequence to acquire the data. A similar simple sequence had been described and discussed previously by Robinson et al.[125] for the measurement of the rate of phosphoryl group transfer catalyzed by phosphoglyceromutase, and has been very recently applied to the study of the $^{23}Na^+$ transport across lipid vesicle membranes in the presence of the ionophore monensin.[126] Figure 22 shows the inversion transfer from $^{23}Na^+$ *out* to $^{23}Na^+$ *in*.

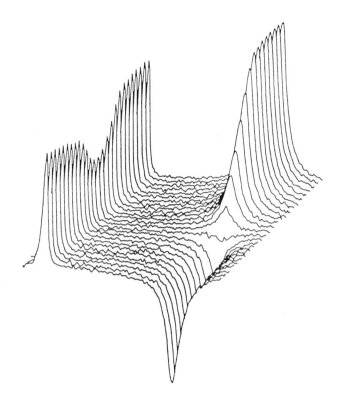

Figure 22. ^{23}Na inversion transfer from Na$^+$ *out* to Na$^+$ *in* through lipid vesicle membranes. (With permission from reference 126.)

These few examples demonstrate the potentialities of the direct NMR observation of the transported cation. Studies on systems as close as possible to *in vivo* systems, should continue to develop considerably in the next few years.

However, as it has been pointed out[127,128] a drawback to this approach is the possibility of losing some signal from intra-cellular cations. Several factors could be at the source of that loss, for example a long-lived cation immobilization on a large entity. The same problem could also occur if only a small population of the cation was bound for a short time on slowly-reorienting macromolecules. Since ^{7}Li, ^{23}Na, and ^{39}K are quadrupolar nuclei having a spin quantum number I = 3/2, their resonance signal is not Lorentzian when the corresponding cations are transiently bound to a slowly-reorienting entity; it consists of the superposition of two Lorentzian lines, corresponding to a double-exponential relaxation.[129] By comparing spectra obtained at different fields, this last hypothesis could be tested and eventually invalidated in some favorable cases, since the linewidth of the narrow component, which is mainly observed, depends on the Larmor frequency.

VII. SOLID STATE NMR

Static and magic angle spinning (MAS) alkali metal NMR spectra of ionophore complexes have been reported only scarcely in the literature. For example, Saitô and his co-workers[130,131] have puplished a series of MAS ^{23}Na NMR spectra of sodium complexed with naturally occurring ionophores in the solid state (see Figure 23).

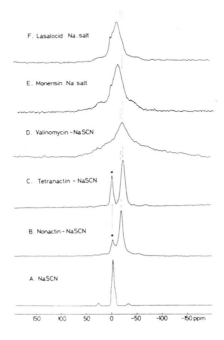

F. Lasalocid Na salt

E. Monensin Na salt

D. Valinomycin – NaSCN

C. Tetranactin – NaSCN

B. Nonactin – NaSCN

A. NaSCN

150   100   50   0   -50   -100   -150 ppm

Figure 23. 79.35 MHz MAS ^{23}Na NMR spectra of the sodium complexes with naturally occurring ionophores in the solid state. The peaks asterisked in B and C are from uncomplexed NaSCN. (With permission from reference 130.)

Typically, the ^{23}NaSCN signals were displaced 20 ppm upfield upon complexation with the ionophores. This shift is similar to the one observed after complexation by crypands! One year later, the same group reported a comparative study of the ^{23}Na chemical shifts in solution and in the solid state for Na$^+$ complexed by several natural and synthetic ionophores.[131] Except for in chloroform, where the solution and solid state chemical shifts were almost identical, the ^{23}Na signals were displaced downfield in solution as compared to the solid state. These shifts could possibly originate from conformational variations of the ligand, from the participation of solvent molecules or of the associated counteranion in the first coordination sphere of the cation.

Dye and co-workers[31,133] have extensively studied the formation of alkali metal anions.

These "alkalides" have been characterized by a number of methods, including alkali metal NMR in solution and in the solid state. Figure 24 shows the non-spinning and the magic-angle spinning ^{23}Na NMR spectra of (Na$^+$-C222,Na$^-$).[132]

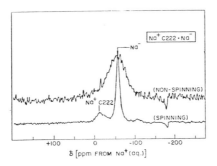

Figure 24. Magic-angle sample spinning ^{23}Na NMR spectrum of (Na$^+$-C222,Na$^-$). Also shown is the nonspinning spectrum. Data was collected with 276 sweeps for spinning. (With permission from reference 132.)

High speed rotation of the sample at the magic angle narrowed the signals in such a way that the two sodium species, Na$^-$ and (Na-C222)$^+$ became observable.[132] Mixed metallic species (M$^+$-L,Na$^-$) in which M is an alkali metal and L a coronand or a cryptand could be synthesized. The characterization of these complexes by alkali metal MAS NMR was particularly helpful as it is demonstrated in Figure 25.

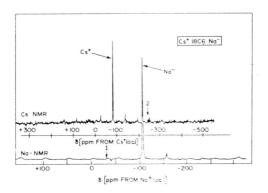

Figure 25. Magic-angle sample spinning ^{23}Na and ^{133}Cs NMR spectra of (Cs$^+$-18C6, Na$^-$). Arrow 1 indicates the chemical shift of Na$^+$ in (Na$^+$-18C6-SCN) while 2 shows the chemical shift of Cs$^-$ in solution. (With permission from reference 132.)

In the case of the complex (Cs^+-18C6, Na^-) only the cesium cation and the sodium anion could be observed by ^{133}Cs and ^{23}Na NMR respectively. This pointed, without any ambiguity, to the formation of (Cs^+-18C6, Na^-).[132] More recently, Rb^- could be detected in crystalline rubidides by ^{87}Rb NMR.[31]

VIII. CONCLUSION

In this review, we have given a brief overview of the many facets through which alkali metal NMR has been applied to the understanding of the complexation of alkali metal cations by ionophores. This field should continue to enjoy popularity in the next years: it provides a very powerful tool for the study of kinetics and mechanisms of reaction, as well as invaluable information on the structures and the microdynamics of the complexes. With the availability of higher field instrumentation and the development of new pulse sequences, broader ranges of kinetic values should be reached. Transport across complex systems, such as biological membranes, should be studied more routinely in the next ten years. In vivo studies are in their embryonic state and we can foresee a dramatic increase of their applications in the near future.

IX. ACKNOWLEDGMENTS

The work from our group, described in this review, was funded by the Natural Science and Engineering Research Council of Canada. H.P.G. wishes to thank FCAR (Fonds pour la formation de Chercheurs et l'Aide à la Recherche) for a postgraduate scholarship.

REFERENCES

1.   F.G. Riddell, S. Arumugam, P.J. Brophy, B.G. Cox, M.C.H. Payne, and T.E. Southon, J. Am. Chem. Soc., 110 (1988) 734.

2.   (a) C.J. Pedersen, J. Am. Chem. Soc., 89 (1967) 7017. (b) C.J. Pedersen, Angew. Chem. Int. Ed. Eng., 27 (1988) 1021.

3.   (a) B. Dietrich, J.M. Lehn and J.P Sauvage, Tetrahedron Lett., (1969) 2885. (b) J.M. Lehn, Angew. Chem. Int. Ed. Eng., 27 (1988) 89. (c) A.I. Popov and J.M. Lehn, "Coordination Chemistry of Macrocyclic Compounds", Chapter 9, G.A. Melson ed., Plenum Press, New York, 1979.

4.   (a) E.P. Kyba, M.G. Siegel, L.R. Sousa, G. Sogah and D.J. Cram, J. Am. Chem. Soc., 95 (1973) 2691. (b) D.J. Cram, Angew. Chem. Int. Ed. Eng., 27 (1988) 1009.

5.   (a) E. Buncel, H.S. Shin, Ng. van Truong, R.A.B. Bannard and J.G. Purdon, J. Phys. Chem., 92 (1988) 4176. (b) E. Buncel, H.S. Shin, R.A.B. Bannard and J.G. Purdon, Can. J. Chem., 62 (1984) 926. (c) E. Buncel, H.S. Shin, R.A.B. Bannard, J.G. Purdon and B.G. Cox, Talanta, 31 (1984) 585. (d) B.G. Cox, E. Buncel, H.S. Shin, R.A.B. Bannard and J.G. Purdon, Can. J. Chem., 121 (1984) 1.

6.   C. Detellier, "NMR of Newly Accessible Nuclei", Vol. 2, Chapter 5, P. Laszlo ed., Academic Press, New York, 1983.

7.   C. Brevard and P. Granger, "Handbook of High Resolution Multinuclear NMR", Wiley, New York, 1981.

8.   J.D. Dunitz, M. Dobler, P. Seiler and R.P. Phizackerley, Acta. Cryst., B30 (1974) 2733.

9.   D. Live and S.I. Chan, J. Am. Chem. Soc., 98 (1976) 3769.

10. J. Homer, J. Magn. Reson., 57 (1984) 171.

11. R.H. Erlich, E. Roach and A.I. Popov, J. Amer. Chem. Soc., 92 (1970) 4989.

12. V. Gutmann and E. Wychera, Inorg. Nulc. Chem. Lett., 2 (1966) 257.

13. A.M. Grotens, J. Smid and E. de Boer, Chem. Comm., (1971) 759.

14. D.H. Haynes, B.C. Pressman and A. Kowalsky, Biochem., 10 (1971) 852.

15. H.P. Graves and C. Detellier, submitted for puplication.

16. H.D.H. Stöver and C. Detellier, J. Phys. Chem., 93 (1989) 3174.

17. P.G. Gertenbach and A.I. Popov, J. Am. Chem. Soc., 97 (1975) 4738.

18. Y.M. Cahen, J.L. Dye, and A.I. Popov, Inorg. Nucl. Chem. Letters, 10 (1974) 899.

19. Y.M. Cahen, J.L. Dye and A.I. Popov, J. Phys. Chem., 79 (1975) 1289-1291.

20. E. Kauffmann, J.L. Dye, J.M. Lehn and A.I. Popov, J. Am. Chem Soc., 102 (1980) 2274.

21. E. Mei, L. Liu, J.L. Dye and A.I. Popov, J. Soln. Chem., 6 (1977) 771.

22. S. Khazaeli, J.L. Dye and A.I. Popov, Spectrochim. Acta, 39A (1983) 19.

23. C. Cambillau and M. Ouvrevitch, J.C.S. Chem. Comm., (1981) 996.

24. A. Cornelis, P. Laszlo and C. Cambillau, J. Chem. Res. (S), (1978) 462.

25. J. Desroches, H. Dugas, M. Bouchard, T.M. Fyles and G.D. Robertson, Can. J. Chem., 65 (1987) 1513.

26. (a) A. Ricard, F. Lafuma and C. Quivoron, Polymer, 23 (1982) 907. (b) A. Ricard and F. Lafuma, Polymer, 27 (1986) 133.

27. F.W. Wehrli, J. Magn. Reson., 25 (1977) 575.

28. J.L. Dye, C.W. Andrews and J.M. Ceraso, J. Phys. Chem., 79 (1975) 3076.

29. M.L. Tinkham and J.L. Dye, J. Am. Chem. Soc., 107 (1985) 6129.

30. R.C. Phillips, S. Khazaeli and J.L. Dye, J. Phys. Chem., 89 (1985) 606.

31. M.L. Tinkham, A. Ellaboudy, J.L. Dye and P.B. Smith, J. Phys. Chem., 90 (1986) 14.

32. A.S. Ellaboudy, N.C. Pyper and P.P. Edwards, J. Am. Chem. Soc., 110 (1988) 1618.

33. P.P. Edwards, A.S. Ellaboudy and D.M. Holton, Nature, 317 (1985) 242.

34. G.C. Levy, R.L. Lichter, and G.L. Nelson, "Carbon-13 Nuclear Magnetic Resonance Spectroscopy" 2nd Ed., Chapter 8, J. Wiley & Son, New York, 1980.

35. D.A. Wright, D.E. Axelson, G.C. Levy, in "Topics in Carbon-13 NMR Spectroscopy", Ed. by G.C. Levy, Vol. 3, Chapter 2, J. Wiley & Son, New York, 1979.

36. R.K. Harris, "Nuclear Magnetic Resonance Spectroscopy", Longman Sc. Tech., Harlow, 1987.

37. D.J. Craik and G.C. Levy "Topics in Carbon-13 NMR Spectroscopy", Ed. by G.C. Levy, Vol. 4, Chapter 9, J. Wiley & Son, New York, 1984.

38. L. Echegoyen, A. Kaifer, H. Durst, R.A. Schultz, D.M. Dishong, D.M. Goli and G.W. Gokel, J. Am. Chem. Soc., 106 (1984) 5100 .

39. B. Eliasson, K.M. Larsson and J. Kowalewski, J. Phys. Chem., 89 (1985) 258.

40. M.C. Fedarko, J. Magn. Resonance, 12 (1973) 30.

41. A.C. Coxon, D.A. Laidler, R.B. Pettman and J.F. Stoddart, J. Am. Chem. Soc., 100 (1978) 8260.

42. P. Laszlo, Angew. Chem. Int. Ed. Engl., 17 (1978) 254.

43. J.P. Kintzinger and J.M. Lehn, J. Am. Chem. Soc., 96 (1974) 3313.

44. C. Hall, R.E. Richards, G.N. Schulz and R.R. Sharp, Mol. Phys., 16 (1969) 529.

45. C. Deverell, Mol. Phys., 16 (1969) 491.

46. A. Delville, C. Detellier, A. Gerstmans and P. Laszlo, J. Magn. Reson., 42 (1981) 14.

47. M. Bisnaire, C. Detellier and D. Nadon, Can. J. Chem., 60 (1982) 3071.

48. K.J. Neurohr, T. Drakenberg, S. Forsén and H. Lilja, J. Magn. Resonance, 51 (1983) 460.

49. C. Detellier and M. Robillard, Can. J. Chem., 65 (1987) 1684.

50. H.D.H. Stöver, L.J. Maurice, A. Delville and C. Detellier, Polyhedron, 4 (1985) 1091.

51. K.M. Brière, H.P. Graves, T.S. Rana, L.J. Maurice and C. Detellier, J. Phys. Org. Chem., in press.

52. J.D. Lin and A.I. Popov, J. Am. Chem. Soc., 103 (1981) 3773.

53. E. Mei, A.I. Popov and J.L. Dye, J. Amer. Chem. Soc., 99 (1977) 6532.

54. G. Rounaghi and A.I. Popov, Polyhedron, 5 (1986) 1935.

55. E. Mei, J.L. Dye and A.I. Popov, J. Am. Chem. Soc., 98 (1976) 1619.

56. E. Mei, J.L. Dye and A.I. Popov, J. Am. Chem. Soc., 99 (1977) 5308.

57. E. Mei, A.I. Popov and J.L. Dye, J. Phys. Chem., 81 (1977) 1677.

58. S. Khazaeli, J.L. Dye and A.I. Popov, J. Phys. Chem., 87 (1983) 1830

59. S. Khazaeli, A.I. Popov and J.L. Dye, J. Phys. Chem., 86 (1982) 5018.

60. R.D. Boss and A.I. Popov, Inorg. Chem., 25 (1986) 1747.

61. R.D. Boss and A.I. Popov, Inorg. Chem., 24 (1985) 3660.

62. M. Shamsipur, G. Rounaghi and A.I. Popov, J. Soln. Chem., 9 (1980) 701.

63. G. Rounaghi and A.I. Popov, Inorg. Chim. Acta., 114 (1986) 145.

64. M. Shamsipur and A.I. Popov, J. Am. Chem. Soc., 101 (1979) 4051.

65. H.D.H. Stöver, M. Robillard and C. Detellier, Polyhedron, 6 (1987) 577.

66. H.D.H. Stöver, A. Delville and C. Detellier, J. Amer. Chem. Soc., 107 (1985) 4167.

67. A.J. Smetana and A.I. Popov, J. Soln. Chem., 9 (1980) 183.

68. A. Hourdakis and A.I. Popov, J. Soln. Chem., 6 (1977) 299.

69. J. Bouquant, A. Delville, J. Grandjean and P. Laszlo, J. Am. Chem. Soc., 104 (1982) 686.

70. J. Grandjean, P. Laszlo, W. Offermann and P.L. Rinaldi, J. Am. Chem. Soc., 103 (1981) 1380.

71. F. Vögtle and E. Weber, "The Chemistry of Ethers, Crown Ethers, Hydroxyl Groups, and their Sulfur Analogs", Chapter 2, S. Patai ed., Springer Verlab Publishers, Berlin, Heidelberg, New York, 1982.

72. R.M. Izatt, J.S. Bradshaw, S.A. Nelson, J.D. Lamb and J.J. Christensen, Chem. Rev., 85 (1985) 271.

73. D. Lamb, R.M. Izatt, J.J. Christensen and D.J. Eatough, "Coordination Chemistry of Macrocyclic Compounds", Chapter 3, G.A. Melson ed., Plenum Press Publishers, New York, 1979.

74. A.V. Bajaj and N.S. Poonia, Coord. Chem. Rev., 87 (1988) 55.

75. P.B. Chock, Proc. Nat. Acad. Sci. USA, 69 (1972) 1939.

76. (a) G.W. Liesegang, M.M. Farrow, N. Purdie and E.M. Eyring, J. Am. Chem. Soc., 98 (1976) 6905. (b) G.W. Liesegang, M.M. Farrow, F. Arce Vazquez, N. Purdie and E. M. Eyring, J. Am. Chem. Soc., 99 (1977) 3240. (c) L.J. Rodriguez, G.W. Liesegang, R.D. White, M.M. Farrow, N. Purdie and E.M. Eyring, J. Phys. Chem., 81 (1977) 2118.

77. (a) H. Farber and S. Petrucci, J. Phys. Chem., 85 (1981) 1396. (b) C.C. Chen and S. Petrucci, J. Phys. Chem., 86 (1982) 2601. (c) C.C. Chen, W. Wallace, E.M. Eyring and S. Petrucci, J. Phys. Chem., 88 (1984) 2541. (d) K.J. Maynard, D.E. Irish, E.M. Eyring and S. Petrucci, J. Phys. Chem., 88 (1984) 729. (e) W. Wallace, E.M. Eyring and S. Petrucci, J. Phys. Chem., 88 (1984) 6353. (f) Wallace, C.C. Chen, E.M. Eyring and S. Petrucci, J. Phys. Chem., 89 (1985) 1357. (g) H. Richman, Y. Harada, E.M. Eyring and S. Petrucci, J. Phys. Chem., 89 (1985) 2373.

78. (a) G.W. Gokel, L. Echegoyen, M. Sook Kim, E.M. Eyring and S. Petrucci, Biophys. Chem., 26 (1987) 225. (b) L. Echegoyen, G.W. Gokel, M. Sook Kim, E.M. Eyring and S. Petrucci, J. Phys. Chem., 91 (1987) 3854.

79. (a) B.G. Cox and H. Scneider, J. Am. Chem. Soc., 99 (1977) 2809. (b) V.M. Loyola, R. Pizer and R.G. Wilkins, J. Am. Chem. Soc., 99 (1977) 7185. (c) B.G. Cox, H. Schneider and J. Stroka, J. Am. Chem. Soc., 100 (1978) 4746. (d) B.G. Cox, J. Garcia-Rosas and H. Schneider, J. Am. Chem. Soc., 103 (1981) 1054. (e) B.G. Cox, Inorg. Chim. Acta., 64 (1982) C263.

80. E. Amat, B.G. Cox and H. Schneider, J. Magn. Reson., 71 (1987) 259.

81. J.M. Lehn and M.E. Stubbs, J. Am. Chem. Soc., 96 (1974) 4011.

82. K.M. Aalmo and J. Krane, Acta Chem. Scand., A36 (1982) 219.

83. K. Ishihara, H. Muira, S.F. Funahashi and M. Tanaka, Inorg. Chem., 27 (1988) 1706.

84. A. Delville, H.D.H. Stöver and C. Detellier, J. Am. Chem. Soc., 107 (1985) 4172.

85. J.L. Dye, Prog. Macrocyclic Chem., 1 (1979) 63.

86. J. Sandström, "Dynamic NMR Spectroscopy", Academic Press Publishers, New York, 1982.

87. D.E. Woessner, J. Chem. Phys., 35 (1961) 41.

88. K.M. Brière and C. Detellier, J. Phys. Chem., 91 (1987) 6097.

89. A. Delville, H.D.H. Stöver and C. Detellier, J. Am. Chem. Soc., 109 (1987) 7293.

90. E. Shchori, J. Jagur-Grodzinski, Z. Luz and M. Shporer, J. Am. Chem. Soc., 93 (1971) 7133.

91. J.M. Lehn, J.P. Sauvage and B. Dietrich, J. Am. Chem. Soc., 92 (1970) 2916.

92. Y.M. Cahen, J.L. Dye and A.I. Popov, J. Phys. Chem., 79 (1975) 1292.

93. M. Shamsipur and A.I. Popov, J. Phys. Chem., 90 (1986) 5997.

94. M. Shamsipur and A.I. Popov, J. Phys. Chem., 91 (1987) 447.

95. B.O. Strasser and A.I. Popov, J. Am. Chem. Soc., 107 (1985) 7921.

96. K.M. Brière, H.D. Dettman and C. Detellier, submitted for puplication.

97. E. Shchori, J. Jagur-Grodzinski and M. Shporer, J. Am. Chem. Soc., 95 (1973) 3842.

98. H.P. Graves and C. Detellier, J. Am. Chem. Soc., 110 (1988) 6019.

99. S.F. Lincoln, A. White and A.M. Hounslow, J.C.S. Faraday Trans. I, 83 (1987) 2459.

100. R.C. Phillips, S. Khazaeli and J.L. Dye, J. Phys. Chem., 89 (1985) 600.

101. B.O. Strasser, K. Hallenga and A.I. Popov, J. Am. Chem. Soc., 107 (1985) 789.

102. P. Szczygiel, M. Shamsipur, K. Hallenga and A.I. Popov, J. Phys. Chem., 91 (1987) 1252.

103. K.M. Brière and C. Detellier, New. J. Chem., 13 (1989) 145.

104. J.M. Ceraso and J.L. Dye, J. Am. Chem. Soc., 95 (1973) 4432.

105. J.M. Ceraso, P.B. Smith, J.S. Landers and J.L. Dye, J. Phys. Chem., 81 (1977) 760.

106. S.F. Lincoln, I.M. Brereton and T.M. Spotswood, J. Chem. Soc., Faraday Trans. 1, 81 (1985) 1623.

107. S.F. Lincoln, I.M. Brereton and T.M. Spodswood, J. Am. Chem. Soc., 108 (1986) 8134.

108. E. Amat, B.G. Cox, J. Rzeszotarska and H. Schneider, J. Am. Chem Soc., 110 (1988) 3368.

109. H. Degani, Biophys. Chem., 6 (1977) 345.

110. M. Shporer and Z. Luz, J. Am. Chem. Soc., 97 (1975) 665.

111. E. Schmidt and A.I. Popov, J. Am. Chem. Soc., 105 (1983) 1873.

112. G. Wipff, P. Weiner and P. Kollman, J. Am. Chem. Soc., 104 (1982) 3249.

113. M. Shamsipur and A.I. Popov, J. Phys. Chem., 92 (1988) 147.

114. B.O. Strasser, M. Shamsipur and A.I. Popov, J. Phys. Chem., 89 (1985) 4822.

115. J. Grandjean and P. Laszlo, Biochem. (Life Sci. Adv.), 6 (1987) 1.

116. R.K. Gupta and P. Gupta, J. Magn. Reson., 47 (1982) 344.

117. M.M. Pike and C.S. Springer, J. Magn. Reson., 46 (1982) 348.

118. S.C. Chu, M.M. Pike, E.T. Fossel, T.W. Smith, J.A. Balschi, and C.S. Springer, J. Magn. Reson., 56 (1984) 33.

119. E. Fernandez, J. Grandjean and P. Laszlo, Eur. J. Biochem., 167 (1987) 353.

120. J.W. Pettegrew, J.F.M. Post, K. Panchalingam, G. Withers, and D.E. Woessner, J. Magn. Reson., 71 (1987) 504.

121. M.C. Espanol and D.M. de Freitas, Inorg. Chem., 26 (1987) 4356.

122 I. Goldberg, H. Shinar and G. Navon, J. Incl. Phen., 5 (1987) 181

123. H. Shinar, G. Navon and W. Klaui, J. Am. Chem. Soc., 108 (1986) 5005.

124. F.G. Riddell, S. Arumugam and B.G. Cox, J.C.S. Chem. Comm., (1987) 1890.

125. G. Robinson, P.W. Kuchel, B.E. Chapman, D.M. Doddrell and M.G. Irving, J. Mag. Res., 63 (1985) 314.

126. D.C. Shungu and R.W. Briggs, J. Magn. Reson., 77 (1988) 491.

127. R.K. Gupta, P. Gupta and R.D. Moore, Ann. Rev. Biophys. Bioeng., 13 1984 221.

128. Y. Boulanger, P Vinay and M. Desroches, J. Biophys., 47 (1985) 553

129. T.E. Bull, J. Mag. Res., 8 (1972) 344.

130. R. Tabeta, M. Aida, and H. Saitô, Bull. Chem. Soc. Jpn., 59 (1986) 1957.

131. H. Saitô and R. Tabeta, Bull. Chem. Soc. Jpn., 60 (1987) 61.

132. A.S. Ellaboudy, M.L. Tinkham, B. Van Eck, J.L. Dye, and P.B. Smith, J. Phys. Chem., 88 (1984) 3852.

Isotopes in the Physical and Biomedical Science, Vol. 2, edited by E. Buncel and J.R. Jones   213
© 1991 Elsevier Science Publishers B.V., Amsterdam

Chapter 5

MULTINUCLEAR AND MULTIPULSE NMR OF ORGANOSILICON COMPOUNDS

ĒRIKS KUPČE AND EDMUNDS LUKEVICS
Institute of Organic Synthesis, Latvian Academy of Sciences,
Riga, 226006, Latvia (USSR)

CONTENTS

214

VII. REFERENCES   278

SYMBOLS AND ABBREVIATIONS

Ad	adamantyl
Bus^t	2,4,6-tri-*tert*-butylphenyl
Cp	cyclopentadienyl
CS	chemical shift
DA	donor-acceptor (bond)
EN	electronegativity
HMPT	$(Me_2N)_3PO$
lp	lone electron pair
Mes	2,4,6-trimethylphenyl (mesityl)
NPT	non-selective polarization transfer
PT	polarization transfer
Py	pyridine
TBP	trigonal bipyramid
THF	tetrahydrofuran
Xyl	2,6-dimethylphenyl

I. INTRODUCTION

Silicon is one of the most abundant elements in nature. It is responsible for the formation of climate and the origin of life on Earth.[1] This element is involved in living processes, it is utilized in industry, electronics and medicine.[2] Silicon occupies a strategically important place in the Periodic System, evoking special interest in both theoretical and experimental chemistry.

Nuclear magnetic resonance (NMR) appears to be the most appropriate, powerful and universal method to study the properties and follow the interconversion of silicon compounds in such different and important fields of human activity. Owing to the advances in the technology of superconducting magnets, the working frequencies and sensitivity of commercial spectrometers have grown significantly. On the other hand, in the last decade considerable progress has been achieved in the development of multipulse NMR techniques. Explosive growth of published works in this field, also called "spin engineering", has led to further growth in sensitivity and resolution of NMR spectrometers. Nowadays this allows the observation of resonances of practically all magnetically active nuclei. The word "multinuclear", introducing many of the current monographs,[3-6] reviews[7,8] and regular papers in NMR, reflects the modern approach and new possibilities in this field or, in other words, "perestroika" and "new thinking" in NMR spectroscopy (see also reviews[9-16]).

The aim of the chapter is to survey the results in NMR of silicon compounds obtained by multipulse and multinuclear approach with emphasis on silicon-29 NMR since 1983. Special attention has been devoted to the "newly accessible" and perspective nuclei in silicon chemistry, ^{15}N and ^{17}O. Another newcomer is the paragraph (III.C.) on isotope effects in NMR of silicon compounds. These topics have been reviewed for the first time and the literature has been fully covered. Isotope effects on chemical shifts are highlighted.

II. APPLICATION OF MULTIPULSE NMR TO SILICON COMPOUNDS

The number of different multipulse techniques in NMR increases continuously. Many are specific not only to the type of nuclei, but also to the class of the compounds studied. The principles of their action and experimental aspects have been described in detail in reviews[11,13,14] and monographs both for chemists[16] and spectroscopists.[15] For these reasons we have presented only the most characteristic examples reflecting the possibilities and specific features of multipulse methods with respect to silicon compounds.

(A) Polarization Transfer

Routine application of ^{29}Si and ^{15}N NMR to silicon compounds became possible after the development of polarization transfer (PT) techniques. The observation of these resonances by conventional techniques is complicated by several factors:

1) relatively low natural abundance of ^{29}Si (4.7%) and ^{15}N (0.36%) nuclei;
2) low sensitivity and negative nuclear Overhauser effects leading to reduced or nulled signals;
3) long spin-lattice relaxation times leading to long data acquisition times.

Systematic studies of long-range ^{29}Si-^{1}H coupling constants have been realized[17] by employing the selective polarization transfer (SPT) technique. Selective irradiation of one of the SiMe$_3$ proton transitions allows the observation of an enhancement of about 5 in the ^{29}Si spectra. The main advantages of this technique are:

1) polarization transfer selectivity in frequency permitting correlation of ^{1}H and ^{29}Si chemical shifts;
2) the overlapping of close ^{29}Si resonances can be avoided;
3) the relative signs of coupling constants, e.g. ^{29}Si-^{15}N, can be determined.[18]

SPT ^{29}Si spectra generally are very complex (90-360 theoretical peaks have been predicted). For their analysis a computer simulation programme has been developed.[19] The SPT technique is limited to molecules in which ^{29}Si satellites can be observed in ^{1}H spectra. It is also impractical for measurement of ^{29}Si chemical shifts (CS) and ^{29}Si couplings to rare-spin nuclei (^{13}C, ^{15}N, ^{29}Si etc.).

Non-selective polarization transfer (NPT) techniques, e.g. INEPT, DEPT and their modifications, permit the excitation of all transitions in ^{1}H spectra and to obtain ^{1}H decoupled spectra as required. For large J values (J>50 Hz) the INEPT and DEPT sequences are both short relative to ^{29}Si or ^{15}N relaxation times, T_2, giving similar enhancements (5 to 9 times in the case of ^{29}Si and about 10 times in ^{15}N spectra). For small J values (J<10 Hz) the INEPT sequence is still short relative to T_2, but the longer DEPT sequence becomes comparable to T_2, giving poorer enhancements, especially when quadrupolar nuclei, e.g. $^{11/10}$B, $^{2}14$N, $^{37/35}$Cl, are bonded to observed nuclei. However, as illustrated by Fig.1, this disadvantage makes the DEPT sequence superior in cases when broad lines prevent the observation of sharp resonances, e.g. ^{15}N satellites. In this case the intensity of the broad central line, e.g. ^{29}Si bonded to the quadrupolar ^{14}N nuclei, is significantly decreased due to loss in transverse magnetization during application of the DEPT pulse sequence, while the ^{15}N satellites remain unaffected by this relaxation mechanism.

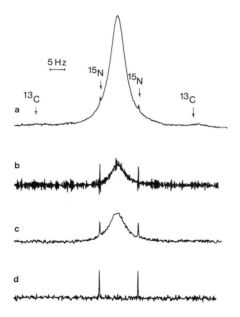

Fig. 1. ^{29}Si spectra of HN(SiMe$_3$)$_2$ (80% in C$_6$D$_6$, 10 mm sample tube) acquired (a) with the INEPT sequence after 16 scans; (b) the same spectrum after Gaussian multiplication of the free indication decay signal (LB=-1 Hz, GB=0.7); (c) with the DEPT sequence after 16 scans; (d) with the INEPT+ HE sequence after 64 scans (τ=35.7 ms, Δ=9.53 ms, T=0.3 s). Spectrometer frequency 71.55 MHz. Reproduced by permission of Academic Press.[24]

Generally, the DEPT sequence offers advantages over INEPT:
a) in coupled spectra (DEPT gives normal multiplet phases and intensities);
b) for samples in which large J variations are present or suspected (DEPT is less likely to suppress a signal than INEPT).

Both the theoretical and practical aspects of the NPT techniques have been explained in detail regarding the [29]Si and ^{15}N nuclei.[20-22]

Substantial enchancement in the sensitivity and decrease in measuring time (up to 300-fold in the case of [29]Si)[20] with NPT techniques offer new fields of application of ^{29}Si and ^{15}N NMR. For example, NPT has been employed to study the stereochemistry and kinetics of the reactions with organosilicon compounds by ^{29}Si NMR.[23] The [15]N-^{1}H coupling constants have been measured in the ^{15}N spectra using DEPT sequence in aminosilanes[24] and silazanes.[25] Application of NPT techniques permits routine measurements of [29]Si coupling to rare spin nuclei, e.g. [15]N, at the natural isotope abundance, which can be used for signal assignment or as a structural probe.[26]

Combination of the refocused INEPT with a selective decoupling (referred to as SPINEPTR) has been proposed as a routine assignment technique for [29]Si spectra of Me$_3$Si derivatives.[27] SPINEPTR spectra appear fully decoupled from protons engaged in polarization transfer but retain all the other couplings resembling single frequency off-resonance decoupled spectra.

The employment of the INEPT sequence in the spin-lattice relaxation time (T_1) measurements for [29]Si and ^{15}N nuclei has yielded a considerable saving of spectrometer time.[28] The combination of the INEPT sequence with Hahn spin-echo technique (INEPT+HE)[24] permits suppression of broad lines in [29]Si spectra (see Fig. 1d.) and can be used in relaxation time T_2 measurements.

(B) Multiquantum Coherence Transfer

The largest sensitivity improvements can be achieved when the resonances of insensitive nuclei are detected indirectly, via the proton resonance. Theoretically this allows one to increase the sensitivity by a factor of $(\gamma_H/\gamma_X)^3$, i.e. by about 128-fold for [29]Si and 961-fold for [15]N. Such results have been obtained in multiquantum coherence transfer experiments initially proposed by Bax et al.[29] This technique has been used to measure long-range [29]Si-^{1}H coupling in silatranes, ethoxysilanes, silylcyclopropenes, silylstyrenes[30] and [15]N-^{1}H couplings in silazanes and silazoxanes.[31]

INADEQUATE sequence creates double quantum coherence in a spin-spin coupled system and then converts it into a detectable transverse nuclear magnetization, revealing the pure satellite spectrum of the coupled spins. This technique is widely used in studies of [13]C-^{13}C coupled systems.[9-16]

Surprisingly, there are only few examples of application of the INADEQUATE technique (in combination with INEPT) to silicon compounds, namely, it has been employed in the [29]Si-[29]Si coupling measurements in cyclodisiloxanes, disilenes and polycyclic polysilanes.[32] The low interest in this technique in the case of silicon compounds is probably due to minor problems with observation of the relatively intense [29]Si satellites.

Heteronuclear version of the INADEQUATE technique has been proposed by Lipp-maa and collaborators.[33] In this experiment the traditional INADEQUATE sequence has been simultaneously applied to the resonance frequencies of both ^{29}Si and ^{13}C nuclei, and ^{13}C signals (coupled to ^{29}Si) have been recorded.

The selective tickling-INADEQUATE modifications of the basic pulse train have been illustrated by studying trimethylsilylated (TMS) sugars. All the studied $^{29}Si-^{13}C$ long-range couplings have been recognized from these spectra. Implementation of these heteronuclear versions of the INADEQUATE sequence requires modification of commercial spectrometers.

(C) Ultrahigh-Resolution NMR

The term "ultrahigh-resolution" (UHR) was introduced in NMR spectroscopy by Allerhand et al. in 1985.[34] UHR has been arbitrarily defined as having been achieved when there is less than 20 mHz instrumental contribution to the line width. This can be realized on commercial spectrometers by eliminating temperature gradients in the sample with (1) a decrease in proton decoupling power using WALTZ-16 technique, (2) an increase in the flow and temperature stability of cooling air. The observed resolution and sensitivity improvements have been equalized with an increase in working frequency of the spectrometer (exceeding the giga Hz level on ^1H) under conventional high-resolution conditions, providing the title for the methodology.

The greatly extended range and precision of measurements of one-bond and long-range spin-spin coupling constants between the rare spin nuclei and the corresponding isotope effects, the isotope shifts induced by minor isotopes having a zero spin etc., permit one to obtain the information which is usually lost in conventional NMR spectra, justifying the rather pretentious title of the methodology. In this way $^{29/28}Si$ and $^{30/28}Si$ isotope effects on ^1H [35] and ^{13}C [34b] CS have been measured for the first time (see also Section IIIC).

Recently, the UHR methodology has been extended to ^{29}Si NMR. Relatively long relaxation times of the ^{29}Si nuclei (typically between 40 and 150 s) result in ^{29}Si line-widths of 22 mHz (including 16 mHz of instrumental broadening) or less. Under these conditions the observation of ^{18}O satellite (natural abundance 0.2%) is feasible in the ^{29}Si spectra (see Fig.2), allowing the first measurement of the $^{18/16}O$ isotope shift.[36] UHR ^{29}Si NMR spectroscopy has been employed in studies of $^{29}Si-^{15}N$ couplings, $^{15/14}N$ isotope shifts,[25] long-range $^{29}Si-^{13}C$ and $^{29}Si-^{29}Si$ couplings, and the corresponding isotope shifts,[25,36,37] as well as $^{18/16}O$ isotope shifts.[37] The advantage of this methodology is its availability on commercial spectrometers after simple and inexpensive modifications. Measurements of long-range couplings between the rare spin nuclei is possible without the use of complicated pulse sequences.

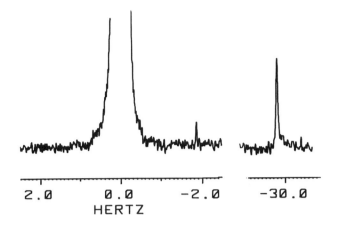

2.0   0.0   -2.0   -30.0

HERTZ

Fig. 2. Ultrahigh resolution ^{29}Si NMR spectrum of (Me$_3$Si)$_2$O (with 20%, v/v,
C$_6$D$_6$ in a 10 mm sample tube) acquired with the INEPT+HE sequence after
16 scans. Upfield from the major resonance an ^{18}O satellite is visible.
One of the ^{13}C satellites is shown for comparison.

(D) Two-Dimensional NMR

During the last decade various techniques of the two-dimensional (2D) NMR
spectroscopy have been widely exploited for spectral assignment and in structural studies. Most of these techniques have been included in the "menu" of commercial spectrometers and are routinely used nowadays. Various aspects of 2D NMR
have been periodically discussed in reviews[9-11,38] and monographs.[15,16] In this
section the most characteristic applications of 2D NMR to the silicon compounds,
reflecting the potential of this technique, are presented.

A	B	B'	A'

The ^{1}H-^{1}H NOESY spectra of 3,7,10-trimethylsilatranes permit the unequivocal
assignment of the signals to the individual rings.[39] The appearance of cross-
peaks from the spatially approaching OCH-protons of neighbouring rings suggests
the existence of RRS isomer in the conformation of type B or B'. These results
provide evidence that the exchange between the enantiomeric "waterweel-like"

conformations in metallatranes proceeds according to a multistep mechanism (A-B-B'-A') rather than a synchronous conversion (A-A').

By means of ^{13}C-^1H and ^{29}Si-^1H COSY experiment at 193 K, eight ^{13}C and four ^{29}Si resonances have been assigned to particular sites in the molecule of compound 1.[40] ^1H-^1H NOESY experiment at 183 K has revealed that exchange process, correlated rotation of all four appendages to the central C atom, is occurring. The enantiotopomerization has been shown to be of ESSS type with free energy of activation of 43.3 kJ/mol.

$$\begin{array}{c} SiMe_3 \\ | \\ ClPh_2Si-C-SiMe_3 \\ | \\ SiMe_2OMe \end{array}$$

1

Structures and ratios of the endo/exo isomers of silyl- and sila-norbornene derivatives were determined by 2D ^1H-^1H COSY and ^{29}Si-^1H correlation spectra.[41] ^{29}Si-^1H COSY has been used for the assignment of signals in the ^1H and ^{29}Si NMR spectra of all four isomers (A-D) of $(ClMeSiO)_4$.[42]

A   B   C   D

$$\begin{array}{ccc} & & Si-O-Si \\ & & | \quad\quad | \\ & = & O \quad\quad O \\ & & | \quad\quad | \\ & & Si-O-Si \end{array}$$

Several papers have been devoted to applications of 2D NMR to trimethylsilylated sugars[43] in place of the empirical rules[44] and the modified INDOR technique[45] employed in the earlier studies. First, the ^1H spectra have been analysed by means of ^1H-^1H COSY technique. Then, the correlation of ^1H and ^{29}Si CS through the ^{29}Si-O-C-^1H couplings provides assignment of the ^{29}Si signals in the ^{29}Si NMR spectra, and allows verification of the analysis of the ^1H spectra (see Fig.3.). Correlation of ^{29}Si and ^1H CS through ^{29}Si-C-^1H coupling permits unambiquous assignment of the $SiCH_3$ signals in the ^1H NMR spectra.

The modified HETCOR and RELAY techniques have been employed by Schraml et al. to obtain ^{13}C-^1H and ^{29}Si-^1H correlated 2D spectra of silylated di- and tri-saccharides.[43c,d] Application of these techniques yielded, in addition to the ^1H, ^{13}C and ^{29}Si line assignment, the ring assignment and unambiguous information about the site of glycosidation. It has been noted that ^{29}Si CS do not follow any simple rule that would allow the empirical assignment. The main difficulties

(A) Chemical Shifts

(1) Theoretical Considerations.

The nuclear screening reflects the electronic environment of a given nucleus and can be described as a sum of local (σ_{loc}) and non-local ($\sigma_{non-loc}$) components. It is usually assumed that [29]Si CS are determined by σ_{loc}, which results from a diamagnetic term (σ_{loc}^d), reflecting the electron density around the nucleus in question, and a paramagnetic term (σ_{loc}^p) arising from electronic circulations around the observed nucleus. The variation in σ_{loc}^d is always small and can be neglected. For instance, going from SiH_4 to Et_3SiH results in variation of σ^d by +2 ppm and σ^p by -50 ppm as calculated by Ando et al. using the EPT method within the CNDO/2 framework.[55] The relative contribution of σ^d slightly increases if there are strongly electronegative (EN) substituents at the Si atom, e.g. going from SiF_4 to Pr_3SiF is related with variation of σ^d by +19 ppm and σ^p by -120 ppm.[55]

Consequently, qualitative treatment of experimental results is possible in terms of the analytical expression for σ_{loc}^p obtained using MO approach:[56]

$$\sigma_{loc}^p = -\frac{2\mu_0\mu_B^2}{3\pi\Delta E}\left[<r^{-3}>_p P_u + <r^{-3}>_d D_u\right] \tag{1}$$

where μ_0 is the permeability of free space, μ_B is the Bohr magneton, ΔE is a mean excitation energy, $<r^{-3}>$ is the inverse of the mean volume of the valence p or d orbitals, and P_u and D_u represent the p and d electron imbalance about the nucleus. The theory of nuclear magnetic shielding has been discussed in detail elsewhere.[57]

(2) Empirical Relationships.

A characteristic feature of the [29]Si shielding pattern is the U-shaped dependence of CS on the sum of FN of the Si substituents, complicating the prediction of [29]Si CS using an empirical approach.[53] This problem can be partially overcome using linear relationships existing in the limited series of compounds.

Linear relationships have been found between group EN sums of ligands bonded to tetravalent Si and [29]Si CS both for P type Si atoms (all ligands have lp available for $d_\pi-p_\pi$ bonding) and S type Si (all ligands have only σ-bonding electrons available).[58]

Two sets of lines of different gradient have been obtained for the dependence of Me→Ph substituent shifts ($\Delta\delta$) on Pauling EN of atoms X in $Me_{2-n}Ph_nSiX_2$.[59] The ligands N, O and F produce smaller $\Delta\delta$ values (deshielding) than would be expected by extrapolation of the lines containing H, C and Cl. This effect has been attributed to $p_\pi-d_\pi$ bonding.[59,60]

The [29]Si CS of $ClCH_2SiRR^1R^2$ (R,R^1,R^2=H, Me, Et, CH_2Ph, CH_2Cl, CH_2CH, Ph, OAlk, OAc, Cl, F) have been correlated with Taft's constants (σ^*) of Si substituents.[61] The positions of the [29]Si signals depend appreciably on noninductive interaction between the Si atom and the substituents bearing the lp. No correlations have been found with the inductive or resonance σ constants of X in Me$_3$SiX.[62]

Polar substituent effects on ^{29}Si CS studied in sterically rigid systems 3 and 4 [63] suggest that the charge dependence of the CS induced by remote substitu-

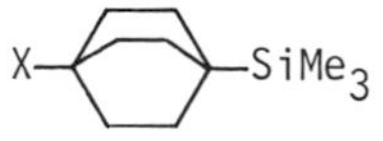

3 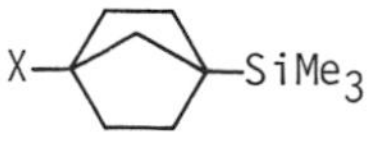   4

ents arises from variation in the $<r^{-3}>_p$ term of Eq.1. Polar field suscepti-bility parameters have also been derived. Linear correlations exist in the se-ries of the isostructural derivatives of C, Si, Ge and Sn relating the CS of corresponding nuclei:[64]

$$\delta^{29}Si = 0.787\delta^{13}C - 61.7, \ r = 0.825, \ n = 43 \tag{2}$$

$$\delta^{73}Ge = 3.32\delta^{29}Si + 39.9, \ r = 0.967, \ n = 29 \tag{3}$$

$$\delta^{119}Sn = 5.119\delta^{29}Si - 18.5, \ r = 0.990, \ n = 48 \tag{4}$$

The slopes of the correlations (3) and (4) correspond fairly closely to the $<r^{-3}>_p$ ratios for the element pairs Sn/Si (4.5) and Ge/Si (3.3), suggesting that σ^p_{loc} term dominates the CS. The poorer agreement for the pair Si/C (1.2-2.0) improves when the O- and F-derivatives are excluded.[64a]

Van Wazer et al. have carried out ab $initio$ calculations of ^{29}Si CS for 28 representative Si compounds both with and without d functions on silicon.[65] An apparent relationship between the ^{29}Si paramagnetic term and the electron-withdrawing power of the substituents on Si was found. A new approach to ana-lysis of the U-shaped dependence of ^{29}Si CS on the sum of group EN of ligands has been proposed by Zhidomirov et al.[66] Results of ^{29}Si CS calculations for 23 organosilicon compounds carried out in terms of the FPT INDO-CS method taking into account the silicon core AO were in good agreement with experimental data. (3) Effect of Ring Size. Considerable and permanent interest has been demonstra-ted in the study of the influence of ring strain on ^{29}Si CS. Some illustrative results obtained recently are presented below.

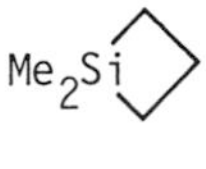

18.4 ref.53a 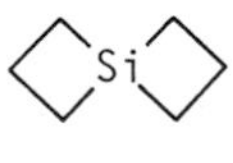   37.2 ref.79 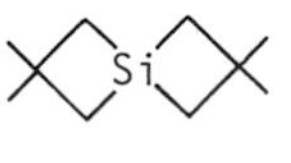   10.7 ref.79 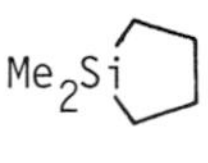   16.8 ref.53a

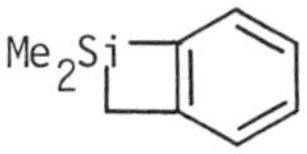

Me$_2$Si〈benzocyclobutene〉

9.2 ref.80

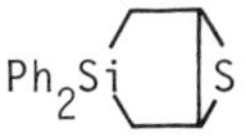

Ph$_2$Si〈S ring〉

11.3 ref.81

R^1-Si-R^2 (dibenzobicyclo structure)

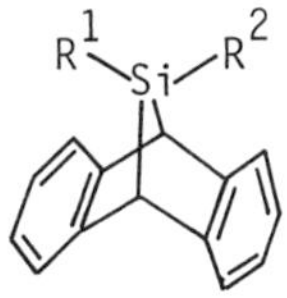

45.4 (R^1=R^2=Mes)

70.1 (R^1=Bu^t, R^2=Mes)

ref.69

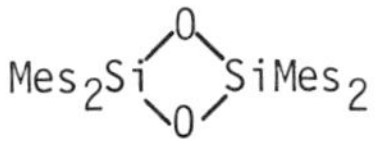

Mes$_2$Si〈O,O〉SiMes$_2$

-3.3 ref.32a

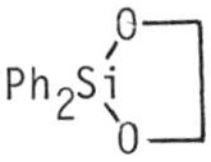

Ph$_2$Si〈O,O ring〉

-7.5 ref.67

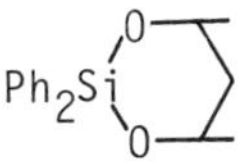

Ph$_2$Si〈O,O ring〉

-29.2 ref.67

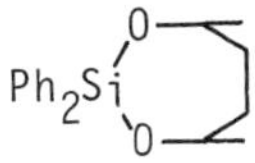

Ph$_2$Si〈O,O ring〉

-29.6 ref.67

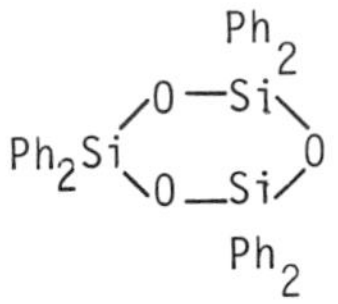

Ph$_2$Si〈O-SiPh$_2$,O-SiPh$_2$〉

-33.8 ref.53a

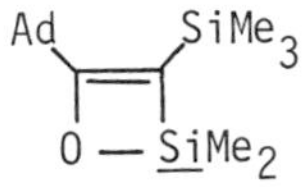

Ad, SiMe$_3$ / O—SiMe$_2$

38.2 ref.82

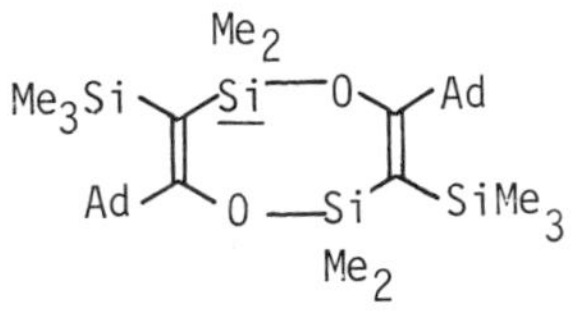

Me$_3$Si, Me$_2$Si—O, Ad / Ad, O—Si, SiMe$_3$, Me$_2$

7.8 ref.82

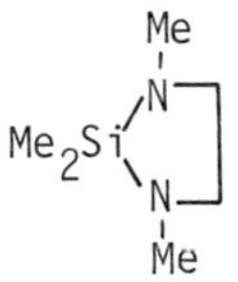

Me$_2$Si〈N(Me),N(Me)〉

9.8 ref.68

Me$_2$Si〈N(Me),N(Me) ring〉

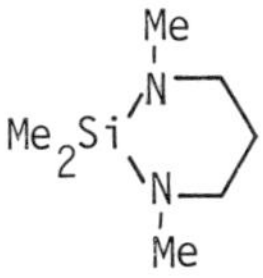

-1.4 ref.68

Me$_2$Si(NEt$_2$)$_2$

-5.5 ref.68

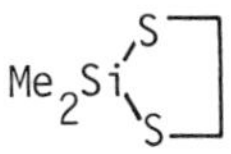

Me$_2$Si〈S,S ring〉

41.7 ref.83

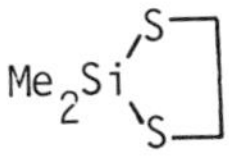

Me$_2$Si〈S,S ring〉

22.5 ref.83

Si〈S,S / S,S〉

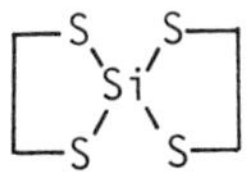

57.5 ref.83

Si(SEt)$_4$

30.7 ref.83

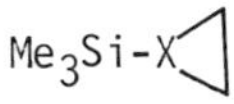

Me$_3$Si-X〈triangle〉

2.6
14.1

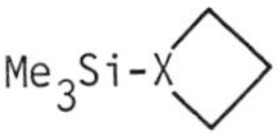

Me$_3$Si-X〈square〉

0.8
1.8

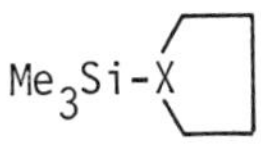

Me$_3$Si-X〈pentagon〉

2.6
2.5

Me$_3$Si-X〈hexagon〉

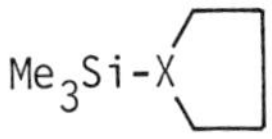

2.3 (X=CH) ref.70
3.9 (X=N) ref.24

Me$_3$Si—⬚

-7.4 ref.71

Me$_3$Si—⬡

-2.5 ref.71

Me$_3$Si-O-C
‖ (CH$_2$)$_n$
Me$_3$Si-O-C

ref.72

n=2	19.48
n=3	17.60
n=4	16.15
n=5	15.92
n=6	14.88
n=7	17.72
n=8	14.62

Cragg et al. [67,68] have demonstrated that cyclization does not significantly affect ^{29}Si CS *per se* but only as a result of strain in the smaller ring. A clear dependence of ^{29}Si CS on ring size for rings of 5-14 atoms have been observed in 2,2-diphenyl-1,3,2-dioxasilacycloalkanes.[67] The large upfield shift which results from an increase in ring sizes 5 to 6 ($\Delta\delta$ = 21.8 ppm) exceeds the total subsequent variation (7.2 ppm), indicating that 5-membered rings are highly strained as supported by their chemical properties and bond angle data. The smaller ^{29}Si CS differences in 5- and 6-membered rings of 1,3,2-diazasilacycloalkanes ($\Delta\delta$ = 9.7 to 11.2 ppm)[68] suggest that N derivatives are considerably less strained as also follows from their chemical properties.

Anomalous ^{29}Si deshielding in dibenzo-7-silanorbornadienes[69] has been explained in terms of an enhanced polarization at the ground state of the molecules associated with σ-π conjugation.

The effect of ring strain on ^{29}Si CS has been observed in Me$_3$Si derivatives of cycloalkanes[70] and cyclic amines.[24] There is a linear correlation between the high-field shift of ^{29}Si resonance and the s-character of Si-C bond in Me$_3$Si derivatives of polycycloalkanes.[71] The strong variation of the ^{29}Si CS in 1- and 1,2-substituted Me$_3$SiO-cycloalkenes has been explained[72] by steric crowding with the nearest CH$_2$ group, becoming more acute with increasing ring size.

The sensitivity of ^{29}Si shielding to steric effects permits the application of ^{29}Si NMR in the conformational analysis of Me$_3$Si derivatives of sugars[43c,d] and steroids.[73] Schraml et al. have demonstrated that ^{29}Si CS of Me$_3$Si steroids fall into two distinct ranges allowing one to distinguish between the equatorial ($\sim$13.2 ppm) and axial ($\sim$14.9 ppm) Me$_3$SiO groups on the cyclohexane ring.[73] Steric interactions are responsible also for an increase in ^{29}Si shielding upon the introduction of Me groups in the ring of monosilacyclobutane derivatives.[74]

Substantial amount of ^{29}Si CS data for small Si containing rings has been published as a result of studies of reactions with Si=E (E=C, N, Si) compounds (δ, ppm):[75-77]

R_2Si—C=$^{Ad}_{OR}$ $R=Me_3Si$ R_2Si<$^{NR'}_{OR}$ (Mes) Bu^t_2Si—N—$SiBu^t_3$
R' $R'=Bu^t$ O—CHPh

-89.5 ($^1J_{SiC}$=38.7 Hz, $^1J_{SiSi}$=73.2 Hz)   -18.6   26.2 ($SiBu^t_2$)   2.4 ($SiBu^t_3$)

ref.76a   ref.76b   ref.77

Boudjouk et al. have reported ^{29}Si CS of siliranes, silirene and cyclotetra-silane (δ, ppm):[78]

Bu^t_2Si (ring with Me, Me) Bu^t_2Si (ring with Me, Me) $Bu^t(H)Si$—$SiBu^t_2$
                                                      Bu^t_2Si—$Si(H)Bu^t$

-43.9 (*trans*)   -86.2   -35.8 (Bu^t,H)

-53.2 (*cis*)   +37.2 (Bu^t_2)

Other examples of the influence of ring strain on ^{29}Si CS can be found in refs. 79-83.

(4) Si-O Compounds. The bulk of the published work dealing with ^{29}Si spectrosco-pic properties of Si-O derivatives can be explained by increased stability and availability of these compounds. Wide ranges of organoxysilanes have been stu-died by ^{29}Si NMR.[84,85] The contribution of each RO group in the ^{29}Si shielding is non-additive and depends on the nature of the other Si-substituents. Varia-tion in the ^{29}Si CS ($\Delta\delta$) resulting from the Me→EtO substitution correlates with the sum of Taft's σ^* constants of Si substituents. Linear correlations have been found for derivatives of type X_3SiOR and inductive and steric constants of the O-substituents. The σ^* and σ_I constants have been estimated from ^{29}Si CS for the series of siloxy-groups.[86]

^{29}Si NMR has been employed to follow the dimerization of several organo-1,3,2-dioxasilacycloheptanes.[87] Dimers can be easily recognized by the lower shielding ($\Delta\delta \sim 7$ ppm) as compared to monomers. The spatial structure of Me_3SiO derivatives of butene has been determined by ^{29}Si NMR.[88]

High polarity of Si-O bonds in $R_3SiOSO_2CF_3$ ($R_3 = H_3$, Me_2H, Me_3, Et_3, Me_2Bu^t, PhMeH, Bu^tPhCl, Cl_3) is accompanied by decreased ^{29}Si shielding (δ between -23.2 ppm for $R_3 = H_3$ and 45.6 ppm for $R_3 = Me_2Bu^t$).[89] The strong ^{29}Si downfield shifts ($\Delta\delta$ between 15.9 and 42.6 ppm) of α-siloxonium and -halonium ions from their precursors diminish on going to β-derivatives (δ, ppm):[90]

Me_3SiOMe	19.1	Et_3SiOH	18.5	Me_3SiCH_2OMe	-1.5
$Me_3Si\overset{+}{O}Me_2$	52.3	$Et_3Si\overset{+}{O}H_2$	51.3	$Me_3SiCH_2\overset{+}{O}HMe$	0.4

A similar effect resulting from coordination of the O atom with $AlEt_3$ decreases with increasing the number of OR groups at Si (δ, ppm):[91]

	Et_3SiOMe	Ph_3SiOMe	$Ph_2Si(OMe)_2$	$PhSi(OMe)_3$	$Si(OMe)_4$
free	22.6	-11.0	-29.0	-54.6	-78.2
x $AlEt_3$	39.7	2.7	-19.9	-51.3	-79.5
$\Delta\delta$	17.1	13.8	9.1	3.3	-1.4

The negative charge on the O atom increases the ^{29}Si shielding (δ, ppm):[92]

$$Bu_2^t Si(NH_2)OH \qquad\qquad Bu_2^t Si(NH)_2 O^- Li^+$$

$$-3.8 \qquad\qquad\qquad\qquad -10.1$$

Oxygen in the ring strongly deshields the ring silicon atom of 1,2-siloxeta-nes 5 (δ between 42 and 65 ppm).[93] Linear relations were found between electron density distribution parameters of compounds 6 (R = Cl, Me, F; R^1 = Cl, Me, NH_2, F; R^2 = H, NO_2, Cl) and ^{29}Si CS.[94] Conjugation was not transmitted through the Si-O moiety in 6. ^{29}Si NMR data of 7 (R,R^1,R^2 = Alk, Ph) were systematically discussed.[95] Functional carboxylic acid silylesters $Bu_2^t(X)SiOCOR$ (X = F, Cl, OLi,

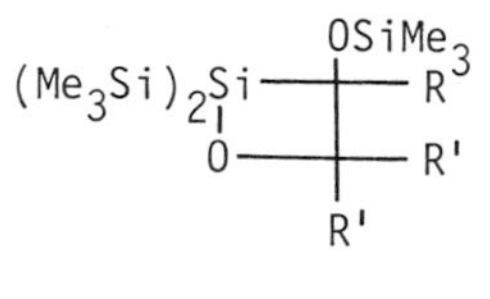

5

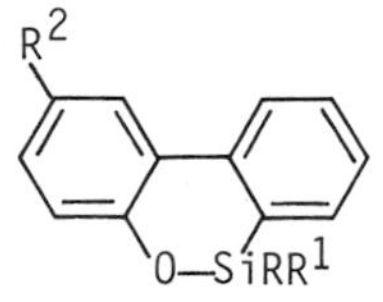

6

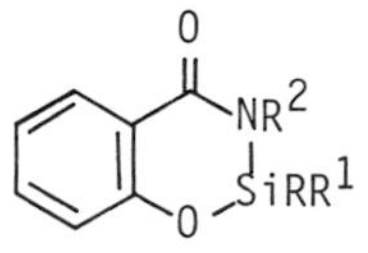

7

OCOR, $NHSi(OH)R_2$, R = Alk, Ar) and $Bu_2^t(Cl)SiOP(O)Cl_2$ were characterized by ^{29}Si (^{19}F, ^{31}P) NMR.[96] ^{29}Si CS of silylated carbazine acids,[97] $p-X-C_6H_4C(OSiMe_3)=$ $NNMe_2$ (X=H, Me, NO_2, MeO) were reported.[98]

The ^{29}Si CS are highly diagnostic, providing unambigous information about ring size and the sequence of O and N atoms in cyclosilazoxanes.[31,99] ^{29}Si NMR has proven particularly useful for assignment of the structure of cyclosiloxanes obtained by the hydrolysis of H_2SiCl_2, $HMeSiCl_2$, Me_2SiCl_2,[100] spirocyclosiloxa-nes obtained by flash vacuum pyrolysis[101] and trisiloxane derivatives of benz-amides (8, 9, X = H, Cl, NO_2, MeO).[102] The signal observed at ca. -20 ppm is

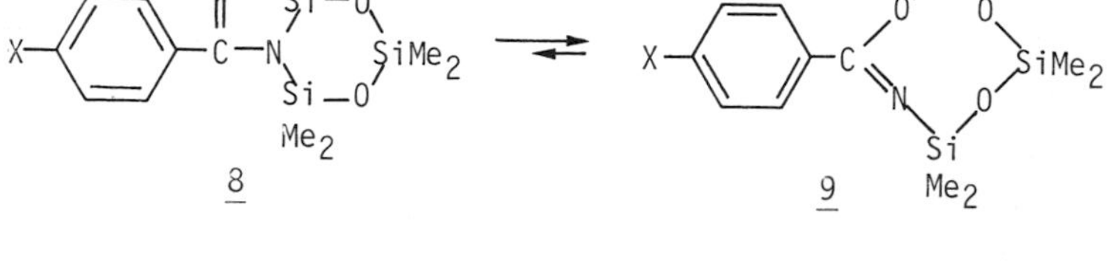

8

9

characteristic of 8-membered siloxanes suggesting the predominance of imidate form ($\underline{9}$).

^{29}Si NMR data of 53 linear and branched siloxanes of the type $R^1R^2R^3SiX$ (R^1,R^2,R^3 = Me, Me_3SiO, $Me_3SiOSiMe_2O$, $(Me_3SiO)_2MeSiO$, $(Me_3SiO)_3SiO$, X = H, Cl, OH, OEt) have been discussed with regard to their dependence on the nature of the substituent X and siloxane network.[103] A linear correlation between ^{29}Si CS and $\sigma^*(X)$ was observed. The relative basicity of (Si)OEt group and the ^{29}Si CS of ethoxysiloxanes were found to be proportional.

^{29}Si NMR has been employed to confirm the structures of disiloxanes contain- ing hydroxyalkyl groups $HO(CH_2)_n[SiMe_2O]_xSiMe_2(CH_2)_nOH$ (n=3-6, x = 1-3 and n=11, x=1) and $Me_3SiOSiMe_2(CH_2)_nOH$ (n=3-6),[104] and organobicyclosiloxanes with an ethylene bridge[105] have been characterized by ^{29}Si NMR. The signal splittings of the ^{29}Si NMR spectra of Me_2N- and Pr(H)N-terminated siloxane oligomers have been assigned to monomer sequences up to the pentad level.[106] The ^{29}Si spectral properties of oligomeric and polymeric Me_3SiO, Cl, Me-siloxanes which are chiral at C and Si,[107] and the stereochemical dependence of ^{29}Si CS and relaxation times T_1 in $Me_3SiO(MeHSiO)_nSiMe_3$ (n=3-8 and 35) have been studied.[108]

^{29}Si NMR has proven a potentially useful technique for the evaluation of com- mercial polysiloxanes,[109] affording new insight into the structures of silylated petroleum fractions, asphaltenes, coal extractions, wood constituents, lignin[110] and sugars[43-45,111] and providing at least semiquantitative structural informa- tion.

^{29}Si NMR has been extensively employed to study the structures of silicates. McCormic et al.[112] have studied the influence of cation size on relaxation rate of ^{29}Si nuclei and rate coefficients for Si exchange between monomer, dimer and cyclic trimer anions. Kinrade and Swaddle[113] have carried out a quantitative study of the dependence of silicate connectivity upon temperature and solution composition which showed that silicate polymerization is favoured by low tempera- ture, low alkalinity and high Si (IV) concentration. ^{29}Si relaxation measurements and detailed analysis of ^{29}Si line broadening provided Si exchange rates. Selec- tive formation of double three-ring, double four-ring and cubic octamer anions were observed at different solution compositions.[114] A cage-like silicic acid Me_3Si ester $[(Me_3Si)_6]Si_6O_{15}$,[115] Me, Et, Me_3Si and Cl derivatives of cyclic silicic acids[116] were characterized by ^{29}Si NMR.

^{29}Si NMR has provided detailed information on the structure of silicate spe- cies which occur in H_2O-D_2O alkali metal silicate solutions,[48,49,117] dynamics of exchange by SiO_4^{4-} anions,[50,118] and on ^{29}Si relaxation mechanisms.[119] Formation of disilicic acid has been followed by ^{29}Si NMR in H^+-$Si(OH)_4$-tropolone system.[120] (5) Si-N and Si-S Compounds. The major impetus for the permanent interest of NMR spectroscopists to Si-N derivatives arises, unquestionably, from the unusual chemical properties and structures of these compounds. Thirthy seven derivatives of the amino-substituted silanes $Me_nSi(NR^1R^2)_{4-n}$, $Me_{3-n}(RO)_nSiNR^1R^2$, and $(RO)_nSi(NR^1R^2)_{4-n}$ (n=0-3, R=Me, Pr^i, Bu^t, o-cresyl, R^1,R^2=H, Me, Et, Pr^i, Bu^n, Bu^t, Ph, p-ClC_6H_4,[121] 50 aminosilanes, silylamides,[24] and various silazanes and cyclosilazanes [25,31] have been studied by ^{29}Si NMR. The Hammett plots of [29]Si CS show both positive and negative slopes with variation of the number of EN atoms attached to Si in $(RO)_nMe_{3-n}SiNHC_6H_4X$.[122]

Protonation of the nitrogen atom in cyclodisilazanes 10-12 is accompanied with the decrease in [29]Si shielding (δ, ppm):[123]

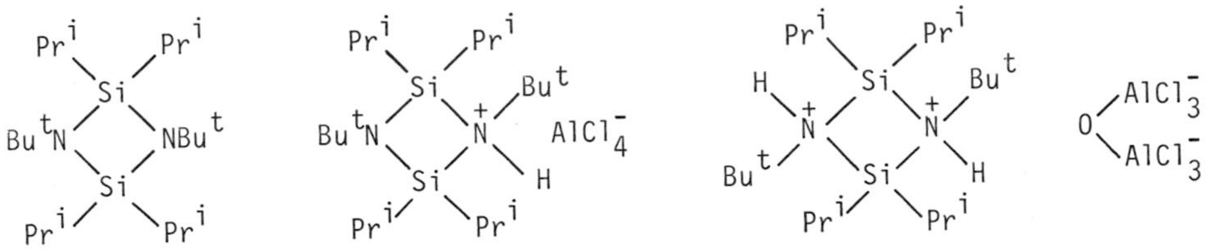

10 7.4   11 30.0   12 30.8

Similar shifts have been observed in bis(trimethylsilyl)imidazolium salts,[124] Me_3Si-guanidinium salts[125] and in the case of the N atom involved in DA bond formation.[126] Opposite effects appears for the negatively charged N atoms (δ, ppm):[25]

$(Me_2N)_2C=NSiMe_3$

-19.5

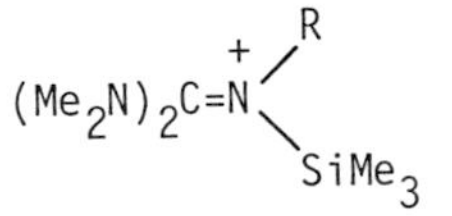

$(Me_2N)_2C=N\diagdown{R \atop SiMe_3}^+$

13.8 (R=H)
14.8 (R=SiMe_3)

ref.125

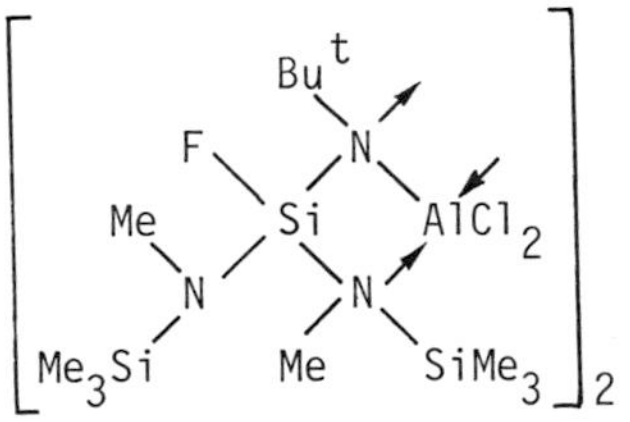

$(Me_3Si)_2NH$

2.1

$(Me_3Si)_2N^-Na^+$

-14.6

ref.25

11.7   25.5

ref.126

The dynamic and equilibrium properties of imidazolium salts have been studied by [29]Si CS titration providing ΔH and ΔS values.[124]

Anderson et al.[127] have studied $Cl(H_2)SiNMe_2$ by ^{29}Si NMR, which was expected to be most sensitive to additional coordination to Si. At low temperatures, which were expected to favour intermolecular N→Si association, the [29]Si signal showed no changes attributable to formation of additional bonds to Si.

The structures of 18 silylsulphonamides $R^1SO_2N(R^2)SiMe_3$ have been determined by ^{29}Si NMR.[128] The three Me_3Si group environments in the model compounds fall into three, non-overlapping CS regions, i.e. δ ca. 30 to 45 (Me_3SiO), ca. 0 to 10 (Me_3SiN) and ca. -3 to -5 ppm ($Me_3SiN=S$), permitting unambiguous interpretation of the spectra. The N-silyl tautomer was the only isomer observed, except, in the cases where strongly electron withdrawing groups, e.g. Cl and NMe_2, were attached to the nitrogen; in such cases the O-silyl tautomer dominates. In a similar way the structures of persilylated urazoles[129] and bis(organosilyl)-amides[130] have been determined.

The ^{29}Si NMR data have shown that silylated cyanides 13 and isocyanides 14 at room temperature equilibrate very slowly with a half-life of about 3 months (δ, ppm):[131]

$$(Bu^tO)_3Si-C\equiv N \qquad\qquad (Bu^tO)_3Si-\overset{+}{N}\equiv\overset{-}{C}$$

$$\underline{13} \quad -104.9 \qquad\qquad \underline{14} \quad -93.2$$

Marsmann et al.[132] have used ^{29}Si NMR in structure determination of the products of photolysis of hexa-Bu^t-cyclotrisilane in the presence of 2,2'-bipyridine. The multistage character of the reaction of $Me_2(Cl)SiCH_2Cl$ with N-trimethylsilyl amides and lactams was shown by ^{29}Si NMR monitoring.[133] ^{29}Si CS of cyclodisilazanes were reported.[134]

Pikies and Wojnowski have made extensive studies of the ^{29}Si spectra of Si-S compounds, e.g. $(RO)_3SiSR'$ ($R=Pr^i$, Bu^i, Bu^t, Am^n, Ar; $R'=H$, Pr, Pr^i, Bu^n, Bu^t, Me_3Si), $(Pr^iO)_{4-n}Si(SEt)_n$, (n=0-4) and cyclic derivatives,[83] providing steric and inductive effects of the Si-substituents. The ^{29}Si CS of X_3SiSH (X=F, Cl, Br, I) have been reported.[135] A "U-shaped" dependence exists between the ^{29}Si CS and the net charge at the Si atom corrected according to the C quantum-chemical model in the trialkoxysilylthio derivatives of permethylpolysilanes $(RO)_3SiS(SiMe_2)_nSi(OR)_3$, (n=2-4,6) and $(RO)_3SiS(SiMe_2)_nMe$, (n=2,4).[136] Hahn and Altenbach have shown that the silylsulphanes $(MeSiPh_2)_2S_n$, (n=2-5) can be unambiguously characterized by ^{29}Si NMR.[137] The individual values of the ^{29}Si CS in Si-S compounds have been published (δ, ppm):[138]

$$(Bu^tO)_3SiSH \qquad (Bu^tO)_3SiSSiMe_3$$

$$-79.2 \qquad\qquad -72.9 \quad 14.8$$

ref.83

$$Ph_2MeSi-(S)_n-SiMePh_2$$
0.9 (n=1)
5.2 (n=2)
6.0-6.6 (n=3-10)

ref.137

-68.5 ($R=OBu^t$)   ref.138a

Recently, Horn and Hemke have systematized the ^{29}Si NMR data of Si-S compounds.[139]

(6) Si-C, Si-Si and Si-Hal Compounds. Al-Mansour et al.[140] have reported for the first time the ^{29}Si data of $(Me_3Si)_3CL$ and $(Me_2Ph)_3SiCL$ where L = $SiR^1R^2R^3$ and R vary widely in EN. ^{29}Si CS exhibit good correlation with the EN of the groups bonded to silicon. Yoder et al. have studied ^{29}Si NMR spectra of nine ortho- and 2,6-disubstituted aryltrimethylsilanes.[141] Highly substituted 1-Bu^t-1-silacyclohexadienes,[142] sila- and silylnorbornene derivatives[41] were characterized by ^{29}Si NMR. ^{29}Si spectra have been involved to study the mechanism of nucleophile-assisted racemization of halosilanes $PhCH(Me)SiMe_2X$ (X=Cl, Br).[143] The difference in the transmission of electronic effects permits one to distinguish the derivatives of 1,2-bis(trimethylsilyl)cyclohexa-1,4-dienes (δ between -6.9 and -7.7 ppm) and corresponding 1,2-bis(trimethylsilyl)benzenes (-2.9 to -5.7 ppm).[144] The transfer of electron density from O, N or S through the aromatic system of phenheterasilines have been studied by ^{29}Si NMR.[145] The ^{29}Si CS data of the Me_3Si and Cl_3Si derivatives of Bu^t_2Cp,[146] the first stable silicon (II) compound decamethylsililicocene,[147] 35 furylsilanes,[148] and trans-1,4-poly-(2,3-bis(trimethylsilyl)-1,3-butadiene) have been reported.[149] The strong deshielding of ^{29}Si is characteristic of 7-silanorbornyl derivatives.[64]

The variation of the ^{29}Si CS in the series of disilanes $R_{3-x}R'_xSiSiR_{3-y}R'_y$ (R,R'=H, Cl, Ph) has been interpreted in terms of the EN of the Si-substituents.[150,151] The ^{29}Si CS data of polycyclic methylsilanes[152] and their Hg derivatives,[153] 1,4-dihydroxyoctaphenyltetrasilane containing intramolecular hydrogen bond,[154] several branched[155] and cyclic[156,157] silanes have been reported (δ, ppm):

-398.0

ref.147

58.2

12.5

ref.157

Contrary to polysilanes composed of asymmetrically substituted silylenes, symmetrically substituted poylsilanes showed a single narrow peak.[158] ^{29}Si CS for these polysilanes decreased with increasing steric bulk of the substituents. ^{29}Si CS of disilanes R_3SiSiR_3 (R = Me, Bu^t, Ph, CH_2Ph),[159] some p-tolyl phenyl substituted di- and trisilanes and of $(Ph_3Si)_2SiH_2$,[160] polycyclic silanes[32b,51,52,161] were reported.

The structures of the products obtained by the redistribution reaction of $Si_2Me_6+Si_2Cl_6$ and $Si_2Me_2Cl_4+Si_2Me_2H_4$,[162] and by the photolysis of $(Me_3Si)_3SiCOR$ (R = Et, $CHMe_2$, CH_2Ph) have been determined[163] by means of ^{29}Si NMR. Substantial increase in the ^{29}Si shielding observed in K-silanides[164] as compared to that in similar Sn (II) derivatives[165] can be explained by the lower excitation energy (ΔE, Eq. 1.) in the latter case (δ, ppm):

	$KSiH_2SiH_3$	$KSiH(SiH_3)_2$	$KSi(SiH_3)_3$	$[(Me_3Si)_3Si]_2Sn$
SiK	-195.4	-238.3	-282.6	-6.2 (SiSn)
SiH_3	-85.7	-83.4	-76.2	-22.1 ($SiMe_3$)

Conjugate addition to methyl cinnamate in the reactions with methylcopper and lithium dimethylcuprate activated either by Me_3SiCl or Me_3SiI has been monitored by ^{29}Si NMR.[166]

(7) Si-Containing Transition Metal Compounds. A large amount of data has been obtained for the Si derivatives of transition metals. ^{29}Si spectra of over 40 iron-group transition-metal-silicon complexes have been studied.[167] In the series $M(SiMe_{3-n}Cl_n)$, (n=0-3) the $SiMeCl_2$ derivative usually gives the ^{29}Si signal to lowest field, whereas an upfield shift is observed on going to the metal lower in the Periodic table (δ, ppm):

$Fe(CO)_4(SiMe_{3-n}Cl_n)$		$M(CO)_4(SiMe_{3-n}Cl_n)$			
n=0	26.6	M=Ru, n=0	2.1	$Fe(CO)_4(SiCl_3)(H)$	46.5
n=1	61.3			$Os(CO)_4(SiCl_3)(H)$	-6.5
n=2	68.5	n=2	56.2	$Mn(CO)_5SiCl_3$	42.2
n=3	41.8	M=Os, n=0	-22.8	$Co(CO)_4SiCl_3$	36.2
		n=2	24.4	$Re(CO)_5SiCl_3$	16.1

^{29}Si spectra have been shown to be useful in structure studies of polysilane-metal complexes because the ^{29}Si CS are strongly dependent upon the location of Si with respect to the metal centre.[168] Thus, compared to the unsubstituted analogues the Si atoms directly bonded to metal (M) exhibit the CS differences ($\Delta\delta = \delta(Si-M) - \delta(Si-CH_3)$, ppm) of ca. 35 to 40 (Fe), 20 (Ru), and -13 to -39 (Re). For Si_β this shift is ca. 10 ppm (Fe, Re, Ru), and for Si_γ ca. 0.0 ppm. When the chain is bonded to Cp ring, $\Delta\delta$ values are as follows, -4.0 (α-Fe), +0.5 (β-Fe) and 0.0 (γ-Fe) ppm.[168]

Very large deshielding of ^{29}Si resonances has been observed in 7-silanor-bornadienyl-iron complexes (δ, ppm):[169]

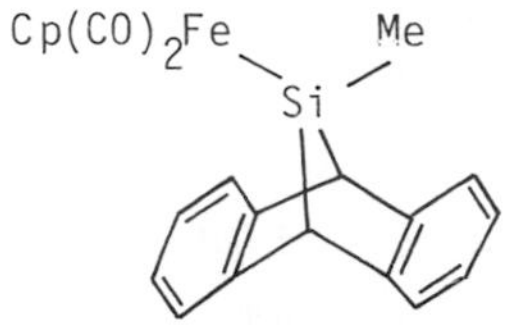

Cp(CO)$_2$Fe — Si — Me

139.0

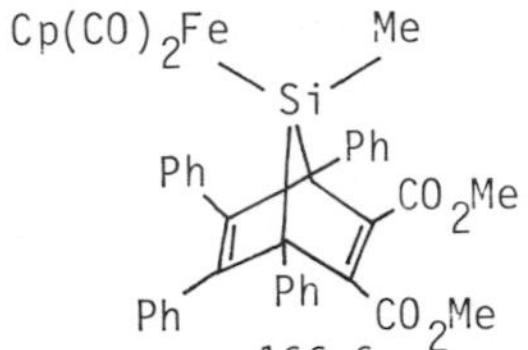

Cp(CO)$_2$Fe — Si — Me, Ph, Ph, Ph, Ph, CO$_2$Me, CO$_2$Me

166.6

Cp(CO)$_2$Fe — Si — Me, Ph, Ph, Ph, Ph

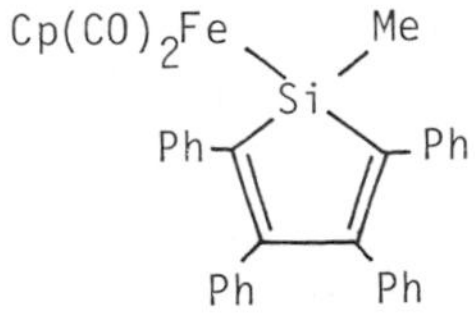

47.6

Products of σ-bond metathesis reactions of Si-H and Si-M (M=Zr, Hf) have been studied by ^{29}Si NMR.[170] ^{29}Si CS of complexes containing Zr-Si bond of the type R(Cp)$_2$Zr(SiHMes$_2$)$_2$[171] and (Me$_3$Si)$_3$SiZr(X)Cp (X=Me, Cl, Cp=C$_5$H$_5$, C$_5$Me$_5$)[172] as well as [C$_5$H$_4$Si(SiMe$_3$)$_3$]$_2$WH and (Cp)$_2$W(H)Si(SiMe$_3$)$_3$,[173] and new silicon mono- and disubstituted (η^4-2,5-di-Ph-silacyclopentadiene)-transition-metal complexes[174] were characterized by ^{29}Si NMR.

The unusually shielded ^{29}Si has been found in the (Cp)$_2$Ti derivatives of si-lacyclobutanes (δ, ppm):[175]

(Cp)$_2$Ti ◁▷ SiMe$_2$

-76.0

(Cp)$_2$Ti ◁▷ Si ◁▷ Ti(Cp)$_2$

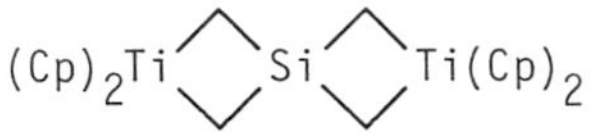

-145.3

^{29}Si resonances of the spirocyclic silylamides M[(NR)$_2$SiMe$_2$]$_2$ (M=Hf, NbMe, NbCl, TaMe, TaCl, R=Bu^t, SiMe$_3$) appear in the range between -33.8 and -41.2 ppm.[176] Nosova et al. have reported the ^{29}Si spectra of 9,9,10,10-tetra-methyl-9,10-disilyl-9,10-dihydroanthracene, (Me$_2$PhSi)$_2$, and their mono- and bis-Cr(CO)$_3$ complexes.[177]

^{29}Si signals of silyl substituted metallocenes with M=V, Cr, Co, Ni cover a range of more than 2200 ppm, yielding the first paramagnetic ^{29}Si NMR shifts.[178] The shifts show a strong temperature dependence. The ^{29}Si data of several dia-magnetic metallocenes have also been reported (δ, ppm):[168,179]

Paramagnetic[178]		Diamagnetic[179]	
(Me$_3$SiCp)$_2$M		(Me$_3$SiCp)$_2$Fe	
M=V	-1364	-4.0	
Cr	-1142	(Me$_3$SiCp)In	
Co	335	-10.6	
Ni	904		

A ca. 30 ppm downfield shift of ^{29}Si resonance in siloxycarbenemanganese complexes reflects the effect of π-complexation of the vinyl group and ring formation (δ, ppm):[180]

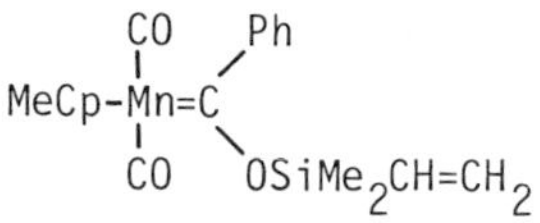

3.2

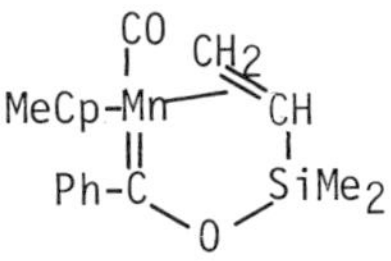

33.6

Complexes containing metal, H, Si 2-electron 3-centre bonds (metal=Mn, Cr, etc.) have been studied by means of ^{29}Si NMR.[181]

(8) Compounds Containing Si=E Bond. Compounds containing the Si=E bond have been obtained only recently. For this reason the present paragraph is new for ^{29}Si NMR. The extremely deshielded ^{29}Si CS values of tricoordinated Si atoms are highly diagnostic and usually fall between 40 and 180 ppm depending on the double bond partner and the substituents at the double bond. The most deshielded ^{29}Si value ever reported for a diamagnetic Si compound has been observed among the Si=P derivatives (δ between 148 and 176 ppm).[182] The ^{29}Si resonances have been found in the ranges between 38 and 144 ppm for Si=C,[76c,183-185] 49 and 95 ppm for Si=Si,[32a,184,186] 60 and 80 ppm for Si=N derivatives (δ, ppm):[187,188]

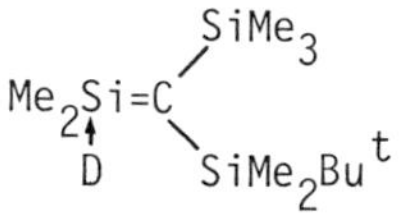

D = -   144.2

THF   52.4

Me$_3$N   34.7

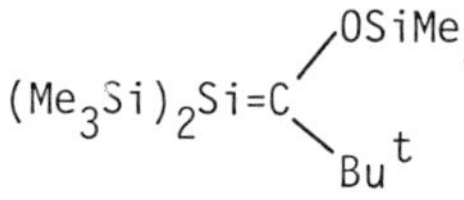

41.4

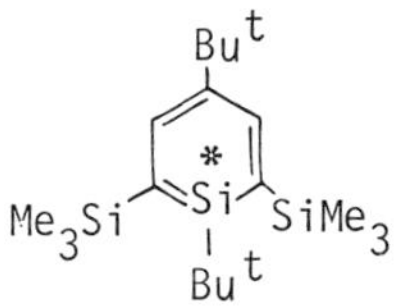

15   26.8

Mes$_2$Si=SiMes$_2$

63.6

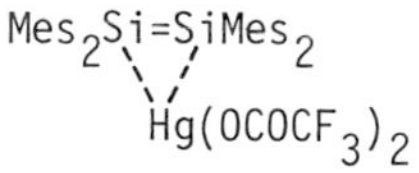

8.5

Mes$_2$Si=PBus^t

151.2

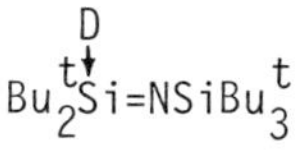

D = THF   1.0

EtMe$_2$N   18.1

Ph$_2$CO   54.2

-         78.4

Märkl and Schlosser[189] have reported ^{29}Si CS of silabenzene 15 which was stable up to ca. -100°C.

The deshielding of ^{29}Si nuclei can be explained in terms of paramagnetic contribution to ^{29}Si CS (Eq. 1) by the involvement of Si in $p_\pi-p_\pi$ bonding which lowers the HOMO-LUMO energy gap ($\pi-\pi^*$).

A characteristic feature of Si=N and Si=C derivatives is the formation of adducts with electron donors (D = THF, Py etc.) which is accompanied by a significant high-field shift of the ^{29}Si resonance.[190] The ^{29}Si data clearly eliminate the possibility of such coordination with phosphasilenes[182] and indicate the formation of π-complex between Hg(OCOCF$_3$)$_2$ and disilenes.[191]

Recently, ^{29}Si data have been reported for compounds containing Si=S and Si=Se bonds (δ, ppm):[192]

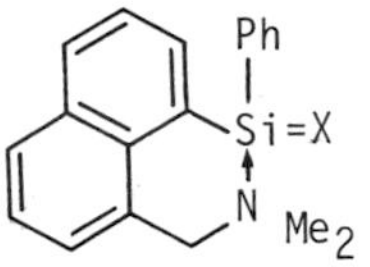

$$X = S \quad 22.3$$
$$Se \quad 29.4 \quad (^1J_{SeSi} = 257 \text{ Hz})$$

The silanediyl complex 16 has been identified by ^{29}Si NMR.[193] Rotation around the Si=Fe bond was unhindered at room temperature. An increase in ^{29}Si shielding is observed on going to Si=Fe and Si=Cr derivatives containing EN substituents at Si (δ, ppm):

$$(OC)_4Fe=SiMe_2$$
$$\uparrow$$
$$D$$
16
92.4 (D = HMPT)

$$(Bu^tO)_2Si=M$$
$$\uparrow$$
$$D$$

M = Fe(CO)$_4$ 7.1 (D = HMPT)

                -9.1 (D = THF)

Cr(CO)$_4$ 12.7 (D = HMPT)

(9) Compounds of 5-, 6- and 7-Coordinated Si. The high sensitivity of ^{29}Si CS to the coordination number of the Si atom permits detection of the formation of weak and labile coordinate bonds to Si. For this reason ^{29}Si NMR has been widely used in studies of Si coordination compounds. Some typical examples are presented below.

5-Coordinated Si Compounds

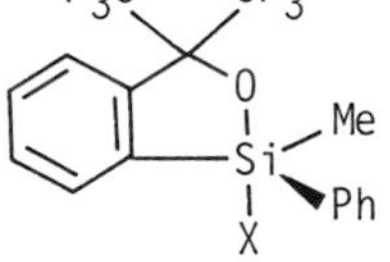

17 X = none   25.0

CN$^-$ -80.1

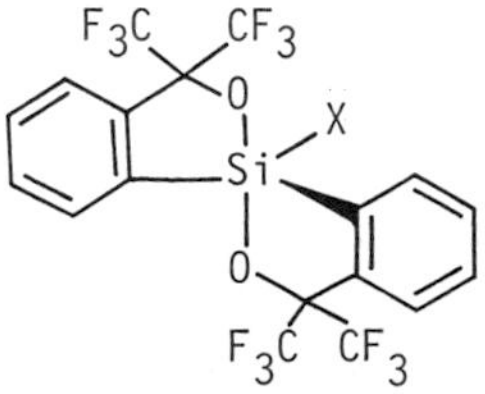

18 X = none   7.9

CN$^-$ -91.5

19 PhSiF₃ −72.9

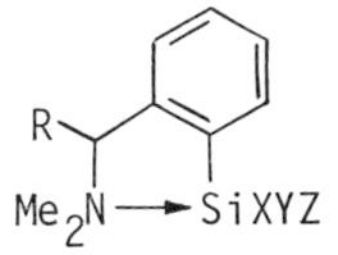

$Me_2N \rightarrow SiXYZ$

−102.3 (XYZ=F₃, R=H)

ref.196

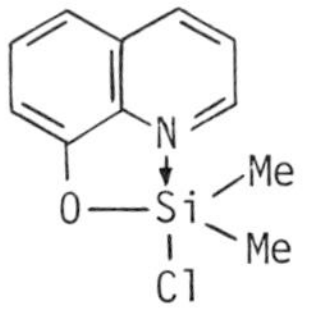

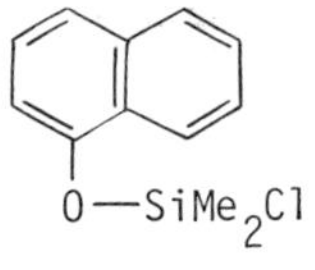

20 −21.0 ref.197

$O\!-\!SiMe_2Cl$ 14.1

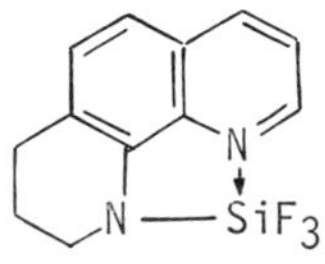

21 −125.2 ref.197

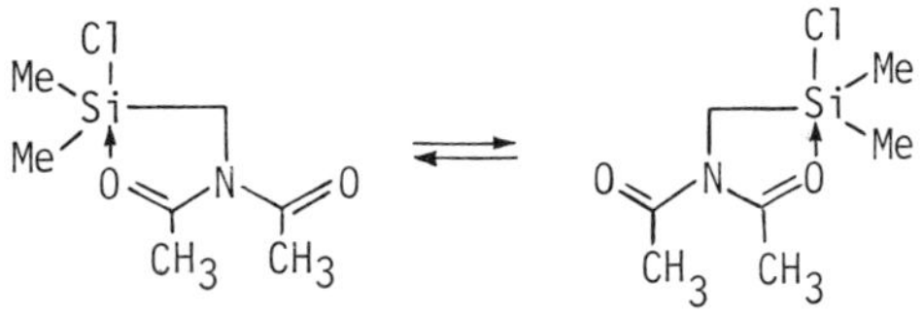

22a ⇌ **22b** −24.2 ref.198a

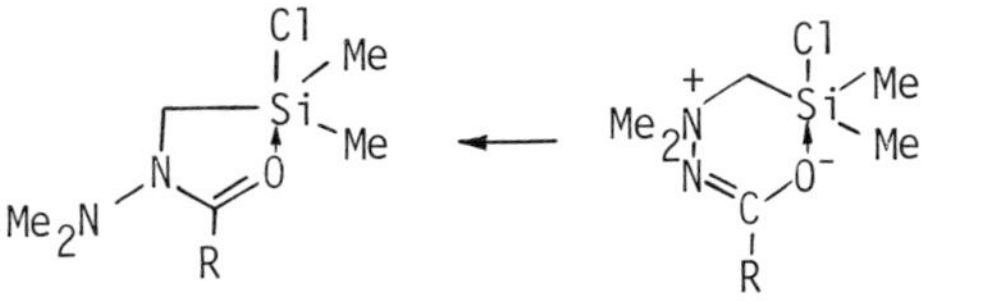

23a −35.9 (R=Me)   **23b** −33.5 (R=Me) ref.198d

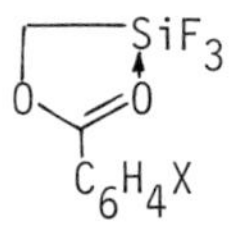

24 −90 to −96 C_6H_4X ref.198b

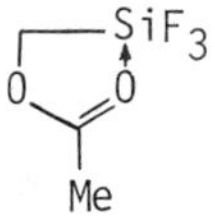

25 −82.1 Me ref.198c

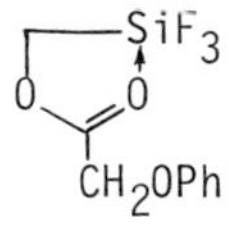

26. −89.0 CH_2OPh ref.198e

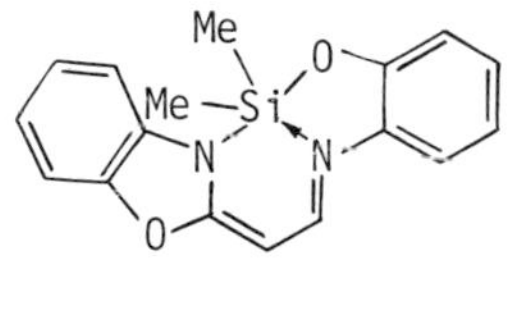

−59.8 ref.213

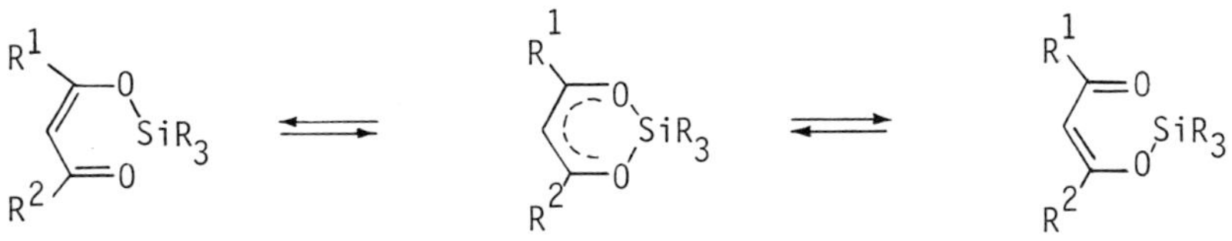

27 R^1, R^2=OMe, Me, Ph; R_3=Me₃, Me₂(CH₂Cl), Me₂(C₆F₅)

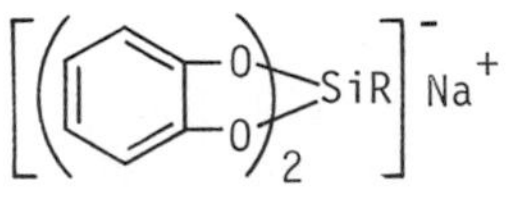

28

-73.8 (R=Me)
-87.4 (R=Ph)
-84.6 (R=1-naphthyl)
ref.200

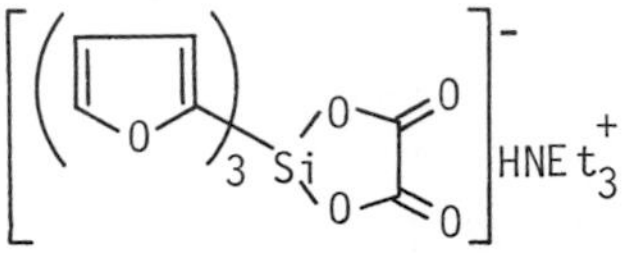

-114.6
ref.216

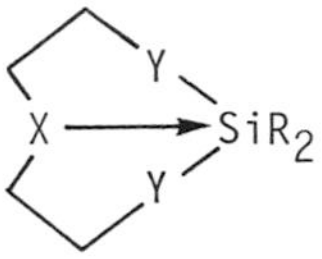

29

a) X=NMe, Y=CH$_2$,
   -2.3 (R=Me) ref.202
b) X=NMe, Y=0.
   -45.7 (R=Ph) ref.201
c) X=Y=0,
   -32.6 (R=Ph) ref.203
d) X=S, Y=0,
   -32.5 (R=Ph) ref.203

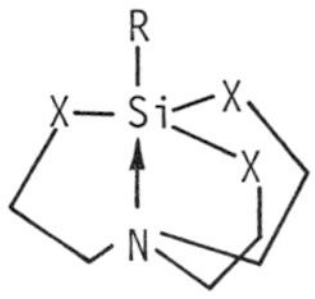

30

a) X=CH$_2$,
   -12.7 (R=Me) ref.202
b) X=NH,
   -68.3 (R=Me) ref.204
c) X=0,
   -69.7 (R=Me) ref.207

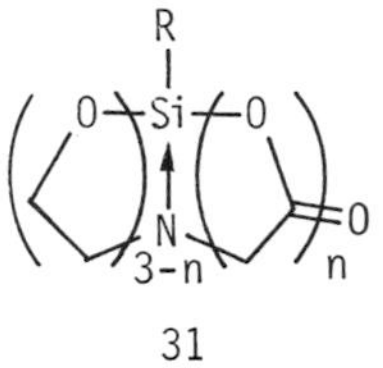

31

-73.2 (R=Me) n=1
-77.2 (R=Me) n=2
-135.8 (R=Me) n=3
ref. 207

6-Coordinated Si Compounds

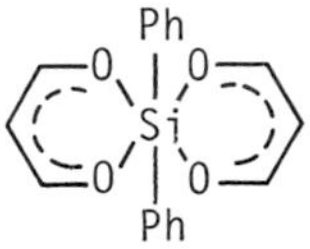

-161.7
ref.214a

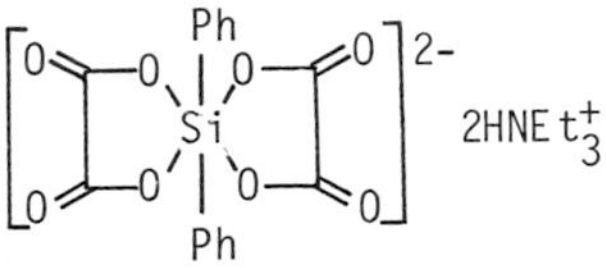

-157.9
ref.216

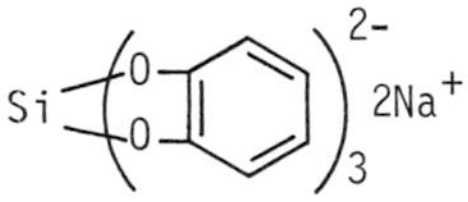

-143.4
ref.214b

7-Coordinated Si Compounds

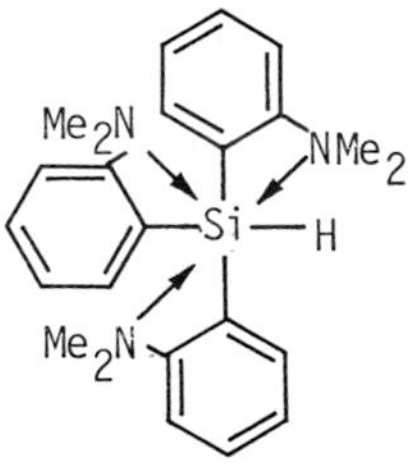

-35.3
ref.215

Ph$_3$SiH

-17.8

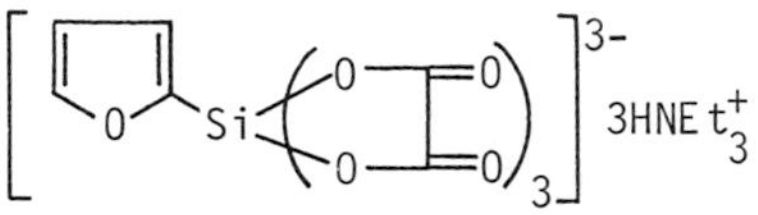

-175.9
ref.216

An increase in [29]Si shielding is characteristic of species containing hyper-valent Si atoms. This allows one to distinguish the formation of four- and five-coordinated silicon salts by [29]Si CS titration of nucleophiles against Me_2HSiX (X=Cl, OSO_2CF_3) (δ, ppm):[124]

$$Nu + Me_2HSiX \rightleftharpoons X^-Nu^+-SiHMe_2$$

$$\underset{\underset{\overset{|}{H}}{Nu^{+-}Si-Nu^+X^-}}{\overset{Me\quad\quad Me}{\diagdown\diagup}}$$

+10 to +24

Nu=pyridine

-44 to -82

N-methylimidazole

4-dimethylaminopyridine

Corriu et al.[194] have reported the [29]Si CS of anionic pentacoordinated hydro-silicates. As expected, upfield shifts of the [29]Si resonances relative to those of the corresponding tetracoordinate silanes were observed (δ, ppm($^1J_{SiH}$, Hz)):[194]

$HSi(OEt)_3$	-58.8	(285)	$HSi(OPh)_3$	-70.5	(320)
$HSi(OEt)_4^-$	-88.1	(223)	$HSi(OPh)_4^-$	-111.9	(296)

Dixon et al.[195] have examined the [29]Si spectra of Me_3SiCN (δ=-11.9 ppm) in the presence of $Bu_4N^+CN^-$. A large upfield shift (δ=-45.0 ppm, broad) was observed due to formation of a substantial amount of $Me_3Si(CN)_2^-Bu_4N^+$. Using IR-determined concentrations [29]Si CS of $Me_3Si(CN)_2^-Bu_4N^+$ was calculated (δ=-138.7 ppm) to be about 127 ppm to high field of the CS for Me_3SiCN. The association constant and thermodynamic parameters have been calculated using ^{29}Si NMR data. A similar shielding effect upon the [29]Si CS was observed in going from tetra- to pentacoordinated cyanosiliconates 17 and 18.[195] In contrast, addition of cyanide ion to Pr_3^iSiCN and $(Bu^tO)_3SiCN$ showed no shift of [29]Si resonance attributable to formation of pentacoordinated siliconates, presumably due to the great bulk of the substituents.

For the compounds 19 which have X=Y=Z=Me or OR, the [29]Si CS are essentially unchanged from those of the reference compounds (PhSiXYZ), suggesting that no N→Si coordination take place.[196] The strong high-field shift of [29]Si resonance observed in 19 (X,Y,Z=H, F or Cl),[196] 20 and 21[197] as compared with model compounds, provides evidence of the formation of an additional DA N→Si bond.

Similar observations suggest a pentacoordinated Si atom is involved in the intramolecular O→Si coordination in compounds 22-26.[198] The temperature dependence of the [29]Si CS in the *cis*-isomer of 27 indicates that the silyl group migration between the O atoms involves formation of the pentacoordinate transition state.[199]

In the series of anionic pentacoordinated Si complexes 28, showing unexpectedly low ^{29}Si shielding,[53b] additional examples have been reported.[200]

Investigations of 1,5-transannular interaction in silacyclooctanes <u>29</u> have continued.[18,201,202] The growth in the strength of the N→Si interaction with increasing donor ability of the heteroatom X and decreasing temperature, is accompanied by high-field shift of the ^{29}Si resonance. The apicophility of the Si-substituents on the TBP has been estimated to increase in the order H<Ph<Me, Pr<OR<Cl.[201,203]

The increase of ^{29}Si shielding in tricarbasilatranes <u>30a</u> (Δδ between -19 and -58 ppm)[202] and triazasilatranes <u>30b</u> (Δδ, -44.2 to -45.5 ppm)[204] with respect to the model compounds, suggests the existence of an intramolecular DA N→Si bond. The decrease in Δδ with increasing EN of the Si-substituent in silatranes <u>30c</u> has been explained by a diamagnetic origin of the high-field shifts.[205] Bellama et al. have used ^{29}Si CS to determine the effective Taft polarity constants for silatrane substituents.[206]

The systematic growth in ^{29}Si shielding with increasing the number (n) of carbonyl groups in silatranones <u>31</u> reflects the appropriate rise in the DA N→Si bond strength.[207] The dramatic increase in high-field shift observed at n=3 has been explained by a further expansion of Si coordination resulting from complexation with a solvent (dimethyl sulfoxide) molecule.

^{29}Si NMR data have provided the most direct evidence for an increase of Si coordination from four to five in 1,2-bis|Me$_2$(X)SiCH$_2$|-1,2-diacetylhydrazines (X=F, Cl, MeO),[208] R$_2$Si|(o-C$_6$H$_4$)CH$_2$|$_2$X (X=S, Se, S$_2$, Se$_2$),[209] Me$_2$XSiCH$_2$N(COCH$_3$)$_2$ (X=Br, I, OCOMe, OCOCF$_3$),[210] Me$_2$(Cl)SiCH$_2$SC=N(CH$_2$)$_3$,[211] and structures depicted below (δ(Δδ), ppm):[212]

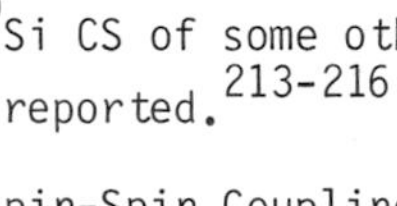

SiF$_3$
NMe$_2$

-100.5 (-36.4)

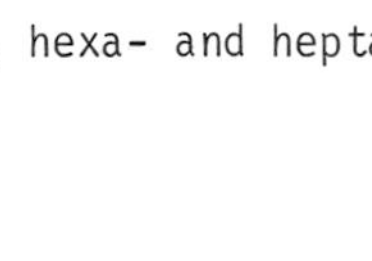

Me$_2$N ⟶ SiH$_2$Ph

-44.1 (-8.5)

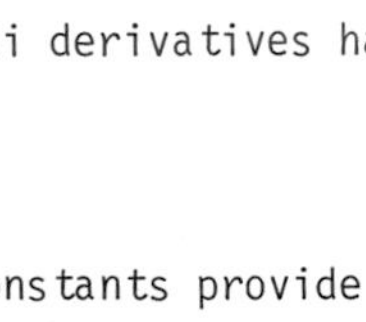

N—SiH$_2$Ph
Me$_2$

-55.5 (-19.9)

^{29}Si CS of some other penta-, hexa- and heptacoordinated Si derivatives have been reported.[213-216]

(B) Spin-Spin Coupling.

(1) Theoretical Considerations. Nuclear spin-spin coupling constants provide valuable information concerning the structure of molecules, the nature of chemical bonds and dynamic processes in liquids. There are three valence electron-mediated nuclear spin coupling mechanisms which are believed to be responsible for the appearance of couplings in the liquid state, i.e. spin-orbital (SO), spin-dipolar (SD) and Fermi-contact (FC) interactions. The last term is generally

assumed to dominate the magnitude of coupling. This allows one to interpret the experimental results in terms of analytical expressions obtained for FC inter-action, e.g. equation derived for one-bond coupling (^1J) within the MO frame-work:[217]

$$^1K^{FC}_{AB} = \frac{4}{9}\mu^2\mu_o^2 s_A^2(0)\ s_B^2(0)\pi_{AB} \tag{5}$$

$$K_{AB} = 4\pi^2 J_{AB}/h\gamma_A\gamma_B \tag{6}$$

where ^1K is the reduced coupling constant which is usually introduced (Eq. 6.) in order to neglect the individual properties of the nuclei, $s_X^2(0)$ is the valence s electron density on the X nucleus and π_{AB} is the mutual polarizability of the A and B s orbitals, while other constants have their usual meanings.

The trend with increasing atomic number down a group and across a row in the Periodic Table is for the π_{AB} term to become more negative, which can lead to a sign change in the reduced coupling from positive to negative. In the absence of lone electron pairs, the π_{AB} term is generally positive and the mean excitation energy (ΔE) approximation is valid, leading to an equation relating ^1K and s-order of the A-B bond ($P_{s_A s_B}$):

$$^1K^{FC}_{AB} = \frac{4}{9}\mu^2\mu_o^2 (\Delta E)^{-1}s_A^2(0)s_B^2(0)P^2_{s_A s_B} \tag{7}$$

The latter equation is extremely simplified and suggests only positive ^1K va-lues. For this reason, its application to systems involving lone electron pairs or heavy atoms may not be justified.

Excellent treatments of the nuclear spin coupling theory have been published elsewhere.[218]

(2) One-Bond Couplings. ^{29}Si-^1H couplings are easily available experimentally and have been studied in detail earlier.[53] Many correlations relating $^1J_{SiH}$ with s-character of the Si bonding orbital, EN and σ^* constants of the Si-substitu-ents, the force constants of the Si-H bond etc., have been found. Illustrative examples of couplings to ^{29}Si obtained recently are collected in Table 1 toge-ther with the most characteristic values reflecting the ranges of these coupl-ings.

In disilanes the magnitude of $^1J_{SiH}$ increases with increasing EN of Si-sub-stituents,[151,162] reflecting a corresponding increase in s-character of the bond Si-H. The lowest $^1J_{SiH}$ values have been observed in K-silanides.[164] A good correlation exists between the ^{29}Si hyperfine coupling constant of a silicon-centred radical and $^1J_{SiH}$ for the corresponding silane.[219] This suggests a li-near relationship between the s-character of the orbital containing the unpaired

Table 1. Examples of ^{29}Si-X Spin-Spin Coupling Constants, Hz

Compounds	$^nJ_{SiX}$	Ref.	Compounds	$^nJ_{SiX}$	Ref.
$^1J(^{29}Si-^1H)$ $(^1K_{SiH}{>}0)$			$(Me_3Si)_2Si{=}\underline{C}(OSiMe_3)Ad$	84.0	184
			$Mes_2Si{=}SiMes_2$	90.0	184
$KSiH_3$	74.8	53a	$Me_3Si\underline{C}{\equiv}CH$	81.79	262
$K\underline{Si}H_2SiH_3$	76.3	164	$(EtO)_3SiC{\equiv}CH$	156.8	322
$KSiH_2\underline{Si}H_3$	148.9	164	$^1J(^{29}Si-^{15}N)$ $(^1K_{SiN}{>}0)$		
H_3SiSiH_3	197.8	151			
Me_3SiH	184.0	53b	$N(CH_2CH_2O)_3\overset{\rightarrow}{Si}Me$	0.7	18
ClH_2SiSiH_2Cl	233.1	151	$N(CH_2COO)_3\overset{\rightarrow}{Si}Me$	8.2	207
$Cl_2HSiSiHCl_2$	285.3	151	$N(CH_2COO)_3\overset{\rightarrow}{Si}CH_2Cl$	10.9	207
$(PrHN)_3SiH$	227.0	204	$Me_2Si{-}SiMe_2$		
$N(CH_2CH_2NH)_3\overset{\rightarrow}{Si}H$	176.6	204	$Pr^iN\quad NPr^i$	5.5	226c
$(EtO)_3SiH$	287.4	204	$SnMe_2$		
$N(CH_2CH_2O)_3\overset{\rightarrow}{Si}H$	280.1	204	$Me_3SiNHOSiMe_3$	6.5	36
F_3SiH	381.7	53b	$(Me_3Si)_3N$	7.6	226c
$H_2Si[Mn(CO)_5]_2{\cdot}4Py$	420	53a	$(Me_3Si)_2NNa$	7.8	25
$^1J(^{29}Si-^6Li)$			$(Me_3Si)_2NH$	13.5	24
			$(Me_2SiNH)_3$	15.4	31
$Ph_2MeSiLi$	16	223b	$(Me_2SiNH)_4$	16.9	31
Ph_3SiLi	17	223b	$Me_3SiNHPh$	15.7	24
$PhMe_2SiLi$	18	223b	$Me_3SiNHBu^t$	17.2	24
$^1J(^{29}Si-^7Li)$			Me_3SiNEt_2	19.2	24
Ph_3SiLi	51	223b	$Me_3SiN{<}\rangle$	13.6	24
$^1J(^{29}Si-^{11}B)$ $(^1K_{SiB}{>}0)$			$Me_3SiNHC(O)Me$	13.4	24
			$Me_3SiN{=}C{=}S$	12.2	24
$(Me_3SiBH_3)Li$	74.0	254b	$Me_3SiN{=}C{=}O$	14.5	24
			$Me_3SiN{=}C{=}NSiMe_3$	16.2	203
$Me_3SiB\langle$ (R,N...N,R ring)	97.0	53a	$(EtO)Me_2SiNHBu^t$	21.5	24
			$(EtO)_2MeSiNHBu^t$	30.9	24
			$(EtO)_3SiNHBu^t$	44.6	24
$^1J(^{29}Si-^{13}C)$ $(^1K_{SiC}{>}0)$			$Si(NHPr)_4$	31.7	24
			$[Si(N{=}C{=}S)_6]^{2-}$ $2K^+$	$(36.1)^a$	203
Pn_3SiLi	10	223c	$Si(NCO)_4$	57.3	203
$Me_3Si\underline{C}(SnMe_3)_3$	30.0	248c	$^1J(^{29}Si-^{19}F)$ $(^1K_{SiF}{<}0)$		
$(Me_3S\underline{i})_3\underline{C}SnMe_3$	37.2	248c			
Me_4Si	50.845	34b	SiF_6^{2-}	108.1	53a
$Me_2Si{<}\rangle Si\underline{Me}_2$	48.29	262	SiF_4	178.6	53a
			$(EtO)_3SiF$	199.1	53a
(CH_2)	35.07	262	Me_3SiF	274.5	53a
$N(CH_2CH_2CH_2)_3Si\underline{Me}$	39.4	202	Cl_3SiF	311.5	53a
(CH_2)	58.2	202	$(Me_3Si)_3SiF$	327.8	53a
Me_3SiNEt_2	56.5	24	SiF_2 ... SiF_2	488	53a
$MeSi(NHPr)_3$	75.3	204			
$N(CH_2CH_2NH)_3Si\underline{Me}$	66.2	204	$^1J(^{29}Si-^{29}Si)$		
$MeSi(OEt)_3$	96.2	53a			
$N(CH_2CH_2O)_3Si\underline{Me}$	107.0	18	$(Me_3Si)_4Si$	52.5	53a
$Me_3Si\underline{C}H{=}CH_2$	64.74	203	$(ClMe_2Si)_4Si$	59.8	155
$Me_3Si\underline{Ph}$	66.02	203	H_3SiSiH_2Cl	88.3	151
$(Me_3Si\underline{Cp})_2Fe$	71.6	179a	$Me_3SiSiMe_2Cl$	94.0	53a
			$Me_3SiSiMe_2F$	98.7	53a
			$(Me_3SiSiMe_2)_2NH$	96.0	53a

Table 1. (Continued)

Compound	$^nJ_{SiX}$	Ref.	Compound	$^nJ_{SiX}$	Ref.
$(Me_3SiSiMe_2)_2O$	103.4	53a	Me_3SiOEt	7.1	53a
$H_3SiSiCl_3$	131.3	151	$MeSi(OEt)_3$	8.3	53a
$Mes_2Si=Si(Mes)Xyl$	158.0	32	Me_3SiCl	7.0	53a
$(MeO)_3SiSiPh_3$	160.0	53a	$MeSiCl_3$	9.0	53a
$Cl_2HSiSiCl_3$	221.0	151	Me_3SiNMe_2	6.5	53a
			$MeSi(NMe_2)_3$	6.9	53a
$^1J(^{31}P-^{29}Si)$ $(^1K_{PSi}<0)$			Me_3SiSMe	6.7	53a
			$MeSi(SMe)_3$	7.3	53a
Me_3SiPH_2	16.2	53a	$Me_3SiSnMe_3$	6.7	53a
Me_3SiPMe_2	20.3	53b	$Me_3SiCH=CH_2$	6.4	241
$(Me_3Si)_3P$	27.3	230	$(PrNH)_3SiCH=CH_2$	6.9	204
$(Me_3Si)_2PSi(SiMe_3)_3$	32.4	230	$(EtO)_3SiCH=CH_2$	7.8	204
$(Me_3Si)_2PSi(SiMe_3)_3$	90.5	230	H_3SiSiH_3	5.2	151
$Mes_2Si=PBu^t$	149	182	ClH_2SiSiH_2Cl	15.4	151
$(H_3Si)_2PLi$	256	53a	$Cl_2HSiSiHCl_2$	36.0	151
$^1J(^{57}Fe-^{29}Si)$			$^2J(^{29}Si-^{13}C)$		
$Si_6Me_{11}Fe(CO)_2Cp$	12.1	235	$Si(OMe)_4$	1.10	37
			$MeSi(OMe)_3$	1.32	37
$^1J(^{77}Se-^{29}Si)$ $(^1K_{SeSi}<0)$			$Me_2Si(OMe)_2$	1.77	37
			Me_3SiOMe	1.77	203
$(H_3Si)_2Se$	110.6	53a	Me_3SiOBu^t	2.34	203
			Me_3SiPh	3.76	203
$^1J(^{103}Rh-^{29}Si)$			H_3SiPh	4.88	257
$CpRh(H)_2(SiEt_3)_2$	16.6	237	$(Me_3SiCp)_2Fe$	5.5	179a
$Cp(C_2H_4)Rh(H)SiEt_3$	22.2	237	$Me_3Si-C\equiv CSnMe_3$	11.6	258
			$Me_3Si-C\equiv CMe$	17.6	258
$^1J(^{119}Sn-^{29}Si)$ $(^1K_{SnSi}>0)$			$Et_3Si-C\equiv CH$	18.6	53a
$(Me_3Sn)_4Si$	220	255	$(EtO)_3Si-C\equiv CH$	31.2	322
$(Ph_3Sn)_2SiPh_2$	515	256			
$Me_3SnSiEt_3$	650	53	$^2J(^{29}Si-^{15}N)$		
$^1J(^{125}Te-^{29}Si)$			$Me_3Si-C\equiv N$	1.6	297c
$Me_3SiTePh$	272	238	$^2J(^{29}Si-^{19}F)$ $(^1K_{SiF}>0)$		
$(Me_3Si)_2Te$	282	238	$(Me_3Si)_3SiF$	17.5	53a
			$Me_3Si(CFCl_2)$	28.4	244
$^1J(^{195}Pt-^{29}Si)$ $(^1K_{PtSi}<0)$			$Me_2Si(CFCl_2)Cl$	33.7	244
$trans$-$PtCl(SiCH_2Cl)(PEt_3)_2$	1600	53a	$MeSi(CFCl_2)Cl_2$	42.6	244
			$Cl_3Si(CFCl_2)$	53.4	244
$^1J(^{199}Hg-^{29}Si)$			Si_2F_6	90.5	53a
$[(Me_3Si)_3Si]_2Hg$	432	240	$^2J(^{29}Si-^{29}Si)$		
$(Me_3Si)_2Hg$	981	240	$(Me_3SiO)_2SiMe_2$	1.32	37
$Me_3SiHgMe$	1367	240	$(Me_3SiO)_4Si$	4.0	53a
$(ClMe_2Si)_2Hg$	1392	240	$(Me_3Si)_2C=C(Et)BEt_2$	8.6	224
$(Cl_2MeSi)_2Hg$	2020	240			
$(Cl_2MeSi)_2Hg\cdot Et_2O$	2395	240	$^2J(^{31}P-^{29}Si)$ $(^2K_{PSi}\lessgtr 0)$		
$(Cl_3Si)_2Hg\cdot Et_2O$	3864	240	$Bu_2^tFSiN=PCl_3$	7.5	246
$^2J(^{29}Si-^1H)(^2K_{SiH}<0)$			$Bu_2^tFSiN=PF_3$	36.0	246
Me_3SiLi	2.8	53a	$Me_3SiNP(NEt_2)_2$	41.5	53a
Me_4Si	6.65	35			

Table 1. (Continued)

Compound	$^nJ_{SiX}$	Ref.	Compound	$^nJ_{SiX}$	Ref.
$^2J(^{109/107}Ag-^{29}Si)$			Me$_3$SiOCH$_2$CH$_3$	2.57	203
			Me$_3$SiOBu^t	1.51	203
[(Bu^{t}O)$_3$SiSAg]$_4$	8.85	138a	(Me$_3$SiCp)$_2$Fe	4.4	179a
$^2J(^{119}Sn-^{29}Si)$			Me$_3$SiPh	5.03	203
			H$_3$SiPh	6.09	257
Me$_3$SiC(SnMe$_3$)$_3$	29.6	248b	Cl$_3$SiPh	9.1	257
(Me$_3$Si)$_3$CSnMe$_3$	38.0	248b			
Me$_3$Si(Me$_3$Sn)C=C(Et)BEt$_2$	92.1/ 93.0	224	$^3J(^{29}Si-^{15}N)$		
$^2J(^{183}W-^{29}Si)$			Ph$_2$Si(NHSiMe$_2$)$_2$NH	2.8	25
			H$_2$NSiMe$_2$N(SiMe$_2$)$_2$NSiMe$_2$NH$_2$	3.1	25
PhMe$_2$SiC(OMe)W(CO)$_5$	11.8	53a	$^3J(^{29}Si-^{19}F)$ ($^3K_{SiF}>0$)		
Ph$_2$MeSiCW(CO)$_2$Cp	55.9	53a			
$^2J(^{199}Hg-^{29}Si)$			(F$_3$Si)$_2$O	<2.5	53a
			Si$_3$F$_8$	15.7	53a
(Me$_3$SiCH$_2$)$_2$Hg	46	252	$^3J(^{29}Si-^{29}Si)$		
(Cl$_3$SiCH$_2$)$_2$Hg	47	252			
[Me$_2$(MeO)Si]$_2$Hg	50	252	B(Me$_3$SiC=CHSiMe$_3$)$_3$ (cis)	6.2	224
[Me(MeO)$_2$Si]$_2$Hg	60	252	Me$_3$SiONHSiMe$_3$	2.45	36
[(MeO)$_3$Si]$_2$Hg	72	252	$^3J(^{31}P-^{29}Si)$		
$^3J(^{29}Si-^1H)$ ($^3K_{SiH}>0$)					
(H$_2$SiCH$_2$)$_3$	0.9	53a	PhP(SiMe$_2$)$_6$	0.9	53a
(EtO)$_4$Si	2.9	30a	P(CH$_2$O)$_3$SiMe	34.0	53a
(EtO)$_3$SiPh	5.6	30a			
(EtO)$_3$Si–C≡CH	5.8	30a	$^3J(^{119}Sn-^{29}Si)$ ($^3K_{SnSi}>0$)		
(EtO)$_3$SiCH$_2$CH$_3$	7.6	30a			
(EtO)$_3$SiCH=CH$_2$ (Cis)	9.6	204	Me$_2$Si(C≡CSnMe$_3$)$_2$	10.6	248a
(Trans)	19.4	204			
Me$_3$Si(BR$_2$)C=CHSiMe$_3$	21.3	224	Me$_2$Si⟨ ⟩SnMe$_2$	15.6	255
$^3J(^{29}Si-^{13}C)$					
Si(OCH$_2$CH$_3$)$_4$	2.35	36			
MeSi(OCH$_2$CH$_3$)$_3$	2.38	203			
Me$_2$Si(OCH$_2$CH$_3$)$_2$	2.46	203			

[a] Calculated by multiplying $^1J(^{29}Si-^{14}N)$ with $\gamma(^{15}N)/\gamma(^{14}N)\approx1.4$, $^1J(^{29}Si-^{14}N)$ = 25.8 Hz.

electron in the radical and s-character of the Si σ-bonding orbital forming the Si-H bond in the parent silane. Correlation between $^1J_{SiH}$ and the sum of σ* constants exists within the limited series of siloxy- and alkoxysilanes.[220] Popowski et al.[103] have observed a linear correlation between $^1J_{SiH}$ and the frequencies of the Si-H bond stretching vibration in siloxanes. Variation of $^1J_{SiH}$ (and $^2J_{SiH}$) with EN of Si substituents in (Me$_3$Si)$_3$CL and (Me$_2$PhSi)$_3$CL (L=Si(H)R^{1}R^2) has been studied.[140] A collection of $^1J_{SiH}$ couplings in arylhydrosilanes have been published.[221]

A decrease in $^1J_{SiH}$ was observed on going from four to five coordinated hydrosilicates (see previous section),[194] suggesting a decrease in s-character of the Si-H bond. Linear relationships exist between $^1J_{SiH}$ and Si-H bond stretching frequencies in silatranes[18] and triazasilatranes.[204] It was assumed that a decrease in $^1J_{SiH}$ reflects a lengthening of the axial Si-H bond with increasing strength of the DA N→Si bond. In contrast, strengthening of the DA N→Si bond in the silocanes containing the equatorially located Si-H bond on TBP is accompanied with a decrease in $^1J_{SiH}$.

Within the series of Si-transition metal complexes the magnitude of J_{SiH} can be correlated with changes in the M-H-Si 3-centre 2-electron bond.[181] Small J_{SiH} values have been observed in the case of short M-Si distances in crystals (J_{SiH}, Hz):[181b,c]

$$
\begin{array}{c}
\qquad\qquad CO \qquad \; H \qquad 70.8 \\
\qquad\qquad | \qquad \diagup \quad \diagdown \\
C_6Me_6-Cr\text{-}\text{-}\text{-}\text{-}\text{-}\text{-}SiPh_2 \\
\qquad\qquad | \qquad\qquad\quad | \\
\qquad\qquad CO \qquad\qquad\quad H
\end{array}
$$

197.0

Lichtenberger and Rai-Chandhuri have attributed this coupling to nonbonded NMR coupling.[222]

The ^{29}Si-^6Li and ^{29}Si-^7Li couplings imply at least a partial covalent contribution to the Si-Li bond in $Me_{3-n}Ph_nSiLi$ (n=1-3).[223] At temperatures above 193 K an averaging of $^1J_{SiLi}$ has been observed due to intramolecular Li exchange.

Earlier studies have revealed a linear relationship between $^1J_{SiC}$ and the s-character of the Si-C bond.[53] Another example has been found within Me_3Si derivatives of polycycloalkanes.[71] This indicates that the systematic increase in $^1J_{SiC}$ observed when EN substituents are introduced at Si in aminosilanes,[24] silazanes and siloxanes,[25,31] alkenylsilanes[30,224] is related to a corresponding increase in the s-character of the Si-C bond. For the apical Si-C bond of tricarbasilatranes 30a[202] and triazasilatranes 30b[204] the value of $^1J_{SiC}$ is considerably decreased as compared to that in model compounds, reflecting a corresponding redistribution of s-electrons. In contrast, increased $^1J_{SiC}$ values in silatranes 30c have been explained by the stereoelectronic effect of the O atom lp.[204] Values of $^1J_{SiC}$ in $Me_3Si(X)C=CHX$ (X=Cl, Br, I) were reported (54.3 to 55.2 (Me) and 55.4 to 62.3 (C_α) Hz).[225] As expected, enhanced values of $^1J_{SiC}$ are characteristic of sp^2-hybridized Si atoms.[184] The $^1J_{SiC}$ value observed in silaimines[187] is surprisingly low ($^1J_{SiC}$, Hz):

$(Me_3Si)_2Si=\underline{C}(OSiMe_3)Ad$	$Mes_2Si=SiMes_2$	$Pr_2^iSi=NBus^t$
84.0 ref.184	90.0 ref.184	55 ref.187

Experimental results[226] and *ab initio* calculations[227] indicate that the values of $^{29}Si-^{15}N$ couplings are dominated by FC-interaction. A systematic increase in $^1J_{SiN}$ occurs with increasing EN of substituents on Si. Linear correlations, $^1J_{SiN}/^1J_{SiC}$ and $^1J_{SiN}/\delta^{29}Si$, have been observed for NPh derivatives of
aminosilanes.[24] $^1J_{SiN}$ shows a marked dependence on the nature of the Si-N
bond, specifically on the extent of $d_\pi-p_\pi$ bonding, nitrogen hybridization and
Si-N bond polarity and length. A linear correlation exists between $^1J_{SiN}$ and
s-character of hybrid orbitals (s_X) involved in Si-N bonding in aminosilanes
(Eq. 8), thus providing a basis for the employment of $^1J_{SiN}$,

$$25\ ^1J_{SiN} = s_{Si}(\%)\ s_N(\%) \tag{8}$$

as a structural probe. A significant decrease in $^1J_{SiN}$ with diminishing ring size in Me_3Si derivatives of cyclic amines is illustrative of the sensitivity of
$^1J_{SiN}$ to nitrogen hybridization. However, relatively small $^1J_{SiN}$ values found
for compounds with a double-bonded nitrogen are indicative of limited applicability of Eq. 8. In cyclosilazanes, $^1J_{SiN}$ increases with growing ring size.[25]
A marked difference between the $^1J_{SiN}$ values in endocyclic and exocyclic Si-N
bonds of compound 32 reflects redistribution of the valence s-electrons
$(^1J_{SiN},$ Hz):

32

This agrees well with the chemical properties and the crystal structures of
these compounds.

Individual values of one-bond $^{29}Si-^{15}N$ couplings in Si-N-Sn and Si-N-Pb
derivatives[228] and $Me_2NSi(Cl)H_2$ (22.6 Hz)[125] have been published.

$^1J_{SiN}$ values were used to predict the DA N→Si bond order and length in silatranes (refs. 18,26). A linear correlation exists between the $^1J_{SiN}$ values and
the free energy of N→Si bond dissociation in silocanes. This allowed one to
estimate the energy of the N→Si bond in silatranes (50 to 70 kJ/mol) from $^1J_{SiN}$
values. $^1J_{SiN}$ data have been systematized elsewhere.[26]

A relatively large body of data regarding $^{29}Si-^{19}F$ couplings has been obtained earlier.[53] However, no general relationships between $^1J_{SiF}$ and structure
could be observed. Usually, EN substituents on Si decrease the absolute values
of $^1J_{SiF}$, but an opposite trend has also been noted.[53] $^1J_{SiF}$ couplings were involved in studies of pentacoordinated Si compounds (19, 21, 24-26).[196-198] An

increase in the DA X→Si (X=O, N) bond strength diminishes the magnitude of $^1J_{SiF}$. For the apical Si-F bond on the TBP of pentacoordinated Si the absolute values of $^1J_{SiF}$ (126.7 to 130.1 Hz) are lower, as compared with equatorial ones (136.0 to 140.9 Hz).

The magnitude of $^1J_{SiSi}$ couplings is governed by the FC interaction.[53] In disilanes, EN substituents on Si increase $^1J_{SiSi}$.[151,155] Kuroda et al.[32b,161] have observed two general trends for ^{29}Si-^{29}Si couplings in highly strained poly-cyclic silanes:

1) with the decrease in ring size from 6 to 3 $^1J_{SiSi}$ values become smaller and $^2J_{SiSi}$ values larger;

2) in strained polycyclic compounds not only are $^1J_{SiSi}$ and $^2J_{SiSi}$ comparable in magnitude, but even two nuclei that are three and four bonds apart and orien-ted in a stereochemically unique manner display a considerable interaction (J_{SiSi}, Hz):[32b]

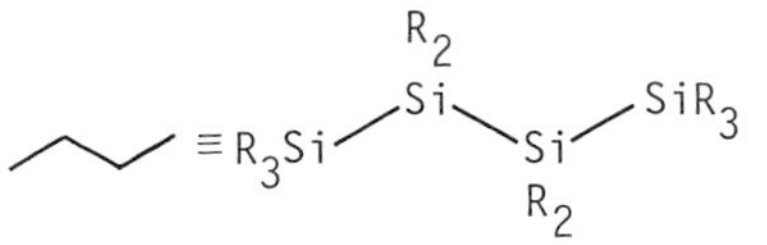

$$\sim \equiv R_3Si- \quad R_3Si-\overset{R_2}{\underset{R_2}{Si}}-\overset{}{\underset{}{Si}}-SiR_3$$

R=H, Alk, Ph

^1J=65.4-68.4 (SiR$_3$)

57.6 (SiR$_2$)

^2J=2.9-4.9

^3J=2.9

^1J=57.4-62.0

^2J=8.2-9.4

^1J=57.6

^2J=9.8

^1J=46.9

^2J=27.3

^1J=24.1

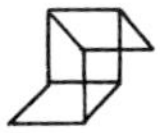

^1J=17.3-42.2

$^{2/3}J$=22.5-23.1

$^{3/4}J$=19.9-24.1

An increase in the magnitude of $^1J_{SiSi}$ with an increase in ring size has been explained by corresponding variation in the s-character of the Si orbitals parti-cipating in the ring formation. Similar values of $^1J_{SiSi}$ (45-60 Hz) and $^2J_{SiSi}$ (8-17 Hz) were observed by West et al.[52] and Hengge and Schrank.[51] Approximately a two-fold increase in $^1J_{SiSi}$ is observed for sp^2-hybridized Si in disilenes ($^1J_{SiSi}$=155 to 158 Hz),[32a] as compared to the values normally observed for the Si-Si bond in organodisilanes (80 to 90 Hz). This is consistent with theoreti-cal predictions, e.g. ab initio calculations gave 183.6 Hz for $H_2Si=SiH_2$.[227] The value of $^1J_{SiSi}$ in disilaoxiranes (99 Hz)[229] is intermediate between that in tetraaryldisilenes and organodisilanes, suggesting that the disilaoxiranes have some of the character of disilene-oxygen π-complexes (A) as well as oxiranes (B).

$$Xyl_3Si \overset{O}{=\!\!\!\uparrow\!\!\!=} SiAr_2$$

A

$$Xyl_2Si \overset{O}{\triangle} SiAr_2$$

B

A relatively bulky collection of [31]P-[29]Si couplings[53] has been supplied with additional data regarding the silaphosphanes $PSi_3Me_{9-x}(SiMe_3)_x$ (x=1-6)[230] and phosphasilenes.[182] A monotonic increase in $^1J_{PSi}$ occurs with increasing n in the series $Me_3SiPSiMe_{3-n}(SiMe_3)_n$ ($^1J_{PSi}$=27.3 to 90.5 Hz).[230] The magnitude of $^1J_{PSi}$ in silaphosphacubane 33,[231] and trisilaphosphanes $PSi_3Me_nPh_{9-n}$ (17.4 to 27.3 Hz)[232] was reported. Decrease in Si-P distance is accompanied with remarkable increase in $^1J_{PSi}$ in compound 34 ($^1J_{PSi}$, Hz):[233]

33 +37.6		ref.233 34 84.1	35

$(^3J_{PSi}$=-12.6)[232]

Relatively large values of $^1J_{PSi}$ were also observed in transition-metal complexes of cyclic organoboron phosphorus compounds 35 (31.8-51.2 Hz).[234] In the case of phosphasilenes, the $^1J_{PSi}$ couplings have unusually high values (149 to 155 Hz), indicating an enhanced s-character of the P=Si bond, as expected for sp^2-hybridization of both elements involved.[182] It is noteworthy that $^1J_{PSi}$ values above 100 Hz have not been observed for single P-Si bonds with the remarkable exception of $LiP(SiH_3)_2$ ($^1J_{PSi}$ = 256 Hz).[53]

Individual values of $^1J(^{57}Fe-^{29}Si)$,[235] $^1J(^{77}Se-^{29}Si)$,[236] $^1J(^{103}Rh-^{29}Si)$[237] and $^1J(^{125}Te-^{29}Si)$[238] couplings have been reported for the first time.

$^{195}Pt-^{29}Si$ couplings of 1775 and 1781 Hz were observed by Caseri and Pregosin[239] in unknown intermediates of Pt-catalysed hydrosilylation reaction of $PhCH=CH_2$ with Et_3SiH.

The magnitude of $^{199}Hg-^{29}Si$ couplings is governed by the FC term and increases with increasing EN of substituents at Hg and Si.[240] Linear correlations were derived between $^1J_{HgSi}$ and $^1J_{SiH}$ of the parent silanes, EN of Si substituents as well as lowest energy UV absorption maxima. Complex formation between $Hg(SiRR^1R^2)_2$ and Et_2O results in further growth of $^1J_{HgSi}$ values.

(3) Two-Bond Couplings. The mechanisms responsible for the magnitudes of two-

bond coupling constants are similar to those governing ^1J couplings. In addition, $^2J_{AB}$ depends on the magnitude of A-X-B valence angle, the properties of the nucleus X involved in the coupling pathway, specifically, on the hybridization of X and the EN of substituents at X.

$^2J_{SiH}$ couplings are of great practical importance, because they are frequently used for enhancing the sensitivity of ^{29}Si spectra by means of PT techniques (see Section II.A.). For this reason, the most characteristic values are collected in Table 1. The magnitude of $^2J_{SiH}$ couplings usually increases with increasing EN of substituents at Si. This has been observed also in the case of disilanes,[151,162] alkenylsilanes,[30,224,225,241,242] disilylmethanes and silacyclopropanes.[17,30,243] Substantial interest in $^2J_{SiH}$ arises from the stereochemical dependence of these couplings. Grignon-Dubois et al. have shown that a linear correlation exists between $^2J_{SiH}$ and the energy of steric hindrance between the silicon and the ring in mono- and disilylated bicyclo[n.1.0]alkanes.[243] A decrease in the valence bond angle Si-C-H is accompanied by a diminishing $^2J_{SiH}$ magnitude, which is zero when silicon is in the exo position or in the case of silacyclopropanes. Some $^2J_{SiH}$ values observed in alkenylsilanes are summarized in Table 2.

Table 2. $^nJ_{SiH}$ for Some Alkenylsilanes, Hz

R_3	$^2J_{(A)}$	$^3J_{cis}(B)$	$^3J_{trans}(C)$	^4J	Ref.
Me$_3$	4.6	(Me)	(Me)	–	241
Me$_3$	6.5	7.5	(Me)	1	241
Me$_3$	8.6	(Me)	12.6	1.4	241
Me$_3$	(Me)	7.3	13.9	–	241
Me$_3$	6.9	6.2	(CH$_2$SiMe$_3$)	2	241
Me$_3$	(BR$_2^!$)	(SiMe$_3$)	21.3	–	224
Me$_3$	4.6	(SiMe$_3$)	(BR$_2^!$)	–	224
Me$_3$	6.4	8.5	15.2	–	241
(Me$_3$SiO)Me$_2$	7.1	9.4	16.2	–	241
(PrHN)$_3$	6.9	8.6	16.7	–	204
N(CH$_2$CH$_2$NH)$_3$	4.6	6.5	13.5	–	204
(EtO)$_3$	7.8	9.6	19.4	–	204
N(CH$_2$CH$_2$O)$_3$	5.8	9.1	18.2	–	204

Direct observation of $^2J_{SiB}$ is complicated by the fast quadrupolar relaxation of ^{11}B nuclei. However, the extent of ^{29}Si line broadening in B containing alkenylsilanes indicates that the absolute values of ^{29}Si-^{11}B couplings across the C=C double bond can be ranged as follows:[224]

$$^2J_{SiB} \approx \, ^3J_{SiB}(cis) < \, ^3J_{SiB}(trans)$$

$^2J_{SiC}$ couplings have received scant attention. Low values of $^2J_{SiC}$ were observed in Me_3Si derivatives of polycycloalkanes (0 to 3.22 Hz),[71] Me_3Si sugars $(2.0\pm0.1$ Hz)[33] and in alkoxysilanes[36,37] (see Table 1.).

An increase in $^2J_{SiF}$ with increasing n in $Me_{3-n}Cl_nSi(CFCl_2)$ has been observed.[244] Values $^2J_{SiF}$ have been reported for F_nSiX_{4-n} (X=Cl, Br, I, $SiHal_3$).[245]

Irregular variation in $^2J_{SiSi}$ with ring size in cyclosiloxanes can be explained by a change in coupling sign ($^2J_{SiSi}$, Hz):[31]

linear   10-membered   8-membered

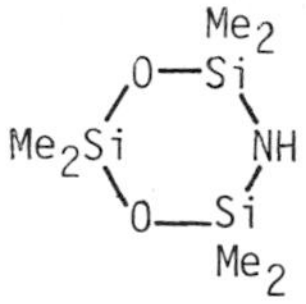

2.1-3.1   1.5   ~0   1.0

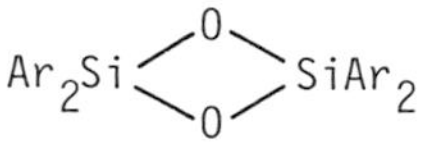

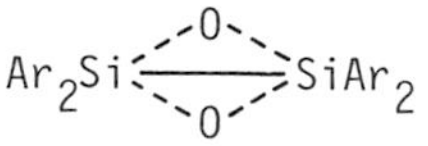

A   3.85-4.02 ref.32   B

In the case of cyclodisiloxanes, the $^2J_{SiSi}$ couplings allow one to differentiate between structures A and B,[32] since J_{SiSi} should be much larger in the case of B (the typical values for directly bonded Si-Si couplings in organodisilanes range from 80 to 90 Hz).[53]

$^2J_{PSi}$ couplings have been studied in silaphosphanes[230] and silylphosphimines $Bu_2^tFSi-N=PX_3$ (X=F, Cl, OR, NR_2).[246] These couplings are particularly useful for diagnostic purposes in the case of silyl derivatives of the phosphoranimines $R_3SiN=PX_2-CH_2SiR_3'$ obtained by *trans*-silylation,[247] because they are quite different for the C-bonded ($^2J_{PSi}$ = 3.3-6.1 Hz) and N-bonded ($^2J_{PSi}$ = 19.5-22.0 Hz) R_3Si groups. Relatively high values of $^2J_{PSi}$ have been observed in HMPT complexes with Si=Fe and Si=Cr compounds (17.5 to 37.2 Hz, see Section III.A.8).[193]

Individual values of $^2J_{SnSi}$ have been reported.[163,224,248-251] Their magnitude depends on the Si-C-Sn valence angle and on the hybridization of the intervening atom. Linear correlation observed between $^2J_{SnSi}$ and $^2J_{SnSn}$ in isostructural Sn and Si derivatives suggests a change in sign of $^2J_{SnSi}$ (Hz):[228]

$(Me_3Sn)_2NSiMe_3$ (-)8.1

$Me_3SnN(SiMe_3)_2$ (+)11.3

$Me_3SnN(Bu^t)SiMe_3$ (+)20.0

$Me_3SnN(BMe_2)SiMe_3$ (+)31.0

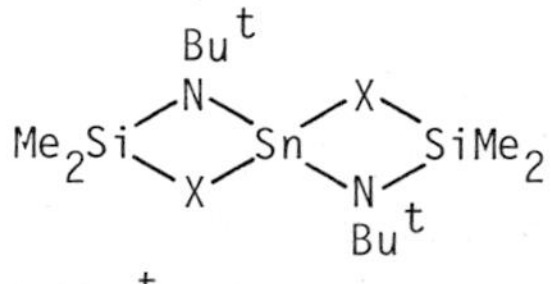

X=NBu^t 8.0

CH_2 81.9

R R'

Me₃Sn Si SnMe₃
       Me₂

19.0-22.0

R R'

Me₃Sn Si SnMe₃
       Me₂

102.3-108.6

Et SiMe₃

Et₂B SnMe₃

93.0

Electronegative substituents at Si increase the value of $^2J_{HgSi}$.[252] The $^2J_{AgSi}$ coupling has been observed for the first time (see Table 1).[138a] Two-bond ^{207}Pb-^{29}Si couplings in diazasilaplumbetidines (20-50 Hz),[249] and ethylene derivatives have been published (^2J, Hz):[251]

R SiMe₃   M = ^{119}Sn ^{207}Pb

R MMe₃   62.1   107.9 (R=Ph)
                  60.5   106.8 (R=CH₂OMe)

(4) Three-Bond Couplings. The magnitude of three-bond couplings depends on bond lengths, hybridization of the atoms of interacting nuclei and those involved in the coupling pathway, and the EN of substituents at these atoms. However, special interest in these couplings arises from the stereochemical dependence of ^3J. The quantitative and qualitative relationships observed between ^3J and the dihedral angle between the given atoms permit one to obtain information concerning the structure of molecules in solutions.

A classical example is the stereochemical dependence of ^3J in ethylene derivatives. As expected, the values of $^3J_{SiH}$ (*trans*) exceed those of $^3J_{SiH}$ (*cis*) approximately by a factor of two (refs. 30,224,225,241,242), decreasing with an increase in EN of the substituents X in the ethylene moiety ($^3J_{SiH}$, Hz):[225]

Me₃Si X

X H

X = Cl   7.1
    Br   7.8
    I   9.6

Me₃Si H

X X

X = Cl   2.9
    Br   3.2
    I   4.1

Cl₃Si SiCl₃

H SiCl₃

25.43 (*cis*)
33.75 (*trans*)
($^2J_{SiH}$=9.29) ref.242

Introduction of silyl groups in the ethylene moiety and substitution of Me by Cl at one or two Si atoms has led to increase in J_{SiH} with chlorinated Si atoms and reduced the values for those Si atoms, where the Me groups were retained.[242] Unfortunately, several values of J_{SiH} reported[242] contradict earlier observations, presumably due to an incorrect assignment of the couplings or structures of the compounds. The magnitude of these couplings also increases with increasing EN of Si-substituents. Representative examples of J_{SiH} in silaethylenes are provided in Table 2.

Grignon-Dubois et al.[17a,b,243] have demonstrated that the magnitude of $^3J_{SiH}$ across the silacyclopropane ring is a function of the dihedral angle between C-H and C-Si bonds suggesting a Karplus-type relationship ($^3J_{SiH}$, Hz):[17b]

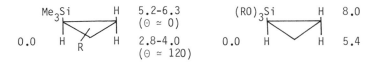

EN substituents on Si increase the magnitude of $^3J_{SiH}$. The higher values of $^3J_{SiH}$ observed in the silatrane cage (4.37 to 5.90 Hz), as compared to those in $RSi(OEt)_3$ (2.64 to 3.52 Hz),[30a] can also be explained in terms of a stereochemical dependence of $^3J_{SiH}$ rather than by $d_\pi - p_\pi$ interaction.

Individual values of $^3J_{SiC}$ have been reported (see Table 1 and references cited therein). Della and Tsanaktsidis have studied the stereochemical dependence of $^3J_{SiC}$ in Me_3Si-polycycloalkanes.[71] The vicinal couplings between ^{29}Si and non-bridgehead γ-carbons in these systems range from 3.2 Hz (SiCCC dihedral angle 135°) to 5.35 Hz (dihedral angle 178°), showing a Karplus-type relationship. The values of $^3J_{SiC}$ involving bridgehead carbons are somewhat enhanced and lie between 6.22 and 10.46 Hz, which has been explained by through-space orbital interactions.

For cyclosilazanes, the values of $^3J_{SiN}$ were observed in the range between 2.8 and 3.1 Hz.[25]

Kowalski has measured magnitudes of $^nJ_{PSi}$ (n=2-4) couplings via CH_2 groups (Hz):[253]

$Ph_2P(CH_2)_xSiMe_2OPh$	$Ph_2(X)P(CH_2)_2SiMe_2OPh$	$Me_3Si(CH_2)_3P(O)Me_2$
18.4 x=1	25.6 (X=O)	2.1
24.2 x=2	29.3 (X=S)	ref.203
0.0 x=3		

(5) Long-Range Couplings. Couplings to ^{29}Si via four or more bonds are rare. Grignon-Dubois et al. reported individual values of $^4J_{SiH}$ in alkenylsilanes ranging between 0 and 2 Hz.[241] Unexpectedly large five-bond $^5J_{SiF}$ couplings (6.59 Hz) have been observed through the bicyclo[1.2.2]octane frame, while $^5J_{SiF}$ is zero in benzene derivative where the elements are similarly disposed.[63] Relatively high values of $^5J_{SnSi}$ have been reported by Wrackmeyer et al. ($^5J_{SnSi}$, Hz):[254]

$$\underset{Me_3Sn}{\overset{Me_3Si}{>}}C=C=C\overset{H}{\underset{\underset{Me_3Sn \quad SiMe_3}{\overset{|}{C-BR_2}}}{<}}$$

22.0/24.4

Individual values of couplings to ^{29}Si have been reported elsewhere (see also Table 1.).[255-258]

C. Isotope Shifts

The feasibility of observing isotope effects in NMR spectra makes it possible to determine the site and degree of isotopic substitution (both by magnetically active or inactive nuclei) in the molecule. This enables one to apply isotope effects to reaction mechanism studies using isotopically enriched compounds. On the other hand, these effects correlate with various structural parameters, such as bond order, bond length, stretching force constants, hybridization parameters, etc., thus making the isotope effects useful for investigation of the electronic structure of molecules.

Recently, the availability and accuracy of isotope effect measurements have been significantly increased as a result of an increase in the magnetic field strength of commercial spectrometers and the introduction of ultrahigh resolution methodology (see Section III.C.). Consequently, one can expect an upsurge in research activity in this field.

(1) Theoretical Considerations. This section focusses on the intrinsic isotope effects on ^{29}Si CS (isotope shifts), because the other effects of isotopic substitution on ^{29}Si NMR parameters have been found insignificant or have not been studied at all. We use here the notation introduced by Gombler[259] for the isotope shift $^{n}\Delta A(^{m'}/^{m}X)$ observed in the NMR spectra of nucleus A upon substitution of the ^{m}X isotope in the molecule with a heavier isotope $^{m'}X$ separated from the A nucleus by n bonds.

There are two factors affecting nuclear shielding upon variation in isotope mass, i.e.: a) the dynamic factor taking into account the mass-related effects on nuclear shielding, which result from differences in averaging over nuclear configuration as the molecule undergoes vibration and rotation (rovibrational averaging); b) the electronic factor - changes in shielding with bond extention or bond angle deformation. In order to eliminate the dynamic factor, the reduced isotope shift $^{n}\Delta R_A$ is usually introduced providing information on electronic properties of the molecule:[260]

$$^{n}_{\Delta}R_A = {}^{n}_{\Delta}A(^{m'}/^{m}X)\left(\frac{m'-m}{m'}\frac{1}{2}\frac{m_A}{m_A+m}\right)^{-1} \tag{9}$$

This expression allows one to compare isotope shifts for a given nucleus upon substitution by isotopes of different elements and to correlate the isotope shifts with various parameters, such as bond length, bond order, substituents EN, CS, coupling constants, etc.

There are several empirical rules which have been derived for experimentally observed isotope shifts:[260,261]

1. Upon substitution with a heavier isotope the NMR signal of the neighbouring nucleus usually shifts towards higher fields, i.e. the isotope shifts are generally negative in sign.

2. The magnitude of the isotope shift is dependent on how remote isotopic substitution is from the observed nucleus. Although there are exceptions, one-bond shifts are larger than two-bond or three-bond isotope shifts.

3. The magnitude of the shift is a function of the observed nucleus and reflects its CS range.

4. The magnitude of the shift is related to the fractional change in mass upon isotopic substitution.

5. The magnitude of the shift is approximately proportional to the number of equivalent atoms which have been substituted by isotopes, i.e. isotope shifts exhibit additivity.

A detailed description of the theory of isotope shifts has been published recently by Jameson and Osten.[260] Extensive compilations of experimental data and empirical treatments of isotope effects have been presented by Hansen.[261]

(2) Isotope Effects on ^{29}Si Shifts. Isotope effects on ^{29}Si CS have been studied since 1985.[226b,262] The deuterium isotope shifts measured by Berchier et al.[263] in several organosilanes and siloxanes show the largest magnitudes and follow the additivity relationship (see Table 3). These shifts exceed the similar effects in ^{13}C spectra (-190 ppb in CH_4),[261] but are smaller compared with those in ^{119}Sn spectra (-403 ppb in SnH_4) (ref. 264). This reflects the approximate ratio of CS ranges of the corresponding nuclei.

For $^{13/12}$C, $^{15/14}$N and $^{18/16}$O isotope shifts there is a tendency to assume less negative values with increasing number of EN substituents on the Si atom (see Table 3). Furthermore, the values of one-bond $^{13/12}$C isotope shifts pass through zero and adopt unusual, positive values if there are several EN substituents on Si. Generally, interpretation of positive isotope shifts meets with certain difficulties, because the theory does not predict positive shifts.[260] Nevertheless, Chesnut has recently shown that positive isotope shifts can appear following an increase in the positive charge on the observed nucleus.[265] An upfield ^{29}Si shift observed[266] with the increasingly positive $^{13/12}$C, $^{15/14}$N and $^{18/16}$O isotope shifts agrees with this conclusion.

The magnitudes of $^{13/12}$C (ref. 262) and $^{15/14}$N (ref. 24) isotope shifts also

Table 3. m'/m_X Isotope Effects on ^{29}Si Chemical Shifts[a]

C o m p o u n d	$^n\Delta^{29}Si$	Ref.	C o m p o u n d	$^n\Delta^{29}Si$	Ref.
$(^{2/1}H)$, n=1			**n=2**		
$PhSiH_2D$	-216	263	Me_3SiOMe	-1.4	266
$PhSiHD_2$	-429(-215)[b]	263	$Me_2Si(OMe)_2$	-0.5	266
$PhSiD_3$	-641(-214)[b]	263	$MeSi(OMe)_3$	-0.3	266
Bu^nSiD_3	-762(-254)[b]	263	$Si(OMe)_4$	-0.8	266
$(Me_2SiD)_2O$	-255	263	$Si(OEt)_4$	-0.7	266
n=2			**n=3**		
$PhSi(CD_3)_3$	-34(-4)[b]	263	Me_3SiOEt	-0.9	266
$(^{13/12}C)$, n=1			$Me_2Si(OEt)_2$	-0.3	266
			$MeSi(OEt)_3$	0.2	266
Me_4Si	-6	262	$Si(OEt)_4$	0.4	36
	-5.6	266	**$(^{15/14}N)$, n=1**		
$Me_2Si\langle\rangle SiMe_2$ (Me)	-4	262	$(Me_3Si)_2NH$	-10.7	24
(CH_2)	-8	262	$Me_3SiNHPr^i$	-14.5	24
Me_3SiCH_2Cl (Me)	-4.0	266	$Me_3SiNHBu^t$	-14.0	24
(CH_2)	-8.8	266	$Me_3SiNHPh$	-16.0	24
Me_3SiPh (Me)	-4.4	266	Me_3SiNEt_2	-11.0	226b
(Ph)	-3.8	266	Me_3SiNPr_2	-9.2	24
$Me_3SiCH=CH_2$ (Me)	-4.3	266	$Me_3SiN(CH_2)_n$ n=6	-11.2	226b
(CH=)	-7.5	266	n=5	-12.0	226b
$Me_3SiC\equiv CH$ (Me)	±1	262	n=4	-11.8	226b
(C≡)	-15	262	n=3	-11.5	226b
$Me_3SiC\equiv CSiMe_3$ (Me)	±1	262	Me_3SiNCO	-15.5	226b
(C≡)	-16	262	$Me_3SiNHOSiMe_3$	-14.3	36
$Me_2Si(C\equiv CH)_2$ (Me)	±1	262	$Me_3SiN(Me)C(O)H$	-9.0	24
(C≡)	-11	262	$Me_3SiN(H)C(O)Me$	-9.3	24
Me_3SiNEt_2	-4.1	266	$Me_2(EtO)SiNHPr$	-9.5	24
$Me_3Si-N\langle$	-2.8	266	$Me_2(EtO)SiNHPh$	-9.5	24
			$Me_2(EtO)SiNPr_2$	-6.4	24
$Me_3SiN=N=N$	-0.5	266	**$(^{18/16}O)$, n=1**		
$(Me_2ClSi)_2NH$	2.1	266			
$(Me_2ClSi)_2O$	5.8	266	Me_3SiOPh	-28.5	37
$(Me_3Si)_2O$	0.2	36	Me_3SiOBu^t	-25.2	37
$Me_3SiONHSiMe_3$ (SiN)	-2.0	36	$Me_3SiOSiMe_3$	-26.2	36
(SiO)	-0.8	36	$Me_3SiONHSiMe_3$	-21.8	36
Me_3SiOMe	-0.8	266	$(Me_3SiO)_2SiMe_2$ (SiMe_3)	-25.6	37
Me_3SiOBu^t	0.1	266	(SiMe_2)	-16.6	37
Me_3SiOPh	-0.1	266	$(Me_2SiO)_4$	-17.3	37
$(Me_3SiO)_2SiMe_2$ (SiMe_3)	0.3	266	$Me_2Si(OMe)_2$	-15.1	37
(SiMe_2)	4.9	266	$MeSi(OMe)_3$	-7.5	37
$(Me_2SiO)_4$	5.2	266	$Si(OMe)_4$	-4.0	37
$Me_2Si(OMe)_2$	3.4	266	$Si(OEt)_4$	-4.6	36
$MeSi(OMe)_3$	4.7	266			

[a] In parts per billion. Negative sign indicates an upfield shift for nucleus bonded to the heavier isotope. Solvent $CDCl_3$ or C_6D_6.

[b] Per one D atom.

depend on the hybridization of the substituted atom, which increases in the following sequence: $Si-X-(sp^3) < Si-X=(sp^2) < Si-X\equiv(sp)$. The correlation between the $^{15/14}N$ isotope shifts and $^1J_{SiN}$ observed in aminosilanes[24] supports the existence of a relationship between $^1\Delta$ and hybridization.

An increase in $^{13/12}C$ isotope shifts in the heavier group 14 nuclei (see Table 4) reflects the appropriate growth in their CS ranges.

Table 4. $^{13/12}C$ Isotope Shifts of Me_4M, M=14 Group Elements, ppb

M	$^1\Delta M(^{13/12}C)$	Ref.
^{29}Si	-5.6	266
^{73}Ge	-14	266
^{119}Sn	-18	262
^{207}Pb	-89	262

The absolute values of isotope shifts (per atom) increase with increasing fractional change in mass: $^{15/14}N$ ($^1\Delta$ = -6.4 to -16 ppb) < $^{13/12}C$ (+5.8 to -16) < $^{18/16}O$ (-4.0 to -28.5) < $^{2/1}H$ (-214 to -255). The corresponding maximum magnitudes of the reduced shifts grow with the diminishing Si-X bond length (r) in the following order: $^{13/12}C$ (r=1.87 Å, $^1\Delta R$=-588 ppb) < $^{15/14}N$ (1.72 Å, -712 ppb) < $^{18/16}O$ (1.63 Å, -796 ppb) < $^{2/1}H$ (1.47 Å, -1055 ppb)[266] in agreement with the general observations.[260,261]

Long-range isotope shifts have been measured only for $^{2/1}H$ and $^{13/12}C$ substitution. Generally, their values are smaller than the corresponding one-bond shifts and they decrease with increasing number of intervening bonds. It is noteworthy that exceptions are possible in the case of $^{13/12}C$ shifts, because one-bond $^{13/12}C$ shifts have either positive or negative sign and can adopt zero value (see Table 3).

Two-bond $^{29/28}Si$ isotope effects on ^{29}Si CS of -0.04 to +0.22 ppm were reported for silicate solutions.[113a] However, these values exceed all two- and even one-bond isotope shifts observed earlier (except those of $^1\Delta^{2/1}H$, see Table 3) and, therefore, are not reasonable. Negligible two-bond $^{2/1}H$ isotope shift (up-frequency) observed in the same silicate solutions[113a] confirm this conclusion. Second order effects are the most plausible source of errors in the $^2\Delta^{29/28}Si$ measurements.

(3) $\underline{^{29/28}Si\text{-Induced Effects on Chemical Shifts of Other Nuclei}}$. These effects are summarized in Table 5.[267,268] Except for ^{199}Hg, their magnitudes are considerably lower than those observed in ^{29}Si spectra. In general, this is related to the smaller mass factor ($m_A/(m_A+m)$, see Eq.9) and is consistent with the general observations, i.e. isotope shifts are predominant in spectra of heavier isotopes.

Table 5. $^{29/28}Si$ Isotope Effects on the Chemical Shifts of
Various Nuclei (A), ppb

Compound	n	A	$^{n}\Delta A(^{29/28}Si)$	Ref.
Me_4Si	2	^1H	-0.06	35
Me_4Si	1	^{13}C	-0.85	34b
			$(-1.7)^a$	
$Bu^tHNSiMe_3$	1	^{15}N	-2.5	24
$(PrHN)_4Si$	1	^{15}N	-4.5	24
$(PhNH)_2Si(OEt)_2$	1	^{15}N	-4.0	24
CH_3SiF_3	1	^{19}F	-9	267
$ClCH_2SiF_3$	1	^{19}F	-14	267
Cl_2CHSiF_3	1	^{19}F	-18	267
Cl_3CSiF_3	1	^{19}F	-23	267
$CH_3CH_2SiF_3$	1	^{19}F	-15	267
$CF_3CH_2CH_2SiF_3$	1	^{19}F	-18	267
$F_2CHCF_2SiF_3$	1	^{19}F	-36	267
$Et_3SiHgEt$	1	^{199}Hg	60	268
$(Et_3Si)_2Hg$	1	^{199}Hg	60	268

a $^{30/28}Si$ Isotope effect.

IV. OXYGEN-17 NMR OF Si-O COMPOUNDS

Historically the ^{17}O NMR spectra of organosilicon compounds ($Si(OMe)_4$ and $Si(OEt)_4$) were recorded for the first time in 1961.[269] However, during the next two decades only three more papers appeared[270-272] reporting ^{17}O CS for seven other Si-O compounds. Systematic studies of ^{17}O NMR spectra of organosilicon compounds started in 1980.

The main reason for the low research activity in this fields is the small natural abundance of the ^{17}O isotope (0.038%). The effectivity of quadrupolar relaxation mechanism of ^{17}O nuclei (spin I=5/2) results in short relaxation times (<10 ms), permitting fast data acquisition. On the other hand, an increase in the ^{17}O linewidth ($\Delta\nu_{1/2}$), which is proportional to the spin-spin relaxation time T_2 ($\Delta\nu_{1/2}= 1/\pi T_2$), complicates the observation of ^{17}O resonances, 1) due to the considerable loss of the signal in the pre-acquisition period, and 2) due to the "acoustic ringing" especially on spectrometers with low working frequences.[5]

Availability of high-field magnets and various multipulse techniques of suppression of "acoustic ringing" allow partially to overcome these problems.[273]

Both the theoretical and practical aspects of ^{17}O NMR have been discussed in detail[273] and only a brief insight is presented in this section.

The rate of quadrupolar relaxation decreases 1) with increasing symmetry of charge distribution around the O atom, and 2) with decrease in reorientation time of the molecules. As a result, decrease in ^{17}O linewidth usually can be observed with increase in the valence bond angle at oxygen, decrease in molecular size, increase in temperature and decrease in solution viscosity, facilitating the observation of ^{17}O resonances.

The value of the ^{17}O CS is governed by the paramagnetic term (σ^p, see Section III.A.1). On the other hand, it has been demonstrated, that in some cases the variation in ^{17}O CS can be determined also by the diamagnetic term (σ^d).[273]

The common feature of various types of Si-O compounds is U-shaped dependence of the ^{17}O CS on the number of OR groups at the silicon atom in the series $Me_{4-n}Si(OR)_n$, R=Alk,[85] Ar[274] or SiR_3.[275] This has been explained in terms of d_π-p_π interaction in the Si-O bond.

Generally the influence of Si-substituents on the ^{17}O CS is smaller as compared to that of O-substituents. On the other hand, the sensitivity of ^{17}O NMR on the nature of Si-substituents can exceed that of ^{29}Si NMR as depicted below (δ, ppm):[31]

	Me_3Si O $SiMe_3$	Me_2ClSi O $SiMe_2Cl$	Me_2ROSi * O $SiORMe_2$
^{17}O	39.0	30.0	71.4
^{29}Si	6.5	6.9	-22

The position of ^{17}O resonances in alkoxysilanes varies over a range of 70 ppm (see Table 6). Their width ($\Delta\nu_{1/2}$=110-680 Hz) increase with larger molecular size, decreasing temperature and increasing viscosity of the solution.

Linear correlations exist between ^{17}O CS in alkoxysilanes and those in corresponding alcohols.[85] The angular dependence shows little variation with different substituents at the silicon atom and is close to 1. Linear correlations have been found between the ^{17}O CS of alkoxysilanes Et_3SiOR and $MePh_2SiOR$ and the OH stretching frequencies of the correspondings alcohols (ROH), whereas no correlation was observed between the total charge on the O atom calculated by CNDO/2 method and ^{17}O CS.[85] A linear correlation has been found between ^{17}O CS and the Taft parameter σ^* of substituent X=F, Cl, Br, I and NO_2, and between ^{17}O and $^{13}C_\alpha(O)$ CS in the series $Me_3Si-O-C_6H_4-X-p$.[274]

Comparison of ^{17}O CS in silatranes[276] and triethoxysilanes[277] reveals a negligible effect of cyclization and transannular N→Si bond formation on ^{17}O CS of equatorial O atoms. In contrast, considerable low-field shift ($\Delta\delta$ = 13.6 to 16.2 ppm) has been observed for the axial O atom in 1-ethoxysilatrane.

The dependence of ^{17}O CS on the conformation of the eight-membered ring in 1,3,2-dioxasilacyclooctanes (29b-d) shows, that ^{17}O NMR can be employed in conformational analysis of organosilicon compounds.[278] A considerable low-field

shift by 7-10 ppm has been observed in compounds existing in a crown conformation relative to their analogues adopting boat-chair type conformations. This
has been attributed to the existence of the anomeric effect in organosilicon
compounds.

A clear distinction between the chain oxygens and those of the end-groups
have been observed in ^{17}O NMR spectra of linear siloxanes (see Fig. 4),[31,279]
which can be employed in structural analysis. On the other hand, ^{17}O CS is practically independent of ring size in cyclosiloxanes.[31] This has been explained
by the opposing action of the factors governing the ^{17}O CS which are balanced.
However, this equilibrium is disturbed in the case of cyclosilazoxanes, as indicated by the tendency of ^{17}O resonances to shift to the higher fields with
decreasing ring size (see Table 6.).

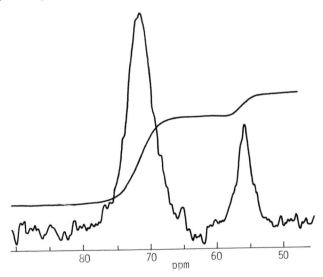

Fig.4. A 48.8 MHz ^{17}O NMR spectrum of $HN[(SiMe_2O)_5SiMe_3]_2$ in $CDCl_3$ showing the
separate resonances for $OSiMe_3$ end-groups and chain oxygens.[31]

There are two signals in ^{17}O NMR spectra of acyloxysilanes ($\delta = 156$ to
182 ppm (SiOC) and $\delta = 370$ to 396 ppm (C=O)) whereas only one signal appears in
the ^{17}O spectra of diacyloxysilanes at 350 K in the intermediate position ($\delta = 270$ to 287 ppm).[280] This reflects the equivalence of both oxygen atoms on the
NMR time scale resulting from their fast exchange as depicted below.

Table 6. Chemical Shifts for Some Representative Si-O Compounds, ppm [a]

Compound	$\delta^{17}O$	Ref.	Compound	$\delta^{17}O$	Ref.
Me₃SiOEt	10.0	270	MeHSi(OEt)₂	27.6	85
Et₃SiOR			MeClSi(OEt)₂	35.7	85
R=Me	-30.8	85	Cl₂Si(OEt)₂	41.6	85
Et	5.2	85	Me₂Si(OPh)₂	94.2	274a
Pr	0.6	85	Me₂Si[OC(O)CH₃]₂	187(393)	280
Prⁱ	33.6	85	(Me₃SiO)₂SiMe₂	53.0	27
Bu	0.8	85	(Me₃Si<u>O</u>SiMe₂)₂O	<u>55.0</u>/71.0	279
Buˢ	29.2	85	(Me₂Si<u>O</u>)₃	72.0	31
CH₂Ph	-0.4	85	Me₂Si(OSiMe₂)₂NH	62.0	31
CH₂CF₃	-20.6	85	Ph₂Si(OSiMe₂)₂NH	56.0	31
R₃SiOEt			Me₂Si(NHSiMe₂)₂O	60.1	31
R₃=Me₂Ph	13.4	85	(Me₂SiO)₄	71.4	31
MePh₂	8.4	85	O(SiMe₂NHSiMe₂)₂O	68.1	31
Ph₃	7.0	85	O(SiMe₂NHSiMe₂O)₂SiMe₂	69.6	31
Me₂Cl	33.0	85	RSi(OEt)₃		
MeCl₂	46.0	85	R=H	27.6	277
Cl₃	54.0	85	Me	21.0	270
Me₃SiOPh	81.0	274b	Et	18.4	277
Me₃SiOC(<u>O</u>)R			CH₂OC(O)CH₃	14.0	271
R=H	182(<u>396</u>)	280	CH₂Cl	18.4	277
Me	179(<u>394</u>)	280	CHCl₂	19.4	277
Et	174(<u>385</u>)	280	Ph	24.2	277
PhCH₂	176(<u>389</u>)	280	CH=CH₂	16.5	277
Ph	163(<u>370</u>)	280	C≡CH	23.3	277
CH₂Cl	172(<u>388</u>)	280	Cl	27.2	85
CHCl₂	165(<u>380</u>)	280	F	11.6	277
CCl₃	156(<u>371</u>)	280	N(CH₂CH₂O)₃SiR		
(Me₃SiO)₃B	68.0	272	R=H	24.4	276
Me₃SiOBR₂ b	132.0	272	Me	22.4	276
(Me₃Si)₂O	43.0	279	Et	17.8	276
(Et₃Si)₂O	15.8	275	CH₂Cl	17.6	276
(Me₂HSi)₂O	22.9	275	Ph	21.0	276
(Me₂PhSi)₂O	33.0	274	EtO	9.1	276
(MePh₂Si)₂O	30.0	274	c	12.5	276
(ClMe₂Si)₂O	73.0	274	Et<u>O</u>	26.0	276
(Cl₂MeSi)₂O	91.0	274	c	28.6	276
Me₂Si(OEt)₂	25.0	274	MeSi(OPh)₃	85.0	274a
Me₂Si(OCH₂CH₂)₂X			(Me₃SiO)₃SiMe	58.0	279
X=NH	20.0	278	(Me₃SiO)₃SiCH₂Cl	47.5	275
NMe	21.4	278	(Me₃SiO)₃Si(CH₂)₃Cl	49.5	275
NBuᵗ	19.6	278	Si(OR)₄		
NPh	26.8	278	R=Me	-26.5	85
S	22.4	278	Et	9.0	270
O	20.5	278	Pr	7.2	85
(X)	(-11.5)	203	Prⁱ	42.6	85
Me₂Si(OCHCH₂)₂NMe RR	47.8	278	CH₂CF₃	-12.6	85
|CH₃ RS	57.8	278	Ph	70.4	274a
Ph₂Si(OEt)₂	18.0	279	SiMe₃	41.7	275

[a] In CDCl₃.

[b] BR₂=9-borabicyclo[3.3.1]nonane, in C₆D₆.

[c] In CD₃CN.

Large differences in ^{17}O CS of amidate and imidate forms permit one to determine the structure of bis(organosilyl)amides in solution.[130] It has been demonstrated that ^{17}O NMR is more sensitive to structural changes of these compounds than ^{14}N and ^{29}Si NMR ($\delta^{17}O$, ppm):

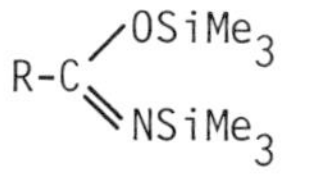

imidate form

170

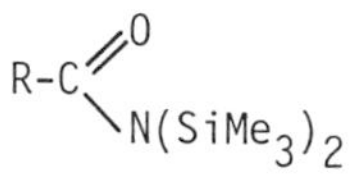

amidate form

393-413

Recently ^{17}O NMR spectra of compounds including Si-O-B and Si-O-Al have been obtained.[281] The ^{17}O CS in Si-O-B derivatives significantly depends on the extent of $(p-p)_\pi$ conjugation in B-O bond. Increase in coordination number of the B atom from 3 to 4 results in considerable high-field shift of ^{17}O resonance ($\Delta\delta \sim -100$ ppm). On the other hand, negligible variation in ^{17}O shielding occurs if the O atom is involved in coordination, e.g. $R_2O \rightarrow AlCl_3$ or $R_2O \rightarrow BR_2$ ($\delta^{17}O$, ppm):

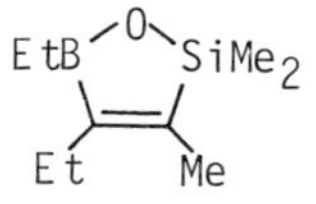

136.5

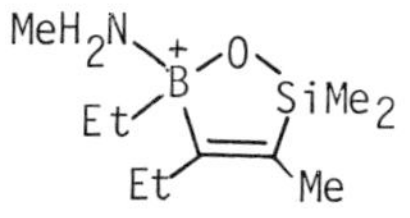

44.8

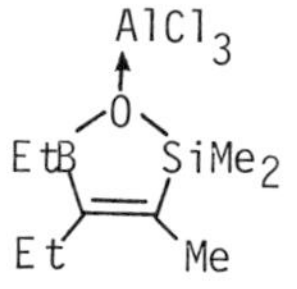

128.2

Natural abundance ^{17}O NMR spectra of aqueous (D_2O) solutions of the oxidized Keggin-structure tungstate $SiW_{12}O_{40}^{4-}$ and its one- and two-electron reduced derivatives $SiW_{12}O_{40}^{5-}$ and $SiW_{12}O_{40}^{6-}$ were recorded with and without the addition of Mn^{2+} to aid in assignment.[282] Well resolved signals between $\delta = 25$ ppm (the internal oxygens) and 76.3 ppm (the terminal oxygen) have revealed differences in spin densities that can be related to intramolecular electron delocalization and permit estimates of intermolecular electron exchange rate limits.

Use of ^{17}O NMR to study silicate solutions offers at least four intrinsic advantages over similar studies using the ^{29}Si nuclei:[283]

1) all silicate species (except monomeric anion) contain two types of oxygen site (bridging and non-bridging), but often only a single silicon site;

2) observation of the ^{17}O nucleus affords insight into solution kinetics;

3) acceptable ^{17}O spectra of ^{17}O enriched silicate solutions may be obtained very rapidly;

4) ^{17}O NMR yields information about other solution species which are not observable by ^{29}Si NMR.

Unfortunately, ^{17}O linewidths of the silicate anions were broad and the CS range was relatively small (ca. 50 ppm). As a result concentrated alkali-metal silicate solutions yielded complex ^{17}O spectra with many overlapping signals. Two spectral regions were distinguished, namely, a 35-55 ppm (terminal Si-O$^-$ or SiOH) and 45-85 ppm (bridging Si-O-Si groups). The authors[283] have concluded that at present ^{17}O NMR spectra of complex silicate solutions may be more useful in providing "fingerprint analysis" rather than in structure elucidation of the individual components.

Kowalewski and Berggren[284] have studied ^{17}O (^{13}C and ^{29}Si) relaxation for neat liquid $(Me_2SiO)_4$ in the temperature range 298 to 373 K. The most interesting chemical information on the overall reorientation of the molecule was obtained in a straightforward manner from the ^{17}O measurements, which were also the easiest to perform.

V. NITROGEN NMR OF Si-N COMPOUNDS

The role of the nitrogen nuclei in NMR spectroscopic studies has significantly increased recently. The series of reviews[22,285] and monographs[286] appeared in the last decade confirm this trial sentence. For this reason we have devoted a special paragraph to this topic with the aim to cover all available data. The low γ values of both magnetically active naturally abundant ^{14}N and ^{15}N nuclei result in low resonance frequencies, low NMR sensitivity, and small observable magnitudes of spin-spin couplings. Because of the high natural abundance (a=99.6%) and effective quadrupolar relaxation of the ^{14}N nuclei (spin I=1), the recording of ^{14}N NMR spectra requires relatively little spectrometer time. On the other hand, extensive quadrupolar line broadening usually lower the spectral resolution and accuracy of CS measurements, as well as restricts the observation of spin-spin couplings. This makes the assignment difficult or even impossible if there are several signals in the ^{14}N spectra.[287]

Very high resolution can be obtained in ^{15}N NMR spectra providing accurate values of CS and coupling constants (see Fig. 5). Unfortunately, long relaxation times, negative NOE, low sensitivity and natural abundance (a=0.365%) of the ^{15}N nuclei (spin I=1/2) make the measurements time consuming. Recent advances in multipulse techniques (see Section II) and growth in magnetic field strengths of modern spectrometers allow one to overcome these problems, increasing the popularity of ^{15}N NMR. On the other hand, ^{14}N NMR is still useful:

a) to evaluate the trends in nitrogen shielding,

b) to predict the ^{15}N CS and coupling constants making the ^{15}N measurements more reasonable,

c) for the solution of structural problems (e.g. amidate-imidate tautomerism) which is possible if the variations in CS exceed the linewidths of the ^{14}N resonances,

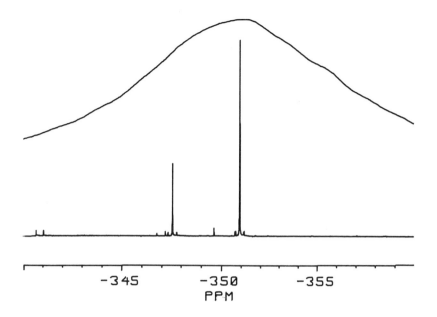

Fig.5. A comparison of the ^{14}N spectrum (the upper trace) and the ^{15}N spectrum (the lower trace) of cyclotrisilazane $Ph_2Si(NHSiMe_2)_2NH$ in $CDCl_3$. Spectrometer frequency 360 MHz (on ^1H).

d) in compounds with high symmetry of charge distribution around the N nuclei, e.g. with linear or tetrahedral structure of the nitrogen atom. In the latter cases sharp lines usually appear in the ^{14}N spectra allowing precise measurements of CS and even the observation of spin-spin couplings (see Fig.6),

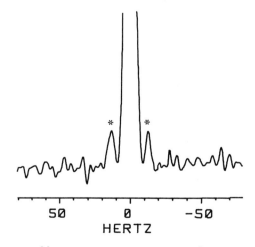

Fig.6. The ^{14}N spectrum of $[Si(NCS)_6]^{2-}2K^+$ in acetone-d_6 showing ^{29}Si satellites (*).

e) if the observation of the ^{15}N resonance fails, e.g. the PT is ineffective or impossible, in the case of small quantities of sample, etc.

For brevity and reasons mentioned above, only ^{15}N data are treated in this chapter in cases when both ^{14}N and ^{15}N NMR data appear in the literature. Representative values are collected in Table 7.

Table 7. Nitrogen Chemical Shifts for Some Representative Si-N Compounds, ppm[a]

Compound	$\delta^{15}N$ $(\delta^{14}N)$	Ref.	Compound	$\delta^{15}N$ $(\delta^{14}N)$	Ref.
Aminosilanes			R_3SiNEt_2		
Et_3SiNH_2	(-377)	287f	$R_3=Me_2H$	-374.5	288
Ph_3SiNH_2	-373.8	25	Me_2Bu^t	-351.4	288
Me_3SiNHR			Me_2Ph	-349.5	288
R=Me	(-377)	287d	$MePh_2$	-342.5	288
Et_i	-345.5	226c	Ph_3	-302.5	288
Bu^i	-353.1	226c	Ph_2Cl	-338.9	288
Bu^s	-333.4	226c	$PhCl_2$	-331.5	288
Bu^t	-321.2	226c	Cl_3	-325.4	288
	-323.9	24	$Cl_2(NEt_2)$	-335.6	203
Ph	-309.5	226c	$Cl(NEt_2)_2$	-342.9	203
	-314.9	24	$(EtO)_3$	-349.9	24
Me_3SiNR_2			$Me_2HSiNMe_2$	-400.5	288
$R_2=Me_2$	-378.5	288	$Me_2Si(NMe_2)_2$	(-374)	287d
Et_2	-351.5	288	$MeSi(NMe_2)_3$	(-371)	287d
	-348.9	203	$Si(NMe_2)_4$	-371.6	203
Pr^i_2	-315.5	288	**Azo compounds**		
$(CH_2)_2$	(-368)	287d			
$(CH_2)_3$	-364.2	203	$Me_3Si-N=X$		
$(CH_2)_4$	(-345)	287d	X=NMe	(302)	287e
Me,Ph	(-326)	287d	NPh	(247)	287e
$R_3SiNHBu^t$			$NSiMe_3$	(614)	287e
$R_3=Me_2Ph$	-325.7	24	$S=NSiMe_3$	-58.7	296
$MePh_2$	-327.3	24	$P-N(SiMe_3)_2$	-67.8	289b
$Me_2(NHBu^t)$	-317.4	24	$P-N(SiMe_3)_2$ 214.0		289b
$Me_2(EtO)$	-317.6	24	$H_3Si-N=C=O$	-364.2	297c
$Me(EtO)_2$	-321.5	24	$H_3Si-N=C=S$	-287.1	297c
$(EtO)_2(NHBu^t)$	-324.9	24	$H_3Si-N=C=Se$	-259.0	297c
$(EtO)_3$	-330.4	24	$Me_3Si-N{\equiv}C=S$	-265.3	24
$R_3SiNHPh$			$Si(NCS)_6^{2-}$	(-231.4)	203
R=Me_2H	-320.4	24	$Me_3SiN=C=NSiMe_3$	-331.1	203
Me_2Ph	-317.7	24	$(Me_3Si)_3CB{\equiv}NSiMe_3$	(-262)	298
$MePh_2$	-319.5	24	$Bu^t_2Si=NSiBu^t_3$	(-234)	190c
$Me_2(NHPh)$	-312.7	24	$THF·Bu^t_2Si=NSiBu^t_3$	(-334)	190c
$Me_2(EtO)$	-310.8	24	**Amides**		
$Me(NHPh)_2$	-313.1	24			
(EtO)(NHPh)Me	-313.5	24	$MeC(O)N(Me)SiMe_3$	(-286)	292
$Me(EtO)_2$	-314.4	24	$MeC(O)NHSiMe_3$	(-264)	292
$(EtO)_2(NHPh)$	-319.2	24	$MeC(O)N(SiMe_2H)_2$	(-253)	130
$(EtO)_3$	-322.3	24	$MeC(O)N(SiMe_3)_2$	(-120)	130
b	-321.5	24	$HN{-}C(O)NHSiMe_3$	-281.3	24
c	-320.7	24			

Table 7. (Continued).

Compound	$\delta^{15}N$ $(\delta^{14}N)$	Ref.	Compound	$\delta^{15}N$ $(\delta^{14}N)$	Ref.
Me₃Si-N (succinimide)	(-181)	287c	BMe₂	(-285)	287a
			B(Me)Br	(-280)	287a
MeC(S)NHSiMe₃	(-226)	287a	B(Cl)N(SiMe₃)₂	(-326)	287a
MeSO₂NHSiMe₃	(-278)	287a	Al[N(SiMe₃)₂]₂	(-314)	287a
p-MeC₆H₄SO₂NHSiMe₃	-283.1	24	Me	(-374)	287a
			Et	(-344)	287a
Silazanes			Pri	(-327)	287a
(H₃Si)₃N	-417.2	297a	But	(-320)	287a
(Me₃Si)₃N	-342.0	289a	Ph	(-291)	287a
(Me₃Si)₂NH	-352.9	286a	Sn(II)	-239.2	289a
(H₃Si)₂NH	-406.3	297a	Pb(II)	-189.1	289a
(Me₂PhSi)₂NH	-357.8	25	HgMe	(-291)	287a
(MePh₂Si)₂NH	-361.5	25	N=CH₂	(-242)	287a
(Me₂ClSi)₂NH	-334.3	25	N=CF₂	(-300)	287a
(Cl₃Si)₂NH	(-297)	287d	N=CCl₂	(-216)	287a
[(MeO)₃Si]₂NH	(-384)	287f	N=NH(SiMe₃)₂	(-250)	287a
(Me₂ClSi)₂NMe	(-338)	287d	N≡C	(-320)	287a
(MeCl₂Si)₂NMe	(-318)	287d	N=C=S	(-266)	287a
			OSiMe₃	(-338)	287a
Cyclosilazanes $(R_2SiNR^1)_n$			N(SiMe₃)₂	(-300)	287a
n=2			SSiMe₃	(-333)	287a
(Me₂Si-NSiX₃)₂			SCF₃	(-372)	299
X₃=Me₂Cl	-322.1	25	Cl	(-346)	287a
Me₂NH₂	-323.8	25	Br	(-310)	287a
Me₂(EtO)	-325.4	25	I	(-412)	287a
n=3					
(Me₂SiNH)₃	-347.3	31	**Miscellaneous**		
(Me₂SiNMe)₃	(-336)	287d	Me₃SiN(SnMe₃)₂	-370.6	226c
(Me₂SiNCl)₃	(-298)	287a	Me₃SiN(But)SnMe₃	-320.3	226c
(Me₂SiNBr)₃	(-269)	287a	[Me₃SiN(But)]₂Sn	-220.9	289b
Me₂Si(NHSiMe₂)₂O	-345.3	31	[Me₃SiN(But)]₂Pb	-140.9	289b
Me₂Si(OSiMe₂)₂NH	-342.9	31	Me₂Si(N-But)₂M M=Sn	-196.3	289c
HN(SiMe₂SiMe₂)₂NH	-356.4	31	Pb	-183.3	289c
n=4					
(Me₂SiNH)₄	-341.7	31			
(PhMeSiNH)₄	-345.8	25			
(SiMe₂OSiMe₂)₂NH	-334.9	31	Me₂Si(N-But)Sn(N-But)SiMe₂	-267.6	289d
n=5					
O(SiMe₂NHSiMe₂O)₂SiMe₂	-335.1	31			
O(SiMe₂OSiMe₂NH)₂SiMe₂	-336.7	31	Me₃SiNHSCF₃	(-372)	299
n=6			Me₃SiN(SCF₃)₂	(-355)	299
(SiMe₂OSiMe₂NH)₃	-334.8	31	Me₃SiNHOSiMe₃	-255.6	36
			H₃SiNHPF₂	-323.2	297c
Trimethylsilylamides (Me₃Si)₂NX*			trans-PtI(PEt₃)₂H₂SiN(SiH₃)₂		
X=Li	(-326)	287a		-401.6	297b
Na	(-341)	287a			
K	(-315)	287a			
BeN(SiMe₃)₂	(-325)	287a			

a Referred to CH₃NO₂, solvent CDCl₃.
b In C₆D₁₂. c In (CD₃)₂SO.

The ^{15}N NMR spectra has been obtained recently for the series of aminosila-nes[24,226,288] and silazanes.[25,31,289] Delocalization of the nitrogen lp due to d_π-p_π conjugation is responsible for decreased shielding in the series $NH_3 > R_3SiNH_2 > (R_3Si)_2NH > (R_3Si)_3N$ (R=Alk), where-as the opposite shielding sequence exists in the case of R=H, suggesting the prevalence of +I effect. Competition of both effects results in a U-shaped de-pendence of ^{15}N CS on the sum of the Si-substituents' EN in aminosilanes (see Fig. 7). A shielding effect of SiPh group is characteristic for both nitrogen and ^{17}O NMR (see Section IV).

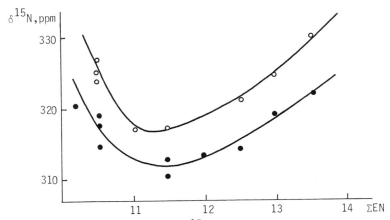

Fig. 7. "U-shaped dependence of the ^{15}N CS on the sum of Si-substituent electro-negativity in aminosilanes X_3SiNHR, R=Ph (●) and R=Bu^t (o).

Upfield shift of the nitrogen resonance indicating strong +I effect of the Me_3Si group can be observed only when the substituents exerting a strong -I ef-fect appear at the nitrogen atom (e.g. NR_2, Cl, Br, I, N≡C, SCN).[287a] A low-field shift indicating delocalization of nitrogen lp appears when p_π-p_π bonding between the nitrogen and X is likely to occur in the series $(Me_3Si)_2NX$ (X=BR_2, AlR_2, BeR, N=Y, P=NR). This shift is considerably larger than downfield shift owing to electronegative ligands (see Table 7).

Substantial decrease in the nitrogen shielding observed in Me_3Si amides of Sn (II) and Pb (II) compared with that in Sn (IV) and Pb (IV) derivatives have been explained by external magnetic field (B_o) induced charge rotation at the nitrogen atoms involving the formally empty $6 p_z$ or $5 p_z$ at Pb or Sn respecti-vely.[289] The low-field shift of ^{15}N resonances in Me_3Si derivatives of pyrrole and pyrazole relative to H and Me substituted parent compounds suggests a weak d_π-p_π interaction (δ^{15}N, ppm):[290]

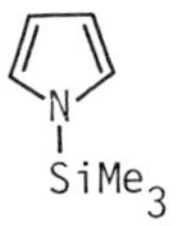

-62.7

-76.5

-216.0 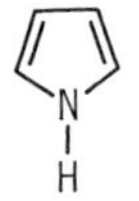 -234.2 -158.7 -180.8

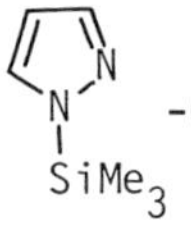

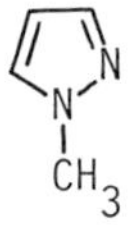

It has been noted that B_o-induced paramagnetic circulations of the $\pi\leftrightarrow\sigma^*$ or $\sigma\leftrightarrow\pi^*$ type in the presence of electropositive Me_3Si group and strongly polarized Si-N bond are reponsible for nitrogen deshielding.

The low-field ^{14}N shift of ~100 ppm observed in $Ph_3SiOClO_3 \cdot Py$ is consistent with complexed pyridine of $Ph_3SiPy^+X^-$ type.[291]

Substantial differences in CS of amidate and imidate forms permit one to employ $^{15/14}N$ NMR for unambiguous determination of the structure of Me_3Si-amides,[130,292] (^{14}N) acylated and carbalkoxylated hydrazines[293] and lactams[294] (^{15}N).

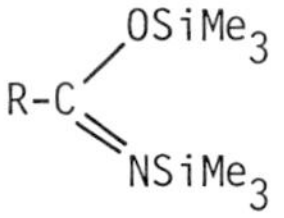

imidate form

$\delta^{14}N$ (ppm): -120 to -177

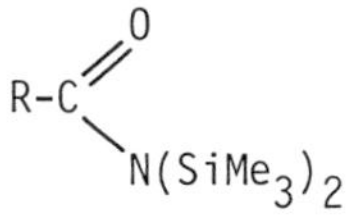

amidate form

-245 to -286

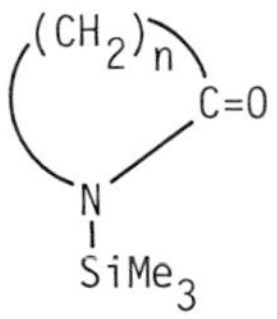

lactam form

$\delta^{15}N$ (ppm): (n=3) -263.9

(n=4) -264.8

(n=5) -259.8

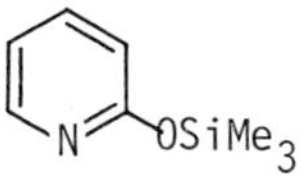

-101.6

^{15}N (and ^{29}Si) NMR has been employed to show that equilibrium between the structures 36 and 37 is shifted toward imidate form 37.[295] The absence of ^{29}Si-^{15}N coupling in the ^{29}Si spectra of ^{15}N labelled derivatives is indicative of fast (on the NMR time scale) silatropic rearrangement between Si-O and Si-N environments due to silyl group exchange.

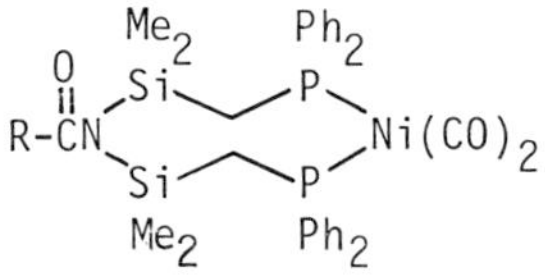

36

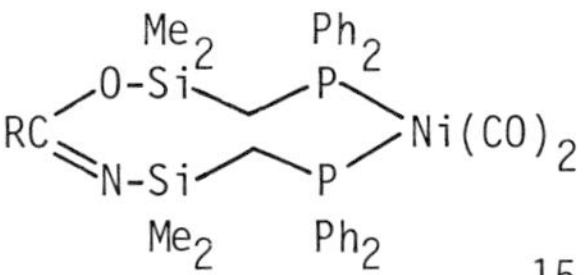

37

$\delta^{15}N$ (ppm):

-135.6 (R = $CH_2=CH$)

-141.0 (R = Ph)

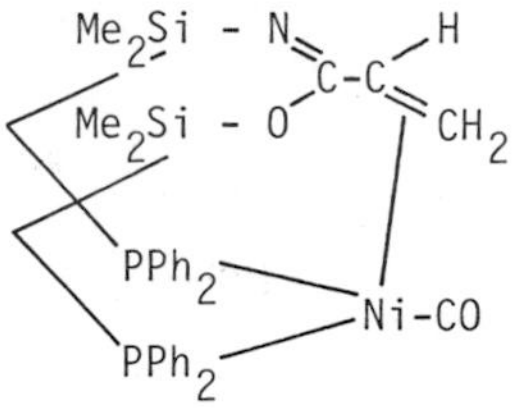

$$\delta, \text{ppm}$$

$$^{15}N = -154.9$$

$$^{29}Si = -13.1 \text{ (SiN)}$$

$$+10.2 \text{ (SiO)}$$

$$^{1}J_{SiN} = 7.5 \text{ Hz}$$

38

Formation of the coordinative linkage in 38 freeze out this process allowing the observation of ^{29}Si-^{15}N coupling.

^{15}N NMR spectrum of the ^{15}N labelled $H_2(Cl)SiNMe_2$ ($\delta = -366.7$ ppm) showed clearly that the Si-N bond is not labile on the NMR time scale, as fully resolved couplings to ^{29}Si and protons at Si ($^2J_{SiH} = 7.26$ Hz) were observed.[125] Wrackmeyer et al.[228] have employed ^{15}N NMR in studies of four- and five-membered Si-N-Sn(II) and Si-N-Sn(IV) rings.

^{14}N NMR has been employed to provide evidence of formation of silaketimines and their adducts with electron donors ($\delta^{14}N$, ppm):[190c]

$$Bu_2^tSi=N-SiBu_3^t$$

-234

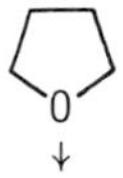

-334

$$Bu_2^tSi=N-SiBu_3^t$$

Negligible changes in ^{15}N shielding of the equatorial N atoms in azasilatranes 30b relative to the model compounds have been explained by weak influence of transannular N→Si bond on the equatorial Si-NH bonds.[204] Individual values of ^{14}N and ^{15}N CS in Si-N compounds have been reported elsewhere[296-301] (see also Table 7).

Some representative values of ^{15}N-^1H couplings in Si-N compounds are collected in Table 8. These couplings provide valuable information on the structure of molecules (see Section III.B) and can be employed in sensitivity enhancement of the ^{15}N spectra using PT or other multipulse techniques (see Section II).

Table 8. ^{15}N-^{1}H Spin-Spin Coupling Constants for Some Representative Si-N Compounds, Hz[a]

C o m p o u n d	$^{n}J_{NH}$	n	Ref.	C o m p o u n d	$^{n}J_{NH}$	n	Ref.
(H$_3$Si)$_3$N	-4.2	2	297a	(Me$_2$ClSi)$_2$NH	68.1	1	31
(H$_3$Si)$_2$NH	71.4	1	300b		1.2	3	203
	-71.3	1	297a	(Me$_2$SiNH)$_3$	69.4	1	31
	-4.3	2	297a	(Me$_2$SiNH)$_4$	67.0	1	31
(Me$_3$Si)$_2$NH	66.9	1	24	(Me$_2$SiOMe$_2$SiNH)$_3$	65.9	1	31
	66.5	1	300b	Me$_2$Si(NHPr)$_2$	75.3	1	24
Me$_3$SiNHBu^t	74.0	1	24	Me$_2$Si(NHBu^t)$_2$	74.0	1	24
Me$_3$SiNHPh	76.5	1	24	Me$_2$Si(NHPh)$_2$	77.4	1	24
	76.0	1	300b	Ph$_2$MeSiNH$_2$	74.0	1	25
Me$_3$SiNHC$\underset{O}{\diagdown}$NH	78.0	1	24	(Ph$_2$MeSi)$_2$NH	66.3	1	24
H$_3$SiNHPF$_2$	-73.1	1	297a	Ph$_2$MeSiNHBu^t	75.0	1	24
	-4.1	2	297a	(EtO)$_3$SiNHBu^t	73.6	1	24
Me$_3$SiNHSCF$_3$	79.2	1	299		2.4	3	203
Me$_3$SiNHOSiMe$_3$	64.8	1	36	(EtO)$_3$SiNHPh	77.8	1	24
Me$_3$SiN(Me)NO$_2$	0.6	3	300c		2.1	3	203
Me$_3$SiN(NO$_2$)COOEt	0.9	3	300c		0.9	4	203
$\underline{11}$	60	1	123		78.7	1	24
$\underline{12}$	65	1	123	$\overline{N(CH_2CH_2O)_3SiH}$[b,c]	0	2	18
				$\overrightarrow{N(CH_2CH_2O)_2Si(Ph)H}$[b,d]	8.4	2	18

[a] In C$_6$D$_6$ or CDCl$_3$. Unsigned numbers represent absolute values of J.
[b] In (CD$_3$)$_2$SO.
[c] Apical H on TBP of Si.
[d] Equatorial H on TBP of Si.

VI. MULTINUCLEAR STUDIES OF SILICON COMPOUNDS

The last section of this review focusses on studies employing nuclei other than ^{1}H, ^{13}C and ^{29}Si published since 1983. The most important parameters of the nuclei, specifically spin (I), natural abundance (a in %) and gyromagnetic ratio (γ in $(rad \cdot T^{-1} \cdot s^{-1})10^{-7}$) are given in parentheses following the name of the nucleus.[5]

Hydrogen-2 (I=1, a=0.015, γ=4.1064). Deuterium labelling allows one to obtain mechanistic information on potential reaction intermediates. The mechanisms of the far-UV photochemical ring opening and cleavage reactions of 1,1-dimethyl-1-silacyclobut-2-ene,[302] formation of silyliron (III) porphyrin complexes,[303] and platinum catalysed hydrosilylation of PhCH=CH$_2$ with Et$_3$SiH have been studied by means of ^{2}H NMR.[239]

Lithium-6 (I=1, a=7.42, γ=3.9366) and lithium-7 (I=3/2, a=92.58, γ=10.3964). ^{7}Li NMR spectra of Me$_{4-n}$Si(NXBu^n)$_n$ and RSi(NXSiMe$_3$)$_3$, X=H, Li ("silazates") permitted the observation of species of different Li coordination numbers in various solvents and when complex-forming ligands are present.[304]

Multinuclear NMR including ^{7}Li spectra have been employed in the study of [Bu$^t_2$SiFLiNBu^t]$_2 \cdot$THF.[305] Fluctuation of Li-N bonds occurs in solution as depicted below:

Buncel et al. have reported ^{6}Li and ^{7}Li NMR parameters for Ph$_n$Me$_{3-n}$SiLi (n=1-3) as a function of solvent, temperature and the presence of crown ether complexing agents.[223] The narrow range of Li CS suggests a significant Si-Li interaction. Below ca. 5°C ^{7}Li and ^{31}P NMR spectra of [LiPPh(SiMe$_3$)tmeda]$_2$ (tmeda = N,N,N',N'-tetramethylethylenediamine) showed scalar ^{31}P-^{7}Li coupling of 40 Hz due to decreased rate of intermolecular Li exchange.[306] A similar value of $^1J_{PLi}$ (49.4 Hz) has been observed for dimer 34 as studied by multinuclear (including ^{7}Li) NMR.[233] The reaction between LiC(SiR$_3$)$_3$ and B(OMe)$_3$ giving (R$_3$Si)$_3$CB(OMe$_2$)$_2$, as well as several further reactions were monitored by ^{7}Li, ^{11}B, ^{19}F and ^{29}Si NMR.[307] Stalke et al. have studied intramolecular Li-F coordination by ^{6}Li, ^{7}Li, ^{19}F and ^{29}Si NMR:[308]

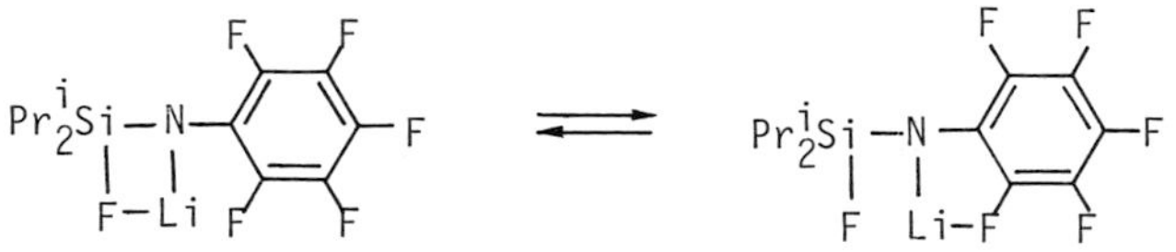

Multinuclear NMR data (including ^{7}Li) have provided direct evidence of cation-anion pairing in alkaline silicate solutions.[112b] At low temperatures formation of a transient dimer [Bu^{t}Li·2Me$_3$SiOMe]$_2$ was observed by ^{6}Li NMR.[309]

Boron-11 (I=3/2, a=80.3, γ=8.5794). Compounds of the 39 type and their transition metal complexes have been characterized by multinuclear NMR including ^{11}B spectra.[234,236,281]

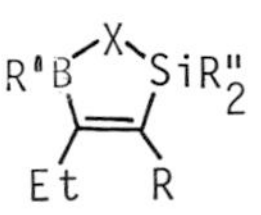

39   X=O, NR, PR, S, Se

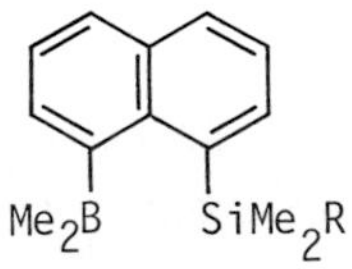

40

a) R = Me
b) R = OEt
c) R = F

A downfield ^{29}Si shift (δ^{29}Si$=20.3$ ppm) and upfield ^{11}B shift (δ^{11}B$=19$ ppm) in 40b relative to that in the model compounds PhSiMe$_2$OEt (δ^{29}Si$=5.1$ ppm) and 1-naphthyl-BMe$_2$ (δ^{11}B$=80$ ppm) provide evidence for the formation of an intra-molecular DA (Si)O→B bond.[310] Similar variations in ^{11}B and ^{29}Si spectra suggest the formation of strongly polarized DA complexes between CF$_3$SO$_3$SiR$_3$ (R=Me, Et) and BX$_3$ (X=Cl, Br).[311] The structure of complexes 40·F$^-$ was determined by means of ^{11}B, ^{19}F and ^{29}Si NMR.[310] An upfield shift of both ^{11}B and ^{29}Si resonances and $J_{SiF}=13.2$ Hz indicates an increase in the coordination number of B(III→IV) and Si(IV→V) atoms.

Upfield shifts of ^{11}B and ^{29}Si resonances in the borasilaindanes 41 as com-pared to the shifts in the model compounds 42 were considered as evidence for a through-bond interaction between both heteroatoms.[312]

R=Me, Cl

41   42

Silicon-containing carboranes were studied with the aid of ^{1}H, ^{11}B, ^{13}C and ^{29}Si NMR.[181e] The small value of J_{SiH} coupling constant (42 Hz) suggests a structure in which one H is involved in the Si-H-B 3-centre 2-electron bond with a silicon atom formally in the 2+ oxidation state, providing the first example of a Si-H-E (E - main group element) bridge bond.

The multinuclear NMR data involving ^{11}B are consistent with planar 6-membered silaborazine rings.[313] ^{11}B CS have been reported for B, Si and Sn containing ethylene derivatives,[224] B containing silacyclopent-3-enes and silacyclopenta-dienes,[248a] and 1,2-dihydro-1,2,5-disilaborepines.[248d]

The ^{11}B and ^{29}Si NMR spectra of Me$_3$Si derivatives of closo-germacarboranes[314] and closo-plumbacarboranes[315] were consistent with apical positions of the Ge(II) and Pb(II) atoms in a pentagonal-bipyramidal cluster, and at least two different kinetic processes in the solution of their 2,2'-bipyridine complexes. Multinuc-lear data including ^{11}B NMR of Me$_3$Si and Alk$_2$B substituted stannacyclopentadie-nes,[250] silyl derivatives of dihydroazaborolyl system,[316] (Me$_3$Si-amino)boranes,[317] Me$_3$Si derivatives of (borylimino)phosphoranes,[318] three- and four-membered rings with ring atom sequence BNSi and BNSiE (E = P, N, As)[319] have been published.

Nitrogen-15 (I=1/2, a=0.36, γ=-2.7116). Unsubstituted and 3,7,10-trimethyl-substituted 1-(4'-tolyl)-silatranes have been studied by multinuclear NMR in-cluding ^{15}N.[320a] The N→Si bond length in 1-halosilatranes (Hal=F, Cl, Br, I) has been calculated from ^{15}N CS.[320b]

^{15}N CS were measured in silocanes 29b[321] and silatranones 31.[207] A downfield shift of ^{15}N resonances with the rising number of C=O groups (n) in the atrane framework of 31 suggests an increase in the DA N→Si bond strength. Multinuclear NMR data for Si-containing aminoalcohols have been systematized recently.[322]

The results of multinuclear study of triazasilatranes 30b suggest the existence of a stronger N→Si bond in 30b, as compared to that in 30c.[204] The shielding of the apical nitrogen increases with the growing N→Si bond strength, whereas only minor variation was noted in the shielding of equatorial nitrogen atoms.

The formation of a N→Si bond in 21 is accompanied by a decrease in shielding of N(10) (δ^{15}N=-34.9 ppm), as compared to the SiMe$_3$ derivative (δ^{15}N=-70.8 ppm), which does not contain the N→Si bond.[197] The ^{15}N CS of 23a,b have been reported.[198d]

Lambert et al.[323a] have employed ^{15}N NMR in studies of complex formation between Ph$_3$SiOClO$_3$ and nitrogen donors. In the presence of Ph$_3$SiOClO$_3$ the ^{15}N CS of CH$_3$CN was essentially unchanged from that of free CH$_3$CN, whereas addition of 1 equivalent of Py led to low-field shift of ^{15}N signal ($\Delta\delta$ = 97 ppm). Thus, ^{15}N NMR indicated very clearly complex formation between Py and Ph$_3$SiOClO$_3$.

Oxygen-17 (I=5/2, a=0.037, γ=-3.6264). The formation of DA O→Si bond in 24 results in an upfield shift of ^{17}O resonances ranging from 52 to 65 ppm for the C=O groups and a downfield shift ranging from 18 to 25 ppm for the OCH$_2$ groups relative to the model compounds.[324] The strength of O→Si coordination decreases with increasing temperature and the acceptor ability of substituent X. In the case of Si(OEt)$_3$ derivatives, O→Si coordination was not observed. Linear correlations exist between ^{17}O CS and σ_p constants of substituents as well as between the ^{17}O CS and C=O bond stretching frequencies.

In the series of furylsilanes the ^{17}O CS are correlated with the magnitude of π-charge on the O atom as calculated by means of the CNDO/2 method.[325]

Fluorine-19 (I=1/2, a=100, γ=25.16). ^{19}F nuclei have been widely used in studies of F-containing compounds due to their extremely favourable NMR properties. An extensive compilation of ^{19}F data has been presented by Wray.[326]

^{19}F NMR allows the observation of enhanced Si coordination. This was employed to provide the evidence for the existence of a 5-coordinated Si in compounds 19,[196] 43,[327] 21 and 44.[328]

R'RN—
R$_2$FSi
\X⁄

43

X=O, NMe
R,R'=H, Me

X— SiFR$_2$

44

X= a) O
    b) NMe
R$_2$=Ph$_2$, F$_2$

A decrease in ^{19}F shielding ($\Delta\delta\sim$10 ppm) and in the absolute values of $^1J_{SiF}$ couplings (ΔJ=34-42 Hz), as compared to those[198b] in the model compound

$ClCH_2SiF_3$, is indicative of O→Si coordination in 24. At low temperatures the $^1J_{SiF}$ couplings show different values for the apical (126.7 to 129.1 Hz) and equatorial (140.4-140.9 Hz) F atoms, suggesting a different character of the corresponding bonds on the TBP of the Si atom. Multinuclear studies involving ^{19}F NMR indicate stronger O→Si coordination in 26, as compared to that in 24.[198e]

The rates of ligand permutation (inversion of Si configuration), as measured by ^{19}F monitoring of site interchange for the geminal CF_3 groups, for spirosilane 45a in weakly nucleophilic media are consistent with the mechanism proceeding by nondissociative Berry-type pseudorotation of the intermediates 45b,c.[329] By contrast, the temperature dependence of the ^{19}F spectrum of a 1:1 mixture of 46+45a indicates that a dissociation/recombination mechanism is operative in the interconversion process within the structure.[330]

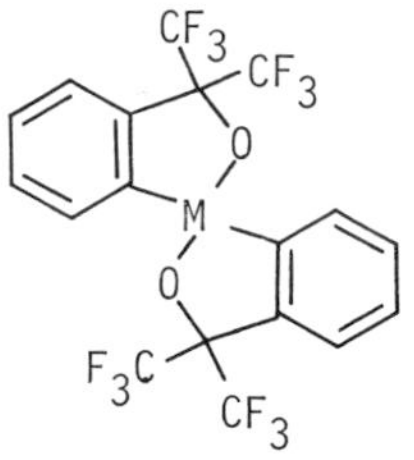

45

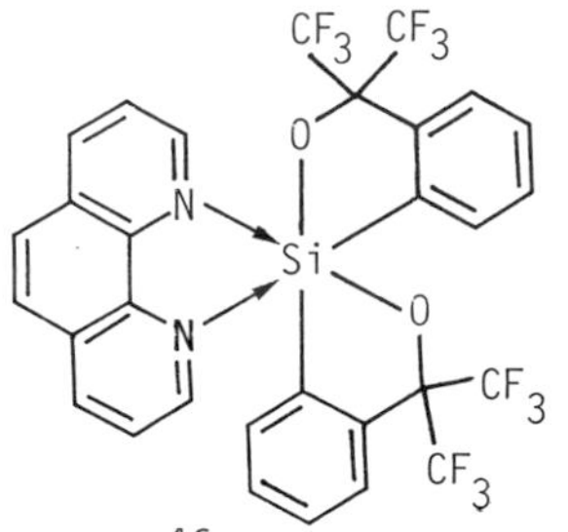

46

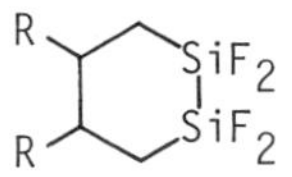

47

a) M=Si

b) M=Si→Nu⁺, Nu = $\overset{O}{\overset{\|}{C}}$—⬡—NMe₂ or MeOH

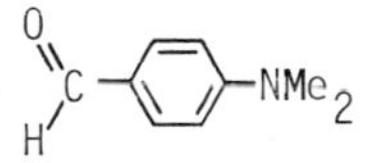

c) M=SiR⁻NMe₄⁺

Activation parameters ($\Delta G^{\neq}$, $\Delta H^{\neq}$ and $\Delta S^{\neq}$) for ring inversion in 47 (R=Cl, Br, Me, H) have been determined with reasonable accuracy by full line-shape analysis of the dynamic ^{19}F NMR spectra of these compounds.[331]

An intermolecular F/Cl exchange occurs with halogenated 1,3-disilazanes, as followed by ^{19}F and ^{29}Si NMR.[332] Corriu et al.[212] have deduced information on the geometry of five-coordinated Si compounds from ^{19}F spectra. Multinuclear NMR (including ^{19}F) data have provided an evidence of Li-F coordination in $Pr^i_2(F)SiN(Li)C_6H_5$.[308] New silicon-monosubstituted and -disubstituted (η^4-2,5-diphenylsilacyclopentadiene) transition-metal complexes,[174] and compound 34[233] were characterized by multinuclear NMR (including ^{19}F) data.

Cycloaddition reactions of tetrafluorodisilacyclobutene and 1,4-disilacyclohexa-2,5-dienes have been studied using ^{19}F NMR.[333] Multinuclear data including

^{19}F NMR have been reported for $(Me_3Si)_3CSiRR'F$ (R,R'=F, Me, Pr^i, Ph, Bu^t, CH_2Ph),[334] $(Me_3Si)_3CSiHRR'$ (R,R'=F, H, Ph, Br, I),[335] $Me_{3-n}SiCl_n(CFCl_2)$ (n=0-3),[244] F_nSiX_{4-n} (X=Cl, Br, I, SiX_3, n=1-4),[245] $(Me_3SiNMe)_2Si(F)N(Bu^t)AlCl_2$,[126] $Bu^t_2Si(F)NHSnMe_3$ and $Bu^t_2Si(F)N(SnMe_3)_2$ showing hindered rotation around the Si-N bond.[336]

Sodium-23 (I=3/2, a=100, γ=7.0761). Cation-anion pairing in alkaline silicate solutions was followed by ^{23}Na NMR.[112b]

Aluminium-27 (I=5/2, a=100, γ=6.9704). ^{27}Al and ^{29}Si NMR have revealed the formation of intramolecular Cl→Al coordination[337] and stereochemically rigid dimers in alumosilazanes[126,338] and compounds of 48 type.[281a]

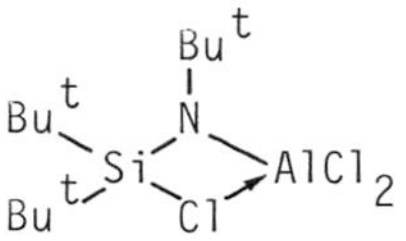

$\delta^{29}Si$=50.4 ppm

$\delta^{27}Al$=106.6 ppm

ref.337

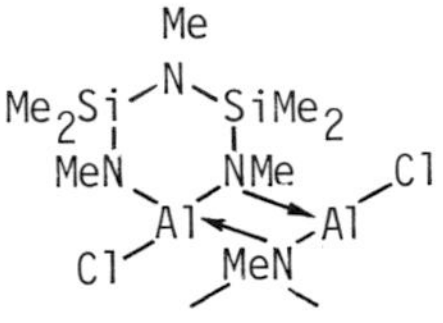

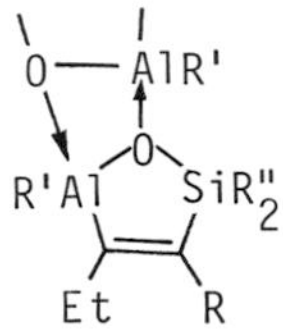

48

The structure of simple alumosilicate anions formed by the reaction of silicate and alumosilicate anions was investigated by ^{27}Al NMR spectroscopy.[112c]

Phosphorus-31 (I=1/2, a=100, γ=10.8289). ^{29}Si, ^{31}P CS and $^2J_{PSi}$ couplings permit to differentiate between the C and N silylated derivatives of 49,[247] between the isomers 50 A and B, and to rule out the presence of the isomer C,[339] to distinguish the structures 51 A and B ($^2J_{PSi}$, Hz):[340]

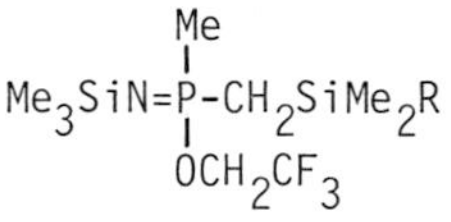

$Me_3SiN=\overset{\underset{|}{Me}}{P}-CH_2SiMe_2R$
$\underset{|}{OCH_2CF_3}$

49

50

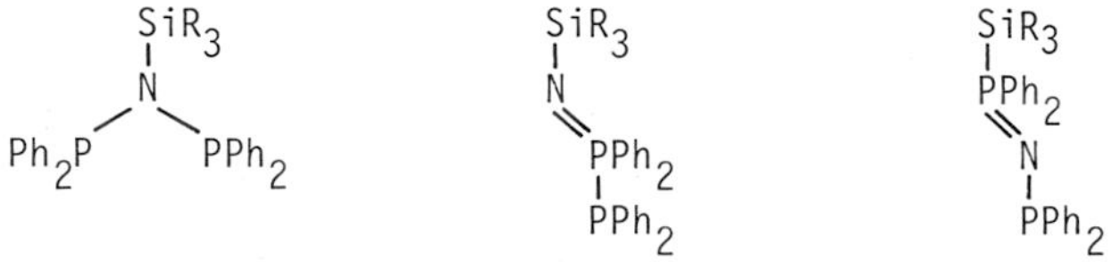

A

3.5-7.6

B

19.1-23.5

C

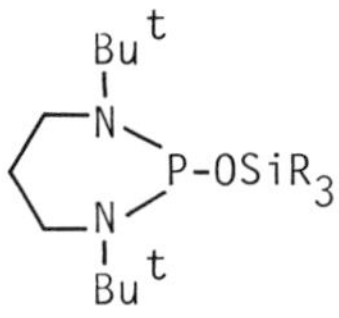

A

0-9.8

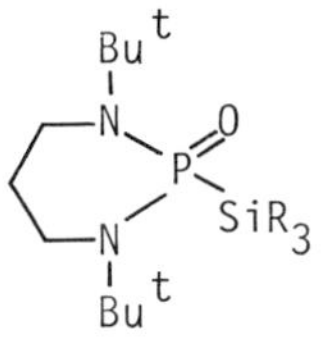

B

15.9-28.6

51

The ^{31}P resonance in phosphasilenes was found at unexpectedly high fields (δ=93.5 to 136.0 ppm), as compared to the P=X compounds where X is a group 15 element (δ=476 to 620 ppm).[182] Surprisingly, the Si-substituent effects are small for [29]Si but rather large for [31]P. [31]P CS in the silaphosphanes $Me_3SiPSiMe_{3-n}(SiMe_3)_n$ (n=0-3) were observed in the range between -251.0 and -303.3 ppm (ref. 230).

Multinuclear data including ^{31}P NMR have been reported for the silylphosphi-mines $Bu_2^tFSi-N=PX_3$ (X=NR_2, OR, F, Cl)[246] and complexes 52 showing smaller $^2J_{PSi}$ couplings in the case of X=O (0 to 7 Hz) as compared to those for X=NH (7 to 10 Hz).[341]

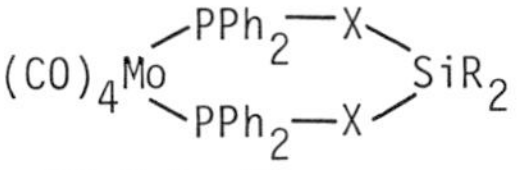

52 X=O, NH

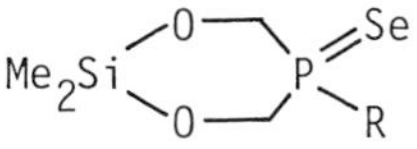

53

^{31}P NMR data have been reported for the silylphosphoranimines $(Me_3Si)_2NP(R,H)$= $NSiMe_3$ (R=N($SiMe_3$)$_2$, Bu^t, Pr^i, $OSiMe_3$).[342] Energy barriers of ca. 12 to 19 kcal/mol were determined for the [1,3]-silyl exchange process. Multinuclear NMR including ^{31}P data suggests the existence of 53 in equilibrium between two enan-thiomeric "chair" conformations in solution.[343] Fritz et al. have reported ^{31}P NMR data for monocyclic and bicyclic silylphosphanes of various ring size.[344] ^{29}Si and ^{31}P NMR data have been obtained for the Me_3Si derivatives of diaza-phosphetidines.[345]

Formation and stereochemistry of phosphite substituted hydridosilyl complexes of iron and derived anionic complexes,[346] silylated di-, tri- and tetraphospha-nes,[347] Os containing metallocycle $CH_2(PPh_2N)_2Os(O)_2(OSiMe_3)_2$,[348] and three different isomeric forms of $Me_3SiNPPh_2CH_2PPh_2$[349] were followed by ^{31}P NMR. A new class of Si-P heterocycles 4-silaphosphorinanes,[350] new metallocycles obtain-ed from $Me_3SiN=PPh_2CH_2PPh_2$ and [RhClX]$_2$ (X=$(CO)_2$ or cyclooctadiene),[351] the first silaphosphacubane (δ=-177 ppm),[231] new amphoteric ligands $(Me_3SiCH_2)_2MPPh_2$ (M=In, Ga),[352] transition-metal (Cr, Mo, W, Fe, Ru, Ni, Co) complexes of cyclic

B-P-Si compounds,[234] transition-metal silyl complexes CpMn((0)$_2$SiMePh$_2$)MX and
Fe(CO)$_3$(PPh$_3$)(SiR$_3$)MX (MX=ZnCl, CdCl or HgBr)[353] M(CO)$_n$L$_2$ and M(CO)$_3$L$_3$ (M=Cr,
Mo, W, L=Me$_3$SiOCH$_2$PMe$_2$ or Me$_2$ViSiOCH$_2$PMe$_2$),[354] P(SiMe$_3$)$_3$-bridged transition-
metal phosphido complexes;[355] [LiPPh(SiMe$_3$)tmeda][306] products of reaction of
(Me$_3$Si)$_3$CLi and (Me$_2$Ph)$_3$SiCLi with PCl$_3$,[356] compound __34__,[223] BNSiP-rings,[319]
trisilylphospnanes PSi$_3$Me$_x$Ph$_{9-x}$,[232] Me$_2$(PhO)Si(CH$_2$)$_n$X (m=1-3, X=PPh$_2$, P(O)Ph$_2$,
P(S)Ph$_2$),[253] Me$_3$Si-phosphaalkenes (Me$_3$Si)$_2$C=PR (R=Cl, F, OBu^t, NR$_2$, Me, Bu^t),[357]
siloxyphosphines,[358] silylated iminomethylenephosphoranes[359] Me$_3$Si derivatives
of (borylimino)phosphoranes,[318] phosphinosilylalanes (Me$_3$Si)$_2$AlPR$_2$ (R=H, Alk,
Ar, Me$_3$Si),[360] and complexes R$_3$P-Au-SiR$_3'$[361] were characterized by ^{31}P NMR
spectra.

   Chlorine-35 (I=3/2, a=75.53, γ=2.6210) and chlorine-37 (I=3/2, a=24.47,
γ=2.1817). Variable temperature (ranging from -75 to +15°C) relaxation time
measurements for ^{13}C, ^{29}Si and ^{35}Cl in Cl$_3$SiH, Cl$_2$MeSiH, ClMe$_2$SiH and Me$_3$SiCl
have provided the group rotation activation energies (ca. 5.1 kcal/mol).[362]
A ^{29}Si and ^{35}Cl NMR study of Ph$_3$SiOClO$_3$ has shown it to be a covalent perchloryl
ester in solution[363] in contrast with earlier claims. No sharp signal, which is
characteristic of ClO$_4^-$ anion, was observed in the ^{35}Cl spectrum, indicating covalent nature of its bonding to Si. Lambert et al.[323] have studied ^{35}Cl and ^{37}Cl
NMR spectra of R$_3$SiOClO$_3$ (R=Me, Ph) solutions in sulpholane and acetonitrile.
Line widths varied from kHz range at 0.1 M concentrations to a few Hz at mili-
molar concentrations (δ from +4.1 to -20.3 ppm, respectively) providing evidence
of a fast two-site exchange between associated (e.g. ion paired) and fully ioniz-
ed forms of the silyl perchlorate. Quantitative analysis of spectra yielded equilibrium constant and percentage of the two forms at each concentration.

   Potassium-39 (I=3/2, a=93.1, γ=1.2483). ^{39}K NMR has been employed in studies
of cation-anion pairing in alkaline silicate solutions.[112b]

   Cobalt-59 (I=7/2, a=100, γ=6.3472). ^{59}Co CS have been reported for the complexes CoCp·39[281a] and Co$_2$(CO)$_6$·Me$_3$SiC≡CSiMe$_3$.[365]

   Germanium-73 (I=9/2, a=7.76, γ=-0.9331). Multinuclear spectra including ^{73}Ge
NMR have been reported for Me$_{4-n}$Si(GeH$_3$)$_n$ (n=1-3) and H$_3$SiGeH$_3$.[365] An increase
in n is accompanied by a decrease in ^{73}Ge shielding (δ^{73}Ge=-315.5 to -277.0 ppm)
and increase in ^{29}Si shielding (δ^{29}Si=-9.4 to -71.7 ppm).

   Selenium-77 (I=1/2, a=7.58, γ=5.1018). Compounds of __53__ type[343] and
Me$_3$SiCH$_2$SeR[366] have been examined using ^{77}Se NMR.

   Organosubstituted 2,5-dihydro-1,2,5-selenasilaboroles and their adducts with
AlCl$_3$, pyridine, PMe$_3$, and π-complexes with Fe(CO)$_3$, Ru(CO)$_3$, CpCO and Ni were
characterized by multinuclear (including ^{77}Se) NMR data.[236] An example presented
below shows the relative sensitivity of various NMR parameters to the structural
changes with __54__ (δ, ppm (^{1}J$_{SeSi}$, Hz)):[236]

Et   Me

EtB〜Se〜SiMe$_2$ 54·AlCl$_3$ 54·Py   54·PMe$_3$ 54·Fe(CO)$_3$ 54·Ru(CO)$_3$ 54·CoCp

54

	54	54·AlCl$_3$	54·Py	54·PMe$_3$	54·Fe(CO)$_3$	54·Ru(CO)$_3$	54·CoCp
^{11}B	79.2	76.3	11.7	-1.4	29.3	25.7	33.1
^{29}Si	24.2	33.0	21.6	22.5	42.2	49.3	18.3
^{77}Se	-95.2	-167.5	-319.6	-464	-549.4	-566.5	-726.6
	(97)	-	(108.6)	(110.3)	(77)	(75.9)	(82.6)

^{77}Se CS of [(Me$_3$Si)$_3$C]$_2$Se$_2$ and (Me$_3$Si)$_3$CSeI were reported.[367]

Rubidium-87 (I=3/2, a=27.85, γ=8.7532). Interactions between Rb$^+$ cations and silicate anions in alkaline silicate solutions were investigated by ^{87}Rb NMR spectroscopy.[112b]

Rhodium-103 (I=1/2, a=100, γ=-0.8520). Wrackmeyer et al. have reported ^{103}Rh CS and J_{RhSi} coupling (1.1 Hz) for complexes 39.[281a] Silicon-containing Rh (III) and Rh (V) hydrides have been studied by multinuclear NMR including ^{103}Rh.[237a]

Cadmium-113 (I=1/2, a=12.26, γ=-5.9328). Adducts of Cd[N(SiR$_3$)$_2$]$_2$ with nitrogen donors have been studied by multinuclear (^{113}Cd, ^{29}Si and ^{13}C) NMR spectroscopy.[368] An increase in the coordination number of Cd from two to three is accompanied by a downfield shift of ^{113}Cd and an upfield shift of ^{29}Si resonances.

Tin-119 (I=1/2, a=8.58, γ=9.9756). ^{119}Sn NMR spectroscopic results have been systematized by Wrackmeyer.[255a] Multinuclear data including ^{119}Sn NMR have confirmed the structures of Me$_3$SnCH$_2$SiMe$_3$ and some halogenated derivatives of this compound,[369] C(SnMe$_3$)$_n$(SiMe$_3$)$_{4-n}$,[248b] Ph$_3$SnSiPh$_2$SnPh$_3$,[256] Me$_3$Sn derivatives of silacyclopentadienes and silacyclopent-3-enes,[187a,c-e] Sn (II) and Sn (IV) silylamides,[226c,289] B, Si and Sn containing ethylene derivatives,[224] Me$_3$Si derivatives of stannacyclopentadienes,[250] R(R')C=C(SiMe$_3$)SnMe$_3$,[251] some four and five-membered Si-N-Sn heterocycles.[228,249]

Tellurium-125 (I=1/2, a=6.99, γ=-8.4398). Products of the reaction of Me$_3$SiCl with PhTeMgBr were identified with the aid of ^{29}Si and ^{125}Te NMR spectroscopy.[328]

Cesium-133 (I=7/2, a=100, γ=3.5087). Cation-anion pairing in alkaline silicate solutions has been investigated by ^{133}Cs NMR.[112b]

Tungsten-183 (I=1/2, a=14.28, γ=1.1145). The Me$_2$SiCp$_2$-bridged binuclear tungsten complexes have been studied by the indirect 2D ^{31}P-^{183}W and ^{1}H-^{183}W shift correlation spectroscopy.[370] New heteropolytungstate anion H$_3$Si$_2$W$_{18}$Zr$_3$O$_{71}^{11-}$ was characterized by ^{29}Si and ^{183}W NMR data.[371]

Platinum-195 (I=1/2, a=33.8, γ=5.7412). ^{195}Pt NMR studies of the platinum catalysed hydrosilylation of PhCH=CH$_2$ with Et$_3$SiH revealed a build-up of a hydride-like intermediate complex.

Mercury-199 (I=1/2, a=16.84, γ=4.7912). The ^{199}Hg NMR spectra were recorded for a series of silylmercury derivatives of the Hg(SiRR'R") $_2$ type.[240] The total range of ^{199}Hg CS observed for these derivatives encompasses 3000 ppm from the least shielded complex reported Li$_2$Hg(SiMe$_3$)$_4$ to quite highly shielded species such as (Cl$_3$Si)$_2$Hg. A linear correlation between the ^{199}Hg CS and the sum of the EN of Si substituents was observed for symmetric species and between ^{199}Hg CS and the lowest energy UV absorption maximum for these derivatives. Limited studies have been reported on the solvent, concentration, and temperature dependence of ^{199}Hg CS for these derivatives. In (Me$_{3-n}$X$_n$SiCH$_2$)$_2$Hg (X=MeO, Cl, n=0-3) an increase in the EN of Si substituents is accompanied by growing ^{199}Hg shielding.[252] Complexes [CpMn(CO)$_2$SiMePh$_2$]HgBr and [Fe(CO)$_3$(PPh$_3$)SiR$_3$]HgBr were characterized by ^{199}Hg NMR.[353]

Lead-207 (I=1/2, a=27.6, γ=5.5797). ^{207}Pb CS were reported on some Pb (II) silylamides.[289] Dynamic behaviour of diazasilaplumbetidines has been interpreted[249] on the basis of multinuclear (including ^{207}Pb) NMR data. ^{207}Pb CS of R(R')C=C(SiMe$_3$)PbMe$_3$ were reported.[251] Me$_3$Si derivatives of closo-plumbacarboranes have been studied by multinuclear (including ^{207}Pb) NMR spectroscopy.[315]

The results of NMR studies of silicon compounds have revealed that multinuclear approach permits one to obtain a more detailed insight into the structure of molecules while multipulse methods increase the sensitivity and resolution of NMR spectroscopy extending the field of its application.

REFERENCES

1. J.F.Kasting, O.B.Toon, J.B.Pollack, Scientific American (Russ.), 4 (1988) 32.
2. a) M.G.Voronkov, G.I.Zelchan and E.Lukevics, "Silizium und Leben", Akademie-Verlag, Berlin, 1975; b) J.Dunogues, La Recherche, 199 (1988) 596.
3. J.Mason, ed., "Multinuclear NMR Spectroscopy", Plenum, New York (1987).
4. J.B.Lambert, F.G.Riddell, eds., "The Multinuclear Approach to NMR Spectroscopy", NATO ASI Series, Reidel, Dordrecht (1983).
5. C.Brevard, P.Granger, "Handbook of High Resolution Multinuclear NMR", Wiley, New York (1981).
6. P.Laszlo, ed., "NMR of Newly Accessible Nuclei", Academic Press, New York (1983).
7. J.Mason, Chem.Rev., 87 (1987) 1299.
8. R.Benn, A.Rufinska, Angew.Chem., Int.Ed.Engl., 25 (1986) 861.

9. H. Kessler, M. Gehrke, C. Griesinger, Angew.Chem., Int.Ed.Engl., 27 (1988) 490.

10. R.R. Ernst, Chimia, 41 (1987) 323.

11. G.A. Morris, Magn.Reson.Chem., 24 (1986) 371.

12. W. McFarlane, D.S. Rycroft in "Annu.Rep. NMR Spectroscopy" G.A. Webb, ed., Academic Press, London (1985), Vol. 16, p. 293.

13. C.J. Turner, Progr. NMR Spectrosc., 16 (1984) 311.

14. R. Benn, A. Günther, Angew.Chem., Int.Ed.Engl., 22 (1983) 350.

15. R. Ernst, G. Bodenhausen, A. Wokaun, "Principles of Nuclear Magnetic Resonance in One and Two Dimensions", Oxford, New York (1987).

16. A.E. Derome, "Modern NMR Techniques for Chemistry Research", Pergamon, Oxford (1987).

17. a) M. Grignon-Dubois, M. Laguerre, B. Barbe, M. Petraud, Organometallics, 3 (1984) 359; b) Ibid., p. 1060; c) G. Deleris, M. Birot, J. Dunogues, B. Barbe, M. Petraud, M. Lefort, J.Organometal.Chem., 266 (1984) 1; d) D. Bennetau, M. Birot, G. Deleris, J. Dunogues, J.P. Pillot, B. Barbe, M. Petraud, Rev.Silicon, Germanium, Tin, Lead Comp., 8 (1985) 327.

18. Ē. Kupče, E. Liepiņš, A. Lapsiņa, I. Urtāne, G. Zelčāns, E. Lukevics, J. Organometal.Chem., 279 (1985) 343.

19. M. Grignon-Dubois, M. Laguerre, Actual Chim., 10 (1983) 21; Comp.Chem., 9 (1985) 279.

20. D.M. Dodrell, D.T. Pegg, W. Brooks, M.R. Bendall, J.Am.Chem.Soc., 103 (1981) 727; b) B.J. Helmer, R. West, Organometallics, 1 (1982) 877; c) J. Schraml, Collect.Czech.Chem.Commun., 48 (1983) 3402; d) T.A. Blinka, B.J. Helmer, R. West, Adv.Organometal.Chem., 23 (1984) 193.

21. G.A. Morris, R. Freeman, J.Am.Chem.Soc., 101 (1979) 160.

22. W. Philipsborn, R. Müller, Angew.Chem., Int.Ed.Engl., 25 (1986) 383.

23. T.H. Chan, M.A. Brook, Tetrahedron Lett., 26 (1985) 2943; U. Scheim, K. Rühlmann, M. Grosse-Ruyken, A. Porzel, J.Organometal.Chem., 314 (1986) 39.

24. Ē. Kupče, E. Lukevics, J.Magn.Reson., 76 (1988) 63.

25. Ē. Kupče, E. Lukevics, Yu.M. Varezhkin, A.H. Mikhailova, V.D. Sheludyakov, Organometallics, 7 (1988) 1649.

26. Ē. Kupče, E. Lukevics, J.Organometal.Chem., 358 (1988) 67.

27. J. Schraml, J.Magn.Reson., 59 (1984) 515.

28. J. Kowalewski, G. Morris, J.Magn.Reson., 47 (1982) 331.

29. A. Bax, R.H. Griffey, B.L. Hawkins, J.Am.Chem.Soc., 105 (1983) 7188.

30. a) E. Liepiņš, I. Birǵele, P. Tomsons, E. Lukevics, Magn.Reson.Chem., 23 (1985) 485; b) Ē. Liepiņš, Yu. Goldberg, I. Iovel, E. Lukevics, J.Organometal.Chem., 335 (1987) 301.

31. Ē. Kupče, E. Liepiņš, E. Lukevics, B. Astapov, J.Chem.Soc., Dalton Trans., (1987) 1593.

32. a) H.B. Yokelson, A.J. Millevolte, B.R. Adams, R. West, J.Am.Chem.Soc., 109 (1987) 4116; b) M. Kuroda, Y. Kabe, M. Hashimoto, S. Masamune, Angew.Chem., Int.Ed.Engl., 27 (1988) 1727.

33. J. Past, J. Puskar, M. Alla, E. Lippmaa, J. Schraml, J.Magn.Reson.Chem., 23 (1985) 1076; J. Past, J. Puskar, J. Schraml, E. Lippmaa, Collect.Czech. Chem.Commun., 50 (1985) 2060.

34. a) A. Allerhand, R.E. Addleman, D. Osman, J.Am.Chem.Soc., 107 (1985) 5809; b) A. Allerhand, M. Dohrenwend, Ibid., p. 6684; c) A. Allerhand, D. Osman, M. Dohrenwend, J.Magn.Reson., 65 (1985) 361; d) S.R. Maple, A. Allerhand, Ibid., 66 (1986) 168; e) A. Allerhand, C.H. Bradley, Ibid., 67 (1986) 173; f) S.R. Maple, A. Allerhand, J.Am.Chem.Soc., 109 (1987) 56; g) S.R. Maple, A. Allerhand, J.Magn.Reson., 72 (1987) 203.

35. F.A.L. Anet, M. Kopelevich, J.Am.Chem.Soc., 109 (1987) 5870.

36. Ē. Kupče, E. Lukevics, J.Magn.Reson., 80 (1988) 359; Latv.PSR Zinat.Akad. Vestis, Kim.Ser., 5 (1987) 633.

37. Ē. Kupče, E. Liepiņš, I. Zicmane, E. Lukevics, J.Chem.Soc., Chem.Commun., (1989) 818.

38. A. Bax, L. Lerner, Science, 232 (1986) 960.

39. Ē. Kupče, E. Liepiņš, E. Lukevics, Khim.Geterotsikl.Soed., (1987) 129.

40. A.G. Avent, P.D. Lickiss, A. Pidcock, J.Organometal.Chem., 341 (1988) 281.

41. N. Auner, J.Organometal.Chem., 353 (1988) 275.

42. H. Jancke, A. Porzel, Z.Chem., 25 (1985) 252.

43. a) R. Nardin, M. Vincendon, J.Magn.Reson., 61 (1985) 338; b) E. Liepiņš, I. Sekacis, E. Lukevics, Magn.Reson.Chem., 23 (1985) 10; c) J. Schraml, E. Petrakova, J. Pelnar, M. Kvicalova, V. Chvalovský, J.Carbohydr.Chem., 4 (1985) 393; d) J. Schraml, E. Petrakova, J. Hirsch, Magn.Reson.Chem., 25 (1987) 75.

44. J. Schraml, J. Vcelak, M. Cerny, V. Chvalovský, Collect.Czech.Chem.Commun., 48 (1983) 2503; J. Schraml, S. Kucar, J. Zeleny, V. Chvalovský, Ibid., 50 (1985) 1176.

45. J. Schraml, M.F. Larin, V.A. Pestunovich, Ibid., 50 (1985) 343.

46. F.C. Schilling, F.A. Bovey, J.M. Zeigler, Macromolecules, 19 (1986) 2309; Polym.Mater.Sci.Eng., 54 (1986) 475.

47. C. Biran, M. Fourtinon, B. Efendene, J. Dunogues, J.Organometal.Chem., 344 (1988) 145.

48. a) R.K. Harris, M.J. O'Connor, E.H. Curzon, O.W. Howarth, J.Magn.Reson., 57 (1984) 115; b) C.T.G. Knight, R.J. Kirkpatrick, E. Oldfield, J.Am.Chem. Soc., 109 (1987) 1632; c) C.T.G. Knight, J.Chem.Soc., Dalton Trans., (1988)

1457.

49. R.K. Harris, C.T.G. Knight, J.Chem.Soc., Faraday Trans. 2, 79 (1983) 1525.

50. C.T.G. Knight, R.J. Kirkpatrick, E. Oldfield, J.Magn.Reson., 78 (1988) 31.

51. E. Hengge, F. Schrank, J.Organometal.Chem., 362 (1989) 11.

52. J.Maxka, B.R. Adams, R. West, J.Am.Chem.Soc., 111 (1989) 3447.

53. a) H. Marsmann in "NMR Basic Principles and Progress", R. Diehl, E. Fluck, R. Kosfeld, eds., Springer-Verlag, Berlin (1981), Vol. 17, p. 65; b) E.A. Williams in "Annu.Rep. NMR Spectroscopy", G.A. Webb, ed., Academic Press, London (1983), Vol. 15, p. 235; c) R.K. Harris, Chapter 16 in ref. 4; d) J. Schraml in "Carbon-Funct. Organosilicon Compounds", V. Chvalovský, J.M. Bellama, eds., Plenum, New York (1984), p. 121; e) B. Coleman, Part 2, p. 197 in ref. 6.

54. R.J.P. Corriu, C. Guerin, J.J.E. Moreau in "Topics in Stereochemistry", E.L. Eliel, N.L. Allinger, eds., Willey, New York (1984), Vol. 15, p. 43.

55. T. Takayama, I. Ando, Bull.Chem.Soc. Japan, 60 (1987) 3125.

56. C.J. Jameson, H.S. Gutowsky, J.Chem.Phys., 40 (1964) 1714.

57. a) I. Ando, G.A. Webb, "Theory of NMR Parameters", Academic Press, London (1983), Chapter 3; b) G.A. Webb, Chapter 2 in ref. 4.

58. N. Janes, E. Oldfield, J.Am.Chem.Soc., 107 (1985) 6769.

59. R.H. Cragg, R.D. Lane, J.Organometal.Chem., 277 (1984) 199.

60. R. Radeglia, Z.Chem., 26 (1986) 147.

61. I.P. Yakovlev, Yu.S. Finogenov, V.A. Gindin, V.P. Feshin, P.A. Nikitin, A.E. Shchegolev, B.A. Ivin, Zh.Obshch.Khim., 55 (1985) 1093.

62. M.S. Samples, C.H. Yoder, J.Organometal.Chem., 312 (1986) 149.

63. W. Adcock, H. Gangodawila, G.B. Kok, V.S. Iyer, W. Kitching, G.M. Drew, D. Young, Organometallics, 6 (1987) 156.

64. a) T.N. Mitchell, J.Organometal.Chem., 255 (1983) 279; b) P.J. Watkinson, K.M. Mackay, Ibid., 275 (1984) 39.

65. J.R. Van Wazer, C.S. Ewing, R. Ditchfield, J.Phys.Chem., 93 (1989) 2222.

66. V.G. Malkin, O.V. Gritsenko, G.M. Zhidomirov, Chem.Phys.Lett., 152 (1988) 44.

67. R.H. Cragg, R.D. Lane, J.Organometal.Chem., 291 (1985) 153.

68. R.H. Cragg, R.D. Lane, J.Organometal.Chem., 294 (1985) 7; R.H. Cragg, C.K. Ma, Ibid., 304 (1986) 87.

69. H. Sakurai, K. Oharu, Y. Nakadaira, Chem.Lett., (1986) 1797; G.K. Henry, R. Shinimoto, Q. Zhou, W.P. Weber, J.Organometal.Chem., 350 (1988) 3.

70. J. Schraml, M. Grignon-Dubois, J. Dunogues, H. Jancke, G. Engelhardt, V. Chvalovský, Collect.Czech.Chem.Commun., 48 (1983) 3396; M. Grignon-Dubois, A. Marchand, J. Dunogues, B. Barbe, M. Petraud, J.Organometal. Chem., 272 (1984) 19.

282

71. E.W. Della, J. Tsanaktsidis, Organometallics, 7 (1988) 1178.
72. J. Schraml, J. Sraga, P. Hrnčiar, Collect.Czech.Chem.Commun., 48 (1983) 2937.
73. J. Schraml, J. Čermák, V. Chvalovský, A. Kasal, C. Bliefert, E. Krahé, J.Organometal.Chem., 341 (1988) C6.
74. N. Auner, J.Organometal.Chem., 336 (1987) 83.
75. G. Gillette, J. Maxka, R. West, Angew.Chem., 101 (1989) 90; N. Auner, Z.anorg.allg.Chem., 558 (1988) 87.
76. a) A.G. Brook, Y.K. Kong, A.K. Saxena, J.F. Sawyer, Organometallics, 7 (1988) 2245; b) A.G. Brook, A.K. Saxena, J.F. Sawyer, Ibid., 8 (1989) 851; c) K.M. Baines, A.G. Brook, R.R. Ford, P.D. Lickiss , A.K. Saxena, W.J. Chatterton, J.F. Sawyer, B.A. Behnam, Ibid., 8 (1989) 693.
77. N. Wiberg, K. Schurz, G. Müller, J. Riede, Angew.Chem., Int.Ed.Engl., 27 (1988) 936.
78. P. Boudjouk, U. Samaraweera, R. Sooriyakumaran, J. Chrisciel, K.R. Anderson, Ibid., 27 (1988) 1355.
79. B.J.J. Van de Heisteeg, M.A.G.M. Tinga, Y. Van der Vinkel, O.S. Akkerman, F.Bickelhaupt, Ibid., 316 (1986) 51.
80. H.J.R. DeBoer, O.S. Akkerman, F. Bickelhaupt, Ibid., 321 (1987) 291.
81. G. Manuel, A. Faucher, P. Mazerolles, Ibid., 327 (1987) C25.
82. G. Maas, K. Schneider, W. Ando, J.Chem.Soc., Chem.Commun., (1988) 72.
83. W. Wojnowski, J. Pikies, Z.anorg.allg.Chem., 508 (1984) 201.
84. J. Pikies, W. Wojnowski, Ibid., 511 (1984) 219.
85. E. Liepiņš, I. Zicmane, E. Lukevics, J.Organometal.Chem., 306 (1986) 72.
86. U. Scheim, H. Grasse-Ruyken, K. Rühlmann, A. Porzel, Ibid., 312 (1986) 27.
87. R.H. Cragg, R.D. Lane, Ibid., 289 (1985) 23; R.H. Cragg, R.D.Lane, Ibid., 270 (1984) 25.
88. J. Schraml, J. Sraga, P. Hrnčiar, Collect.Czech.Chem.Commun., 48 (1983) 3097.
89. A. Bassindale, T.Stout, J.Organomet.Chem., 271 (1984) C1.
90. G.A. Olah, H. Doggweiler, J.D. Felberg, S. Frohlich, J.Org.Chem., 50 (1985) 4847.
91. P. Sormunen, E. Liskola, E. Vähäsarja, T.T. Pakkanen, T.A. Pakkanen, J.Organometal.Chem., 319 (1987) 327.
92. O. Graalmann, U. Klingebiel, W. Clegg, M. Haase, G.M. Sheldrick, Angew. Chem., Int.Ed.Engl., 23 (1984) 891.
93. A.G. Brook, W.J. Chatterton, J.F. Sawyer, D.W. Hughes, K.Vorspohl, Organometallics, 6 (1987) 1246.
94. M.G. Kuznetsova, T.L. Abronin, T.L. Krasnova, A.V. Kisin, A.V. Bochkarev, N.V. Alekseev, A.M. Mosin, Zh.Obshch.Khim., 57 (1987) 1715.

95. M. Wang, D. Zhang, G. Li, Q. Xie, Gaodeng Xuexiao, Huaxue Xuebao, 9 (1988)
    410; Chem.Abstr. 110 (1989) 114922x.

96. K. Dippel, O. Graalmann, U. Klingebiel, Z.anorg.allg.Chem., 552 (1987) 195.

97. V.D. Sheludyakov, A.B. Lebedeva, A.V.Kisin, I.S.Nikishina, A.V.Lebedev,
    A.D.Kirilin, Zh.Obshch.Khim., 58 (1988) 393.

98. I.D. Kalikhman, O.B. Bannikova, L.I.Volkova, B.A. Gostevskii, T.I. Yushmano-
    va, V.A. Lopirev, M.G. Voronkov, Izv.Akad.Nauk SSSR, Ser.Khim., (1988) 460.

99. B.D. Lavrukhin, B.A. Astapov, A.V. Kisin, A.A. Zhdanov, Izv.Akad.Nauk SSSR,
    Ser.khim., 5 (1983) 1059.

100. D. Seyferth, C. Prud'homme, G.H. Wiseman, Inorg.Chem., 22 (1983) 2163.

101. C.D. Juengst, W.P. Weber, G. Manuel, J.Organometal.Chem., 308 (1986) 187.

102. B. Dejak, Z. Lasocki, H. Jancke, Bull.Pol.Acad.Sci., Chem., 33 (1985) 275.

103. H. Jancke, J. Schulz, E. Popowski, H. Kelling, J.Organometal.Chem., 354
     (1988) 23; E. Popowski, J. Schulz, H. Kelling, H. Jancke, Z.anorg.allg.Chem.,
     547 (1987) 100; E. Popowski, J. Schulz, H. Jancke, K. Feist, H. Kelling,
     Ibid., 558 (1988) 206.

104. F. Braun, L. Willner, M. Hess, R. Kosfeld, J.Organometal.Chem., 332 (1987)
     63; Ibid., 366 (1989) 53.

105. A.A. Zhdanov, T.V. Astapova, B.D. Lavrukhin, Izv.Akad.Nauk SSSR, Ser.Khim.,
     (1988) 657.

106. P.J.A. Brandt, R. Subramanian, P.M. Sormani, T.C. Ward, J.E. McGrath,
     Polym.Prepr., 26 (1985) 213.

107. Y.M. Pai, K.L. Servis, W.P. Weber, Organometallics, 5 (1986) 683.

108. Y.M. Pai, W.P. Weber, K.L. Servis, J.Organometal.Chem., 288 (1985) 269.

109. F.N. Stone, B.N. Ranganathan, R.F. Evilia, Anal.Lett., 17 (1984) 2243;
     D.A. Laude Jr., R.W. Lee, C.C. Wilkins, Anal.Chem., 57 (1985) 1286.

110. J. Schraml, V. Blechta, M. Kvicalova, L. Nondek, V. Chvalovsky, Anal.Chem.,
     58 (1986) 1892; J.M. Dereppe, B. Parbhoo, Ibid., 56 (1984) 2740; R.Brezny,
     J. Schraml, M. Kvicalova, J. Zeleny, V. Chvalovsky, Holzforschung, 39
     (1985) 297.

111. J. Schaml, J. Vcelak, M. Cerny, V. Chvalovsky, Collect.Czech.Chem.Commun.,
     48 (1983) 2503.

112. a) A.V. McCormic, A.T. Bell, C.J. Radke, J.Phys.Chem., 93 (1989) 1737;
     b) Ibid., p.1733; c) Ibid., p.1741.

113. a) S.D. Kinrade, T.W. Swaddle, Inorg.Chem., 27 (1988) 4253; b) Ibid.,
     p.4259; c) Ibid., 28 (1989) 1952.

114. F. Schlenkrich, O. Rademacher, H. Scheler, Z.anorg.allg.Chem., 547 (1987)
     109; O. Rademacher, S. Busse, Ibid., 564 (1988) 104; I. Hasegawa, S. Sakka,
     Y. Sugahara, K. Kuroda, C. Kato, J.Chem.Soc., Chem.Commun. (1989) 208.

115. D. Hoebbel, G. Engelhardt, A. Samoson, K. Ujszaszy, Yu.I. Smolin, Z.anorg.allg.Chem., 552 (1987) 236.

116. H. Marsmann, E. Meyer, Z.anorg.allg.Chem., 548 (1987) 193; H.C. Marsmann, E. Meyer, Makromol.Chem., 188 (1987) 887; E. Bertling, H.C. Marsmann, Phys.Chem.Minerals, 16 (1988) 295.

117. R.K. Harris, J. Jones, C.T.G. Knight, R.H. Newman, J.Mol.Liq., 29 (1984) 63.

118. G. Engelhardt, D. Hoebbel, J.Chem.Soc., Chem.Commun., (1984) 514.

119. C.T.G. Knight, R.K. Harris, Magn.Reson.Chem., 24 (1986) 872; S.D. Kinrade, T.W. Swaddle, J.Am.Chem.Soc., 108 (1986) 7159.

120. S. Sjoberg, N. Ingri, A.M. Nenner, L.O. Oehman, J.Inorg.Biochem., 24 (1985) 267.

121. J. Pikies, W. Wojnowski, Z.anorg,allg.Chem., 521 (1985) 173.

122. J. Pikies, A. Herman, W. Wojnowski, A. Meller, Ibid., 498 (1983) 218.

123. U. Kleibisch, U. Kleingebiel, D. Stalke, G.M. Sheldrick, Angew.Chem., Int.Ed.Engl., 25 (1986) 915.

124. A.R. Bassindale, T. Stout, J.Chem.Soc., Perkin Trans. 2, (1986) 221; Ibid., 227; J.Chem.Soc., Chem.Commun., (1984) 1387.

125. S.C. Chaudhry, D. Kummer, J.Organometal.Chem., 339 (1988) 241.

126. W. Clegg, M. Haase, U. Klingebiel, J. Neemenn, G.M. Sheldrick, Ibid., 251 (1983) 281.

127. D.G. Anderson, J. Armstrong, S. Cradcock, J.Chem.Soc., Dalton Trans., (1987) 3029.

128. J. Iley, A.R. Bassindale, P. Patel, J.Chem.Soc., Perkin Trans. 2, (1984) 77.

129. K. Mitteilung, G. Faleschini, E. Nachbaur, Monatsh.Chem., 119 (1988) 457.

130. M.S. Samples, C.H. Yoder, J.Organometal.Chem., 332 (1987) 69.

131. W.R. Hertler, D.A. Dixon, E.W. Matthews, F. Davidson, F.G. Kitson, J.Am.Chem.Soc., 109 (1987) 6532.

132. M. Weidenbruch, A. Schäfer, H. Marsmann, J.Organometal.Chem., 354 (1988) C12.

133. I.D. Kalikhman, A.I. Albanov, O.B. Bannikova, L.I. Belousova, M.G. Voronkov, V.A. Pestunovich, A.G. Shipov, E.P. Kramarova, Yu. I. Baukov, Ibid., 361 (1989) 147.

134. S. Bartholmei, U. Klingebiel, G.M. Sheldrick, D. Stalke, Z.anorg. allg. Chem., 556 (1988) 129.

135. H.G. Horn, M. Hemke, Chem.Ztg., 109 (1985) 1.

136. A. Herman, B. Dreczewski, W. Wojnowski, Z.anorg.allg.Chem., 551 (1987) 196.

137. J. Hahn, K. Altenbach, Z.Naturforsch., Teil B., 41B (1986) 675.

138. a) W. Wojnowski, M. Wojnowski, K. Peters, E.-M. Peters, H.G. von Schnering, Z.anorg.allg.Chem., 530 (1985) 79; b) J. Pikies, W. Wojnowski, Ibid., 511 (1984) 219; c) W. Wojnowski, W. Bochenska, K. Peters, E.-M. Peters, H.G. von Schnering, Ibid., 539 (1986) 165.

139. H.G.Horn, M. Hemke, Chem.Ztg., 110 (1986) 18.

140. A.I. Al-Mansour, S.S. Al-Showiman, I.M. Al-Najjar, Spectrochim.Acta, 44A (1980) 643; Inorg.Chim.Acta, 134 (1987) 275.

141. J.A. Albanese, D.E. Gingrich, C.D. Schaeffer, Jr., S.M. Coley, J.C. Otter, M.S. Samples, C.H. Yoder, J.Organometal.Chem., 365 (1989) 23.

142. P. Jutzi, M. Meyer, Chem.Ber., 121 (1988) 1393.

143. A.R. Bassindale, J.C.-Y. Lau, P.G. Taylor, J.Organometal.Chem., 341 (1988) 213.

144. K. Mach, H. Antropiusova, L. Petrusova, F. Turecek, V. Hanus, P. Sedmera, J. Schraml, Ibid., 289 (1985) 331.

145. V.O. Reikhsfeld, S.V. Nesterova, N.K. Skvortsov, Ē. Kupče, E. Lukevics, Zh.Obshch.Khim., 56 (1986) 1306.

146. S.T. Abu-Orabi, P. Jutsi, J.Organometal.Chem., 329 (1987) 169.

147. P. Jutzi, U. Holtmann, D. Kanne, C. Krüger, R. Blom, R. Gleiter, I. Hyla-Kryspin, Chem.Ber., 122 (1989) 1629.

148. E. Lukevics, N.P. Erchak, I. Kastro, J. Popelis, A.K. Kozyrev, V.I. Anosh-kin, I.F. Kovalev, Zh.Obshch.Khim., 55 (1985) 2062.

149. Y.-X. Ding, W.P. Weber, J.Organometal.Chem., 341 (1988) 267.

150. R. Radeglia, G. Engelhardt, Ibid., 254 (1983) C1.

151. H. Söllradl, E. Hengge, Ibid., 243 (1983) 257.

152. K. Kumar, M.H. Litt, R.K. Chadha, J.E. Drake, Can.J.Chem., 65 (1987) 437; F.K. Mitter, E. Hengge, J.Organometal.Chem., 332 (1987) 47.

153. F.K. Mitter, G.I. Pollhammer, E. Hengge, Ibid., 314 (1986) 1.

154. L. Parkanyi, H. Strüger, E. Hengge, Ibid., 333 (1987) 187.

155. K. Mitteilung, K. Hassler, Monatsh.Chem., 117 (1986) 613.

156. H. Sakurai, S. Hoshi, A. Kamiya, H. Hosomi, C. Kabuto, Chem.Lett., (1986) 1781.

157. A. Sekiguchi, S.S. Zigler, R. West, J. Michel, J.Am.Chem.Soc., 108 (1986) 4241.

158. A.R. Wolff, J. Maxka, R. West, J.Polym.Sci., Part A: Polym.Chem., 26 (1988) 713.

159. A. Fürstner, H. Weidmann, J.Organometal.Chem., 354 (1988) 15.

160. K. Hassler, Monatsh.Chem., 119 (1988) 1051.

161. Y. Kabe, M. Kuroda, Y. Honda, O. Yamashita, T. Kawase, S. Masamune, Angew.Chem., Int.Ed.Engl., 27 (1988) 1725.

162. H. Schmölzer, E. Hengge, J.Organometal.Chem., 260 (1984) 31.

286

163. K.M. Baines, A.G. Brook, Organometallics, 6 (1987) 693.

164. F. Feber, M. Krander, Z.Naturforsch, Part B, 40b (1985) 1301.

165. A.M. Arif, A.H. Cowley, T.M. Elkins, J.Organometal.Chem., 325 (1987) C11.

166. M. Bergdahl, E.-L. Lindstedt, T. Olsson, J.Organometal.Chem., 365 (1989) C11.

167. R. Krentz, R.K. Pomeroy, Inorg.Chem., 24 (1985) 2976.

168. K.H. Pannell, S.P. Vincenti, R.C. Scott III, Organometallics, 6 (1987) 1593; Ibid., 5 (1986) 1056; K.H. Pannell, J.M. Rozell, W.-M. Tsai, Ibid., 6 (1987) 2085; K.H. Pannell, J.M. Rozell, J. Lii, S.-Y. Tien-Mayr, Ibid., 7 (1988) 2524; K.H. Pannell, L.-J. Wang, J.M. Rozell, Ibid., 8 (1989) 550; U. Wachtler, W. Malisch, E. Kolba, J. Matreux, J.Organometal.Chem., 363 (1989) C36.

169. A. Marinetti-Mignani, R. West, Organometallics, 6 (1987) 141.

170. H.-G. Woo, T.D. Tilley, J.Am.Chem.Soc., 111 (1989) 3757.

171. D.M. Roddick, R.H. Heyn, T.D. Tilley, Organometallics, 8 (1989) 324.

172. F.H. Elsner, T.D. Tilley, A.L. Rheingold, S.J. Geib, J.Organometal.Chem., 358 (1988) 169.

173. U. Schubert, A. Schenkel, Chem.Ber., 121 (1988) 939.

174. F. Carre, R.J.P. Corriu, C. Guerin, B.J.L. Henner, W.W.C. Wong Chi Man, Organometallics, 8 (1989) 313.

175. B.J.J. Van de Heisteeg, G. Schat, O.S. Akkerman, F. Bickelhaupt, Ibid., 5 (1986) 1749; Tetrahedron Lett., (1984) 5191.

176. O.J. Brauer, H. Büger, G.R. Liewald, J. Wilke, J.Organometal.Chem., 310, (1986) 317.

177. V.M. Nosova, A.V. Kisin, N.V. Alekseev, Yu.E. Zubarev, Zh.Obshch.Khim., 57 (1987) 1541.

178. F.H. Köhler, W.A. Geike, J.Magn.Reson., 53 (1983) 297.

179. a) F.H. Köhler, W.A. Geike, N. Hertkorn, J.Organometal.Chem., 334 (1987) 359; b) P. Jutzi, W. Leffers, G. Müller, Ibid., 333 (1987) C24; c) U. Schubert, A. Schenkel, Chem.Ber., 121 (1988) 939.

180. M.J. McGeary, J.L. Templeton, J.Organometal.Chem., 323 (1987) 199.

181. a) U. Schubert, K. Bahr, J. Müller, Ibid., 327 (1987) 357; b) U. Schubert, J. Müller, H.G. Alt, Organometallics, 6 (1987) 469; c) C. Schubert, G. Scholz, J. Mueller, K. Ackermann, B. Woerle, R.F.D. Stansfield, J.Organometal.Chem., 306 (1986) 303; d) M.J. Fernandez, P.M. Bailey, P.O. Bentz, J.S. Ricci, T.F. Koetzle, P.M. Maitlis, J.Am.Chem.Soc., 106 (1984) 5458; e) U. Sirivardane, M.S. Islam, T.A. West, N.S. Hosmane, J.A. Maguire, A.H. Cowley, Ibid., 109 (1987) 4600; f) E. Matarasso-Tchiroukhine, G. Jaouen, Can.J.Chem., 66 (1988) 2157.

182. C.N. Smit, F. Bickelhaupt, Organometallics, 6 (1987) 1156.

183. G. Raabe, F. Michl, Chem.Rev., 85 (1985) 419.

184. R. West, Pure and Appl.Chem., 56 (1984) 163.

185. a) N. Wiberg, G. Wagner, G. Müller, Angew.Chem., Int.Ed.Engl., 24 (1985)
     229; b) A.G. Brook, K.D. Safa, P.D. Lickiss, K.M. Baines, J.Am.Chem.Soc.,
     107 (1985) 4338.

186. R. West, Angew.Chem., Int.Ed.Engl., 26 (1987) 1201.

187. a) M. Hesse, U. Klingebiel, Angew.Chem., Int.Ed.Engl., 25 (1986) 649;
     b) N. Wiberg, K. Schurz, G. Reber, G. Müller, J.Chem.Soc., Chem.Commun.,
     (1986) 591.

188. N. Wiberg, K. Schurz, J.Organometal.Chem., 341 (1988) 145.

189. G. Märkl, W. Schlosser, Angew.Chem., Int.Ed.Engl., 27 (1988) 963.

190. N. Wiberg, G. Wagner, G. Reber, J. Riede, G. Müller, Organometallics, 6
     (1987) 35; b) N. Wiberg, G. Preiner, P.Karampatses, C.-K. Kim., K. Schurz,
     Chem.Ber., 120 (1987) 1357; c) N. Wiberg, K. Schurz, Ibid., 121 (1988) 581.

191. C. Zybill, R. West, J.Chem.Soc., Chem.Commun., (1986) 857.

192. P. Arya, J. Boyer, F. Carré, R. Corriu, G. Lanneau, J. Lappasset,
     M. Perrot, C. Priou, Angew.Chem., Int.Ed.Engl., 28 (1989) 1016.

193. a) C. Zybill, D.L.Wilkinson, C. Leis, G. Müller, Angew.Chem., Int.Ed.Engl.,
     28 (1989) 203; b) C. Zybill, G. Müller, Organometallics, 7 (1988) 1368.

194. B. Becker, R. Corriu, C. Guerin, B. Henner, Q. Wang, J.Organometal.Chem.,
     359 (1989) C33.

195. D.A. Dixon, W.R. Hertler, D.B. Chase, W.B. Farnham, F. Davidson,
     Inorg.Chem., 27 (1988) 4012.

196. B.J. Helmer, R. West, R.J.P. Corriu, M. Poirier, G. Royo, A. DeSaxce,
     J.Organometal.Chem., 251 (1983) 295; R.J.P. Corriu, A. Kpoton, M. Poirier,
     G. Royo, J.Y. Corey, Ibid., 277 (1984) C25.

197. G. Klebe, K. Hensen, J. Von Jouanne, Ibid., 258 (1983) 137.

198. a) I.D. Kalikhman, O.B. Bannikova, B.A. Gostevski, O.V. Vyazankina,
     N.S. Vyazankin, V.A. Pestunovich, Izv.Akad.Nauk SSSR, Ser.khim., (1985)
     1688; b) A.I. Albanov, L.I. Gubanova, M.F. Larin, V.A. Pestunovich,
     M.G. Voronkov, J.Organometal.Chem., 244 (1983) 5; c) M.G. Voronkov,
     S.V. Basenko, R.G. Mirskov, E.I. Brodskaya, D.D. Toryashinova, V.A. Pestu-
     novich, M.F. Larin, V.Yu. Vitkovski, Izv.Akad.Nauk SSSR, Ser.khim., (1985)
     2224; d) I.D. Kalikhman, V.A. Pestunovich, B.A. Gostevskii, O.B. Bannikova,
     M.G. Voronkov, J.Organometal.Chem., 338 (1988) 169; e) I.D. Kalikhman,
     O.V. Bannikova, L.P. Petukhov, V.A. Pestunovich, M.G. Voronkov, Dokl.
     Akad. Nauk SSSR, 287 (1986) 870; f) M.G. Voronkov, G. Dolmaya, A.I. Alba-
     nov, S.G. Shevchenko, E.I. Dubinskaya, Zh.Obshch.Khim., 56 (1986) 371.

199. I.D. Kalikhman, O.V. Bannikova, L.I. Belosonova, O.A. Vyazankina, N.S. Vyazankin, Izv.Akad.Nauk SSSR, Ser.khim., 6 (1986) 1424; I.D. Kalikhman, O.B. Bannikova, L.I. Belousova, B.A. Gostevskii, O.A. Vyazankina, N.S. Vyazankin, Zh.Obchsh.Khim., 54 (1984) 2609.

200. A. Boudin, G. Cerveau, C. Chuit, R.J.P. Corriu, C. Reye, Bull.Chem.Soc. Japan, 61 (1988) 101.

201. Ē. Kupče, Ē. Liepiņš, E. Lukevics, J.Organometal.Chem., 248 (1983) 131.

202. K. Jurschat, C. Mugge, J. Schmidt, A. Tzschach, Ibid., 287 (1985) C1.

203. Ē. Kupče, unpublished results.

204. Ē. Kupče, E. Liepiņš, A. Lapsiņa, G. Zelčāns, E. Lukevics, J.Organometal. Chem., 333 (1987) 1.

205. V.F. Sidorkin, V.A. Pestunovich, M.G. Voronkov, Magn.Reson.Chem., 23 (1985) 491.

206. J.M. Bellama, J.D. Nies, N. Ben-Zvi, Ibid., 24 (1986) 748.

207. Ē. Kupče, E. Liepiņš, A. Lapsiņa, G. Zelchan, E. Lukevics, J.Organometal. Chem., 251 (1983) 15.

208. I.D. Kalikhman, O.B. Bannikova, B.A. Gostevskii, M.G. Voronkov, V.A. Pestunovich, Izv.Akad.Nauk SSSR, Ser.Khim., (1989) 492.

209. S.N. Tandura, Yu.A. Strelenko, S.I. Androsenko, E.E. Masterov, N.V. Alekseev, V.I. Rakitskaya, O.G. Rodin, V.F. Traven, Zh.Obshch.Khim., 58 (1988) 398.

210. I.I. Belousova, B.A. Gostevskii, I.D. Kalikhman, O.A. Vyazankina, O.B. Bannikova, N.S. Vyazankin, V.A. Pestunovich, Ibid., 58 (1988) 407.

211. M.G. Voronkov, A.I. Albanov, A.E. Pestunovich, V.N. Sergeev, V.A. Pestunovich, I.I. Kandrov, Yu.I. Baukov, Metallorg.Khim., 1 (1988) 1435.

212. J. Boyer, C. Brelière, F. Carré, R.J.P. Corriu, A. Kpoton, M. Poirier, G. Royo, J. Colin Young, J.Chem.Soc., Dalton Trans., (1989) 43.

213. S. Kerschl, B. Wrackmeyer, J.Organometal.Chem., 332 (1987) 25.

214. a) K.M. Taba, W.V. Dahlhoff, Ibid., 280 (1985) 27; b) A. Boudin, G. Cerveau, C. Chuit, R.J.P. Corriu, C. Reye, Organometallics, 7 (1988) 1165.

215. C. Brellère, F. Carré, R.J.P. Corriu, G. Royo, Ibid., 7 (1988) 1007.

216. N.P. Erchak, E. Lukevics, in "International Symposium on Furan Chemistry" Abstracts, Riga (1988) p.3.

217. C.J. Jameson, H.S. Gutowsky, J.Chem.Phys., 40 (1964) 1714.

218. J. Kowalewski in "Annu.Rep. NMR Spectroscopy", G.A. Webb, ed., Academic Press, London (1985), Vol. 14, p. 62; G.A. Webb, Chapter 3 in ref. 4; I. Ando, G.A. Webb, Chapter 4 in ref. 54a.

219. A. Hudson, R.A. Jackson, C.J. Rhodes, A.L. DelVeechio, J.Organometal.Chem., 280 (1985) 173.

220. E. Popowski, N. Holst, H. Kelling, Z.anorg.allg.Chem., 543 (1986) 219.

221. E. Lukevics, V.F. Matorykina, Izv.Akad.Nauk Latv.SSSR, Ser.khim., (1983) 643.

222. D.L. Lichtenberger, A. Rai-Chaudhuri, J.Am.Chem.Soc., 111 (1989) 3583.

223. a) E. Buncel, T.K. Venkatachalam, U. Edlund, Can.J.Chem., 64 (1986) 1674; b) U. Edlund, T. Lejon, T.K. Venkatachalam, E. Buncel, J.Am.Chem.Soc., 107 (1985) 6408; c) U. Edlund, T. Lejon, P. Pyykkö, T.K. Venkatachalam, E. Buncel, Ibid., 109 (1987) 5983; d) E. Buncel, T.K. Venkatachalam, U. Edlund, B. Eliasson, J.Chem.Soc., Chem.Commun., (1984) 1476; J.Am.Chem. Soc., 107 (1985) 303.

224. B. Wrackmeyer, Polyhedron, 5 (1986) 1709.

225. K. Kamienska-Trela, L. Kania, J. Sitkowski, E. Bednarek, J.Organometal. Chem., 364 (1989) 29.

226. a) Ē. Kupče, E. Liepiņš, O. Pudova, E. Lukevics, J.Chem.Soc., Chem.Commun., (1984) 581; b) Ē. Kupče, E. Liepiņš, E. Lukevics, Angew.Chem., Int.Ed. Engl., 7 (1985) 568; c) B. Wrackmeyer, S. Kerschl, C. Stader, K. Horchler, Spectrochim. Acta., 42A (1986) 1113.

227. V. Galasso, G. Fronzoni, Chem.Phys., 103 (1986) 29.

228. C. Stader, B. Wrackmeyer, D. Schlosser, Z.Naturforsch., 43B (1988) 707.

229. H.B. Jokelson, A.J. Millevolte, G.R. Gillette, R. West, J.Am.Chem.Soc., 109 (1987) 6865.

230. K. Hassler, J.Organometal.Chem., 348 (1988) 33.

231. M. Boudler, G. Scholz, K.-F. Tebbe, M. Fehér, Angew.Chem., Int.Ed.Engl., 28 (1989) 339.

232. K. Hassler, Monatch.Chem., 119 (1988) 851.

233. D. Stalke, M. Meyer, M. Andrianarison, U. Klingebiel, G.M. Sheldrick, J.Organometal.Chem., 366 (1989) 281.

234. R. Köster, G. Seidel, R. Boese, B. Wrackmeyer, Chem.Ber., 121 (1988) 1941.

235. E. Hengge, M. Eibl, F. Schrank, J.Organometal.Chem., 369 (1989) C23.

236. R. Köster, G. Seidel, R. Boese, B. Wrackmeyer, Chem.Ber., 121 (1988) 1955.

237. S.B. Duckett, D.M. Haddleton, S.A. Jackson, R.N. Perutz, M. Poliakoff, R.K. Upmacis, Organometallics, 7 (1988) 1526.

238. C.H.W. Jones, R.D. Sharma, J.Organometal.Chem., 268 (1984) 113.

239. W. Caseri, P.S. Pregosin, J.Organometal.Chem., 356 (1988) 259.

240. M.J. Albright, T.F. Schaaf, A.K. Hovland, J.P. Oliver, Ibid., 259 (1983) 37.

241. M. Grignon-Dubois, M. Laguerre, Organometallics, 7 (1988) 1443.

242. H. Schmidbaur, J. Ebenhöch, Z.Naturforsch., 42B (1987) 1543.

243. M. Grignon-Dubois, M. Ahra, M. Laguerre, J.Organometal.Chem., 348 (1988) 157.

244. R. Josten, I. Ruppert, Ibid., 329 (1987) 313.

245. B.S. Suresh, J.C. Thompson, J.Chem.Soc., Dalton Trans., (1987) 1123.

246. U. Kleibisch, U. Klingebiel, J.Organometal.Chem., 314 (1986) 33.

247. U.G. Wettermark, P. Visian-Neilson, G.M. Sheide, R.H. Neilson, Organo-metallics, 6 (1987) 959.

248. a) B. Wrackmeyer, J.Organometal.Chem., 310 (1986) 151; T.N. Mitchell., R. Wickenkamp, Ibid., 291 (1985) 179; c) C. Stader, B. Wrackmeyer, Ibid., 321 (1987) C1; d) A. Sebald, P. Seiberlich, B. Wrackmeyer, Ibid., 303 (1986) 73; e) S. Kerschl, B. Wrackmeyer, Ibid., 338 (1988) 195.

249. B. Wrackmeyer, K. Horchler, H. Zhou, M. Veith, Z.Naturforsch., 44B (1989) 288.

250. B. Wrackmeyer, J.Organometal.Chem., 364 (1989) 331.

251. T.N. Mitchell, Magn.Reson.Chem., 25 (1987) 1019.

252. M.F. Larin, P.V. Gendin, V.A. Pestunovich, L.T. Ribin, O.A. Vyazankina, N.S. Vyazankin, Izv.Akad.Nauk SSSR, Ser.Khim., 9 (1983) 2139.

253. J. Kowalski, J. Chojnowski, J.Organometal.Chem., 356 (1988) 285.

254. a) B. Wrackmeyer, C. Bihlmayer, M. Schilling, Chem.Ber., 116 (1983) 3182; b) W. Biffar, H. Nöth, H. Pommerening, R. Schewerthöffer, W. Storch, B. Wrackmeyer, Chem.Ber., 114 (1981) 49.

255. B. Wrackmeyer, in "Annu.Rep. NMR Spectroscopy" G.A. Webb, ed., Academic Press, London (1985), Vol. 16, p. 73.

256. S. Adams, M. Dräger, J.Organometal.Chem., 323 (1987) 11.

257. V. Wray, L. Ernst, T. Lund, H.J. Jakobsen, J.Magn.Reson., 40 (1980) 55.

258. K. Kamienska-Trela, Z. Biedrzycka, R. Machinek, B. Knieriem, W. Lüttke, Org.Magn.Reson., 22 (1984) 317.

259. W. Gombler, J.Am.Chem.Soc., 104 (1982) 6616.

260. C.J. Jameson, H.J. Osten in "Annu.Rep. NMR Spectroscopy", G.A. Webb, ed., Academic Press, London (1986), Vol. 17, p. 1.

261. P.E. Hansen in "Annu.Rep. NMR Spectroscopy", G.A. Webb, ed., Academic Press, London (1983), Vol. 15, p. 105; Progr. NMR Spectrosc., 20 (1988) 207.

262. S. Kerschl, A. Sebald, B. Wrackmeyer, Magn.Reson.Chem., 23 (1985) 514.

263. F. Berchier, Yo-Ming Pai, W.P. Weber, K.C. Servis, Magn.Reson.Chem., 24 (1986) 679.

264. K.L. Leighton, R.E. Wasylishen, Can.J.Chem., 65 (1987) 1469.

265. D.B. Chesnut, Chem.Phys.Lett., 110 (1986) 415.

266. Ē. Kupče, E. Lukevics, in preparation.

267. J. Dyer, J. Lee, Spectrochim.Acta, Part A, 26A (1970) 1045.

268. Yu.K. Grishin, Yu.A. Ustinyuk, Zh.Strukt.Khim., 5 (1982) 163.

269. H.A. Christ, P. Diehl, H.R. Schneider, H. Dahn, Helv.Chim.Acta, 44 (1961) 865.

270. R.K. Harris, B.J. Kimber, Org.Magn.Reson., 7 (1975) 460.

271. J. Schraml, Nguen-Duc-Chuy, P. Novák, V. Chvalovský, M. Mägi, E. Limpaa, Collect.Czech.Chem.Commun., 43 (1978) 3202.

272. W. Biffar, H. Nöth, H. Pommerening, B. Wrackmeyer, Chem.Ber., 113 (1980) 333.

273. J.-P. Kintzinger, in "NMR Basic Principles and Progress", P. Diehl, E. Fluck, R. Kosfeld, eds., Springer-Verlag, Berlin (1981) Vol. 17; Part 2, p. 79 in ref. 6.

274. a) E. Liepiņš, I.A. Zicmane, E. Lukevics, E.I. Dubinskaya, M.G. Voronkov, Zh.Obshch.Khim., 53 (1983) 1092; b) G.A. Kalabin, D.F. Kushnarev, R.B. Valeyev, B.A. Trofimov, M.A. Fedotov, Org.Magn.Reson., 18 (1982) 1.

275. E.E. Liepiņš, I.A. Zicmane, I.A. Sleikša, E. Lukevics, Latv. PSR Zinat. Akad.Vestis, Kim.Ser., (1988) 726.

276. E.E. Liepiņš, I.A. Zicmane, G.I. Zelchan, E. Lukevics, Zh.Obshch.Khim., 53 (1983) 245.

277. I.A. Zicmane, E.E. Liepiņš, E. Lukevics, Latv. PSR Zināt.Akad.Vēstis, Kīm.Sēr., (1982) 91.

278. Ē. Kupče, E. Liepiņš, I. Zicmane, E. Lukevics, Magn.Reson.Chem., 25 (1987) 1084.

279. K. Rühlmann, U. Scheim, S.A. Evans, Jr., J.W. Kelly, A.R. Bassindale, J.Organometal.Chem., 340 (1988) 19.

280. A. Lyčka, J. Holeček, K. Handlir, J. Pola, V. Chvalovský, Collect.Czech. Chem.Commun., 51 (1986) 2582.

281. a) R. Köster, G. Seidel, R. Boese, B. Wrackmeyer, Chem.Ber., 121 (1988) 597; b) R. Köster, G. Seidel, R. Boese, B. Wrackmeyer, Ibid., 121 (1988) 709; c) R. Köster, G. Seidel, S. Kerschel, B. Wrackmeyer, Z.Naturforsch., Teil B, 42B (1987) 191.

282. K. Piepgrass, J.N. Barrows, M.T. Pope, J.Chem.Soc., Chem.Commun., (1989) 10.

283. C.T.G. Knight, A.R. Thompson, A.C. Kunwar, H.S. Gutowsky, E. Oldfield, R.J. Kirkpatrick, J.Chem.Soc., Dalton Trans., (1989) 275.

284. J. Kowalewski, E. Berggren, Magn.Reson.Chem., 27 (1989) 386.

285. J. Mason, Chem.Rev., 81 (1981) 205.

286. a) G.J. Martin, M.L. Martin, J.-P. Gouesnard in "NMR Basic Principles and Progress", P. Diehl, E. Fluck, R. Kosfeld, eds., Springer-Verlag, Berlin, (1981) Vol. 18; b) M. Witanowski, L. Stefaniak, G.A. Webb in "Annu.Rep. NMR Spectroscopy", G.A. Webb, ed., Academic Press, London (1986), Vol. 18, p. 1.

287. a) K. Barlos, G. Hübler, H. Nöth, P. Wanninger, N. Wiberg, B. Wrackmeyer, J.Magn.Reson., 31 (1974) 363; b) H. Nöth, W. Tinhof, T. Taeger, Chem.

Ber., 107 (1974) 3113; c) H. Nöth, W. Storch, Ibid., 109 (1976) 884; d) H. Nöth, W. Tinhof, B. Wrackmeyer, Ibid., 107 (1974) 518; e) J. Kroner, W. Schneid, N. Wiberg, B. Wrackmeyer, G. Ziegleder, J.Chem.Soc., Faraday Trans.2, 74 (1978) 1909; f) K.A. Andrianov, V.F. Andronov, V.A. Drozdov, D.Ya. Zhinkin, A.P. Kreschkov, M.M. Morgunova, Dokl.Akad.Nauk SSSR, 202 (1972) 583.

288. M.L. Filleux-Blanchard, N.D. An, Org.Magn.Reson., 12 (1979) 12.

289. a) B. Wrackmeyer, J.Magn.Reson., 61 (1985) 536; b) C. Stader, B. Wrack-meyer, Z.Naturforsch., Teil B, 42B (1987) 1515; c) C. Stader, B. Wrack-meyer, J.Magn.Reson., 72 (1987) 544; d) C. Stader, B. Wrackmeyer, J.Organo-metal.Chem., 321 (1987) C1.

290. B. Wrackmeyer, Ibid., 297 (1985) 265.

291. J.B. Lambert, J.A. McConnell, M.J. Schulz, Jr., J.Am.Chem.Soc., 108 (1986) 2482.

292. A.R. Bassindale, T.B. Posner, J.Organometal.Chem., 175 (1979) 273.

293. B.N. Khasapov, A.V. Kalinin, A.Ya. Shteinshneider, A.A. Blumenfeld, Izv. Akad.Nauk SSSR, Ser.Khim., 6 (1984) 1296.

294. P.R. Srinivasan, S.P. Gupta, S.Y. Chen., J.Magn.Reson., 46 (1982) 163.

295. M.D. Fryznk, P.A. MacNeil, J.Am.Chem.Soc., 106 (1984) 6993.

296. T. Chivers, R.T. Oakley, O.J. Scherer, G. Wolmershäuser, Inorg.Chem., 20 (1981) 914.

297. a) D.W.W. Anderson, J.E. Bentham, D.W.H. Rankin, J.Chem.Soc., Dalton Trans., (1973) 1215; b) E.A.V. Ebsworth, J.M. Edward, D.W.H. Rankin, Ibid., (1976) 1673; c) D.E.J. Arnold, S. Cradcock, E.A.V. Ebsworth, J.D. Murdoch, D.W. Rankin, R.K. Harris., B.J. Kimber, Ibid., (1981) 1349.

298. M. Haase, U. Klingebiel, Angew.Chem., Int.Ed.Engl., 24 (1985) 324.

299. A. Haas, M. Willert-Porada, Z.anorg.allg.Chem., 545 (1987) 24.

300. a) E. Randall, J.J. Ellner, J.J. Zuckerman, J.Am.Chem.Soc., 88 (1966) 622; E.W. Randall, J.J. Zuckerman, Ibid., 90 (1968) 3167; b) A.H. Cowley, J.R. Schweiger, Ibid., 95 (1973) 4179; c) S.L. Ioffe, A.L. Blumenfeld, A.S. Shashkov, Izv.Akad.Nauk SSSR, Ser.Khim., (1978) 246.

301. M. Herberhold, S.M. Frank, B. Wrackmeyer, Z.Naturforsch., 43B (1988) 985; O.A. Miller, A.V. Zibarev, M.A. Fedotov, G.G. Furin, Zh.Obshch.Khim., 59 (1989) 586.

302. M.G. Steinmetz, B.S. Udayakumar, M.S. Gordon, Organometallics, 8 (1989) 530.

303. Y.O. Kim, H.M. Goff, J.Am.Chem.Soc., 110 (1988) 8706.

304. a) B.J. Aylett, C.-F. Liaw, J.Organometal.Chem., 325 (1987) 91; b) D.J. Brauer, H. Buerger, G.R. Liewald, J. Wilke, Ibid., 283 (1985) 305.

305. D. Stalke, U. Klingebiel, G.M. Sheldrick, Ibid., 344 (1988) 37.

306. E. Hey, C.L. Raston, B.W. Skelton, A.H. White, J.Organometal.Chem., 362 (1989) 1.

307. S.S. Al-Juaid, C. Eaborn, M.N.A. El-Kheli, P.B. Hitchcock, P.D. Lickiss, M.M.E. Molla, J.D. Smith, J.A. Zora, J.Chem.Soc., Dalton Trans., (1989) 447.

308. D. Stalke, U. Klingebiel, G.M. Sheldrick, Chem.Ber., 121 (1988) 1457.

309. T.F. Bates, R.D. Thomas, J.Organometal.Chem., 359 (1988) 285.

310. H.E. Katz, J.Am.Chem.Soc., 108 (1986) 7640.

311. G.A. Olah, K. Laali, O. Faroog, Organometallics, 3 (1984) 1337.

312. W. Schacht, D. Kaufmann, J.Organometal.Chem., 331 (1987) 139.

313. E. Henecker, H. Nöth, Z.Naturforsch., Teil B, 40b (1985) 717.

314. N.S. Hosmane, M.S. Islam, B.S. Pinkston, U. Siriwardane, J.J. Banewicz, J.A. Maguire, Organometallics, 7 (1988) 2340.

315. N.S. Hosmane, U. Siriwardane, H. Zhu, G. Zhang, J.A. Maguire, Ibid., 8 (1989) 566.

316. G. Schmid, D. Zaika, J. Lehr, N. Augart, R. Boese, Chem.Ber., 121 (1988) 1873.

317. J.F. Janik, C.K. Narula, E.G. Gulliver, E.N. Duesler, R.T. Paine, Inorg.Chem., 27 (1988) 1222.

318. B.-L. Li, P. Mukherjee, R.H. Neilson, Ibid., 28 (1989) 605.

319. K.-H. van Bonn., P. Schreyer, P. Paetzold, R. Boese, Chem.Ber., 121 (1988) 1045.

320. a) L. Parkanyi, P. Hencsei, L. Bihatsi, I. Kovacs, A. Szöllösy, Polyhedron, 4 (1985) 243; b) V.A. Pestunovich, B.Z. Shterenberg, L.P. Petukhov, V.I. Rakhlin, V.P. Baryshok, R.G. Mirskov, M.G. Voronkov, Izv.Akad.Nauk SSSR, Ser.Khim., 8 (1985) 1935.

321. E.E. Liepiņš, I.S. Birǵele, G.I. Zelchan, I.P. Urtāne, E. Lukevics, Zh.Obshch.Khim., 53 (1983) 1076.

322. E.E. Liepiņš, Ē.L. Kupče, I.S. Birǵele, in "Organosilicon Derivatives of Aminoalcohols" (Russ.), E. Lukevics, ed., Zinātne, Riga (1987) Chapter 2.

323. a) J.B. Lambert, W.J. Schulz, Jr., J.A. McConnell, W. Schif, J.Am.Chem. Soc., 110 (1988) 2201; b) J.B. Lambert, W. Schilf, Ibid., 110 (1988) 6364; c) J.B. Lambert, J.A. McConnell, W. Schilf, W.J. Schulz, Jr., J.Chem.Soc., Chem.Commun., (1988) 455.

324. E.E. Liepiņš, I.A. Zicmane, L.M. Ignatovich, E. Lukevics, L.I. Gubanova, M.G. Voronkov, Zh.Obshch,Khim., 53 (1983) 1789.

325. Yu. Popelis, E. Liepiņš, N.P. Erchak, E. Lukevics, Latv. PSR Zināt.Akad. Vēstis, Ķīm.Sēr., (1985) 111.

326. V. Wray in "Annu.Rep. NMR Spectroscopy", G.A. Webb, ed., Academic Press, London, (1980) Vol. 10B.

327. R. Krebs, D. Schomburg, R. Schmutzler, Z.Naturforsch., Teil B, 40B (1985) 282.

328. a) G. Klebe, R. Hensen, J.Chem.Soc., Dalton Trans., (1985) 5; b) G.Klebe, M. Nix, K. Hensen, Chem.Ber., 117 (1984) 797.

329. W.H. Stevenson III, J.C. Martin, J.Am.Chem.Soc., 107 (1985) 6352; W.H. Stevenson III, S. Wilson, J.C. Martin, W.B. Farnham, Ibid., 107 (1985) 6340.

330. W.B. Farnham, J.F. Whitney, Ibid., 106 (1984) 3992.

331. C. Yu, S.C. Chang, C. Liu, J.Org.Chem., 48 (1983) 5228.

332. E. Helmers, M. Hesse, U. Klingebiel, Z.anorg.allg.Chem., 565 (1988) 81.

333. C. Lin, C. Lee, C. Liu, Organometallics, 6 (1987) 1861; 1869; 1878; Y. Heng, C. Lin, C. Lee, C. Liu, Ibid., 1882.

334. S. Pillmann, U. Klingebiel, J.Organometal.Chem., 321 (1987) 1.

335. D.B. Azarian, C. Eaborn, P.D. Lickiss, Ibid., 330 (1987) 1.

336. D. Stalke, U. Klingebiel, G.M. Sheldrick, Ibid., 341 (1988) 119.

337. W. Clegg., U. Klingebiel, J. Neeman, G.C. Sheldrick, Ibid., 249 (1983) 47.

338. U. Wannagat, T. Blumental, D.J. Brauer, H. Bürger, Ibid., 249 (1983) 33.

339. H. Schmidbaur, S. Lauteschläger, F.H. Köhler, Ibid., 271 (1984) 173.

340. E.E. Nifantiev, N.S. Vyazankin, S.F. Sorokina, L.A. Vorobieva, O.A. Vyazan-kina, D.A. Bravo-Zhivotovsky, Ibid., 277 (1984) 211.

341. G.M. Gray, C.S. Kraihanzel, Ibid., 421 (1983) 201; G.M. Gray, K.A. Redmill, Ibid., 280 (1985) 105.

342. H.R. O'Neal, R.H. Neilsen, Inorg.Chem., 23 (1984) 1372.

343. I.I. Patsanovskii, E.A. Ishmaeva, E.N. Strelkova, A.V. Ilyasov, T.A. Zyab-likova, I.E. Ishmaev, N.M. Kudyakov, M.G. Voronkov, A.N. Pudovik, Zh. Obshch.Khim., 54 (1984) 1738.

344. G. Fritz, P. Anann, Z.anorg.allg.Chem., 535 (1986) 106; G. Fritz, R. Bias-toch, Ibid., 95; G. Fritz, R. Biastoch, W. Hönle, H.G. Schnering, Ibid., 86; G. Fritz, D. Hanke, Ibid., 537 (1986) 17; G. Fritz, R. Biastoch, Ibid., 535 (1986) 63.

345. K. Dastal, J. Sikola, M. Meisel, H. Grunze, Ibid., 543 (1986) 199.

346. M. Knorr, U. Schubert, J.Organometal.Chem., 365 (1989) 151; G. Bellachioma, G. Cardaci, E. Colomer, R.J.P. Corriu, A. Vioux, Inorg.Chem., 28 (1989) 519.

347. G. Fritz, T. Vaahs, Z.anorg.allg.Chem., 552 (1987) 7; Ibid., p.18; G. Fritz, T. Vaahs, J. Härer, Ibid., p.11; G. Fritz, H. Fleischer, W. Hönle, H.G. van Shnering, Ibid., p.34; G. Fritz, H. Fleischer, Ibid., 570 (1989) 67.

348. K.V. Kathi, H.W. Roesky, M. Rietzel, Ibid., 553 (1987) 123.

349. K. Katti, R.G. Cavell, Inorg.Chem., 28 (1989) 413.

350. M.L.J. Hackney, R.C. Haltiwanger, P.F. Brandt, A.D. Norman, J.Organometal. Chem., 359 (1989) C36.

351. K.V. Katti, R.G. Cavell, Organometallics, 7 (1988) 2236.

352. O.T. Beachley, Jr., J.P. Kopasz, H. Zhang, W.E. Hunter, J.L. Atwood, J.Organometal.Chem., 325 (1987) 69.

353. E. Kunz, U. Schubert, Chem.Ber., 122 (1989) 231.

354. P. Aslanidis, G.E. Manoussakis, Z.anorg.allg.Chem., 555 (1987) 169.

355. H. Schäfer, W. Leske, Ibid., 550 (1987) 57; Ibid., 552 (1987) 50; H. Schäfer, D. Binder, Ibid., 546 (1987) 55.

356. S.S. Al-Juaid, S.M. Ohaher, C. Eaborn, P.G. Hitchcock, C.A. McGeary, J.D. Smith, J.Organometal.Chem., 366 (1989) 39.

357. D. Gudat, E. Niecke, W. Sachs, P. Rademacher, Z.anorg.allg.Chem., 546 (1987) 7.

358. W. Urbaniak, B. Marciniec, Synth.React.Inorg.Met.-Org.Chem., 18 (1988) 695.

359. D.A. DuBois, R.H. Nielson, Inorg.Chem., 28 (1989) 899.

360. J.F. Janik, E.N. Duesler, M.F. McNamara, M. Westerhausen, R.T. Paine, Organometallics, 8 (1989) 506.

361. J. Meyer, J. Willnecker, U. Schubert, Chem.Ber., 122 (1989) 223.

362. W. Storek, Chem.Phys.Lett., 98 (1983) 267; K.M. Larson, J. Kowalewski, U. Henrikson, J.Magn.Reson., 62 (1985) 260.

363. G.K.S. Prakash, S. Keyaniyan, R. Aniszfeld, L. Heiliger, G.A. Olah, R.C. Stevens, H.K. Choi, R. Rau, J.Am.Chem.Soc., 109 (1987) 5123.

364. P. Galow, A. Sebald, B. Wtackmeyer, J.Organometal.Chem., 259 (1983) 253.

365. S.P. Foster, K.F. Leung, K.M. Mackay, R.A. Thomson, Austral.J.Chem., 39 (1986) 1089; A.L. Wilkins, P.J. Watkinson, K.M. Mackay, J.Chem.Soc., Dalton Trans., (1987) 2365.

366. J.M. Chehayber, J.E. Drake, Inorg.Chim.Acta, 111 (1986) 51.

367. W.-W. DuMont, I. Wagner, Chem.Ber., 121 (1989) 2109.

368. E.C. Alyea, K.J. Fisher, Polyhedron, 5 (1986) 695.

369. D.W. Hawker, P.R. Wells, Org.Magn.Reson., 22 (1984) 280.

370. R. Benn., H. Brenneke, J. Heck, A. Rufinska, Inorg.Chem., 26 (1987) 2826.

371. R.G. Finke, B. Rapko, T.J.R. Weakley, Ibid., 28 (1989) 1573.

Isotopes in the Physical and Biomedical Science, Vol. 2, edited by E. Buncel and J.R. Jones
© 1991 Elsevier Science Publishers B.V., Amsterdam

Chapter 6

ONE-BOND $^{13}C-^{13}C$ COUPLING CONSTANTS IN STRUCTURAL STUDIES

K.KAMIENSKA-TRELA

Polish Academy of Sciences, Institute of Organic Chemistry,

Warsaw, Poland

CONTENTS

I.INTRODUCTION

The magnetically active isotope of carbon, ^{13}C, belongs to rare ones, since its natural abundance is 1.1 per cent only. The natural content of doubly ^{13}C-labelled isotopomers, those where the $^{13}C-^{13}C$ spin-spin interaction can take place, is still smaller by a factor of 100. Therefore, for a long time a ^{13}C enrichment had been an absolute requirement in ^{13}C-^{13}C spin-spin coupling studies. Spin-spin couplings between carbon nuclei were measured for the first time in the early sixties, for ^{13}C-enriched ethane, ethene and acetylene.[1,2] About twenty years later, with the advent of the FT pulse technique and superconducting solenoid magnets, it became possible to measure CC couplings at the natural abundance of the ^{13}C isotope.[3-5] A breakthrough took place when the INADEQUATE pulse sequence was introduced.[6-8] The latter technique suppresses signals of singly-labelled molecules and allows one to observe the ^{13}C NMR spectra of doubly-labelled isotopomers. This reduces the determination of ^{13}C-^{13}C spin-spin coupling constants to a fairly routine experiment, but for relatively small molecules only. As far as large molecules are concerned, those of biological interest (proteins, carbohydrates, etc.), ^{13}C labelling is needed still. The number of papers devoted to CC spin-spin coupling studies continously increases. Since the early eighties, hundreds of $^1J_{CC}$ couplings have been recorded. The earliest reviews on one-bond CC spin-spin coupling constants are those by Maciel,[9] Llinas et al.,[10] Stothers,[11] and Bystrov;[12] the latter author collected and discussed the data for amino acids. Wasylishen included ^{13}C-^{13}C coupling constants in his review on coupling constants between ^{13}C and the first row elements.[13] Major reviews and data collections have been published by Hansen and Wray,[14-17] and by Krivdin and Kalabin.[18] In a number of books, special chapters or comments were devoted to one-bond CC spin-spin coupling constants.[19-21] Theoretical aspects of $^1J_{CC}$ coupling have been discussed by Ellis and Ditchfield,[22] and in two reviews by Kowalewski.[23,24] A recent review by Buddrus and Bauer was devoted to the INADEQUATE experiment.[25]

The present chapter is not intended to present a collection of all $^1J_{CC}$ values available. We shall rather focus our attention on the main factors which determine the magnitudes of $^1J_{CC}$s. These include orbital hybridization, substituent electronegativities, intermolecular forces and steric effects. A more thorough understanding of the influence of these factors on $^1J_{CC}$ is of great interest, as spin-spin couplings have turned out to be a sensitive probe for the electronic structure of chemical bonds. The couplings between ^{13}C carbon nuclei are also proving useful in a variety of situations which involve the elucidation of biosynthetic pathways and of unknown molecular structures.

II. MAIN FACTORS CONTROLLING THE MAGNITUDE OF INDIRECT ONE-BOND CC SPIN-SPIN COUPLING CONSTANTS. THEORETICAL CONSIDERATIONS

A general theory of spin-spin coupling was developed by Ramsey,[26,27] who showed that indirect magnetic coupling between nuclear spins A and B results from three types of interactions:

$$J_{AB} = J_{AB}^{OD} + J_{AB}^{SD} + J_{AB}^{FC} \qquad (1)$$

those involving nuclear spin and electron orbital magnetic moment (J_{AB}^{OD}); dipole-dipole interactions between nuclear and electron spins (J_{AB}^{SD}); and Fermi contact interactions between the latter spins (J_{AB}^{FC}) . Thus, electrons are involved in all of them as a transmitting medium for the indirect coupling between nuclear spins.

Applications of Ramsey's theory to larger molecules we owe to McConnell,[28,29] Pople and co-workers,[30-32] and Blizzard and Santry.[33,34] Though the theory was applied at an approximate level only, involving among others the so-called average excitation energy approximation, it afforded a succesful interpretation of the FC, SD and OD contributions in terms of a commonly used characteristics of the ground electronic states of the molecules under study.

Thus, it has been shown that the contact contribution depends upon the product $S^2(A).S^2(B)$ of the s electron densities at the coupled nuclei; the orbital and dipolar interactions are proportional to the product of the one-centre integrals, $\langle r_A^{-3} \rangle$ and $\langle r_B^{-3} \rangle$ for the coupled nuclei, where $\langle r^{-3} \rangle$ is the expectation value of the inverse cube of the valence p electron radii. The first large sets of theoretical one-bond CC spin-spin coupling constants have been published in early 70's by Maciel et al., [35-38] who calculated the SCF INDO values of $^1J_{CC}$s for a variety of organic molecules. In spite of the fact that the computations have been performed for the Fermi contact term only, the agreement between the theory and experiment was, in most cases, satisfactory. The values of non-contact terms for CC spin-spin coupling have been obtained by Blizzard and Santry[34] also from SCF perturbation theory calculations at the INDO level.

Four important conclusions follow from these studies:

(1) The Fermi contact contribution is invariably the dominant factor, and in most cases this sole term can reasonably reproduce experimental $^1J_{CC}$ values.

(2) The orbital and dipolar terms are not negligible for multiple CC bonds. However, for double CC bonds, the non-contact terms are of mutually opposite signs and partly cancel each other. Only in the case of the triple CC bond, the orbital as well as the dipolar terms are large and positive.

(3) The values of spin-spin coupling constants in unsubstituted hydrocarbons can be related to the concept of hybridized orbitals; the values increase in the order: $^1J_{Csp^3Csp^3} < ^1J_{Csp^3Csp^2} < ^1J_{Csp^2Csp^2}$ (single bond) $\approx ^1J_{Csp^2Csp^2}$(aromatic bond) $< ^1J_{Csp^2Csp^2}$ (double bond) $< ^1J_{Csp^2Csp} < ^1J_{CspCsp}$(single bond) $< ^1J_{CspCsp}$ (triple bond).

(4) In substituted compounds the magnitude of $^1J_{CC}$ is governed by the electronegativity of the substituents concerned.

The mechanism of spin-spin coupling is still a subject of continuing interest and numerous studies on this topic, many of them performed at a rather sophisticated level of theory, have recently been published.[39-66] Among others, new methods of evaluating and electron contributions to coupling constants have been applied to coupling across multiple CC bonds,[67-71] and in a paper by Pyykko,[72] relativistic effects have been taken into account.

The most relevant theoretical data for the simplest molecular models, ethane, ethene, ethyne and bicyclobutane, have been collected in Table 1. Those for substituted compounds are given in Tables 2-4, and shown in Figs. 1 and 2. Though the values of the calculated spin-spin coupling constants and those of the OD, SD and FC terms do depend on the method employed, the general trends observed at the early stage of CC coupling mechanism studies have

TABLE 1

Contributions from different mechanisms to the carbon-carbon coupling constants in ethane, ethene and ethyne. All values are given in Hz.

	contact	dipolar	orbital	total	Ref.
CH_3-CH_3					
SCPT INDO	35.6	0.7	-2.9	33.4	34
SCF INDO	41.5	-	-	41.5	35
LMO INDO	27.2	0.2	-1.3	26.1	44
ab initio	17.4	0.55	-0.04	17.91	39
EHMO	30.24	0.18	0.04	30.46	55
FOPPA INDO/MCI	43.24	-1.45	0.59	42.38	57
Experimental				34.6	1
$H_2C=CH_2$					
SCPT INDO	70.6	3.9	-18.6	55.9	34
SCF INDO	82.2			82.2	35
LMO INDO	71.6	0.5	-3.2	68 9	44
ab initio	69.2	3.67	-4.83	68.04	39
EHMO	74.90	1.64	-3.97	72.57	55
FOPPA INDO/MCI	84.22	2.09	-6.96	79.35	57
FP MC+CI [a]	81.9		-10.0	71.9	54
ab initio SOPPA[a,b]	98.6	2.5	-9.3	91.8	59
SOPPA[a,c]	78.8	2.3	-9.0	72.1	59
ab initio EOM	86.99	1.85	-6.48	82.37	61
Experimental				67.6	1
HC≡CH					
SCPT INDO	140.8	8.3	23.6	172.7	34
SCF INDO	163.6			163.6	35
LMO INDO	141.1	2.2	1.4	144.7	44
ab initio	190.7	0.35	1.89	192.94	39
EHMO	157.13	5.30	10.32	172.75	55
FOPPA/INDO/MCI	164.49	3.79	1.09	169.37	57
FP MC+CI[a]	173.81		15.3	189.1	54
ab initio SOPPA[a,b]	194.2	6.4	1.8	202.4	59
SOPPA[a,c]	186.7	6.6	2.5	177.8	59
SOPPA[b]	183.34	7.24	5.82	196.4	63
E-CCDPPA	180.83	7.25	5.40	193.5	63
EOM	210.05	5.20	1.74	216.99	61

TABLE 1 contd.

Experimental				170.6	2
				171.5	1

bicyclobutane (central CC bond)					
			-0.92		48
	-1.4	-1.3	-2.9	-5.6	40
EOM	-7.00	-0.50	-0.27	-7.77	62

Experimental (in 2,2,4,4-tetramethylbicyclobutane) -17.49 74

a) re-calculated from the reduced coupling constants K

b) DZ basis

c) DZP basis

SCPT self-consistent perturbation theory

SCF self consistent field theory

LMO localized molecular orbitals

EHMO extended Huckel molecular orbital

FOPPA INDO/MCI theory the first-order polarization propagator
approach (multicenter integrals)

FPMC+CI finite perturbation-multiconfiguration SCF method +

SOPPA second-order polarization propagator approximation

ECCDPPA extended coupled-cluster doubles polarizatien propagator
approximaton

EOM equations of motion

been fully confirmed. The Fermi contact term invariably remains the main contribution to the spin-spin coupling mechanism including aromatic,[5,42] and open-chain conjugated systems.[51,58,61] The spin-spin coupling constant between the bonded bridgehead carbons in the highly strained molecule of bicyclobutane seemed to be an exception, with its large negative orbital term $^1J_{CC}^{OD}$ as a possible dominating contributor to the negative $^1J_{CC}$ value.[40,43] However, more recent *ab initio* calculations performed by Galasso[62] have shown that the orbital contribution for the central CC bond of bicyclobutane actually resembles closely that for a CC single bond, particularly such as found in cyclopropane,[48] and that also in the case of this unique bonding, the coupling mechanism is controlled by the Fermi contact interaction whose sign is exceptionally negative. This leads to a negative value of $^1J_{CC}$ coupling constant, in accordance with the experimental findings.[73,74] INDO data obtained by Wray[5] for derivatives of benzene, and by Kamienska-Trela *et al.*, for substituted ethenes,[65] allenes,[65] mono- and disubstituted acetylenes[56] and diacetylenes,[58] where the substituents were varied systematically

TABLE 2

Calculated total values of $^1J_{C≡C}$ coupling constants (in Hz) from INDO FPT method in mono- and disubstituted acetylenes and diacetylenes.

Substituent	$^1J_{C≡C}$		
X	$XC≡CH$ [56]	$XC≡CX$ [56]	$XC≡C-C≡CX$ [58]
Li	61.4	37.2	43.2
BeH	95.3	51.3	93.5
BH_2	136.4	112.8	137.5
CH_3	175.6	174.1	178.9
NH_2	193.8	219.4[a]	201.9[a]
		217.3[b]	202.1[b]
OH	206.0	252.3[a]	216.4[a]
		250.0[b]	216.5[b]
F	215.9	293.1	233.3
Range of OD	4.9	0.7	3.0
	to	to	to
contribution	9.8	11.1	10.1
Range of SD	4.7	3.5	4.1
	to	to	to
contribution	6.1	6.4	6.3

a) NH_2, OH groups *cis* arranged
b) NH_2, OH groups *trans* arranged

along the first and second rows of the Periodic Table, have shown that in all these cases the electronegativity of substituent is,

304

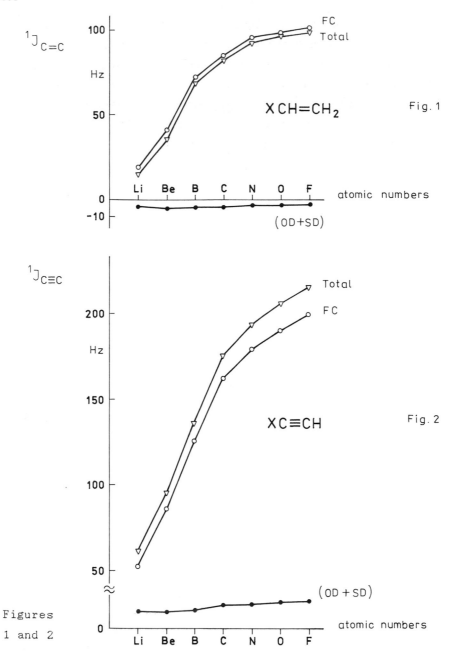

$^1J_{C=C}$

100

Hz

50

XCH=CH$_2$ Fig. 1

FC
Total

Li Be B C N O F   atomic numbers

0

-10   (OD+SD)

$^1J_{C\equiv C}$

200

Hz

150

XC≡CH   Fig. 2

Total

FC

100

50

(OD + SD)

0

Li Be B C N O F   atomic numbers

Figures
1 and 2

The plots of the total, the Fermi contact and the sum of orbital and dipolar terms vs. atomic numbers of the first atom of the substituent for $^1J_{CC}$ coupling constants in substituted ethenes (Fig. 1) and in substituted ethynes (Fig. 2).

as was earlier demonstrated by Maciel *et al.*,[35] an important factor governing the magnitude of $^1J_{CC}$ spin-spin coupling. An increase in substituent electronegativity always brings an increase in the $^1J_{CC}$ values and *vice versa.* A similar dependence of $^1J_{CC}$ vs. electronegativity has been revealed by *ab initio* calculations performed by Fronzoni and Galasso[64] for monoheteroanalogs of cyclopropane and cyclopropene with the bridgehead heteroatoms within the set: B, C, N, O, Si, and S.

TABLE 3

Calculated total values of $^1J_{CC}$ coupling (in Hz) in monosubstituted ethanes, ethenes, allenes and benzenes.

X /Method/	$^1J_{CC}$			
	XCH_2CH_3 [49] /SCPT INDO/	$XCH=CH_2$ [a] /FPT INDO/	$X\overset{*}{C}H=C=\overset{*}{C}H_2$ [a] /FPT INDO/	C_6H_5X [5] /MO INDO/
Li	–	14.6	27.3	–
BeH	–	35.4	55.1	–
BH_2	–	68.5	75.7	–
CH_3	34.8	82.1	107.5	69.2
NH_2	37.8	93.3	120.1	77.6
OH	40.6	96.8^b	128.8^b	82.3
		95.0^c	125.4^c	
F	40.7	99.6^{65}	130.8	85.8
Range of OD contribution	-1.3 to -1.6	-5.4 to -6.5	-3.7 to -4.4	-4.0 to -4.6
Range of SD contribution	0.9 to 1.0	1.6 to 2.9	1.1 to 2.8	1.0 to 1.4

a) K. Kamieńska-Trela, P. Gluziński and B. Knieriem, unpublished data

b) OH group *s-cis* arranged

c) OH group *s-trans* arranged

In substituted ethanes, ethenes and ethynes, the largest calculated values have been found for fluorosubstituted compounds and the smallest for the derivatives of lithium. The calculations reveal that only the Fermi contact term reflects changes in the electronegativity of substituents. Although orbital-dipole and spin-dipole terms are not negligible, especially for multiple CC bonds, they remain practically constant within a given bonding pattern (see Figs. 1 and 2), and in most cases they together account for less than 15 % of the total $^1J_{CC}$ value.

It is noteworthy that the sensitivity of CC spin-spin coupling constants towards substitution is strongly related to the s-characters of the bonds involved and diminishes in the order $^1J_{C\equiv C} > {}^1J_{C=C} >> {}^1J_{C-C}$. Thus, for example, the calculated spin-spin couplings across triple CC bonds vary from 61.4 Hz in $HC\equiv CLi$ to 215.9 Hz in $HC\equiv CF$, and from 14.6 Hz in $H_2C=CHLi$ to 99.6 Hz in $H_2C=CHF$, while changes of only few hertz have been found in derivatives of ethane. As will be shown in further sections of the present chapter, these theoretical trends have been fully confirmed by experiment.

TABLE 4

Calculated total INDO FPT[65] values of $^1J_{C=C}$ in fluorethenes. For comparison the FOPPA[57] INDO/MCI results are included.

Compound	$^1J_{C=C}$	
	FOPPA INDO/MCI[57]	INDO FPT[65]
$H_2C=CF_2$	116.4	119.2
FCH=CHF *cis*	109.8	113.5 a
FCH=CHF *trans*	121.3	129.4
FCH=CF$_2$	–	146.6 a
F$_2$C=CF$_2$	–	196.4
Range of OD	-5.6	-3.6
contribution	to	to
	-6.2	-4.8
Range of SD	2.4	2.5
contribution	to	to
	2.8	3.3

a) K. Kamieńska-Trela and P. Gluziński, unpublished data

The INDO calculations of CC spin-spin coupling constants predict for them a slight dependence on both conformation and configuration of the compounds. Thus, variations by as much as 5-6 Hz were noted by Barfield and co-workers[75] in the calculated results for one-bond CC coupling constants as the dihedral angle Φ was varied in butane, butanol and butanoic acid. Substantially greater $^1J_{CC}$ values were found for *trans* 1,2-difluoroethene than for the *cis*-difluoro isomer.[57,65]

III. HYBRIDIZATION AND ONE-BOND CC COUPLING CONSTANTS.
    UNSUBSTITUTED HYDROCARBONS

In the absence of strongly influencing factors such as complexation, electron-donating or -attracting substituents one can expect that the relationship predicted by theory, between hybridization of bonding carbon orbitals and CC coupling constants, will dominate the magnitude of the latter. The measurements performed for ethane, ethene and ethyne[1,2] revealed that an increase of $^1J_{CC}$ values follows this sequence: 34.6 Hz in C_2H_6, 67.6 in C_2H_4 and 171.5 Hz in C_2H_2. An empirical equation connecting the $^1J_{CC}$ values measured for unsubstituted hydrocarbons and the product of the S_A and S_B characters of intervening carbon atoms was proposed by Frei and Bernstein in 1963.[76]

$$^1J_{CC} = 576 \ S_A \cdot S_B - 3.4 \tag{2}$$

Since then, several slightly differing values of numerical coefficients entering this equation have been proposed.[13,20,41,50,77-79] Among others, Newton, Schulman and Manus[41] found the following linear relationship between the experimental $^1J_{CC}$ couplings and the relevant products of s-densities as calculated by the INDO method:

$$^1J_{CC} = 621 \ S_A \cdot S_B - 10.2 \tag{3}$$

On the basis of equation (3), negative values for the $^1J_{CC}$ between bridgehead carbon atoms in benzvalene and bicyclobutane were predicted.

The existence of a correlation between the magnitude of the coupling constants and the "s-character" of the carbon atoms involved was further confirmed experimentally for a variety of compounds[36,37,80-83] including carbocations[84-86] (see also Table 5).

TABLE 5

Examples of CC spin-spin coupling constants for all possible variations of formal CC hybridization. All values are given in Hz.

34.6^{78}

$CH_3-CH_2-CH_3$

$68.8 \quad 53.7^{87}$

$CH_2=CH-CH=CH_2$

$67.6^{1,2}$

$CH_2=CH_2$

$41.9, \ 70.0^{37}$

$CH_3-CH=CH_2$

$71.1^{89} \ 83.9, \ 175.6^{88}$

$H_2C=CH-C\equiv CH$

98.7^{90}

$H_2C=C=CH_2$

$67.4 \ 175.0^{88,91}$

$CH_3-C\equiv CH$

$194.1 \ 154.8^{92}$

$HC\equiv C-C\equiv CH$

171.5^{1}

$HC\equiv CH$

Various versions of the Frei-Bernstein equation have been used in many cases for the estimation of the electronic structure of CC bonds. Most often, they were combined with an analogous equation obtained by Muller and Pritchard[93] for one-bond carbon-hydrogen coupling constants.

Particularly interesting results were obtained for cyclopropane and its derivatives. The one-bond CC coupling constant in cyclopropane is much smaller than those in open-chain hydrocarbons or in cyclohexane, 12.4 Hz in C_3H_6, 32.7 Hz in C_6H_{12} and 34.6 Hz in C_6H_{14}.[78] The same pattern is observed when the coupling constants across endocyclic CC bonds are compared with the couplings in other cyclic compounds such as cyclobutane and cyclopentane (see data in Table 6). The couplings across exocyclic CC bonds are, on the contrary, much larger in alkyl and methylene substituted cyclopropanes than those for four, five, and six-membered rings (Table 7).

Using the Muller-Pritchard[93] equation with $^1J_{CH}$ of 160.4 Hz and the sum rule for the s-character, Luttke and co-workers[78] calculated the s-character for cyclopropane as being equal to 0.1792. An insertion of this S_{CC} value into the modified equation (2)[78]

$$^1J_{CC} = 658 \ S_A S_B - 7.9 \qquad (4)$$

gave 13.2 Hz for $^1J_{CC}$, which is close to the experimental value of 12.4 Hz.[78] The small difference has been interpreted by the authors as a measure of the non-contact contributions to the $^1J_{CC}$ in cyclopropane. A similar interpretation of the data was given also by Stocker[98] who, in addition, pointed out that the result is in an excelent agreement with the hybridization model of cyclopropane proposed by Walsh.[99,100] The Walsh model predicts an unusually low s-character for each of the endocyclic carbon orbitals, 16.67% , and a correspondingly high s-character for the exocyclic bonds, i.e. 33.33% (sp^2 hybridization) for each exocyclic carbon orbital in dimethylcyclopropane and 50% (sp hybridization) in methylenecyclopropane. This leads to a value of 101.6 Hz for $^1J_{C=C}$ in the latter compound which is in good agreement with the experimental value of 95.2 Hz.[80] The same interpretation can be given for the remaining CC coupling constants in methyl- and methylenecyclopropanes. This provides support for the Walsh model of the cyclopropane molecule.

However, not all of the results obtained for formally unsubstituted hydrocarbons can be explained so simply. First of all, an alternative interpretation can be given for the observed low $^1J_{CC}$ value in cyclopropane and its derivatives. Indirect spin-spin interaction between any two carbons in a cyclic compound can be transferred *via* different, independent pathways. Thus, $^1J_{CC}$ between two adjacent carbon atoms in cyclopropane can be presented as a sum of one-bond and geminal constants, and in cyclobutane as a sum of one-bond and vicinal CC couplings. It has been shown by Marshall and co-workers[101] and confirmed by others[102-104] that the measured coupling constant amounts to the algebraic sum of the expected J values for the pathways that contribute to the coupling constant in question. INDO calculated coupling constants, J_{CC} for

310

TABLE 6

CC coupling constants in saturated cycloalkanes. All values in Hz.

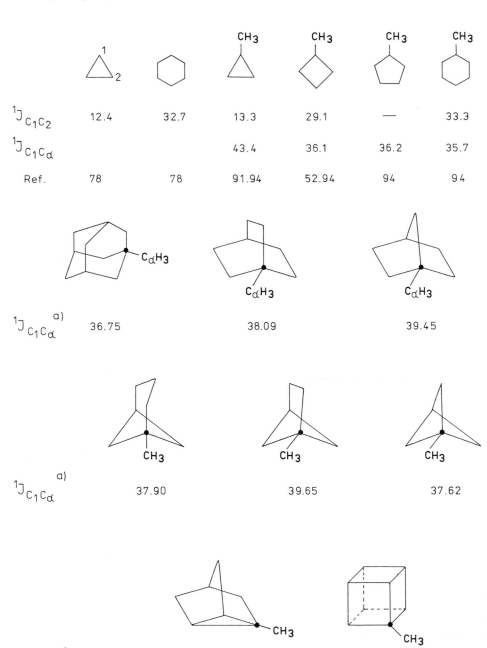

$^1J_{C_1C_2}$	12.4	32.7	13.3	29.1	—	33.3
$^1J_{C_1C_\alpha}$			43.4	36.1	36.2	35.7
Ref.	78	78	91.94	52.94	94	94

$^1J_{C_1C_\alpha}$ $^{a)}$ 36.75   38.09   39.45

$^1J_{C_1C_\alpha}$ $^{a)}$ 37.90   39.65   37.62

$^1J_{C_1C_\alpha}$ $^{a)}$ 48.48   37.02

a) ref.[95]

TABLE 7

CC coupling constants, $^1J_{CC}$, in open-chain and cyclic monoenes and dienes. All values are given in Hz.[a]

	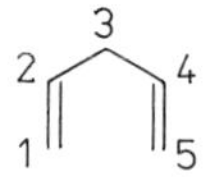	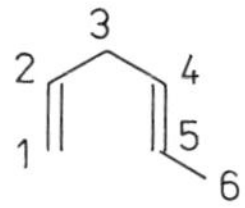	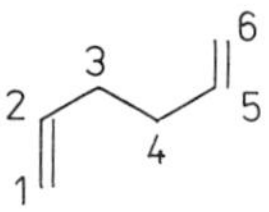
$^1J_{C_1C_2}$	70.0	70.0	70.0
$^1J_{C_2C_3}$	41.8	43.3	42.2

	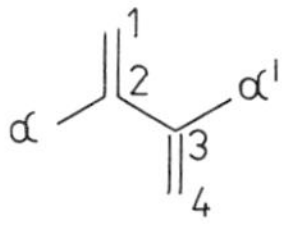	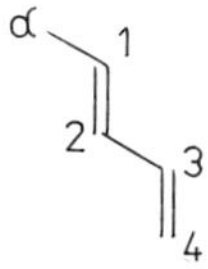	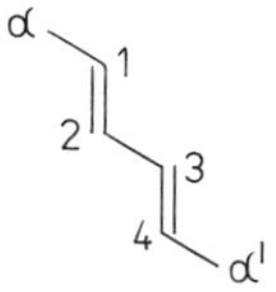	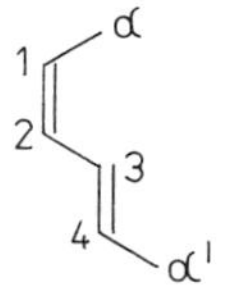
$^1J_{C_1C_2}$	71.4	71.7	71.9	70.0
$^1J_{C_2C_3}$	b)	54.9	b)	56.0
$^1J_{C_\alpha C_1 \text{ (or } C_2)}$	43.0	43.6	44.4	42.7
$^1J_{C_3C_4}$	71.4	74.4	71.8	73.0

	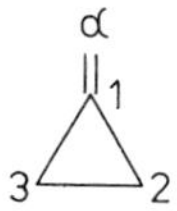	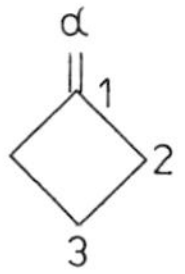	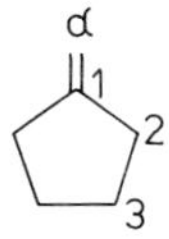	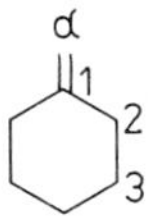	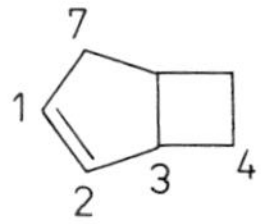	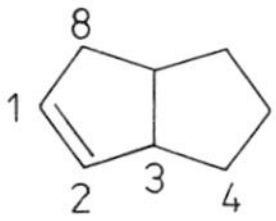
$^1J_{C_1C_\alpha}$	95.2 [80]	73.1	73.8	72.3	1,2 68.7[97]	1,2 60.0[97]
$^1J_{C_1C_2}$	23.2 [80]	33.6	39.6	39.7	2,3 40.0[97]	2,3 40.0[97]
$^1J_{C_2C_3}$		28.5	33.3	31.9	1,7 38.9[97]	1,8 38.4[97]

	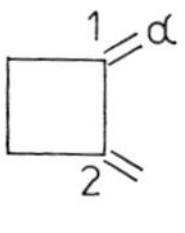	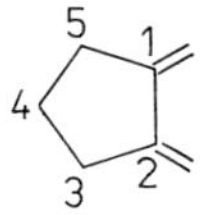	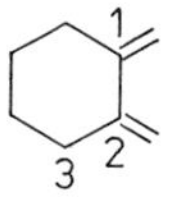	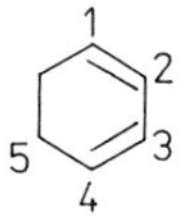	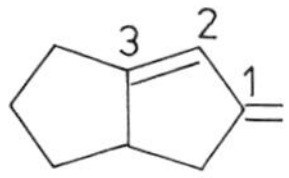	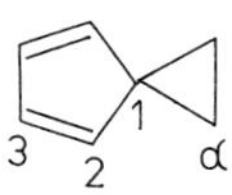
$^1J_{C_1C_\alpha}$	72.8	73.5	72.4	1,2 67.6	72.3[97]	9.6[97]
$^1J_{C_2C_3}$	34.2	40.1	40.2	4,5 40.1	74.1[97]	65.8[97]
$^1J_{C_1C_2}$					50.9[97]	51.4[97]

a) All data in the Table are, if not otherwise stated, from ref. 96; b) could not be measured, because of magnetic equivalence of the carbons.

TABLE 8

Calculated coupling constants, J_{CC}, for a two- and three-carbon unit and for cyclopropane (in Hz); reproduced with permission.[98]

		Fermi-contact term	Orbital term	Dipolar term	Total
$^2J_{CC}$		-19.34	-0.71	-0.03	-20.08
$^1J_{CC}$		35.63	-2.91	0.73	33.45
$^1J_{CC}$		16.65	-3.87	-0.31	12.47

[a] CCC bond angle $60°$.

two- and three- carbon units and for cyclopropane, those given in Table 8,[98] lead to the carbon-carbon coupling in cyclopropane equal to 33.45 + (-20.08) = 13.37 Hz, which is in accord with the calculated and experimental values for the cyclopropane molecule (12.47 and 12.4 Hz, respectively). A similar interpretation can be applied to the bicyclobutane molecule, where two large negative two-bond contributions should result in a negative $^1J_{CC}$ value of the coupling between the bridgehead carbon atoms, in agreement with the negative experimental $^1J_{CC}$ value determined for some derivatives of bicyclobutane, by Finkelmeier and Luttke,[74] and by Pomerantz and co-workers.[73] In light of the above facts one could expect that the coupling constants between the ring C atoms in tetra-tert-butyltetrahydrane,[105] ($^1J_{CC}$ = 9.2Hz) and between the bridgehead carbon atoms in 1-cyanobicyclobutane[106,107] ($^1J_{CC}$ = 16.0Hz) are also negative. However, this problem still remains to be solved by experiment. The positive $^1J_{C1C2}$ value of 17.36 Hz has been calculated by Galasso for tetrahydrane.[62]

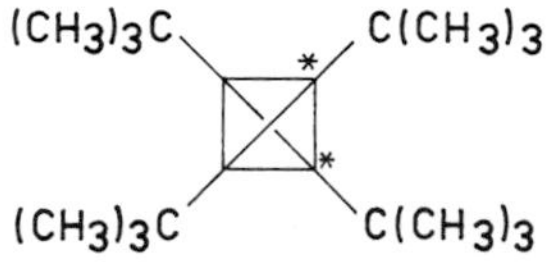

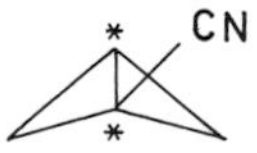

J_{CC} =   9.2 Hz   16.0 Hz

TABLE 9

One-bond spin-spin coupling constants, $^1J_{CC}$, in alkyl and phenyl substituted ethanes, ethenes and ethynes (in Hz).

$CH_3-CR_1R_2R_3$				
Substituents			$^1J_{CC}$	Ref.
R^1	R^2	R^3		
H	H	H	34.6	1
H	H	CH_3	34,6	78
H	CH_3	CH_3	34.7	108
CH_3	CH_3	CH_3	35.5	105
H	H	Ph	33.8	14

$R^1R^2C = CR^3R^4$						
Substituents				$^1J_{C=C}$	$^1J_{C-C}$	Ref.
R^1	R^2	R^3	R^4			
H	H	H	CH_3	70.0	41.9	37
H	H	CH_3	CH_3	72.6	–	37
H	CH_3	CH_3	CH_3	74.7	–	14
H	Ph	H	Ph	72.9[a]	57.0[a]	109
Ph	Ph	H	H	72.3	–	65
Ph	Ph	Ph	Ph	76.9	54.4	109

$R^1R^2R^3CC \equiv CR^4$						
Substituents				$^1J_{C \equiv C}$	$^1J_{C-C}$	Ref.
R^1	R^2	R^3	R^4			
H	H	H	H	175.0	67.4	88,111
CH_3	CH_3	CH_3	H	168.1	66.1	88
				168.7	–	112
H	CH_3	CH_3	CH_3	174.9	–	113

314

TABLE 9 contd.

CH$_3$	CH$_3$	CH$_3$	CH$_3$	173.9	68.4	88
H	H	H	Ph	181.2		(b)
H	H	CH$_3$	Ph	177.6		114

a) *trans*; b) K. Kamieńska-Trela and Z. Biedrzycka, unpublished data

TABLE 10

CC spin-spin coupling constants (in Hz) in some selected aromatic hydrocarbons.

55.3 ± 0.5

ref. 115

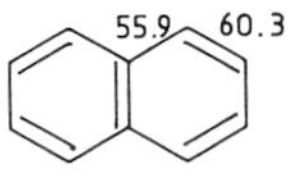

55.9   60.3

ref. 110

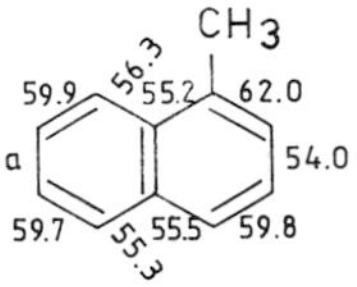

CH$_3$
59.9   56.3   55.2   62.0
a   54.0
59.7   55.5   59.8
55.3

ref. 117

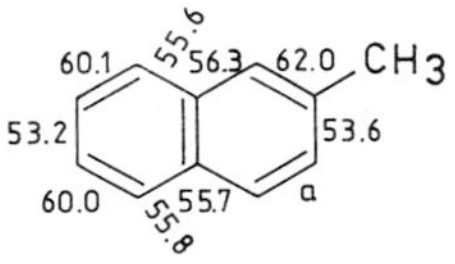

60.1   55.6   56.3   62.0 CH$_3$
53.2   53.6
60.0   55.8   55.7   a

ref. 117

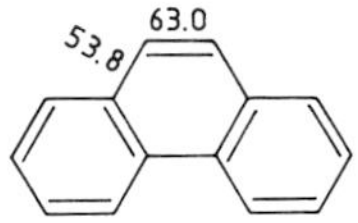

53.8   63.0

ref. 110

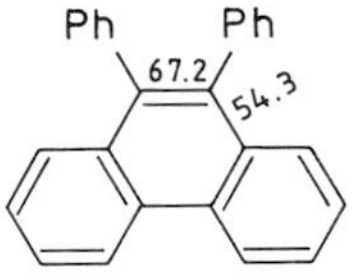

Ph   Ph
67.2   54.3

ref. 110

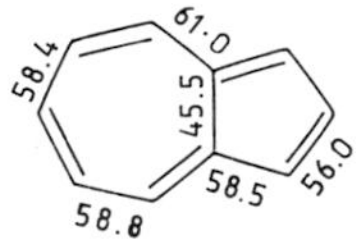

58.4   61.0
45.5
58.8   58.5   56.0

ref. 117

Furthermore, a survey of the $^1J_{CC}$ values collected in Tables 7, 9 and 10 reveals that the substitution of a hydrogen with methyl groups at double (both ethylenic and aromatic) and triple bonds leads to an increase in the $^1J_{CC}$ values. The methyl increment is about 5 Hz in acetylenes and about 2.5 Hz in ethylenes and aromatic compounds. Effects of similar range are observed when a hydrogen atom is replaced with the $- CH=CH_2$ and $-Ph$ groups (Tables 5,9), and a large increase by 22 Hz was found upon passing from acetylene to diacetylene[92] (Table 5). These results can hardly be explained in terms of conjugation effects, since in such a case, a decrease in spin-spin coupling constants is predicted.

On the other hand, it is well known that the electronegativity of vinyl and acetylenic moieties is greater than that of hydrogen (see *Coulson's Valence*, p.166).[118] Therefore, it is reasonable to assume that the main effect exerted by these substituents on the magnitudes of spin-spin $^1J_{CC}$ coupling is concerned with a change in the electronegativity, with respect to a hydrogen atom. The increase observed in the $^1J_{CC}$ values in substituted acetylenes $RC \equiv CH$ in the order R = t-Bu< i-Pr< Et< Me ≈ Ph ≈ $HC=CH_2$ << $-C \equiv CH$ suggests an analogous sequence of the electronegativities involved. A similarity of the effects exerted by methyl, vinyl and phenyl groups indicates that the electronegativities of these substituents are very close to each other.

It should be mentioned at this point that in several papers a general correlation between $^1J_{CC}$ coupling constants in aromatic compounds and the relevant HMO π bond orders has been postulated.[110,116,119,120] In particular, Hansen *et al.*,[110,119] noted that $^1J_{CC}$ values in a series of polyclic aromatic hydrocarbons plotted against the corresponding bond orders or bond lengths gave two parallel lines, one corresponding to couplings between methine and quaternary carbons, and the other to couplings involving only methine carbons. The corresponding diagram was presented by Kalinowski *et al.*,[20] who used Hansen's numerical data. A similar analysis has been perfomed for benzo[α]pyrene by London and co-workers,[120] who plotted the corresponding $^1J_{CC}$ values as a function of theoretical bond length, and also confirmed qualitatively Hansen's conclusions. However, more recent data seem to indicate that the validity of these conclusions is rather limited to closely related groups of compounds. Thus, for example, an attempt to correlate $^1J_{CC}$ coupling constants determined in azulene against HMO π order values failed.[117]

Furthermore, as has been stressed by Berger,[116] the existence of the correlation does not necessarily mean that the spin-spin coupling constants are transmitted *via* π - electrons. It has been pointed out by Pawliczek and Gunther[121] that the π -framework only reflects the changes in the sigma system, and this yields the apparent correlations between spin-spin coupling constants and HMO parameters.

There have been only few studies of steric effects on $^1J_{CC}$ couplings in unsubstituted hydrocarbons, but some interesting results have been published.[113,122] One of the most crucial results in this respect has been obtained by Booth and Everett,[122] who measured the CC coupling constants in [^{13}C-1-methyl]-*cis*-1,4-dimethylcyclohexane and its *trans* isomer at about 180 K.

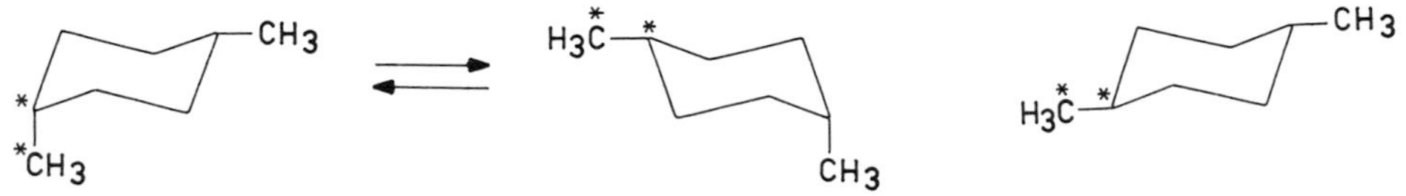

cis

T = 180 K

$^1J_{CC(ax)}$ = 34.64 Hz $^1J_{CC(eq)}$ = 35.40 Hz

T = 300 K

$^1J_{CCav.}$ = 35.55 Hz

trans

T = 173 K

$^1J_{CC}$ = 35.64 Hz

T = 300 K

$^1J_{CC}$ = 35.89 Hz

This allowed them to determine $^1J_{CC}$ coupling constants for *equatorial* and *axial* methyl groups separately, $^1J_{CC}$ (*eq*) of 35.40 Hz and $^1J_{CC}$ (*ax*) of 34.64 Hz . Two effects have been involved for explanation of the difference observed: (i) a greater electronegativity of the equatorially oriented methyl as compared with the *axial* group;[123] (ii) a slight flattening of the ring. It should be added that a similar stereospecifity is revealed by $^1J_{CH}$ coupling constants.[124] The difference between $^1J_{CH}(eq)$ and $^1J_{CH}(ax)$ is as should be expected, about four times greater[124] than that for the relevant $^1J_{CC}$ constant, ($^1J_{CH}(eq)$ = 126.44 Hz and $^1J_{CH}(ax)$ = 122.44 Hz).

In an interesting study Luttke and co-workers[113] have analysed $^1J_{CC}$ coupling constants in two highly strained molecules cyclooc-tyne and 1-sila-4-cycloheptyne. The $^1J_{CC}$ values in both compounds

(166.0 and 159.4 Hz, respectively) are lower than in the corresponding open-chain dialkylacetylenes (174.9 Hz in 4-methylpentyne-2 and 177.0 Hz in 2,2-dimethyl-5-methyl-hexyne-3). The experimental value of $^1J_{CC}$ for octyne was compared with those calculated by means of INDO method for 2-butyne where geometry of the C-C≡C-C system was forced out of linearity into both planar and non-planar structures. The authors came to the conclusion that the experimental $^1J_{CC}$ value corresponds to a distorted model of 2-butyne with an in-plane *cis* distortion combined with a subsequent twist from planarity. This result has been confirmed by gas electron diffraction measurements. The ϑ C-C≡C-C dihedral angle was estimated to be ca. 40°, while the < C-C≡C: angle amounted to about 155°. In this way cyclooctyne represents the first reported example of a highly bent and twisted CC triple bond.

Thus it can be concluded that in unsubstituted hydrocarbons, hybridization plays an important role, and may be considered as a main factor which determines the magnitude of $^1J_{CC}$ coupling constants. However, even in the case of such "simple" compounds, other effects such as electronegativity of substituents, steric hindrance and multi-path coupling mechanisms have to be taken into account if subtle differences in the couplings are considered. The role of these effects increases, when electron-donating or -attracting groups are introduced.

IV SUBSTITUENT, COMPLEXATION AND STERIC EFFECTS ON ONE-BOND CC SPIN-SPIN COUPLING CONSTANTS

The electronegativity of substituents is a second important factor which, according to theoretical calculations, determines the magnitude of $^1J_{CC}$ coupling. A large number of one-bond CC spin-spin coupling constants, determined experimentally for variously substituted ethanes, ethenes, allenes, aromatic compounds and ethynes, have already been published. It has been found that sensitivity towards substitution is quite different for couplings across single, double and triple CC bonds. It is therefore convenient to discuss each of the groups separately, and we start the discussion with the most sensitive coupling i.e. that across a triple bond. In addition to substituent electronegativity effects, the influence of complexation and steric effects will be discussed in this section.

A *Triple CC bonds*

There have already been numerous studies devoted to spin-spin coupling constants across triple CC bonds. These include variously substituted acetylenes,[56,66,111,112,114,125-133] diacetylenes[92,134,135] as well as cobalt[136] and tungsten[137] complexes of acetylene. The coupling constant of 230.4 Hz has been found in *meta*-CH$_3$OPhC≡COCH$_3$ and this has so far been the largest CC coupling determined experimentally.[132] The $^1J_{C≡C}$ values diminish very quickly along with the decreasing electronegativity of substituents, and $^1J_{CC}$ of 81 Hz has been found in (C$_4$H$_9$)$_3$SnC≡CSn(C$_4$H$_9$)$_3$ (E$_{Sn}$ = 1.7)[125] and $^1J_{C≡C}$ of 56.8 Hz[56] in (C$_2$H$_5$)$_3$SiC≡CLi, (E$_{Si}$ = 1.9; E$_{Li}$= 0.98).

The dependence of $^1J_{C≡C}$ couplings on the electronegativity of the atom attached directly to the triple bond is illustrated by the data collected in Table 11 and shown in Fig. 3, where the $^1J_{C≡C}$ values, taken from Table 11, have been plotted against the products of Pauling's electronegativities, E$_x$.E$_y$. The plot comprises acetylenes substituted with Cl, Br, I, Li, OR, SR, NR$_2$, PR$_2$, CR$_3$, C$_6$H$_5$, SiR$_3$ and GeR$_3$, where R denotes, with few exceptions, alkyl groups. In order to avoid second-order effects, substituents like HOCH$_2$- , BrCH$_2$-, Cl$_3$Si- and similar ones have been excluded. All electronegativity values have been taken from ref.[118] A least-squares fit yielded the following equation:

$$^1J_{XC≡CY} = 23.23(\pm0.51)E_xE_y + 15.45(\pm3.32) \qquad (5)$$
$$n= 27; \; r= 0.99; \; s.d.= 4.2 \text{ Hz}$$

Since the substituents involved in the correlation represent a wide range of electronic interactions: mesomeric (π-π; p-π; d-π), hyperconjugation and inductive effects, one can expect that the validity of Eq. (5) should be quite general. In consequence, the equation can be used either for the estimation of unknown spin-spin coupling constants or the other way around, for calculations of substituent electronegativities (see Table 12 and comments in ref.[114]).

TABLE 11

CC coupling constants, $^1J_{C≡C}$, (in Hz) in XC≡CY acetylenes.[114]

Substituents X / Y	No	Phenyl $^1J_{C≡C}$	No	$(CH_3)_3C$ a $^1J_{C≡C}$	No	$(CH_3)_3Si$ a $^1J_{C≡C}$
OR		–	11	224.3 b R = C_2H_5	20	166.7 R = CH_3
Cl	1	216.0	12	204.8	21	155.3
$N(C_2H_5)_2$	2	204.3	13	204.0 b		–
Br	3	202.5	14	190.6	22	143.2
Phenyl	4	185.0		–		–
SCH_3	5	184.2	15	175.0	23	134.2
C_2H_5	6	177.6	16	170.7		–
I	7	179.7	17	169.5	24	126.6
$SeCH_3$	8	173.2		–		–
PR_2	9	154.1 R = n-C_4H_9		–	25	115.2 c R = C_6H_5
$Ge(CH_3)_3$		–	18	131.5		–
SiR_3	10	136.9 R = CH_3	19	130.6 R = CH_3	26	101.4 c R = C_2H_5
Li		–		–	27	56.8 c

a) If not stated otherwise; b) Y = CH_3; c) Y = $(C_2H_5)_3Si$

The Table 11 and the Figure 3 reproduced with permission.[114]

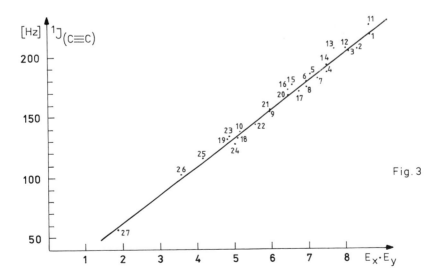

Fig. 3 Plot of CC couplings in disubstituted acetylenes against products of Pauling's E_XE_Y electronegativities.

It is tempting to consider a possible physical interpretation of the intercept of the $^1J_{CC}$ axis (15.45. Hz) in Eq.(5). Its numerical value is equal to $^1J_{C \equiv C}$ in the instance where the electronegativity of either substituent is zero, i.e. when a substituent at the triple bond is replaced by a lone electron pair. This will null the Fermi contact term, and in consequence the coupling constant in the anion will be approximately equal to the sum of the OD and SD terms. For most of the compounds studied so far (see Table 2) the latter ranges from 10 to 16 Hz and only rarely, for some electron-donating substituents, it falls below 10 Hz.[56] The value of 15.45 Hz seems to be indeed a good estimate of the participation of OD and SD interactions.

It is of interest to note that the results obtained for substituted acetylenes have been invoked to interpret the trends observed in the one-bond couplings across CP and CN triple bonds. It has been found that $^1J_{C \equiv N}$s decrease in the order: nitriles (one free electron pair) > $(CH_3)_3$ SiC≡N (polar Si-C≡ bond) > cyanide anion (two free electron pairs).[138]

Summarizing the results one can expect that the smallest $^1J_{CC}$ values will be found in the anions and the largest in the corresponding cations. Moreover, this trend seems to be quite general and valid for other types of CC bonds. Thus, a large increase in $^1J_{C-C^+}$ for α-cyanodiarylmethyl and 1,1-diaryl-2-butynyl cations in comparison with the $^1J_{CC}$ in the corresponding alcohols or hydrocarbons observed by Olah and co-workers does not seem to be "unusual".[108]

The results presented above include effects of α-substituents only. The influence of electronegative substituents in the β-position with respect to the triple bond has been studied by Krivdin et al.,[131] (Table 13). They found that there is no simple correlation between the properties of substituents and their influence on the $^1J_{CC}$ value concerned. Thus, an average increase of 3 to 5 Hz in $^1J_{CC}$ is observed upon the introduction of one chlorine or bromine atom, but a decrease of a similar magnitude is found when hydroxy and phenoxy groups are introduced. A small but systematic decrease in $^1J_{CC}$ has been observed upon the introduction of subsequent β-methyl groups in silyl acetylenes.[111]

TABLE 12

Spin-spin coupling constants, $^1J_{C \equiv C}$, in substituted acetylenes
and electronegativity values, E_A, of H, Hg, Sn, Pb and Te.
calculated by means of Eqn. (5).[114] All $^1J_{C \equiv C}$s are given in Hz.

Compound	$^1J_{C \equiv C}$	E_A
$HC \equiv COCH_3$	216.5	2.54±0.19
$HC \equiv CC_6H_5$	175.9	2.51±0.16
$HC \equiv CC(CH_3)_3$	168.7	2.57±0.15
$HC \equiv CGe(C_2H_5)_3$	132.5	2.51±0.13
$HC \equiv CSi(C_2H_5)_3$	131.8	2.64±0.13
$HC \equiv CH$	171.5	2.59±0.15
Weighted mean value of E_H		2.56±0.02
$(CH_3)_3SnC \equiv COC_2H_5$	151.6	1.72±0.13[a]
$(C_2H_5)_3SnC \equiv CC(CH_3)_3$	120.5	1.77±0.11
$(CH_3)_3SnC \equiv CSi(CH_3)_3$	94.0	1.78±0.11
$(n-C_4H_9)_3SnC \equiv CSn(n-C_4H_9)_3$	81.0	1.68±0.10
Weighted mean value of E_{Sn}		1.74±0.02[a]
$(C_2H_5)_3PbC \equiv CH$	113.0	1.64±0.11
$CH_3HgC \equiv CH$	125.7	1.85±0.12
$CH_3TeC \equiv CC_6H_5$[b]	154.4	2.18±0.14

[a] Individual statistical deviations were used as statistical
weights in the calculations of weighted mean values.
[b] In ref.[114] erroneously $CH_3TeC \equiv CH$ was given.

TABLE 13

Influence of β substitutents on $^1J_{C \equiv C}$ coupling constants in
acetylenes.[132] All values are in Hz.

Compounds	$^1J_{C \equiv C}$	Compounds	$^1J_{C \equiv C}$
$HOCH_2C \equiv CH$	169.3	$CH_2ClC \equiv CH$	178.6
$(CH_3)_2C(OH)C \equiv CH$	166.3	$CH_2BrC \equiv CH$	179.3
$(CH_3)_2C(OH)C \equiv CCl$	202.2	$Cl_3SiC \equiv CH$	146.0
$(CH_3)_2C(OH)C \equiv CBr$	188.4	$Cl_3GeC \equiv CH$	150.6
$C_6H_5OCH_2C \equiv CH$	173.1	$C_5F_{11}C \equiv CH$	188.5[a]

a) Ref.126

It is noteworthy that the $^1J_{C\equiv C}$ values depend only slightly on the number of acetylenic groups attached to a heteroatom.[126] A rather small increase in the order: $i\text{-}Pr_3SnC\equiv CH$ (119.8 Hz), $i\text{-}Pr_2Sn(C\equiv C)_2$ (125.3 Hz) and $i\text{-}PrSn(C\equiv CH)_3$ (127.7 Hz) can be interpreted in terms of β-effects rather than from the point of view of changes in the electronegativities of the heteroatoms involved.

Only few CC spin-spin coupling constants have been reported for diacetylene derivatives.[92,134,135] (Table 14) However, even the sparse data show that also in this group of compounds the electronegativity of substituent remains the main factor which determines the magnitude of the $^1J_{C\equiv C}$.

TABLE 14

One-bond CC spin-spin coupling constants, $^1J_{CC}$, in derivatives of diacetylene $XC\equiv C\text{-}C\equiv CX$. All values are in Hz.

X	$^1J_{C\equiv C}$	$^1J_{\equiv C\text{-}C\equiv}$	Ref.	X	$^1J_{C\equiv C}$	$^1J_{\equiv C\text{-}C\equiv}$	Ref.
$(CH_3)_3C$	188.3	a	134	$(C_2H_5)_3Si$	146.4	137.2	92
$(C_2H_5)_3Ge$	146.8	137.7	135	$(CH_3)_3Sn$	134.5	a	134

a) not observed

A comparison of the experimental spin-spin coupling constants given in Tables 11 and 14 (or those calculated by means of eq. (5) with those calculated by means of the INDO method (Table 2) shows that the latter reproduce quite well the experimental values. Since, as was shown, the $^1J_{CC}$ changes are reflected mainly by the corresponding changes in the contact Fermi contribution, which in turn is related to the s characters of the CC bonds involved, the observed variations of CC spin-spin coupling constants can be interpreted in terms of the Walsh rule, which states that "if group X attached to carbon is replaced by a more electronegative group Y, then the carbon valency towards Y has more p character than it has towards X."[139,140] A similar interpretation has been invoked[141] to account for the nearly additive changes in the $^1J_{CH}$ coupling constants in the CHXYZ substituted methanes.[142]

Of great interest are the results obtained for cobalt[136] and tungsten[137] complexes of acetylene. $^1J_{CC}$s of 55.9 Hz and 21 Hz were measured in Co complexes in which the acetylene is bonded to two $(Co_2(CO)_6(HC{\equiv}CH))$ (28) and four $(Co_4(CO)_{10}(HC{\equiv}CH)$ (29) cobalt atoms, respectively, and $^1J_{CC} = 15.8$ Hz was found in the tungsten complex $W_2(O\text{-}t\text{-}Bu)_6py(\mu^*C_2H_2)$ (30).[137] The corresponding

TABLE 15
One-bond carbon-carbon and carbon-hydrogen coupling
constants in acetylene ligand; all values in Hz.

Complex	$^1J_{CC}$	$^1J_{CH}$	Ref.
28	55.9	209.8	136
29	21.	174.7	136
30	15.8	192	137

carbon-hydrogen spin-spin coupling constants, $^1J_{CH}$ are 209.8, 174.7 and 192 Hz. These are the smallest $^1J_{CC}$ and $^1J_{CH}$ values, determined so far for acetylene derivatives. The dramatic decrease in $^1J_{CC}$ in the acetylenic ligand of the cobalt complexes as compared with free acetylene, has been explained by Aime et al.,[136] on basis of CNDO calculations, in terms of a decrease in the Fermi contact contribution. By analogy, a similar explanation can be offered for the tungsten complex.

B *Double CC bonds*

More than one hundred spin-spin coupling constants across double CC bond have already been published.[65,89,143-150] Most of them have been determined for monosubstituted ethenes.[89] Some selected data for this group of compounds are given in Table 16. The $^1J_{C=C}$ values vary from 54.7 Hz in $H_2C{=}CHSn(C_4H_9)_3$ to 82.2 Hz in $H_2C{=}CHOC(CH_3)_3$,[89] and are linearly related to the

TABLE 16

Spin-spin coupling constants across double CC bond, $^1J_{C=C}$, in some selected monosubstituted ethenes $H_2C=CHX$. All values are given in Hz.

Substituent	$^1J_{C=C}$	Ref.	Substituent	$^1J_{C=C}$	Ref.
OC_6H_5	81.8	89	Br	76.3	89
p- $OC_6H_4OCH_3$	81.4	89	SCH_3	72.9	89
p- $OC_6H_4NO_2$	81.4	89	SCH_2CH_3	72.8	89
p- $OC_6H_4OCH_3$	81.4	89	$S(CH_2)_2CH_3$	72.9	89
p- OC_6H_4Cl	82.1	89	$SCH(CH_3)_2$	72.5	89
p- OC_6H_4Br	81.7	89	$SC(CH_3)_3$	72.4	89
$OC(CH_3)_3$	82.2	89	$SeCH_3$	72.3	89
$OCH(CH_3)_2$	79.6	89	$C(CH_3)_3$	70.0	153
$O(CH_2)_2CH_3$	78.6	89	CH_3	70.0	37
OCH_2CH_3	78.5	89	$TeCH=CH_2$	69.2	89
OCH_3	78.6	89	$Ge(C_2H_5)_3$	60.6	89
Cl	77.6	147	$Si(CH_3)_3$	58.8	153
N⟨ 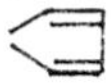	77.5	146	$Sn(C_4H_9)_3$	54.7	89

TABLE 17

Spin - spin coupling constants across double CC bond in *trans* and *cis* ethenes. All values are given in Hz.

Compound	$^1J_{C=C}$ *trans*	$^1J_{C=C}$ *cis*	Ref.
FCH = CHBr	96.0	87.9	a
ClCH = CHCl	91.9	84.5	148
BrCH = CHBr	86.4	82.2	150
ICH =CHI	78.3	78.7	143
$(CH_3)_3SiCCl = CHCl$	77.5	67.5	150
$(CH_3)_3SiCBr = CHBr$	72.4	64.9	150
$(CH_3)_3SiCI = CHI$	66.3	62.5	150
BrCH = $CHCH_3$	78.6	77.7	a
p-$CH_3OC_6H_4CH=CHCN$[b]	74.0	73.8	108

a) K. Kamieńska-Trela and Z. Biedrzycka, unpublished data
b) configuration was not assigned.

electronegativity of substituents:[146]

$$^1J_{C=C} = 14.4\ E_X + 33.1 \qquad\qquad (6)$$

A rather poor correlation with Swain-Lupton's parameter F has also been published:[89]

$$^1J_{C=C} = (32.4 \pm 4.1)F + (69.2 \pm 0.9) \qquad\qquad (7)$$

$$r = 0.903, \quad \text{standard deviation} = 2.2\ Hz, \quad n = 15$$

TABLE 18

Spin-spin coupling constants across double CC bond in some halogeno non-symmetrically substituted ethenes (in Hz)

Compounds	$^1J_{C=C}$	Ref.
$C_6H_5CH=CF_2$	115.3	65
$Cl_2C=CHCl$	99.5	144
	103.0	147
$Cl_2C=CH_2$	90.5	144
	91.0	147
$CH_3CBr=CH_2$	79.0	a)
$ClCH=CClLi$ *trans*	37.0	145
$CH_3CCl=CCl_2$	104.8	147
$ClCH_2CCl=CCl_2$	105.8	147
$Cl_2CHCCl=CCl_2$	106.8	147
$Cl_3CCCl=CCl_2$	110.8	147

a) K. Kamieńska-Trela and Z. Biedrzycka, unpublished data

Along with the increasing number of substituents attached to the double CC bond, further variations in $^1J_{CC}$ follow. The coupling constant of 37.0 Hz has been determined in 1-lithium-*trans*-1,2-dichloroethene,[145] and that of 115.3 Hz in 1,1-difluoro-phenylethene.[65] These are, respectively, the smallest and the largest $^1J_{C=C}$ coupling constants determined so far. The observed changes are in excellent agreement with the trend observed in INDO computed values of $^1J_{C=C}$ (Table 3). Still larger values can be anticipated for tri- and tetrafluoroethenes, and smaller ones for double CC bonds which bear electron-donating substituents.

A full analysis of spin-spin coupling constants across double CC bond requires that steric effects, i.e. those of configuration and conformation of substituents, be taken into consideration.

It has been shown by Koole et al.,[151] (see also references quoted therein) that vinyl ethers exist at room temperature as mixtures of two forms. The more stable isomer has the s-cis form, while the other is either s-trans or gauche. In methyl vinyl ether the s-cis form prevails but the equilibrium shifts towards the other isomer along with increasing bulkiness of the alkoxy groups. In t-butyl vinyl ether, the s-trans form predominates. Vinyl thioethers exist as s-cis isomers only. An inspection of Newman projections of both conformers (see below) reveals that an

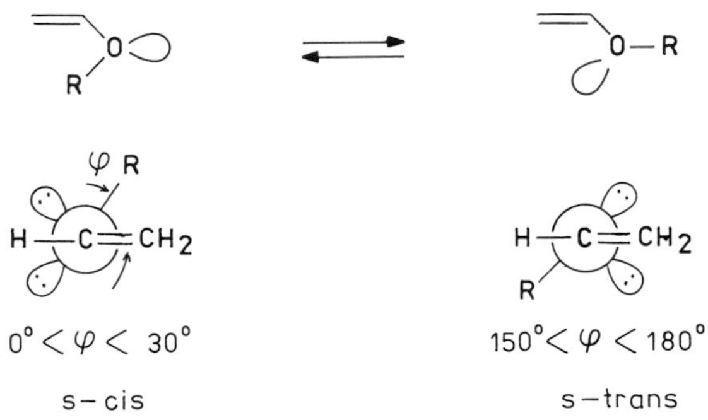

$$0° < \varphi < 30°$$

s-cis

$$150° < \varphi < 180°$$

s-trans

interaction between the π electrons of a double CC bond and a lone electron pair at the oxygen atom can take place in the s-trans conformer only, which in turn may influence the magnitude of the spin-spin coupling constant across the C=C bond. A thorough study of conformational effects on the $^1J_{C=C}$ coupling constants in vinyl ethers as well as in vinyl thioethers has been performed by Krivdin et al..[89,149] An increase in the value of $^1J_{C=C}$ has been observed in the order methyl ≈ ethyl < iso-propyl < tert-butyl vinyl ethers (Table 16). The difference between the $^1J_{CC}$ values in methyl and tert-butyl vinyl ethers (about 4 Hz) was assigned by the authors to a positive contribution of the oxygen electron lone pair to the $^1J_{CC}$ value in the s-trans conformer. It has also been concluded that the actual contribution is most probably somewhat larger than the value quoted, since the two compounds are not single conformers but contain some admixture of other conformers. An almost constant (within 1 Hz) $^1J_{C=C}$ value has been found, as could be expected, in a series of vinyl thioethers[89] (Table 16).

An analysis of the $^1J_{C=C}$ data, obtained recently for *cis* and *trans* 1,2-dihalogenoethenes and for *cis* and *trans*-1-trimethyl-silyl-1,2-dihalogenoethenes[150] (Table 17), showed that: (i) the magnitude of $^1J_{C=C}$ decreases monotonically with the decreasing electronegativity of the halogen; (ii) the value of $^1J_{C=C}$ *trans* is generally larger than that of $^1J_{C=C}$ *cis*, but the difference diminishes in both the groups studied with decreasing electronegativity of the halogen. The difference is negligible between *cis* and *trans* 1,2-diiodoethene; $^1J_{C=C}$ in the *cis* isomer is larger by only 0.4 Hz with respect to the *trans* isomer. The differences in $^1J_{C=C}$ values tend to reflect differences in energy between *cis* and *trans* isomers. Available data indicate that *cis*-1,2-difluoroethene is more stable than the *trans* isomer by 3.9 kJ/mol.[152] The difference decreases in the order: 1,2-difluoro-, 1,2-dichloro- and 1,2-dibromoethenes (3.9, 2.7 and 1.3 kJ/mol, respectively), and is close to zero for 1,2-diiodoethenes, *cis*-1,2-diiodoethene being less stable than its *trans* analog. No theoretical justification could be offered for the observed relationship between $\Delta^1J_{C=C}$ and ΔE. However, since $^1J_{C=C}$ coupling constants depend upon the relevant contributions of the s-electrons to the corresponding bonding orbitals, the observed decreases in $\Delta^1J_{C=C}$ (*trans* - *cis*) and in ΔE (*trans* - *cis*) can therefore be tentatively associated with the s-electron densities at the bonding orbitals of the *cis* and *trans* isomers.

In striking contrast to the behaviour of one-bond hydrogen-carbon coupling constants, whose values remain practically unchanged on passing from free alkenes to their complexes,[153-155] the $^1J_{C=C}$ values have proven to be very sensitive towards the geometry and electronic structure of a given complex. They have been measured for a series of η^2-alkene complexes of rhodium(I),[153] η^5-C_p-η,η^2-alkenylnickel(II)[156] complexes, for tetracarbonyl-iron(0) and tetracarbonylruthenium(0) complexes of acrolein,[157] and for a considerable number of η^4-butadiene,[156-159] and

η^3-allyl[160] complexes. Invariably, the $^1J_{C=C}$ values in the complexes are considerably smaller than those in the uncoordinated compounds. The magnitude of the decrease varies upon the nature of the metal involved and depends on the geometry of a given complex,[159] but remains constant for a given series. In the studies of a series of closely related η^2-alkenerhodium (I) complexes $^1J_{C=C}$ values have been found to be related to the electronegativity of the substituent in coordinated alkenes $CH_2=CHX$ (X = Si, C, O) and a constant decrease of about 22 Hz has been observed on coordination of all alkenes.[153] Thus, the $\Delta^1J_{C=C}$ rather than $^1J_{C=C}$ itself may be the important parameter in discussing bonding and hybridization changes in coordinated alkenes.

Interesting results have been obtained by Fox and Shultz[161] for the dipotassium salts of tetraphenylethylene dianions which exist as contact ion pairs (31^{2-} and 32^{2-}). Relatively small changes have

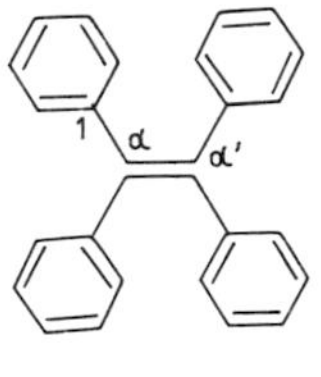

31 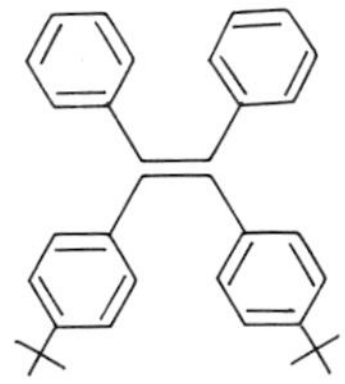   **32**

been found in the carbon-hydrogen, $^1J_{CH}$, (by an average of 3.6%) and carbon-carbon $^1J_{C(arom)C(arom)}$, coupling constants in the phenyl rings of **31** upon reducing it to 31^{2-} (Table 19). A large decrease in $^1J_{C\alpha C\alpha'}$ and an increase in $^1J_{C\alpha C1}$ (an average of 21.1%) have been observed in 32^{2-} by comparison with the parent compound **32**. The $^1J_{CC}$ values obtained have been related to the π orders of the corresponding bonds and a linear relationship has been found. However, in light of the results presented in the previous parts of this section, it seems reasonable to assume that the observed $^1J_{CC}$ changes result from corresponding changes of the σ framework of the compounds studied.

TABLE 19

One-bond CH and CC coupling constants   in tetraphenylethylenes 31
and 32, and in their dianions. All values are in Hz.[161]

C-H coupling constants			$C_{arom}-C_{arom}$ coupling constants			
C-H	31	31^{2-}	$C_{arom}-C_{arom}$	31	31^{2-}
2	159.0	153.7	1-2	58.3	55.7
3	158.8	149.5	2-3	56.2	57.7
4	160.7	158.0	3-4	55.6	54.6

CC coupling constants across double bond					
	32		32^{2-}		$32 - H_2$
$\alpha - \alpha'$	76.6	$\alpha - \alpha'$	60.6	$\alpha - \alpha'$	33.9
$\alpha - 1$	54.6	$\alpha - 1$	64.5	$\alpha - 1$	42.2

C *Substituted allenes*

Only few CC spin-spin coupling constants   have   been   published
for substituted allenes and all the data presented in this section
are taken from the paper by Krivdin *et al.,*[162] who   measured $^1J_{CC}$
couplings for a series of alkoxy-substituted allenes. As could   be
expected (see INDO data in Table 3), effects   of   substituents   on
$^1J_{CC}$ across an unsubstituted CC allenyl bond are rather weak (less
than 4 Hz). Spin-spin coupling constants across a CC bond with   an
alkoxy substituent attached are, on the other   hand,   much   larger
(by about 15 Hz)   than   the   coupling   constant   in   unsubstituted
allene (98.7 Hz[90]). The couplings do not change within the   series
methoxy, ethoxy and *iso*-propoxyallene (113.7-114.0 Hz) but a sharp
increase occurs on going to *tert*-butoxyallene (^1J =   118.2 Hz). A
positive contribution, of about 4 Hz, to $^1J_{C=C}$ has been   predicted
for a lone electron pair of oxygen in the *s-trans* isomers in vinyl
ethers (see section 4 B).   It has therefore been concluded by   the
authors that a shift of the *s-cis* ⇌ *s-trans* equilibrium towards
the *s-trans* isomer takes place in *tert*-butoxyallene. However, it

TABLE 20

CC spin-spin constants in alkoxyallenes $RO\overset{1}{C}H=\overset{2}{C}H=\overset{3}{C}H_2$. All values are in Hz.[162]

R	$^1J_{C=C}$		R	$^1J_{C=C}$	
	$^1J_{1,2}$	$^1J_{2,3}$		$^1J_{1,2}$	$^1J_{2,3}$
CH_3	113.8	101.7	$CH(CH_3)_2$	113.7	101.4
CH_2CH_3	114.1	101.8	$C(CH_3)_3$	118.2	102.2
$CH_2CH_2CH_3$	113.7	101.4	$CH=CH_2$	117.8	101.9
$CH_2CH(CH_3)_2$	114.0	101.9	C_6H_5	117.7	102.1
$CH_2CH=CH_2$	114.8	101.1			

should be noticed, that the results concerning the s-cis/s-trans ratios in alkoxyallenes, those based on carbon-hydrogen spin-spin coupling constants measured at various temperatures,[151] electron diffraction[163] and photoelectron spectra[164] data are rather contradictory and do not corroborate Krivdin's assumption. Thus, the analysis of $^1J_{CH}$ values performed at various temperatures by Koole et al.,[151] indicates that at room temperature ethoxy and propoxyallenes contain about 20% of less stable conformer, whose percentage considerably increases in iso-propoxyallene; only tert-butoxyallene exists as a pure s-trans conformer. According to photoelectron data[164] methoxyallene contains a considerable admixture of s-trans isomer. Only in the electron diffraction spectrum obtained for methoxyallene the s-trans isomer was not detected.[163] Moreover, there is no agreement concerning the nature of the less stable conformer and both s-trans and gauche forms are considered by various authors.[151]

D Substituted aromatic and heteroaromatic compounds.

One-bond CC spin-spin coupling constants in mono- and disubstituted benzenes, in condensed aromatic systems and in some heteroaromatic compounds have been a subject of extensive studies.

Monosubstituted benzenes have been studied by Wray et al.,[5] Marriott et al.,[165] Wrackmeyer,[128] and Krivdin and Kalabin.[166-170] $^1J_{CC}$s in disubstituted benzenes have been analysed by Sandor and Radics,[171] Krivdin et al.,[172] and by Olah and co-workers.[173-175] The data for substituted condensed aromatic systems have been published by Berger and Zeller.[176] A large set of CC coupling constants in 2-, 3- and 4-substituted pyridines have been obtained by Denisov et al.,[177,178] and for heteroanalogs of fluorenone by Krivdin et al.,[79,179] Very precise (within 0.001 Hz) $^1J_{CC}$ values have been determined by Maple and Allerhand[180] in pyridine and 2,6-lutidine. The CC spin-spin coupling constants in ortho-, metha- and para- aminobenzoic acids, benzoic acids and aniline have been measured by Berger[181] for the neutral compounds and for the corresponding anions and cations.

TABLE 21

CC spin-spin coupling constants, $^1J_{CC}$, in monosubstituted benzenes. All values in Hz.

X (at C1)	$^1J_{1,2}$	$^1J_{2,3}$	$^1J_{3,4}$	Ref.
F	70.8	56.6	56.2	5
NO_2	67.4	56.1	55.3	5
OCH_3	67.0	57.8	56.1	5
OH	65.6	57.7	56.1	5
Cl	65.2	55.8	56.1	5
Br	63.6	54.9	56.1	5
$N(CH_3)_2$	62.8	59.0	56.1	5
SH	60.1	-	56.0	5
SCH_3	59.8	56.5	56.0	169
CH_3	57.0	56.4	56.0	169
$C(CH_3)_3$	57.8	56.7	55.8	5
$SeCH_3$	58.3	-	56.3	169
$TeCH_3$	56.5	-	-	169
$P(C_6H_5)_2$	55.0	55.3	-	5
SiH_3	49.5	54.7	55.4	5
$Pb(C_6H_5)_3$	51.1	54.0	55.5	128
$Bi(C_6H_5)_2$	51.0	53.5	55.6	128
HgC_6H_5	49.5	53.4	-	128

$\Delta J(X_1-X_2)$ $\Delta\,^F_{Hg}$ 21.3 Hz $\Delta\,^N_{Hg}$ 5.6 Hz $\Delta\,^F_{Hg}$ 0.6 Hz

The most striking feature of one-bond CC spin-spin coupling constants in all aromatic compounds studied so far is, that the influence of substitutent decays quickly with the increasing distance from the CC bond concerned. Only the couplings across $C_{ipso}C_{ortho}$ bond show the usual $^1J_{CC}$ vs. electronegativity substituent dependence[5,165] (Table 21). Thus, for example, in monosubstituted benzenes the coupling constants across $C_{ipso}C_{ortho}$

TABLE 22

One-bond CC coupling constants, $^1J_{CC}$, (in Hz) in pyridine and pyridine N-oxide derivatives.

Substituents	$^1J_{2,3}$	$^1J_{3,4}$	$^1J_{4,5}$	$^1J_{5,6}$	Ref
\multicolumn{6}{c}{Pyridines}					
2-F	75.4	55.7	54.4	56.0	177
OCH=CH$_2$	72.3	57.7	–	56.8	179
OCH$_3$	70.7	57.6	53.6	57.0	177
Cl	67.4	54.1	54.3	55.3	177
Br	64.9	53.3	54.4	55.4	177
CH$_3$	56.2	54.7	54.2	54.7	177
3-F	70.3	68.8	54.2	54.2	177
OCH=CH$_2$	67.2	65.7	–	54.8	179
OCH$_3$	66.1	64.6	55.2	54.9	177
Cl	62.4	63.0	53.5	54.3	177
Br	59.9	61.4	52.9	54.3	177
CH$_3$	55.4	55.1	54.3	54.6	177
4-OCH$_3$	56.2	64.2	64.2	56.2	177
CH$_3$	54.9	54.7	54.7	54.9	177
Cl	53.6	62.5	62.5	53.6	177
Br	52.7	60.8	60.8	52.7	177
H	54.3	53.7	53.7	54.3	20[a]
\multicolumn{6}{c}{Pyridine N-oxides}					
X = 2-CH$_3$	62.5		54.2	60.9	b
4-CH$_3$	60.5	55.2	55.2	60.5	b
H	60.2	54.8	54.8	60.2	b

a) very precise $^1J_{CC}$ values (within 0.001 Hz) are reported by Maple and Allerhand[180] for pyridine and 2,6-dimethylpyridine;
b) Z.Biedrzycka and K. Kamieńska-Trela, unpublished data.

bond vary from 49.5 Hz in silyl to 70.8 Hz in fluorobenzene,[5] while the corresponding, $^1J_{C(ortho)C(meta)}$s, are of 54.7 Hz and 56.64 Hz,[5] respectively ($^1J_{CC}$ in benzene is of 55.37 ± 0.5 Hz).[115] The influence of substitutents upon the couplings across C_{meta}-C_{para} bonds is smaller than one 1 Hz. A similar pattern has been found in all substituted pyridines[177] (Table 22) and heteroanalogs of fluoronenone.[79,179]

A large set of CC spin-spin coupling constants determined for ortho-, meta- and para-substituted anilines has been analysed by Sandor and Radics[171] from the point of view of additive effects of substituents on $^1J_{CC}$s. The experimental $^1J_{CC}$ values were compared with $^1J_{CC}$ (cald.), the hypothetical coupling constants calculated upon an assumption of additivity of the substituent coupling increments (Table 24). The differences $\Delta = {}^1J_{CC}(exp.) - {}^1J_{CC}(calcd.)$ are generally smaller than 1 Hz in meta and para substituted anilines and only in few cases they are slightly larger than that. In ortho-anilines, non-additivity of substituent effects is more pronounced. In particular, all experimental coupling constants across $C_{ipso}C_{ortho}$ bonds are much smaller than the calculated ones. This was explained by the authors in terms of a partial mutual compensation of the inductive effects by the substituents themselves. A full additivity of substituent effects, including ortho-substituted compounds (deviations from additivity were generally smaller than 1 Hz) has also been observed by Krivdin et al.,[172] who analysed a large set of CC coupling constants in benzenes containing, among others, SCH_3, OH, OCH_3, F, Cl and Br substituents. Deviations from additivity (about 4 to 7 Hz) have been observed only in such cases where either hydrogen bond was present (33) or a large steric hindrance occurred (34).

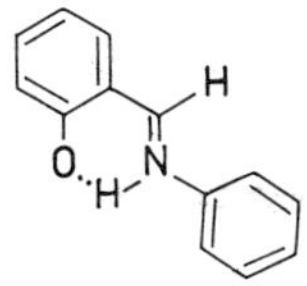

33

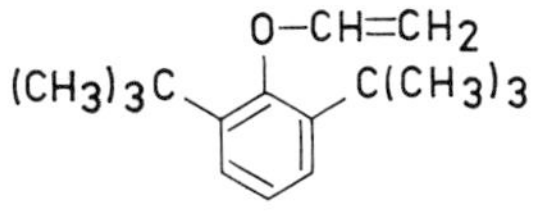

34

TABLE 23

CC coupling constants (in Hz) in p-substituted benzenes across C1C2 bonds.

Substitutents							
X	Y	CHO[a]	COCH$_3$[a]	COOCH$_3$[a]	COCl[a]	CN[b]	SCH$_3$[c]
F		70.5	70.6	70.7	70.3	70.7	71.6
OCH$_3$		66.4	66.5	66.5	66.3	66.5	67.8
Cl		64.8	65.0	64.8	64.6	64.7	d
Br		63.1	63.6	63.4	d	63.0	64.3
CN		60.4	60.4	d	d	d	d
CH$_3$		56.1	56.5	56.5	56.1	56.2	57.3
H		55.4	55.4	55.5	55.2	54.7	56.0

a) ref. 175; b) ref. 174; c) ref. 172 ; d) not measured

TABLE 24

CC coupling constants (in Hz) in *ortho-*, *meta-*, and *para-* disubstituted benzene derivatives

Substitutents X(at C1) Y		$^1J_{1,2}$	$^1J_{2,3}$	$^1J_{3,4}$	$^1J_{4,5}$	$^1J_{5,6}$	$^1J_{1,6}$	Ref.
		ortho						
NH$_2$	F	73.4	74.7	57.0	56.8	59.2	62.6	171
NH$_2$	NO$_2$	70.9	68.5	59.0	53.0	60.7	58.4	171
NH$_2$	OCH$_3$	70.0	70.4	57.4	57.1	58.4	64.4	171
NH$_2$	Cl	69.4	68.8	56.5	55.9	59.7	61.5	171
NH$_2$	Br	68.3	67.2	55.8	55.9	59.7	60.7	171
NH$_2$	NH$_2$	-	64.5	58.1	-	58.1	64.5	171
NH$_2$	I	66.0	64.2	55.4	55.8	59.9	60.2	171
NH$_2$	CN	64.8	62.2	58.1	54.5	59.3	60.6	171
NH$_2$	CH$_3$	61.1	60.6	56.6	55.7	59.0	62.2	171
OCH$_3$	CH$_3$	67.2	60.2	55.8	56.9	58.0	67.6	172
SCH$_3$	Cl	67.9	66.0	55.5	-	-	60.1	172
SCH$_3$	Br	66.8	65.1	55.6	56.7	56.7	59.8	172
SCH$_3$	CH$_3$	58.9	58.4	56.7	-	56.3	61.8	172
		meta						
NH$_2$	NH$_2$	64.7	64.7	60.9	58.8	58.8	60.9	171
NH$_2$	OCH$_3$	63.9	70.9	66.9	57.9	59.2	61.1	171
NH$_2$	F	62.8	74.8	71.1	57.1	59.8	61.3	171

TABLE 24 contd.

NH$_2$	NO$_2$	62.3	71.6	67.8	56.6	59.2	60.3	171
NH$_2$	CN	62.3	63.8	60.1	56.8	58.8	60.2	171
NH$_2$	CH$_3$	61.6	60.3	57.2	56.7	58.9	61.2	171
NH$_2$	Cl	61.0	69.3	65.5	55.9	59.4	61.2	171
NH$_2$	Br	60.0	67.9	63.9	55.3	59.6	61.1	171
NH$_2$	I	59.7	65.0	61.4	54.6	59.5	61.2	171
OCH=CH$_2$	OCH$_3$	69.8	68.6	68.6	58.3	58.3	69.2	172
OCH=CH$_2$	CH$_3$	68.8	58.5	58.0	57.1	57.2	67.8	172
OCH=CH$_2$	NO$_2$	68.7.	69.4	67.9	56.6	57.2	67.7	172

para

NH$_2$	OCH$_3$	62.5	60.1	68.3	–	–	–	171
NH$_2$	NH$_2$	62.5	d	62.5				171
NH$_2$	F	62.2	59.3	72.2				171
NH$_2$	Cl	61.8	59.0	66.4				171
NH$_2$	CH$_3$	61.6	59.1	57.6				171
NH$_2$	Br	61.5	58.1	64.8				171
NH$_2$	I	61.4	57.8	61.7				171
NH$_2$	CN	59.7	60.6	60.2				171
NH$_2$	NO$_2$	59.4	60.7	67.3				171
OCH=CH$_2$	OCH$_3$	69.1	–	68.1				172
OCH=CH$_2$	Cl	68.8	57.8	66.4				172
OCH=CH$_2$	Br	68.8	57.1	64.7				172
OCH=CH$_2$	CH$_3$	68.3	58.0	57.6				172
OCH=CH$_2$	NO$_2$	68.1	–	68.9				172

E *Single CC bonds*

These include spin-spin coupling constants across Csp^3-Csp^3, Csp^3-Csp^2, Csp^3-Csp, Csp^2-Csp^2, Csp^2-Csp and Csp-Csp bonds; among these, Csp^3-Csp^3 couplings have been most thoroughly studied (Table 25). The measurements have been carried out for substituted open-chain alkanes,[36,38,125,182-185] adamantanes,[186] and diadamantanes,[187] cyclopropanes,[91,98,188] bicycloalkanes,[189] and saturated heterocycles.[190,191] A large series of measurements has been done for mono- and disaccharides.[192-201]

The sensitivity of Csp^3Csp^3 coupling constants towards substitution is smaller than that of any other type of coupling, but they show the expected variation with substituent electronegativity. In monosubstituted hydrocarbons the variation is within about 20 Hz. Thus, for example, $^1J_{CC}$ in $(CH_3)_3CF$ attains 40.3 Hz and 28.4 Hz in $(CH_3)_3CLi$.[184] In 1-chlorocyclopropane $^1J_{CC}$

TABLE 25

Coupling constants across one Csp^3Csp^3 bond. All values are in Hz.

Compound	$^1J_{C-C}$	Ref.	Compound	$^1J_{C-C}$	Ref.
CH_3CH_2X			CH_3CH_2X		
$X=OC(CH_3)_3$	39.3	185	$X=TeC_2H_5$	35.4	185
OCH_3	39.0	36	SC_2H_5	35.4	185
F	38.2	36	CHO	35.4	15
$N(C_2H_5)_2$	38.1	15	C_6H_5	34.2	38
OH	37.7	36	$Pb(C_2H_5)_3$	33.8	128
Cl	36.1	36	HgC_2H_5	33.6	128
Br	36.0	36	$Sn(C_2H_5)_3$	33.1	128
SeC_2H_5	35.9	185	$B(C_2H_5)_3$	33.0	128
I	35.8	36	CN	33.0	38
NH_2	35.8	36	$Ge(C_2H_5)_3$	32.8	185
NO_2	35.7	36	$SiH(C_2H_5)_2$	31.7	15
FCH_2CH_2OH	40.8	15	$(CH_3)_3CF$	40.3	184
$BrCH_2CH_2Br$	38.9	14	$(CH_3)_3CLi$	28.4	184
CH_3CBr_3	38	183	$(CH_3)_2CHLi$	22.9	184
CH_3CHBr_2	37	183	CH_3CLiBr_2	8	183

equals to 13.9 Hz[91] and is close to zero (<0.5 Hz) in 1-lithium-2-phenylcyclopropane.[188] Similarly small changes upon substitution have been observed in monosubstituted bicycloalkanes,[189] adamantanes,[186] and diadamantanes.[187]

Only few systematic studies have been performed on the other types of $^1J_{CC}$ coupling across single bonds. These include the analysis of $^1J_{Csp^3-Csp^2}$ couplings in 2-substituted propenes,[37] and chloropropenes,[147] in aliphatic carbonyl compounds,[202] and in a series of ^{13}C enriched aldono-1,4-lactones.[203] $^1J_{Csp^2(ar)-C(O)X}$ have been analysed in aromatic carbonyl compounds,[204,205] and particularly valuable results have been obtained by Olah et al., who studied $^1J(C_{ar}C(O)^+)$ coupling constants in substituted benzoyl cations.[171,172] The coupling constants between Csp^2 and Csp in a series of acetylenic acids and esters have been determined by Chaloner.[206] The most relevant data are displayed in Tables

TABLE 26

$^{1}J_{CC}$ coupling constants (in Hz) across Csp^3-Csp^2, $Csp^3-Csp^2(O)$, $Csp^3-Csp^2(N)$, Csp^3-Csp^2 (aromatic) bonds.

Compound	$^{1}J_{C-C}$	Ref.	Compound	$^{1}J_{C-C}$	Ref.
Csp^3-Csp^2			$Csp^3-Csp^2(O)$		
			$CH_3-CO\ X$		
$Cl_3C-CCl=CCl_2$	68.6	147	$X = OC_2H_5$	58.8	202
$Cl_2CH-CCl=CCl_2$	60.8	147	OH	56.7	202
$ClCH_2-CCl=CCl_2$	55.2	147	Cl	56.1	202
$CH_3-CCl-CCl_2$	50.3	147	Br	54.1	202
$CH_3-CCl=CH_2$	48.5	37	$N(CH_3)_2$	52.2	202
$CH_3-C(OC_2H_5)=CH_2$	51.8	37	I	46.5	202
$HOCH_2-CH=CH_2$	45.4	53	C_6H_5	43.3	202
$CH_3-C(CN)=CH_2$	44.9	37	CH_3	40.1	202
$CH_3-C(C_6H_5)=CH_3$	42.9	37	H	39.4	202
$CH_3-C(CH_3)=CH_2$	41.8	37	C_2H_5	38.4	202

					Csp^3-Csp^2(aromatic)		
$Csp^3-Csp^2(N)$							
$CH_3-CH=N-OH$	anti	48.4	209[a])	$C_6H_5-CCl_2C_6H_5$	51.6	110
$CH_3-CH=N-OH$	syn	40.5	209	$C_6H_5-CH_2F$	49.3	209
$(CH_3)_2C=N-OH$	anti	49.3	209	$C_6H_5-CH_2Cl$	47.8	82
	syn	41.4		$C_6H_5-CH_2OH$	47.7	82
$(CH_3)_2C=N-OH\ HCl$	anti	42.5	212	$C_6H_5-C(CH_3)_2OH$	47.3	9
	syn	42.3		$C_6H_5-CH_2CH_3$	45.5	209

a) Further examples have been published in refs. 210, 211, 219

26-29. In most cases the usual increase in $^{1}J_{CC}$ coupling constants along with the increasing electronegativity of substitutent has been observed. This includes also substituent effects transferred trough a triple CC bond on $^{1}J_{CC}$'s across single Csp^3-Csp and Csp-Csp bonds (see Tables 27 and 14, respectively). However, some exceptions have been noted. Thus, no relationship between $^{1}J_{Csp^3-Csp}$ and the nature of X has been found in $XCH_2C\equiv CH$ acetylenes (X = OH, OC_6H_5, Cl, Br).[127] Also the data for $^{1}J_{Cipso-Csp}$ couplings in $C_6H_5C\equiv CX$ are rather difficult to rationalize in terms of substituent electronegativity effects only, though a regular decrease is observed within each group of

TABLE 27

$^1J_{CC}$ coupling constants (in Hz) across with Csp^3-Csp bonds.

Compound	$^1J_{C-C}$	Ref.	Compound	$^1J_{C-C}$	Ref.
$CH_3-C≡COC_2H_5$	74.8	126	$(CH_3)_3C-C≡CCl$	68.5	a
$CH_3-C≡CN(C_2H_5)_2$	70.0	126	$(CH_3)_3\overset{*}{C}-\overset{*}{C}≡CC_2H_5$	68.2	56
$CH_3-C≡CSi(CH_3)_3$	63.5	56	$(CH_3)_3C-C≡CSCH_3$	67.1	131
$CH_3-C≡CSn(CH_3)_3$	62.2	126	$(CH_3)_3C-C≡CBr$	67.0	56
$CH_3-C≡CPb(CH_3)_3$	59.0	126	$(CH_3)_3C-C≡CI$	65.1	56
$BrCH_2-C≡CH$	78.0	130	$(CH_3)_3C-C≡CSi(CH_3)_3$	62.0	56
$ClCH_2-C≡CH$	77.5	130	$(CH_3)_3C-C≡CGe(CH_3)_3$	62.5	a
$HOCH_2-C≡CH$	71.5	130	$(CH_3)_3C-C≡CSn(C_2H_5)_3$	61.4	a

a) K. Kamieńska-Trela and Z. Biedrzycka, unpublished data

TABLE 28

$^1J_{CC}$ coupling constants (in Hz) across single Csp^2-Csp^2 bonds.

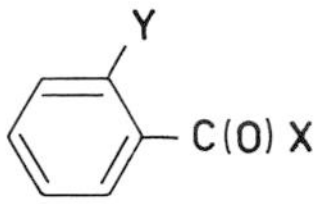

Substituents X	Y	$^1J_{Csp^2-Csp^2}$	Ref.	Substituents X	Y	$^1J_{Csp^2-Csp^2}$	Ref.

F	CH_3	81.6^a	205	OCH_3	H	74.8	82
F	C_2H_5	80.8	205	Cl	H	74.3	76,82
F	$(CH)_3C$	80.6	205	OH	H	71.9	14
OCH_3	CH_3	74.6	204	ONa	H	65.9	82
OH	CH_3	72.0	204	C_6H_5	H	54.8	14
OH	$(CH_3)_3C$	72.0	205	H	H	53.2	14
OH	C_2H_5	71.9	205	CH_3	H	52.5	14

YCH=CH-COX

OCH_3	C_6H_5	76.5 trans	15	OH	C_6H_5	73.9 trans	15
		75.5 cis				73.5 cis	
OH	H	70.4	14	CH_3	H	51.3	15

a) meta derivative

TABLE 29

$^1J_{CC}$ coupling constants (in Hz) across Csp^2-Csp bonds.

Compound	$^1J_{Csp^2-Csp}$	Ref.	Compound	$^1J_{Csp^2-Csp}$	Ref.
$XC{\equiv}C-C_6H_5$			$C_6H_5C{\equiv}C-COOCH_3$	127.7	206
X = $N(C_2H_5)_2$	95.8	66	$C_6H_5C{\equiv}C-COOH$	119.3	206
Cl	93.6	56	$CH_3C{\equiv}C-COOCH_3$	127.5	13
Br	92.0	56	$C_6H_{13}C{\equiv}C-COOCH_3$	126.2	206
SCH_3	91.2	131	$C_6H_{13}C{\equiv}C-COOH$	123.0	206
CH_3	91.2	a	C_6H_5CN	80.4	82
C_6H_5	91.1	14			
C_2H_5	90.9	a			
SeC_2H_5	89.7	131			
I	89.5	56			
$P(C_4H_9)_2$	87.4	66			
$TeCH_3$	87.2	131			
$P(C_6H_5)_2$	86.8	66			
$Si(CH_3)_3$	84.5	a			

a) K. Kamieńska-Trela and Z. Biedrzycka, unpublished data

substituents, such as halogens and chalcogens,[56,131] considered separately. Mesomeric effects have been invoked to interpret the data for benzoyl cations.[171,172]

In an interesting paper[207] a CC coupling constant of ca. 20 Hz has been observed in an adduct of rhodium octaethyl porphyrin dimer and two CO units, which provided clear evidence of C-C formation. The relatively small value of this coupling constant, combined with the large value of $^2J_{CRh}$, has been explained in terms of a multicentered interaction between the (OEP)Rh and the C_2O_2 unit.

Steric effects play an important role in this group of couplings . A stereospecifity of couplings across axial and equatorial CC bonds, analogous to that in 1,4-dimethylcyclohexane, ($^1J_{CC(eq)}$ > $^1J_{CC}$ (ax)) has been observed by Barna and Robinson,[190] in 4-tert-butyl-2-methylpiperidines and 4-tert-butyl-2-methyl cyclohexanones, but it was attributed to the proximity of non-bonding electrons or a double bond.

A relationship between $^1J_{C1C2}$ and configuration of substituents has been observed in aldofuranose anomers.[198] $^1J_{C1C2}$ couplings in compounds with *trans*-O1/O2 are consistently greater (46.2±0.7 Hz) Hz) than in those with *cis* O1/O2 (42.7 ± 0.7 Hz). This correlation can be considered as a reliable indication in structural studies on furanoses,[197-199] but is not generally valid and cannot be applied authomatically to other saccharides. Thus, in a marked contrast to the behaviour of $^1J_{C1,C2}$ in furanoses, this coupling is not a reliable probe for anomeric configuration in aldopyranoses.[198]

α-furanose   β-furanose

D-ribose

A dependence on configuration has been observed for $^1J_{CC}$ in a series of mono- and di-substituted cyclopropanes, where greater $^1J_{CC}$ values have been found in *trans* substituted compounds.[208]

$^1J_{CC}$ couplings have been determined by Wray and Ernst[209] in a large series of *syn* and *anti* oximes, and invariably $^1J_{CC}(anti)$ greater than $^1J_{CC}(syn)$ has been found. Further examples have been published by Krivdin *et al.*,[210,211] who attributed the observed

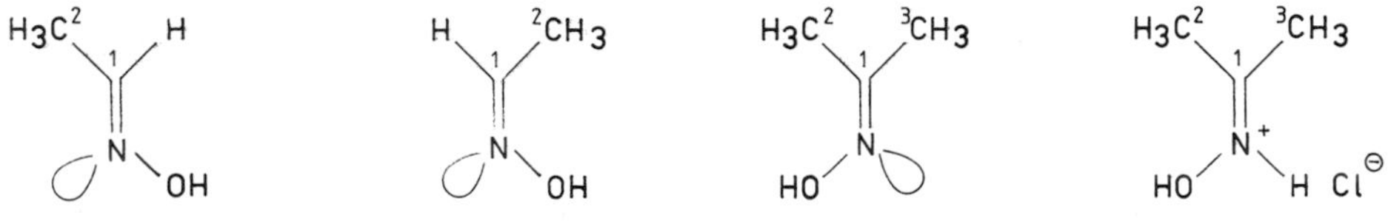

^1J	C1C2	C1C2	C1C2	C1C2
	48.4 Hz	40.5 Hz	41.4 Hz	42.3 Hz
			C1C3	C1C3
			49.3 Hz	42.5 Hz
ref.	209	209	209	212

difference to the interaction between the CC bond involved and a lone electron pair on nitrogen. The conclusion has been corroborated by the results obtained for protonated acetoxime, where very similar $^1J_{CC}$ values have been found for the couplings across both *anti* and *syn* bonds.[212] The correlation seems to be of a general character and can be applied as a valuable tool in the determination of configuration of imines.[213]

V. EMPIRICAL CORRELATIONS

Additivity is one of the most widely applied and fruitful concepts in physical organic chemistry. It allows one to rationalize substitutent effects and to estimate unknown values of the parameters involved. It has been tacitly assumed that the additivity approach should be valid also in the case of one-bond CC coupling constants. In particular, Bauer *et al.*,[147] analysed $^1J_{CC}$ couplings in various chloro substituted compounds and estimated the Cl-substituents increments for Csp^3-Csp^3, Csp^3-Csp^2 and Csp^2-Csp^2 (double) bonds. A similar analysis was carried out by Krivdin *et al.*,[132] for $^1J_{CC}$ couplings across triple bonds. However, the differences between the estimated J values and those determined experimentally are rather large (within 1.6 to 14.9 Hz) for most typical substituents, such as halogens and silyl and stannyl groups. Although the authors consider the observed deviations to be "the typical ones" for a given substituent, the validity of the additive scheme does not seem unquestionable in the case of one-bond CC couplings. An alternative, multiplicative approach has been proposed by Egli and Philipsborn,[214] who considered $^1J_{CC}$ as the product of two empirical factors, I(C1) and I(C2). These factors can be derived either as the square roots of $^1J_{CC}$'s in symmetrically substituted molecules (key factors) or from unsymmetrical molecules of any kind by dividing the experimental CC coupling constant by the corresponding key factor. In other words:

$$^1J_{C1C2} = I(C1)I(C2) \qquad (8)$$

The following example illustrates this approach. The $^1J_{CC}$

couplings in $HC\equiv CH$, $Et_3SiC\equiv CSiEt_3$ and $Bu_3SnC\equiv CSnBu_3$ are 171.5 Hz[1], 101.4 Hz and 81.0 Hz[125], respectively, and this yields $I(CH\equiv)$ = 13.09 $Hz^{1/2}$, $I(Et_3SiC\equiv)$ = 10.07 $Hz^{1/2}$ and $I(Bu_3SnC\equiv)$ = 9.0 $Hz^{1/2}$. This leads to $^1J_{CC}$ couplings of 131.8 Hz in $Et_3SiC\equiv CH$ and of 117.8 Hz in $Bu_3SnC\equiv CH$, which is in a good agreement with the experimental $^1J_{CC}$,s of 130.9 and 119.8 Hz,[125] respectively. Further examples can be found in papers by Egli and Philipsborn[214] and others.[56]

In the case of ethane and ethene derivatives, a further step can be made, and each of the I factors can in turn be separated into two components, according to the equations:

$$I(XYZC) = i_x + i_y + i_z \text{ (in ethane)} \tag{9}$$

and
$$I(XYC) = i_x + i_y \qquad \text{(in ethene),} \tag{10}$$

where i_x, i_y and i_z are associated with substituents X, Y and Z, respectively. An analogous additivity scheme was already applied by Malinowski for predicting carbon-hydrogen couplings.[142] A limitation of the proposed scheme lies in the fact that it does not account for any steric interactions between substituents. The paucity of the relevant data does not allow one to perform a full analysis of the problem. Nevertheless, a combination of additive and multiplicative approaches seems to be more proper than a simple additive scheme in the case of one-bond CC couplings .

As was shown in the previous sections of the present chapter, one-bond CC coupling constants depend almost always on the electronegativities of substitutents. The same has been observed for various carbon-heteroatom couplings. It is therefore not surprising that linear correlations have been established for couplings between geometrically equivalent CC and CX (X = H, CN) pairs of atoms. Thus, Weigert and Roberts[91] found that $^1J_{CC}$ couplings are linearly related to $^1J_{CH}$. The equations relating $^1J_{CC}$ to $^1J_{CH}$, and to $^2J_{CH}$ were obtained by Kalabin et al.,[130] for derivatives of acetylene:

$$^1J_{C\equiv C} = (2.87 \pm 0.15)\ ^1J_{CH} - (545.6 \pm 37.1) \tag{11}$$
$$n = 16, \quad \text{standard deviation} = 4.37\ Hz, \quad r = 0.98$$

$$^1J_{C\equiv C} = (5.67 \pm 0.25)\,^2J_{CH} - (109.6 \pm 12.0) \tag{12}$$
$$n = 18, \quad \text{standard deviation} = 3.70 \text{ Hz}, \quad r = 0.99$$

Runge and Firl[217] have compared the reduced coupling constants $^1K_{CC}$ with $^1K_{CN}$ in structurally related compounds:

$$^1K(^{15}N^{13}C) = 0.51\,^1K(^{13}C^{13}C) - 3.19 \times 10^{20} \tag{13}$$

The ^{13}C-^{29}Si couplings over one, two, and three bonds have been found to show the same magnitude changes as those observed for the corresponding CC couplings.[5,218]

A large set of equations relating various types of $^1J_{CC}$'s have been derived by Krivdin et al.,[89,166,132,169,219] These include correlations of $^1J_{CC}$ couplings across double[89], aromatic[169] and triple[132] CC bonds with the CC couplings in all possible remaining systems. A few examples are given below:

$$^1J_{1,2}(C_6H_5) = (2.14 \pm 0.14)\,^1J_{CC}(CH_3-CH_2) - (16.8 \pm 5.2) \tag{14}$$
$$(n=17, \text{ standard deviation} = 1.2 \text{ Hz}, \text{ r=0.968})$$

$$^1J_{1,2}(C_6H_5) = (0.91 \pm 0.03)\,^2J_{CC}(CH_2=CH) - (5.7 \pm 1.9) \tag{15}$$
$$(n=30, \text{ standard deviation} = 0.8 \text{ Hz}, \text{ r=0.988})$$

$$^1J_{CC}(H_2C=CHX) = (0.20 \pm 0.02)\,^1J(C\equiv CX) + (33.64 \pm 3.24) \tag{16}$$
$$(n=7, \text{ standard deviation} = 1.6 \text{ Hz}, \text{ r=0.969})$$

In most of the remaining equations of this type the correlations are even worse.

A correlation between one-bond CC couplings and the sum of the chemical shifts of the spin-spin coupled carbons was observed by Dhavan and Goux[200] in perbenzylated methyl furanosides.

In several studies the relationship between $^1J_{CC}$ and bond length, r_{CC}, was analysed.[80,120,220]

Equations relating $^1J_{CC}$ vs. stretching force constants, K_{CC}, were derived for single, double and triple CC bonds.[111,221]

$$K_{CC} = 0.0344 \ ^1J_{CC} + 3.25 \ (\text{single CC bonds}) \qquad (17)$$

$$K_{CC} = 0.180 \ ^1J_{CC} - 3.25 \ (\text{double CC bonds}) \qquad (18)$$

$$K_{CC} = 0.0295 \ ^1J_{CC} + 11.04 \ (\text{triple CC bonds}) \qquad (19)$$

A relatively small number of papers have been devoted to the influence of pH upon one-bond CC couplings. The first systematic studies were performed by Berger[181] who measured $^1J_{CC}$'s in *ortho-*, *meta-* and *para*-aminobenzoic acids as a function of pH. The spin-spin coupling constants determined for protonated and deprotonated species were plotted against the electron charge at the carboxyl carbon atom calculated using INDO method. A very rough linear correlation was observed between the charge and $^1J_{C_{ipso}C(O)}$ coupling.

Few studies have been on $^1J_{CC}$ variations with pH in various aminoacids.[222-225]

VI EXPERIMENTAL METHODS

Spin-spin coupling between carbon nuclei can be observed only in the spectra of isotopomers containing at least a pair of carbon-13 atoms. Since the natural abundance of such isotopomers is very low (ca. 0.01% of total), the corresponding spectra appear as doublets of low intensity, located around the strong signals of non-coupled carbon atoms. The satellite lines are about 200 times weaker than the main signals of a ^{13}C spectrum. This combined with an unfavourable sensitivity of ^{13}C nucleus towards NMR detection caused a formidable challenge for ^{13}C-^{13}C spin-spin coupling studies, since for a long time an expensive and often difficult experimentally ^{13}C enrichment was an absolute requirement.

With the introduction of high-field magnets and pulsed Fourier transform (FT) technique, it became feasible to record ^{13}C-^{13}C satellite spectra at the natural abundance of ^{13}C isotope[3-5] (Fig. 4). An analysis of such a spectrum is straightforward since the satellites represent invariably AB spin systems, and the coupling constants can be extracted by inspection. The population of ^{13}C-^{13}C-^{13}C isotopomers (ABC or ABX spin systems) is a hundred times lower than that of doubly labelled ones and the corresponding satellite lines never appear in the spectra registered at the natural concentration of ^{13}C .

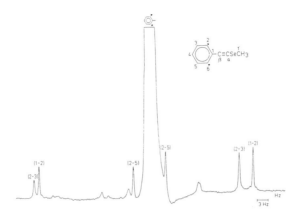

Fig.4 ^{13}C satellite spectra observed in the proton   decoupled ^{13}C FT NMR spectrum for C-2 (or C-6) of 1-methylseleno-2-phenylethyne. The difference between lines denoted 1-2 corresponds to $^{1}JC1C2$, 2-3 to $^{1}JC2C3$ and 2-5 to $^{2}JC2C5$.

Using the INADEQUATE (Incredible Natural Abundance Double QUAntum Transfer Experiment) pulse sequence, developed by Bax   and Freeman,[6-8] which uses double quantum coherence transitions it   is possible to suppress the signals corresponding to singly   labelled isotopomers and to record the satellite spectrum only.   The theoretical background of the method and of its experimental details is given   in a few   excellent   treatises[226-228]   and reviews,[25,229,230] and only some of the   most   important   features will be briefly discussed here.

The pulse sequence which generates the   suitable   double quantum-coherence between two   coupled   spins   in   the   INADEQUATE experiment is

$\pi/2$ - t_D - π -t_D - $\pi/2$, where

$t_D = (4J)^{-1}$.

J_{CC} values, in most common natural compounds such as   steroids, carbohydrates and antibiotics, are within a range of 30 to 80   Hz. However, it should be kept in mind before starting   an   experiment that the actual value of $^{1}J_{CC}$ may strongly deviate   from   "typical ones", depending upon hybridization, substitution and complexation (see sections III and IV of the present chapter).

Two examples illustrate the method. In Fig. 5, a spectrum of a mixture of *cis* and *trans* isomers of 1-bromopropene is shown. In this case, all satellite spectra represent AB spin systems only. In Fig. 6 a spectrum of 1,1-difluoro-2-phenylethene is displayed. In this case the carbon signals are additionally split owing to a coupling to two ^{19}F atoms (ABMX spin system).

In the two-dimensional version of the INADEQUATE method,[231,232] the coupled spins are identified by means of a different criterion – the frequency of the double-quantum coherence, which is equal to the sum of the chemical shifts of the carbon sites, measured with respect to the transmitter frequency. The projection of the contours of a two-dimensinal INADEQUATE spectrum onto the X-axis of a coordinate system (F_2 domain) gives the chemical shifts of the satellite signals, that onto the y-axis (F_1 domain) the double-quantum frequencies of the different AX spin system. The advantage over a one-dimensional spectrum is that no overlap occurs, even when $^1J_{CC}$ values are equal, since the double doublets of each of the different AX spin systems appear in different rows. Owing to the straightforward interpretation of 2D INADEQUATE spectra, this is in general the preferred technique if the full connectivity information is required.

The INADEQUATE technique does not provide sensitivity enhancement in comparison with conventional pulse sequences. Actually, rather large samples and high molar concentrations are required in order to keep experiments within reasonable time limits. It is therefore not surprising that numerous variants of the original INADEQUATE method have been designed with a goal to enhance the sensitivity of the method.

Just a simple combination of the one-dimensional INEPT (*IN*sensitive *E*nhancement by *P*olarization *T*ransfer) technique with the INADEQUATE method[233] provides a rather small (by a factor of 4/3) but useful intensity increase for the signals of protonated carbons. A similar result is obtained if INADEQUATE is combined with DEPT (*D*istortionless *E*nhancement by *P*olarization *T*ransfer).[234] Further improvements based on 2D-versions of the HCC

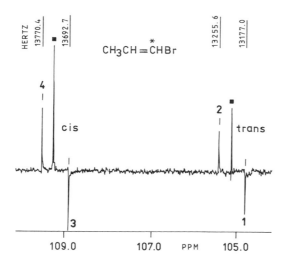

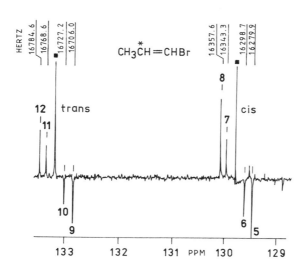

Fig. 5 ^{13}C satellite spectra observed in the proton-decoupled INADEQUATE ^{13}C FT NMR spectrum of a mixture of cis and trans isomers of 1-bromopropene. The difference between lines 1,2 and 9,12 corresponds to $^{1}J_{C=C}$ in trans isomer; 3,4 and 5,8 to $^{1}J_{C=C}$ in cis one; 6,7 to $^{1}J_{CCH_3}$ in cis and 10,11 to $^{1}J_{CCH_3}$ in trans isomers, respectively. Not suppressed to the end peaks of mono-^{13}C-isotopomers are marked with ∎

348

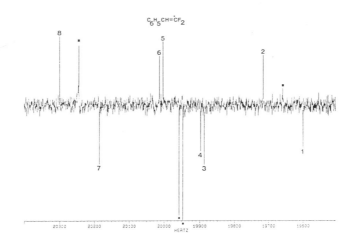

Fig. 6 The part of ^{13}C NMR proton decoupled INADEQUATE spectrum
of 1,1-difluoro-2-phenylethene corresponding to the
carbon **1**; the difference between lines 1,2; 3,5; 4,6 and
7,8 corresponds to $^1J_{C=C}$. Not suppressed to the end peaks
of mono-^{13}C-isotopomer are marked with ■ . Published with
permission.[65]

relay experiments have been proposed by Kessler *et al.*,[235] and
Morris and co-workers.[236]

    If the resolution of a 2D-INADEQUATE spectrum is insufficient
then the SEMINA-1 and SEMINA-2 (*S*ubspectral *E*diting using a
*M*ultiple quantum trap for *INA*dequate) techniques can be
applied.The SEMINA method divides the spectra according to the sum
of the H atoms in the CH$_n$-CH$_m$ fragment: all satellite signals for
which the sum n+m is even appear in one spectrum, while those for
which the sum is odd appear in a second spectrum.[237-239]

    In overcrowded spectra, some particular multiplets (usually AX
spin systems) can be selected by means of the method called

DOUBTFUL (*DOU*ble quantum *T*ransition for *F*inding *U*nresolved *L*ines). The method published by Kaptein et al.,[240] was designed originally with a goal to analyse complex ^1H NMR spectra of large molecules.[241,242] Furthermore, it has been adapted to the analysis of ^{13}C NMR spectra and actually applied to those of alkaloids.[243]

The SLAP pulse sequence (*S*ign *L*abelled *P*olarization Transfer) published by Sorensen and Ernst,[244] which is based on the proton-to-carbon polarization transfer and the creation of ^{13}C zero and double-quantum coherences allows one to establish the sign relations between $^1J_{CC}$, $^2J_{CC}$ and $^3J_{CC}$ couplings for a given molecule.[244,245]

In a further attempt to make the technique more sensitive, a generally applicable frequency-selective version of INADEQUATE (SELINQUATE) which makes 2D recordings superfluous, was designed,[246] and a combination of the SEMUT and DEPT spectral editing techniques with the INADEQUATE and SLAP techniques has been proposed.[247]

The detection of CC connectivities in the solid state is of great interest though it is a rather difficult task. This can be accomplished *via* the observation of ^{13}C NMR spectra of solids rotated at the magic angle. The spectra exhibit intensity and line width contributions due to homonuclear dipolar couplings. Such residual direct dipolar couplings, those across ^{13}C-^{13}C bonds, can be exploited in a double-quantum experiment in order to establish carbon-carbon connectivities in the solid state. Such an experiment has been reported by Griffin and co-workers[248] for a solid sample of doubly labelled glycine at 90% of ^{13}C enrichment.

Very recently, however, the INADEQUATE method has also been applied to a solid sample of camphore whose cross-polarization (CP) -magic angle spinning (MAS) INADEQUATE spectrum was recorded by Benn et al.[249] This provided the first two-dimensional solid state NMR spectrum from which the CC connectivities have been derived using the homonuclear indirect CC scalar couplings.

CC and (CN) scalar couplings have been totally unexploited in the NMR spectroscocpy of large molecules until now. A few papers have been published recently where two dimensional ^{13}C (^{13}C) double quantum (DQC) correlation NMR experiments were developed and applied to proteins.[250,251] In this case, however, uniform or selective ^{13}C enrichment was necessary in order to increase sensitivity of experiments and to introduce a readily observable level of $^{13}C-^{13}C$ couplings.

VII.   ILLUSTRATIVE   EXAMPLES   OF   STRUCTURE   SOLVING   WITH ^{13}C ISOTOPOMERS

Structure elucidation (CC connectivity). One of the most useful applications of both one- and two-dimensional INADEQUATE experiments concerns the determination of carbon-carbon connectivities of the carbon network. The number of structures solved grows rapidly, including studies on new antibiotics,[252-256] steroid derivatives,[257] di-[258] and triterpenes,[259] sesquiterpenes,[260] lipids in the house fly,[261] silk fibroin,[262] tricyclic antidepressants,[263] poly(vinyl)alcohol,[264] two rotenoid analogs, boeravinone A and B,[265] isolated from the roots of *B. diffusa*, bharangin, a novel diterpenoidquinonemethide,[266] and the adducts of diazocyclopentadiene derivatives with dimethyl-acetylenedicarboxylate.[267]

The elucidation of the structure of a new alkaloid, amphimedine, isolated from an *Amphimedon sp.* of sponge collected at Guam Island provides an elegant example of the applicability of CC coupling constants to structural analysis.[268]

Its general formula, $C_{19}H_{11}N_3O_2$, was established using high-resolution mass spectroscopy, and the presence of two intense $\nu(CO)$ bands at 1690 and 1640 cm^{-1} in the infrared spectrum was attributed to α,β-unsaturated ketone and amide functionalities, respectively. A full analysis of ^1H and ^{13}C spectra gave information regarding the presence of the fragments shown below:

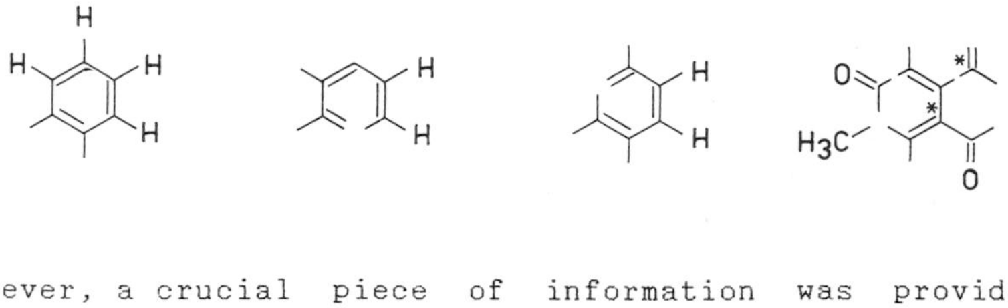

However, a crucial piece of information was provided by the two-dimensional INADEQUATE experiment. Because of the small amount of the compound available, not all of the CC couplings could be detected, but those observed allowed one to determine the connectivity of the most important fragments of the molecule. Thus one-bond connectivities were established between the following pairs of carbons: 1-2, 1-13a, 2-3, 3-4, 4a-4b, 4a-13a, 4b-5, 4b-12c, 5-6, 7a-12c, 8a-9, 12b-12c, 12c-7a, 12c-4b, 13a-1, 13a-4a. This yielded the final structure, **35**, which is in agreement with IR, and UV data as well as ^{1}H and ^{13}C chemical shifts and carbon-hydrogen spin-spin couplings.

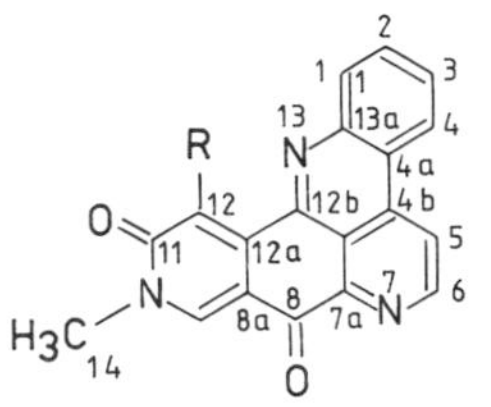

35

Biosynthetic studies. CC spin-spin coupling constants were widely applied in tracing down various courses of biological processes. Usually enriched samples were studied. The older results have already been reviewed by Wray,[14] Marshall,[19] Buddrus and Bauer,[25] and Krivdin and Kalabin[18] and only a few recent examples will be presented here. A novel diterpenoid aldehyde from the fungus *Cercospora travesiana*,[269] which is infecting fenugreek fields, was isolated and identified. Its structure was established *via* ^{1}H and ^{13}C NMR studies performed for the natural product and for the ^{13}C-labelled material obtained by [1,2-^{13}C$_2$]acetate incorporation experiments. Various methods including two-dimensional ^{13}C-^{13}C correlation, COSY, and INADEQUATE, provided the crucial information on the carbon-carbon connectivities, and led to the structure shown below:

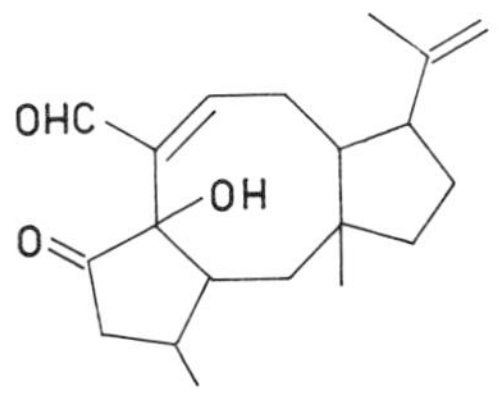

36

Further examples include the elucidation of the structure of tetracycline,[270] and that of the polyether ionophore antibiotic cationomycin which is produced by *Actinomodura azurea*.[271] The synthesis of isotopomers of L-tryptophan[272] *via* a combination of organic synthesis and biotechnology, and the biosynthesis of chromomycin-A_3,[273] which is a typical member of the aureolic acid family of antitumor antibiotics, were followed by means of CC couplings.

One-bond CC coupling constants as probes for homoaromatic structures. Homoaromatic structures were postulated for many carbocations, carbanions and neutral species, but they still remain a subject of controversy. CC coupling constants seem to be a promising tool in the analysis of such problems. They have been applied by Jonsaell and Ahlberg in studies on the structure of doubly ^{13}C-labelled 1,4-bishomotropylium cation,[86] and bicyclo[3.2.2]nonatrienyl anion.[274]

The doubly labelled ^{13}C-labelled alcohol, [2,3-[13]C_2]bicyclo[3.2.2]-nona-3,6,8-trien-2-ol **37a** and its 2-methyl derivative **37b** give under the influence of superacids, the doubly labelled 9-barbaryl cations, which in turn undergo further rapid rearrangements. Two structures have been proposed for the final products: bis-homotropylium ions **38a** and **38b**, and dihydroindenyl cation, **39**. The CC coupling constants measured for the final products (Table 30) were compared with the CC couplings

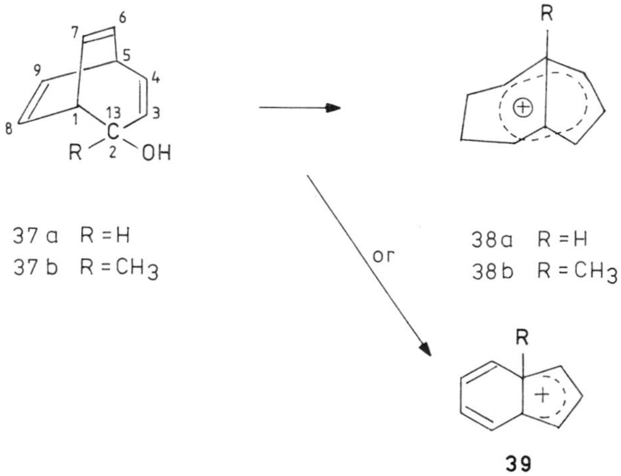

37 a R = H
37 b R = CH3

38 a R = H
38 b R = CH3

39

in the corresponding model compounds such as cyclohexadiene, methylcyclopropane, naphtalene, neopentane and others. In particular, the high value of $^1J(C1CH_3)$ (42 Hz) in 38b, and the rather low ones for those across the C1C2 and C1C9 bonds (28 Hz) were interpreted in terms of a substantial re-hybridization of the bridgehead atoms, that approaching the sp^2-hybridization

TABLE 30
CC coupling constants (in Hz) in 1,4-bis-homo-tropylium ion 38a and its 9-methyl derivative 38b.

carbons	$^1J_{CC}$	
coupled	38a	38b
C_1C_2	28.1	29.3
C_1C_9	28.7	29.3
C_2C_3	60.4	–
C_8C_9	51.9	–
C_1C_6	a	38.1
C_1CH_3	–	42

a) not observed

characteristic for the bonds in three membered rings. Also the $^1J(C5C4)$ of 60.4 Hz seems to be too low if the cation 39 were the final product ($^1J_{CC}$ of 68.0 Hz was measured in cyclohexadiene). It was therefore concluded by the authors that the homoaromatic bis-homotropylium cations 38a and 38b are better models for the cations studied, than cation 39.

In the case of bicyclo[3.2.2]nonatrienyl anion $(C_9H_9)^-$, the lack of an observable isotopic perturbation effect on the ^{13}C

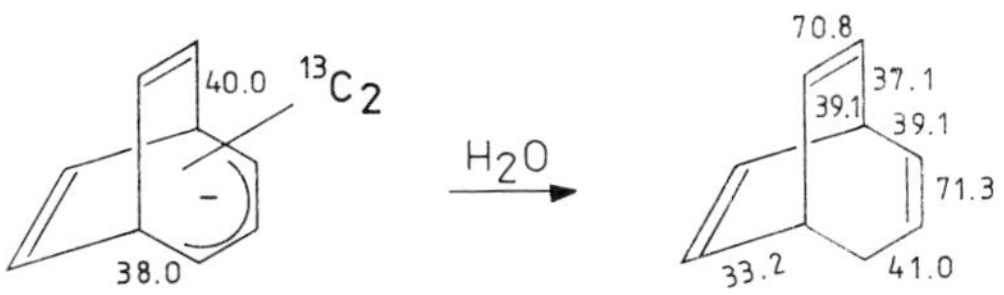

40   41

chemical shifts was used as an argument in the assignment of the symmetric structure **40** to $(C_9H_9)^-$. In the light of the CC-coupling measurements another possibility which assumes an equilibrium of the appropriate homoaromatic ions was rejected.

CC coupling constants as a probe for mesomeric and tautomeric structures. Vitamin C exists in solution dominantly as tautomer **42**, which upon a titration with NaOH forms the anion **43**, and then dianion **44**.[275] For **43** and **44**, mesomeric structures **43a/43b** and **44a/44b** can be written:

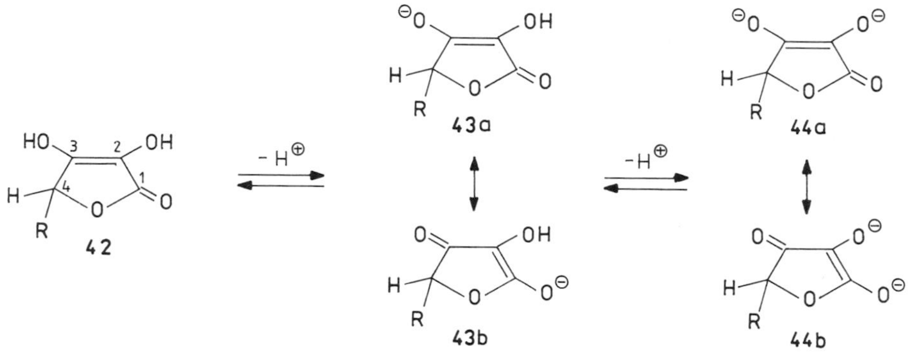

$^1J(C1C2)$ and $^1J(C2C3)$ couplings are highly sensitive to the amount of NaOH added to the initially neutral solution. They vary from 84.9 Hz and 91.8 Hz, respectively, in neutral solution, to 92 Hz and 84 Hz respectively, upon addition of 1 mole of NaOH, and to 85 Hz and 83.5 Hz when another mole of NaOH was added. This indicates that for the monoanion, **43a** rather than **43b** predominates, while for the dianion, **44a** and **44b** make roughly comparable contributions.

1-Phenylazo-2-naphtol and 2-phenylazo-1-naphthol are characteristic representatives of tautomeric *ortho*-hydroxyazo compounds. The ^{13}C chemical shifts of the phenyl ring carbons were assigned, but the assignment of the ring carbons of the naphthalene moiety was difficult because of the lack of suitable reference data and the lability of the equilibrium. Carbon-carbon coupling constants were measured for both compounds using the

SEMINA-1 pulse sequence.[239] These data provided an unambigous assignment of the ^{13}C chemical shifts in both compounds and were found to reflect clearly their tautomeric equilibria.

Further examples of applications of one-bond CC couplings involve the analysis of highly symmetric carbocyclic compounds,[276] of the structure of cyclopentadiene and its derivatives[277], and of endo-dicyclopentadiene.[278] Finally $^1J_{CC}$'s were used to verify spectral assignments and gain insight into details on structure of nortricyclylnorbornenyl cation[279] and 2-norbornyl-^{13}C$_2$ chloride.[280]

ACKNOWLEDGMENTS The author expresses her gratitude to Prof. M. Witanowski for a careful reading of the manuscript and valuable comments. This work was supported by the Polish Academy of Sciences under project CPBP 01.13.1.16.

REFERENCES
1 R. M. Lynden-Bell and N. Sheppard, Proc. R. Soc., London, Ser. A 269 (1962) 385.
2 D. M. Graham and C. E. Holloway, Can. J. Chem., 41 (1963) 2114.
3 P. S. Nielsen, R. S. Hansen and H. J. Jakobsen, J. Organometal. Chem., 114 (1976) 145.
4 H. J. Jakobsen, T. Lund and S. Sorensen, J. Magn. Reson., 33 (1979) 477.
5 V. Wray, L. Ernst, T. Lund and H. J. Jakobsen, J. Magn. Reson., 40 (1980) 55.
6 A. Bax, R. Freeman and S. P. Kempsell, J. Am. Chem. Soc., 102 (1980) 4849.
7 A. Bax, R. Freeman and S. P. Kempsell, J. Magn. Reson., 41 (1980) 349.
8 A. Bax and R. Freeman, J. Magn. Reson., 41 (1980) 507.
9 G. E. Maciel, in NMR Spectroscopy of Nuclei other than Protons, eds. T. Axenrod and G. A. Webb, Wiley-Interscience, New York (1974).
10 J.- R. Llinas, E.-J. Vincent and G. Peiffer, Bull. Soc. Chim. Fr., (1973) 3209.
11 J. B. Stothers, Carbon-13 NMR Spectroscopy, Academic Press, New York (1972).
12 V. F. Bystrov, in Progress in NMR Spectroscopy, 10 (1976) 1, ed. J. W. Emsley, J. Feeney and L. H. Sutcliffe.
13 R. E. Wasylishen, in Annual Reports on NMR Spectroscopy, 7 (1977) 245; ed. G. A. Webb.
14 V. Wray, in Progress in NMR Spectroscopy, 13 (1979) 177;

356

eds.J. W. Emsley, J. Feeney and L. H. Sutcliffe.

15 V. Wray and P. E. Hansen, in Annual Reports on NMR
   Spectroscopy 11A (1981) 99; ed. G. A. Webb.

16 P. E. Hansen, ibid., 11A (1981) 65.

17 P. E. Hansen and V. Wray, Org. Magn. Reson., 15 (1981) 102.

18 L. B. Krivdin and G. A. Kalabin, in Progress in NMR
   Spectroscopy, 21 (1989) 293; eds. J. W. Emsley, J. Feeney
   and L. H. Sutcliffe.

19 J. L. Marshall, Carbon-carbon and carbon-proton NMR couplings:
   Applications to Organic Stereochemistry and Conformational
   Analysis; ed. A. P Marchand, Deerfield Beach, Verlag
   Chemie Int. 1983.

20 H.- O. Kalinowski, S. Berger and S. Braun, ^{13}C-NMR
   Spectroskopie, Georg Thieme Verlag Stuttgart, New York, 1984.

21 E. Breitmaier and V. Voelter, "Carbon-13 NMR Spectroscopy",
   3rd edition, VCH Gesellschaft, 1987.

22 P. D. Ellis, R. Ditchfield, in Topics in Carbon-13 NMR
   Spectroscopy, Vol.2 (1976) 434; ed. G. C. Levy.

23 J. Kowalewski, in Progress in NMR Spectroscopy, 11 (1977) 1;
   eds. J. W. Emsley, J. Feeney and L. H. Sutcliffe.

24 J. Kowalewski, in Annual Reports on NMR Spectroscopy, 12
   (1982) 81; ed. G. A. Webb.

25 J. Buddrus and H. Bauer, Angew. Chem.,Int. Ed. Engl., 99
   (1987) 625.

26 N. F. Ramsey and E. M. Purcell, Phys. Rev., 85 (1952) 143.

27 N. F. Ramsey, Phys. Rev., 91 (1953) 303.

28 H. M. McConnell, J. Chem. Phys., 23 (1955) 760, 2454.

29 H. M. McConnell, J. Chem. Phys., 24 (1956) 460.

30 J. A. Pople and D. P. Santry, Mol. Phys., 8 (1964) 1.

31 J. A. Pople, J. W. McIver, Jr. and N. S. Ostlund,
   Chem. Phys. Lett., 1 (1967) 465.

32 J. A. Pople, J. W. McIver, Jr., and N. S. Ostlund,
   J. Chem. Phys., 49 (1968) 2960, 2965.

33 A. C. Blizzard and D. P. Santry, Chem. Commun., (1970) 87.

34 A. C. Blizzard and D. P. Santry, J. Chem. Phys., 55 (1971)
   950; ibid., 58 (1973) 4714.

35 G. E. Maciel, J. W. McIver, Jr., N. S. Ostlund and
   J. A. Pople, J. Am. Chem. Soc., 92 (1970) 11.

36 V. J. Bartuska and G. E. Maciel, J. Magn. Reson., 5 (1971) 211.

37 V. J. Bartuska and G. E. Maciel., J. Magn. Reson., 7 (1972) 36.

38 K. D. Summerhays and G. E. Maciel, J. Am. Chem. Soc.,
   94 (1972) 8348.

39 C. Barbier, H. Faucher and G. Berthier, Theor. Chim. Acta,
   21 (1971) 105.

40 J. M Schulman and M. D. Newton, J. Am. Chem. Soc.,
   96 (1974) 6295.

41 M. D. Newton, J. M. Schulman and M. M. Manus,
   J. Am. Chem. Soc., 96 (1974) 17.

42 P. Lazzeretti, F. Taddei and R. Zanasi, J. Am. Chem. Soc.,
   98 (1976) 7989.
43 J. M. Schulman and T. J.Venanzi, Tetrahedron Lett.,
   18 (1976) 1461.
44 K. Hirao and H. Kato, Bull. Chem. Soc. (Japan) 50 (1977) 303.
45 J. M. Andre, J. B. Nagy, E. G. Derouane, J. G. Fripiat and
   D. P. Vercauteren, J. Magn. Reson., 26 (1977) 317.
46 S. A. T. Long and J. D. Memory, J. Magn. Reson.,
   29 (1978) 119.
47 W. S. Lee and J. M. Schulman, J. Am. Chem. Soc.,
   101 (1979) 3182.
48 W. S. Lee and J. M. Schulman, J. Magn. Reson., 35 (1979) 451.
49 T. Khin and G. A. Webb, Org. Magn. Reson., 12 (1979) 103.
50 C. Van Alsenoy, H. P. Figeys and P. Geerlings,
   Theor. Chim. Acta, 55 (1980) 87.
51 K. Kamieńska-Trela and B. Knieriem, J. Organometal. Chem.,
   198 (1980) 25.
52 M. Stocker, Org. Magn. Reson., 16 (1981) 319.
53 M. Stocker, Acta Chem. Scand., 37B (1983) 166.
54 A. Laaksonen, J. Kowalewski and V. R. Saunders, Chem. Phys.,
   80 (1983) 221.
55 F. A. A. M. de Leeuw, C. A. G. Haasnoot and C. Altona,
   J. Am. Chem. Soc., 106 (1984) 2299.
56 K. Kamieńska-Trela, Z. Biedrzycka, R. Machinek, B. Knieriem
   and W. Luttke, Org. Magn. Reson., 22 (1984) 317.
57 J. C. Facelli and M. Barfield, J. Am. Chem. Soc.,
   106 (1984) 3407.
58 K. Kamieńska-Trela and P. Gluziński, Croatica Chem. Acta,
   59 (1986) 883.
59 G. E. Scuseria, Chem. Phys. Lett., 127 (1986) 236.
60 J. Geersten and J. Oddershede, Chem. Phys., 104 (1986) 67.
61 V. Galasso and G. Fronzoni, J. Chem. Phys., 84 (1986) 3215.
62 V. Galasso, Chem. Phys., 117 (1987) 415.
63 J. Geersten, Chem. Phys. Lett., 134 (1987) 400.
64 G. Fronzoni and V. Galasso, J. Magn. Reson., 71 (1987) 229.
65 K. Kamieńska-Trela and Z. Biedrzycka, Bull. Pol. Acad. Sci.,
   Chem., 36 (1988) 285; K. Kamieńska-Trela, P. Gluziński and
   B. Knieriem, ibid., to be published.
66 K. Kamieńska-Trela, Z. Biedrzycka, R. Machinek and W. Luttke,
   Bull. Pol. Acad. Sci., Chem., 36 (1988) 105.
67 H. Fukui, T. Tsuji and K. Miura, J. Am. Chem. Soc., 103
   (1981) 3652.
68 H. Fukui, K. Miura. K. Ohta and T. Tsuji, J. Chem. Phys.,
   76 (1982) 5169.
69 A. R. Engelmann, G. E. Scuseria and R. H. Contreras,
   J. Magn. Reson., 50 (1982) 21.
70 M. F. Tufro and R. H. Contreras, Z. Phys. Chem., Leipzig,
   267 (1986) 873.
71 H. Fukui, K. Miura and T. Sakurai, J. Chem. Phys.,

88 (1988) 7040.

72 P. Pyykko and L. Wiesenfeld, Mol. Phys., 43 (1981) 557.

73 M. Pomerantz, R. Fink and G. A. Gray, J. Am. Chem. Soc., 98 (1976) 291.

74 H. Finkelmeier and W. Luttke, J. Am. Chem. Soc., 100 (1978) 6261.

75 M. Barfield, I. Burfitt and D. Doddrell, J. Am Chem. Soc., 97 (1975) 2631.

76 K. Frei and H. J. Bernstein, J. Chem. Phys., 38 (1963) 1216.

77 Z. B. Maksic, M. Eckert-Maksic and M. Randic, Theor. Chim. Acta, (Berl.) 22 (1971) 70.

78 J. Wardeiner, W. Luttke, R. Bergholz and R. Machinek, Angew. Chem., Int. Ed. Engl., 21 (1982) 872.

79 L. B. Krivdin, S. V. Zinchenko. V. V. Shcherbakov, S. N. Elovskii and G. A. Kalabin, Zh. Org. Khim., 23 (1987) 1420.

80 H. Gunther and W. Herrig, Chem. Ber., 106 (1973) 3938.

81 R. D. Bertrand, D. M. Grant, E. L. Allred, J. C. Hinshaw and A. B. Strong, J. Am. Chem. Soc., 94 (1972) 997.

82 A. M. Ihrig and J. L. Marshall, J. Am. Chem. Soc., 94 (1972) 1756.

83 G. W. Buchanan. J. Selwyn and B. A. Dawson, Can. J. Chem., 57 (1979) 3028.

84 G. A. Olah and P. W. Westerman, J. Am. Chem. Soc., 96 (1974) 2229.

85 R. P. Kirchen and T. S. Sorensen, J. Am. Chem. Soc., 99 (1977) 6687.

86 G. Jonsaell and P. Ahlberg, J. Am. Chem. Soc., 108 (1986) 3819.

87 G. Becher, W. Luttke and G. Schrumpf, Angew. Chem., 85 (1973) 357.

88 W. Luttke and R. Machinek, personal communication.

89 L. B. Krivdin, V.V. Shcherbakov and G. A. Kalabin, Zh. Org. Khim., 23 (1987) 2070.

90 G. Gray, Ph. D. Thesis, University of California, Davis, California (1967) quoted in ref. 36.

91 F. J. Weigert and J. D. Roberts, J. Am. Chem. Soc., 94 (1972) 6021.

92 K. Kamieńska-Trela, Org. Magn. Reson., 14 (1980) 398.

93 N. Muller and D. E. Pritchard, J. Chem. Phys., 31 (1959) 1471.

94 W. Luttke, G. Becker, R. Machinek, J. Wardeiner and R. B. Bergolz, private communication cited in H. Egli and W. von Philipsborn, Tetrahedron Lett., (1979) 4265.

95 E. W. Della and P. E. Pigou, J. Am. Chem. Soc., 106 (1984) 1085.

96 R. Hubers, M. Klessinger and K. Wilhelm, Org. Magn. Reson., 24 (1986) 1016.

97 R. Benn, H. Butenschon, R. Mynott and W. Wisniewski, Magn. Reson. Chem., 25 (1987) 653.

98 M. Stocker, Org. Magn. Reson. 20 (1982) 175.

99 A. D. Walsh, Trans. Faraday Soc., 45 (1949) 179.

100 A. D. Walsh, Nature (London) 159 (1947) 165, 712.

101 J. L. Marshall, L. G. Faehl and R. Kattner, Org. Magn. Reson., 12 (1979) 163.

102 M. Klessinger, H. van Megen and K. Wilhelm, Chem. Ber., 115 (1982) 50.

103 M. Barfield, E. W. Della and P. E. Pigou, J. Am. Chem. Soc., 106 (1984) 5051.

104 A. G. Swanson, Tetrahedron Lett., 24 (1983) 1833.

105 T. Loerzer, R. Machinek, W. Luttke, L. H. Franz, K.-D. Malsh and G. Maier, Angew. Chem., 95 (1983) 914.

106 M. Pomerantz and D. F. Hillenbrand, Tetrahedron, 31 (1975) 217.

107 M. Pomerantz and D. F. Hillenbrand, J. Am. Chem. Soc., 95 (1973) 5809.

108 V. V. Krishnamurthy, G. K. S. Prakash, P. S. Iyer and G. A. Olah, J. Am. Chem. Soc., 108 (1986) 1575.

109 P. E. Hansen, O. K. Poulsen and A. Berg, Org. Magn. Reson., 8 (1976) 632.

110 P. E. Hansen, O. K. Poulsen and A. Berg, Org. Magn. Reson., 12 (1979) 43.

111 K. Kamienska-Trela, J. Mol. Struct., 78 (1982) 121.

112 B. Wrackmeyer, J. Organometal. Chem., 166 (1979) 353.

113 M. Traetteberg, W. Luttke, R. Machinek, A. Krebs and H. J. Hohlt, J. Mol. Struct., 128 (1985) 217.

114 Z. Biedrzycka and K. Kamienska-Trela, Spectrochim. Acta, 42A (1986) 1323; and references therein.

115 P. Diehl, H. Bosiger and J. Jokisaari, Org. Magn. Reson., 12 (1979) 282.

116 S. Berger, Org. Magn. Reson., 22 (1984) 47.

117 S. Berger and K.-P. Zeller. J. Org. Chem., 49, (1984) 3725.

118 R. McWeeny, Coulson´s Valence, Oxford University Press, (1979) 163, 166.

119 P. E. Hansen, Org. Magn. Reson., 12 (1979) 109.

120 C. J. Unkefer, R. E. London, T. W. Whaley and G. H. Daub, J. Am. Chem. Soc., 105 (1983) 733.

121 J. B. Pawliczek and H. Gunther, Tetrahedron, 26 (1970) 1755.

122 H. Booth and J. R. Everett, Can. J. Chem., 58 (1980) 2709.

123 H. Booth and P. R. Thornburrow, Chem. Ind., (1968) 685.

124 V. A. Chertkov and N. M. Sergeyev, J. Am. Chem. Soc., 99 (1977) 6750.

125 K. Kamienska-Trela, J. Organometal. Chem., 159 (1978) 15.

126 A. Sebald and B. Wrackmeyer, Spectrochim. Acta, 37A (1981) 365.

127 A. Sebald and B. Wrackmeyer, Spectrochim. Acta, 38A (1982) 163.

128 B. Wrackmeyer, Spectr. Int. J., 1 (1982) 201.

129 A. Sebald, B. Wrackmeyer and W. Beck, Z. Naturforsch.,

38b (1983) 45.

130 G. A. Kalabin, L.B. Krivdin, A. G. Projdakov and
D. F. Kushnarev, Zh. Org. Khim., 19 (1983) 476.

131 K. Kamieńska-Trela, Z. Biedrzycka, R. Machinek, B. Knieriem
and W. Luttke, J. Organometal. Chem., 314 (1986) 53.

132 L. B. Krivdin. A. G. Projdakov, B. N. Bazhenov, S. V.
Zinchenko and G. A. Kalabin, Zh. Org. Khim., 24 (1988) 1595.

133 L. B. Krivdin, V. V. Shcherbakov, A.I. Gritsa, A. G. Malkina
and Iu. M. Skvortsov, Izv. AN SSSR, Ser. Khim., (1988) 1287.

134 F. Holzl and B. Wrackmeyer, J. Organometal. Chem.,
179 (1979) 397.

135 K. Kamieńska-Trela, H. Ilcewicz, H. Barańska and
A. Łabudzińska, Bull. Pol. Acad., Sci., Chem., 32 (1984) 143.

136 S. Aime, D. Osella, E. Giamello and G. Granozzi,
J. Organometal. Chem., 262 (1984) C1.

137 M. H. Chisholm, K. Folting, D. M. Hoffman and J. C. Huffman,
J. Am. Chem. Soc., 106 (1984) 6794.

138 B. Wrackmeyer, Z. Naturforsch., 43b (1988) 923.

139 A. D. Walsh, Discuss. Faraday Soc., 2 (1947) 18.

140 H. A. Bent, Chem. Rev., 61 (1961) 275.

141 C. Juan and H.S. Gutowsky, J. Chem. Phys., 37 (1962) 2198.

142 E. R. Malinowski, J. Am. Chem. Soc., 83 (1961) 4479.

143 G. E. Maciel, P. D. Ellis, J. J. Natterstad and
G. B. Savitsky, J. Magn. Reson., 1 (1969) 589.

144 D. Ziessow and E. Lippert, Ber. Bunsenges. Phys. Chem.,
74 (1970) 568.

145 D. Seebach, H. Siegiel, J. Gabriel and R. Hassig,
Helv. Chim. Acta, 63 (1980) 2046.

146 L. B. Krivdin, D. F. Kushnarev, G. A. Kalabin and
A. G. Proidakov, Zh. Org. Khim., 20 (1984) 949.

147 H. Bauer, J. Buddrus, W. Auf der Heyde and W. Kimpenhaus,
Chem. Ber., 119 (1986) 1890.

148 S. Berger, J. Magn. Reson., 66 (1986) 555.

149 L. B. Krivdin, V. V. Shcherbakov, V. M. Brzozowski and
G. A. Kalabin, Zh. Org. Khim., 22 (1986) 972.

150 K. Kamieńska-Trela, L. Kania, J. Sitkowski and E. Bednarek,
J. Organometal. Chem., 364 (1989) 29.

151 N. J. Koole, M. J. A. de Bie and P. E. Hansen,
Org. Magn. Reson., 22 (1984) 146.

152 N. C. Craig, L. G. Piper and V. L. Wheeler, J. Phys. Chem.,
75 (1971) 1453.

153 J. W. Fitch, E. B. Ripplinger, B. A. Shoulders and
S. D. Sorey, J. Organometal. Chem., 352 (1988) C25.

154 C. A. Tolman, A. D. English and L. E. Manzer, Inorg. Chem.,
14 (1975) 2353.

155 P. W. Jolly and R. Mynott, Adv. Organometal. Chem.,
19 (1981) 257 and references cited therein.

156 R. Benn and A. Rufińska, J. Organometal. Chem.,
238 (1982) C27.

157 S. Zobl-Ruh and W. von Philipsborn, Helv. Chim. Acta,
    64 (1981) 2378.
158 R. Benn and A. Rufińska, J. Organometal. Chem.,
    323 (1987) 305.
159 H. Yamamoto, H. Yasuda, K. Tatsumi, K. Lee, A. Nakamura,
    J. Chen, Y. Kai and N. Kasai, Organometallics, 8 (1989) 105.
160 R. Benn and A. Rufińska, Organometallics, 4 (1985) 209.
161 M. A. Fox and D. Shultz, J. Org. Chem., 53 (1988) 4386.
162 L. B. Krivdin, V. V. Shcherbakov, A. G. Projdakov,
    G. A. Kalabin, B. A. Trofimov, O. A. Tarasova and S. V.
    S. V.Amosova, Zh. Org. Khim., 24 (1988) 1023.
163 J. M. J. M. Bijen and J. L. Derissen, J. Mol. Struct.,
    14 (1972) 229.
164 P. C. Burgers, C. W. Worrell and M. P. Groenewege,
    Spectr. Lett., 13 (1980) 381.
165 S. Marriott, W. F. Reynolds, R. W. Taft and R. D. Topsom,
    J. Org. Chem., 49 (1984) 959.
166 G. A. Kalabin, L. B. Krivdin and B. A. Trofimov,
    Izv. AN SSSR, Ser. Khim., (1982) 113.
167 L. B. Krivdin and G. A. Kalabin, Zh. Org. Khim., 21 (1985) 521.
168 L. B. Krivdin, G. A. Kalabin, R. G. Mirskov and
    S. P. Solovieva, Izv. AN SSSR, Ser. Khim., (1982) 2038.
169 L. B. Krivdin and G. A. Kalabin, Zh. Org. Khim.,
    25 (1989) 690.
170 B. A. Trofimov, L. B. Krivdin, V. V. Shcherbakov and
    H. A. Aliev, Izv. AN SSSR, Ser. Khim., (1989) 64.
171 P. Sandor and J. Radics, Magn. Reson. Chem., 24 (1986) 607.
172 L. B. Krivdin, V. V. Shcherbakov, I. A. Aliev and
    G. A. Kalabin, Zh. Org. Khim., 23 (1987) 569.
173 V. V. Krishnamurthy, G. K. S. Prakash, P. S. Iyer and
    G. A. Olah, J. Am. Chem. Soc., 106 (1984) 7068.
174 G. A. Olah, P. S. Iyer, G. K. S. Prakash and
    V. V. Krishnamurthy, J. Am. Chem. Soc., 106 (1984) 7073.
175 P. S. Iyer, V. V. Krishnamurthy and G. A. Olah,
    J. Org. Chem., 50 (1985) 3059.
176 S. Berger and K.-P. Zeller, Org. Magn. Reson., 11 (1978) 303.
177 A. Iu. Denisov, V. I. Mamatiuk and O. P. Shkurko,
    Khim. Geterotsikl. Soed., (1985) 1383.
178 A. Iu. Denisov, V. I. Mamatiuk and O. P. Shkurko,
    Izv. AN SSSR, Ser. Khim., (1986) 2825.
179 L. B. Krivdin and G. A. Kalabin, Zh. Org. Khim.,
    24 (1988) 2268.
180 S. R. Maple and A. Allerhand, J. Am. Chem. Soc.,
    109 (1987) 56.
181 S. Berger, Tetrahedron, 42 (1986) 2055.
182 W. M. Litchman and D. M. Grant, J. Am. Chem. Soc.,
    89 (1967) 6775.
183 H. Siegel, K. Hiltbrunner and D. Seebach, Angew. Chem.,
    91 (1979) 845.

184 T. Spoormaker and M. J. A. de Bie, Rec. Trav. Chim. Pays-Bas, 98 (1979) 380.

185 L. B. Krivdin, V. V. Shcherbakov, N. G. Gluchich and G. A. Kalabin, Zh. Org. Khim., 24 (1988) 2276.

186 V. V. Krishnamurthy, P. S. Iyer and G. A. Olah, J. Org.Chem., 48 (1983) 3373.

187 V. V. Krishnamurthy, J. G. Shih and G. A. Olah, J. Org. Chem., 50 (1985) 1161.

188 D. Seebach, R. Hassig and J. Gabriel, Helv. Chim. Acta, 66 (1983) 308.

189 M. Barfield, J. C. Facelli, E. W. Della and P. E. Pigou, J. Magn. Reson., 59 (1984) 282.

190 J. C. J. Barna and M. J. T. Robinson, Tetrahedron Lett., (1979) 1459.

191 D. Seebach, J. Gabriel and R. Hassig, Helv. Chim. Acta, 67 (1984) 1083.

192 T. E. Walker, R. E. London, T. W. Whaley, R. Barker and N. A. Matwiyoff, J. Am. Chem. Soc., 98 (1976) 5807.

193 A. S. Serianni, E. L. Clark and R. Barker, Carbohydr. Res., 72 (1979) 79.

194 D. Gagnaire, H. Reutenauer and F. Taravel, Org. Magn. Reson., 12 (1979) 679.

195 A. S. Serianni, J. Pierce and R. Barker, Biochemistry, 18 (1979) 1192.

196 A. Neszmelyi and G. Lukacs, J. Am. Chem. Soc.,104 (1982) 5342.

197 A. S. Serianni and R. Barker, J. Org. Chem., 49 (1984) 3292.

198 M. J. King-Morris and A. S. Serianni, J. Am. Chem. Soc., 109 (1987) 3501.

199 J. R. Snyder and A. S. Serianni, Carbohydr. Res., 163 (1987) 169.

200 S. N. Dhawan and W. J. Goux, Carbohydr. Res., 183 (1988) 47.

201 J. R. Snyder, E. R. Johnston and A. S. Serianni, J. Am. Chem. Soc., 111 (1989) 2681.

202 G. A. Gray, P. D. Ellis, D. D. Traficante and G. E. Maciel, J. Magn. Reson., 1 (1969) 41.

203 T. Angelotti, M. Krisko, T. O'Connor and A. S. Serianni, J. Am. Chem. Soc., 109 (1987) 4464.

204 P. E. Hansen, O. K. Poulsen and A. Berg, Org. Magn. Reson., 11 (1977) 649.

205 P. E. Hansen, A. Berg and K. Schaumburg, Magn. Reson. Chem., 25 (1987) 508.

206 P. A. Chaloner, J. Chem. Soc., Perkin Trans., II (1980) 1028.

207 V. L. Coffin, W. Brennen and B. B. Wayland, J. Am. Chem. Soc., 110 (1988) 6063.

208 M. Stocker and M. Klessinger, Liebigs Ann. Chem., (1979) 1960.

209 V. Wray and L. Ernst, unpublished results in ref. 14.

210 L. B. Krivdin, G. A. Kalabin, R. N. Niestierenko and B.A. Trofimov, Zh. Org. Khim., 20 (1984) 2477.

211 G. A. Kalabin, L. B. Krivdin, V. V. Shcherbakov and

B. A. Trofimov, J. Mol. Struct., 143 (1986) 569.

212 S. V. Zinchenko, V. V. Shcherbakov, G. E. Salnikov,
L. B. Krivdin and G. A. Kalabin, Zh. Org. Khim.,
25 (1989) 684.

213 V. M. S. Gil and W. von Philipsborn, Magn. Reson. Chem.,
27 (1989) 409.

214. H. Egli and W. von Philipsborn, Tetrahedron Lett.,
(1979) 4265.

215 L. Lunazzi, D. Macciantelli and F. Taddei, Mol. Phys.,
19 (1970) 137.

216 R. M. Hammaker, J. Chem. Phys. 43 (1965) 1843.

217 W. Runge and J. Firl, Z. Naturforsch., 31b (1976) 1515.

218 B. Wrackmeyer and W. Biffar, Z. Naturforsch., 34b (1979) 1270.

219 L. B. Krivdin V. V. Shcherbakov and G. A. Kalabin,
Zh. Org. Khim., 22 (1986) 342.

220 V. Wray, A. Gossauer, B. Gruning, G. Reifenstahl and
H. Zilch, J. Chem. Soc., Perkin Trans II (1979) 1558.

221 K. Kamieńska-Trela, Spectrochim. Acta, 36A (1980) 239.

222 S. Fermandjian, S. Tran-Dinh, J. Savrda, E. Sala,
R. Mermet-Bouvier, E. Bricas and P. Fromageot,
Biochim. et Biophys. Acta, 399 (1975) 313.

223 R. E. London, T. E. Walker, V. H. Kollman and N. A. Matwiyoff,
J. Am. Chem. Soc., 100 (1978) 3723.

224 W. Haar, S. Fermandijan, J. Vicar, K. Blaha, and P. Fromageot,
Proc. Nat. Acad. Sci. USA, 72 (1975) 4948.

225 R. E. London, J. Chem. Soc., Chem. Comm., (1978) 1070.

226 J. K. M. Sanders and B. K. Hunter, Modern NMR Spectroscopy,
A Guide for Chemists, Oxford University Press, 1987.

227 G. E. Martin and A. S. Zektzer, Two-dimensional NMR methods
for establishing molecular connectivity, VCH Publishers.

228 A. Bax, Two-dimensional Nuclear Magnetic Resonance in Liquids
(Delft University Press), D. Reidel Dordrecht, 1982.

229 R. Benn and H. Gunther, Angew. Chem., 95 (1983) 381.

230 H. Kessler, M. Gehrke and Ch. Griesinger,
Angew. Chem. Int. Ed. Engl., 27 (1988) 490.

231 A. Bax, R. Freeman and T. A. Frenkiel, J. Am. Chem. Soc.,
103 (1981) 2102.

232 A. Bax, R. Freeman, T. A. Frenkiel and M. H. Levitt,
J. Magn. Reson., 43 (1981) 478.

233 O. W. Sorensen, R. Freeman, T. A. Frenkiel, T. H. Mareci and
R. Schuck, J. Magn. Reson., 46 (1982) 180.

234 S. W. Sparks and P. D. Ellis, J. Magn. Reson., 62 (1985) 1.

235 H. Kessler, W. Bermel and C. Griesinger, J. Magn. Reson.,
62 (1985) 573.

236 K. W. Lee and G. A. Morris, Magn. Reson.Chem., 25 (1987) 176.

237 O. W. Sorensen, U. B. Sorensen and H. J. Jakobsen,
J. Magn. Reson., 59 (1984) 332.

238 U. B. Sorensen, H. J. Jakobsen and O. W. Sorensen,
J. Magn. Reson., 61 (1985) 382.

239 P. E. Hansen and A. Lycka., Magn. Reson. Chem., 24 (1986) 772.

240 P. J. Hore, E. R. P. Zuiderweg, K. Nicolay, K. Dijkstra and
    R. Kaptein, J. Am. Chem. Soc., 104 (1982) 4286.

241 P. J. Hore, R. M. Scheck, A. Volbeda and R. Kaptein,
    J. Magn. Reson. 50 (1982) 328.

242 A. D. Bain, D. W. Hughes, J. M. Coddington and R. A. Bell,
    J. Magn. Reson., 58 (1984) 490.

243 S. R. Johns and R. I. Willing, Magn. Reson. Chem.,
    23 (1985) 16.

244 O. W. Sorensen and R. R. Ernst, J. Magn. Reson.,
    54 (1983) 122.

245 J. Lambert, K. Wilhelm and M. Klessinger, J. Magn. Reson.,
    63 (1985) 189.

246 S. Berger, Angew. Chem., Int. Ed. Engl., 27 (1988) 1196.

247 N. C. Nielsen, H. Bildsoe, H. J. Jakobsen and O. W. Sorensen,
    J. Magn. Reson., 78 (1988) 223.

248 E. M. Menger, S. Vega and R. G. Griffin, J. Am. Chem. Soc.,
    108 (1986) 2215.

249 R. Benn, H. Grondey, C. Brevard and A. Pagelot, J. Chem. Soc.,
    Chem. Commun., (1988) 102.

250 W. M. Westler, M. Kainosho, H. Nagao, N. Tomonaga and
    J. L. Markley, J. Am. Chem. Soc., 110 (1988) 4093.

251 B. J. Stockman, W. M. Westler, P. Darba and J. L. Markley,
    J. Am. Chem. Soc., 110 (1988) 4095.

252 J. M. Bulsing, E. D. Laue, F. J. Leeper, J. Staunton,
    D. H. Davies, G. A. F. Ritchie, A. Davies, A. B. Davies and
    R. P. Mabelis, J. Chem. Soc., Chem. Commun., (1984) 1301.

253 N. Otake, Nippon Kagaku Kaishi, (1986) 1579.

254 M. Kakushima, Y. Sawada, M. Nishio, T. Tsuno and T. Oki,
    J. Org. Chem., 54 (1989) 2536.

255 H. W. Fehlhaber, H. Kogler, T. Mukhopadhyay,
    E. K. S. Vijaykumar and B. N. Ganguli, J. Am. Chem. Soc.,
    110 (1988) 8242.

256 S. Rajan and G. W. Stockton, Magn. Reson. Chem.,
    27 (1989) 437.

257 A. Neszmelyi, W. E. Hull, G. Lukacs and W. Voelter,
    Z. Naturforsch., 41b (1986) 1178.

258 T.G. Dekker, T. G. Fourie, E. Matthee, F. O. Snyckers and
    W. Ammann, S. Afr. J. Chem., 40 (1987) 74.

259 M. Tori, R. Matsuda, M. Sono and Y. Asakawa,
    Magn. Reson. Chem., 26 (1988) 581.

260 R. Faure, E. M. Gaydou and O. Rakotonirainy, J. Chem. Soc.
    Perkin Trans, II (1987) 341.

261 P. P. Halarnkar, C. R. Heisler and G. J. Blomquist,
    Arch. Insect. Biochem. Physiol., 5 (1987) 189.

262 T. Asakura, A.Tateno, Y. Kawaguchi, K. Hamano and
    F. Mukaiyama, Nippon Sanshigaku Zasshi, 56 (1987) 38.

263 D. J. Craik, J. G. Hall and S. L. A. Munro,
    Chem. Pharm. Bull.,35 (1987) 188.

264 K. Hikichi and M. Yasuda, Polym. J. (Tokio)., 19 (1987) 1003.

265 S. Kadota, N. Lami, Y. Tezuka and T. Kikuchi,
     Chem. Pharm. Bull., 36 (1988) 834.

266 A. V. B. Sankaram, M. M. Murthi, K. Bhaskaraiah,
     G. L. Narasimha Rao, M. Subrahmanyam and J. N. Shoolery,
     Tetrahedron Lett., 29 (1988) 245.

267 S. Braun, V. Sturm and K. O. Runzheimer, Chem. Ber.,
     121 (1988) 1017.

268 F. J. Schmitz, S. K. Agarwal, S. P. Gunasekera, P. G. Schmidt
     and J. N. Shoolery, J. Am. Chem. Soc., 105 (1983) 4835.

269 A. Stoessl, G. L. Rock, J. B. Stothers and R. C. Zimmer,
     Can. J. Chem., 66 (1988) 1084.

270 I. K. Wang, L. C. Vining, J. A. Walter and A. G. McInnes,
     J. Antibiot., 39 (1986) 1281.

271 M. Ubukata, J. Uzawa and K. Isono, J. Am. Chem. Soc.,
     106 (1984) 2213.

272 E. M. M. van den Berg, A. U. Baldew, A. T. J. W. Goede,
     J. Raap and J. Lugtenburg, Rec. Trav. Chim. Pays-Bas,
     107 (1988) 73.

273 A. Montanari and J. P. N. Rosazza, Tetrahedron Lett.,
     29 (1988) 5513.

274 G. Jonsaell and P. Ahlberg, J. Chem. Soc. Perkin Trans II
     (1987) 461.

275 S. Berger, J. Chem. Soc., Chem. Commun., (1984) 1252.

276 J. Buddrus, H. Bauer and H. Herzog, Chem Ber., 121 (1988) 295.

277 G. I. Borodkin, Ie. R. Susharin, M. M. Shakirov, V. G. Shubin,
     Iu. A. Ustyniuk, E. I. Lazko and V. Amman, Zh. Org. Khim.,
     25 (1989) 132.

278 J. Blumel and F. H. Kohler, J. Organometal. Chem.,
     340 (1988) 303.

279 M. Saunders, R. M. Jarret and P. Pramanik, J. Am. Chem. Soc.,
     109 (1987) 3735.

280 R. M. Jarret and M. Saunders, J. Am. Chem. Soc.,
     109 (1987) 647.

Isotopes in the Physical and Biomedical Science, Vol. 2, edited by E. Buncel and J.R. Jones
© 1991 Elsevier Science Publishers B.V., Amsterdam

Chapter 7

NMR OF ^{6}LI-ENRICHED ORGANOLITHIUM COMPOUNDS

RUTHANNE D. THOMAS

Center for Organometallic Research and Education, Department of
Chemistry, University of North Texas, Denton, TX   76203   USA

CONTENTS

I.   INTRODUCTION

Organolithium compounds remain among the most widely used class
of reagents for organic and organometallic syntheses.[1-5] They are
also interesting in their own right due to their unusual structure
and bonding properties.[6] Much of the current understanding of

organolithium compounds comes from single crystal x-ray structures.[7-9] However, it has become increasingly clear that the structures in solution are not necessarily the same as those in the solid state. To understand the reaction mechanisms of these compounds, it is therefore critical to have methods for studying lithium compounds in solution.

Historically NMR spectroscopy has been the method of choice for studying organolithium compounds in solution. Much of the early work centered on ^{7}Li NMR, as developed by Brown.[10-12] In the late 1970's Wehrli introduced the use of ^{6}Li NMR.[13-15] Lithium NMR has since been the subject of a number of reviews.[16-18] More recently, Günther reviewed the various NMR techniques now available for the study of organolithium compounds.[19]

Of particular impact on the study of organolithium compounds in the last ten years has been the use of ^{6}Li-enriched compounds. Originally reported by Fraenkel,[20-21] and soon followed by additional examples from Seebach and coworkers,[22-24] the use of ^{6}Li-enriched organolithium compounds has enabled the determination of solution structures and provided for detailed studies of reaction mechanisms. This chapter emphasizes the practical considerations of doing NMR studies with ^{6}Li-enriched organolithium compounds and reviews some of the many experiments now possible using these compounds. Although emphasis is on compounds containing carbon-lithium bonds, "organolithium compounds" will be used in a broad sense to include compounds containing lithium bonded to a variety of elements.

(A) Properties of selected nuclei

There are at least four NMR active nuclei available in organolithium compounds: ^{1}H, ^{13}C, ^{6}Li, and ^{7}Li. The properties of these nuclei are summarized in Table 1. Additional nuclei of importance in selected compounds are also included in Table 1.

The natural isotopic abundances of ^{7}Li and ^{6}Li are 92.6% and 7.4%, respectively. However, the isotopic composition of lithium in commercially available compounds may vary widely due to undisclosed isotopic separation. One author notes[25] that lithium salts with ^{6}Li abundance as low as 3.75% have been marketed. If necessary, the lithium isotopic abundance of a sample can be conveniently determined by comparing the ^{6}Li and ^{7}Li NMR signal intensities of an unknown sample to that of a sample with known isotopic composition.[26]

TABLE 1

Properties of selected nuclei

	Spin	Natural Abundance (%)	Magnetogyric Ratio (rad T^{-1} s^{-1})	NMR Frequency (MHz/2.35T)	Quadrupole Moment (10^{-28} m^2)	Relative Receptivity
^1H	1/2	99.985	26.7510 x 10^7	100.000	---	5.68 x 10^3
^{13}C	1/2	1.108	6.7263 x 10^7	25.145	---	1.00
^6Li	1	7.42	3.9366 x 10^7	14.716	-8 x 10^{-4}	3.58
^7Li	3/2	92.58	10.396 x 10^7	38.864	-4.5 x 10^{-2}	1.54 x 10^3
^{29}Si	1/2	4.70	-5.3141 x 10^7	19.867	---	2.09
^{15}N	1/2	0.37	-2.7107 x 10^7	10.137	---	2.19 x 10^{-2}
^{31}P	1/2	100	10.829 x 10^7	40.481	---	3.77 x 10^2

The receptivities in Table 1 are relative to ^{13}C and assume natural isotopic abundances for all nuclei as defined by eq. 1, where γ is the magnetogyric ratio, I is the spin quantum

$$R_c = (3.95 \times 10^{-24})\gamma^3 I(I+1)P \qquad (1)$$

number, and P is the percent isotopic abundance. As seen from the table, ^7Li is very sensitive at natural abundance levels and even ^6Li is over three times more sensitive than ^{13}C (ignoring Overhauser effects). For ^6Li NMR of compounds with 95% ^6Li isotopic abundance, typical of the ^6Li-enriched samples discussed below, R_c increases by almost 13 times to 45.8. At this isotopic composition, ^7Li still has a receptivity of 83 relative to ^{13}C. Both lithium isotopes are thus readily observable in compounds enriched to 95% ^6Li isotopic abundance.

Both ^6Li and ^7Li are quadrupolar nuclei, with spins of 1 and 3/2, respectively. However, the quadrupole moments of the two nuclei are quite small, resulting in relatively inefficient quadrupolar relaxation even in unsymmetrical environments. Line-broadening is not a serious problem for either nucleus, although ^6Li linewidths are typically much narrower. Despite the lower resonance frequency of ^6Li, the narrower linewidths often lead to an effective increase in resolution of ^6Li NMR vs ^7Li NMR. However, for some lithium compounds, the relaxation is so inefficient that the ^6Li T_1 relaxation times are tens or even hundreds of seconds long, leading to prohibitively long experiment times. As noted below, this is a greater problem for ^6Li-enriched compounds, as opposed to natural abundance samples.

There is also a significant difference in the NMR spectra of nuclei bonded to ^{6}Li vs that for nuclei bonded to ^{7}Li. Quadrupolar relaxation of the ^{7}Li nucleus is sufficiently rapid in most cases to decouple nuclei coupled to the ^{7}Li nucleus, while ^{6}Li quadrupolar relaxation is not. This is the basis for the most important advantage of ^{6}Li isotopic enrichment: observation of coupling between ^{6}Li and directly bonded nuclei such as ^{13}C, ^{29}Si, or ^{15}N.

(B) Preparation of ^{6}Li-enriched compounds

To our knowledge there is currently no commercial source of ^{6}Li enriched organolithium compounds. However, ^{6}Li metal ($\geq 95\%$ isotopic enrichment) is readily available. The metal can then be used to produce lithium compounds either directly from the lithium metal or via an intermediate lithium compound. A recent monograph by Wakefield[27] reviews many of the synthetic reactions and includes practical suggestions for handling organolithium compounds.

^{6}Li isotopic enrichment of $\geq 95\%$ is recommended, particularly for compounds forming larger aggregates. Even at 95% enrichment, only 81% of tetramers, $(0.95)^4 = 0.814$, and 74% of hexamers, $(0.95)^6 = 0.735$, contain all ^{6}Li nuclei. The relatively large percentage of isotopomers containing one or more ^{7}Li nuclei causes interference and possible misinterpretation of results in some of the NMR experiments described in this chapter. (For example, see footnote 55 of ref. 28).

We have observed in several reactions of solid lithium, that ^{6}Li metal is much less reactive than corresponding commercial lithium (primarily ^{7}Li). We suspect this is due to small amounts of sodium contained in some commercial lithium. Sodium can be added, the reaction times increased, or other measures taken to allow for this decreased reactivity.

(C) NMR hardware and isotopic enrichment requirements

Most of the experiments described in this chapter require no special NMR hardware beyond normal multi-nuclear capabilities and normal gating of the observe radiofrequency. Most of the experiments also require low temperature capabilities to slow interaggregate exchange.

Several of the experiments described in this chapter do require triple resonance capabilities (See Table 2). The custom hardware is for either gated ^{13}C or gated ^{6}Li decoupling, in addition to ^{1}H

TABLE 2

NMR experiments and required special experimental conditions

NMR Experiment	Required Enrichment	Special NMR Hardware Required	Section
ONE-DIMENSIONAL EXPERIMENTS			
^{13}C (or other nuclei directly bonded to lithium)	^6Li	none	IIB-IIC
^6Li	none	none	IIIA
^6Li J-modulated spin echo	^{13}C	^{13}C decoupler	IID1
^{13}C J-modulated spin echo	^6Li	^6Li decoupler	IID2
$^{13}C\{^6Li\}$ INEPT	^6Li	^6Li decoupler	IIE1
$^6Li\{^1H\}$ NOE	none	none	IIIC1
^{13}C CPMAS	none	CPMAS hardware	IVB
TWO-DIMENSIONAL EXPERIMENTS			
^{13}C 2D-J-resolved	^6Li	^6Li decoupler	IID3
^6Li 2D-J-resolved	^{13}C	^{13}C decoupler	IID3
$^6Li, ^6Li$ 2D-COSY	^6Li	none	IIID
^6Li 2D Exchange	none	none	IIIE
$^6Li, ^{13}C$ 2D-HETCOR			
a. ^6Li decouple	^6Li	^6Li decoupler	IIE2
b. ^{13}C decouple	^{13}C	^{13}C decoupler	IIE2
$^6Li\{^1H\}$ 2D-HOESY	none	none	IIIC2

decoupling and an observe frequency. $^6Li\{^{13}C\}\{^1H\}$ triple resonance experiments have been discussed by Günther.[19,29] $^{13}C\{^6Li\}$ experiments have been described by Fraenkel[30] and by Thomas.[31] With the increased interest in inverse detection NMR,[32-33] and in particular ^{13}C decoupling, it is likely such hardware capabilities will become increasingly available. In addition, the ^6Li resonance frequency is very close to the resonance frequency of ^2H. As discussed by several authors,[19,30,34] the normal ^2H lock coil in most NMR probes can readily be altered for use with ^6Li. It should be noted, however, that some pulse sequences may need to be modified to take into account the fact that ^6Li has a spin of one. Moreover, some decoupling schemes (e.g. WALTZ) will not work for spin-1 nuclei.

Also included in Table 2 is the isotopic enrichment *required* for each type of experiment. Although no isotopic enrichment is required for the determination of approximate chemical shifts of nuclei directly bonded to lithium, ^6Li enrichment is usually required for resolution of individual aggregates (when multiple aggregates are present) and for resolution of $^6Li-^{13}C$ (or ^6Li-X) coupling. Table 2 thus lists ^6Li-enrichment as a requirement for one-dimensional ^{13}C NMR. Alternatively, experiments using ^6Li as

the observe nucleus generally do *not* require [6]Li isotopic
enrichment. However, such additional enrichment of the observe
nucleus improves sensitivity and significantly decreases
experiment times.

II.   ASSIGNMENT OF AGGREGATION STATES FROM SCALAR SPIN-SPIN
COUPLING

(A)   RLi aggregation states

     Organolithium compounds typically exist as aggregates,
$(RLi)_n$,[1,35] where n = 2,3,4,6,8, or 9.   This is true in the solid
state, solution, and the gas phase.   Much of the work in
organolithium chemistry has concentrated on the determination of
aggregation states and the study of the effects which different
aggregation states have on reactivity.[9,36-38]   The structure of
the organolithium compound in solution, and in large part the
aggregation state, may be the source of the so-called "proximity
effects" attributed to some organolithium compounds.[39]

     The structures of some representative organolithium aggregates
are shown in Figure 1.   In the presence of Lewis bases, such as
ethers or amines, each lithium atom is typically coordinated to
one to three bases to achieve a total coordination of four.

     The aggregation state of a particular compound depends on the
R-group, solvent, concentration, and temperature.   Although it is
difficult to predict *a priori* the aggregation state of a

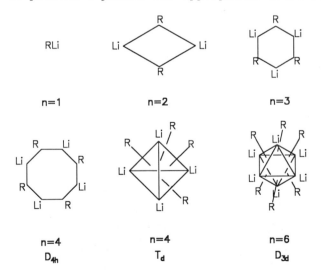

Fig. 1.   Typical structures of organolithium aggregates.

particular compound under a certain set of conditions, a number of generalizations can be made.[35] Increasing the steric requirements of the organic group tends to reduce the aggregate size. Coordinating solvents likewise tend to decrease the aggregate size relative to non-coordinating solvents, with stronger bases leading to smaller aggregates. Increased concentration of the organolithium compound can shift the equilibria to larger aggregates if multiple aggregation states are present. Similarly, temperature can also change the distribution among different aggregates, although the actual effect depends on the nature of the solvent. Decreasing the temperature in non-coordinating solvents tends to increase the aggregation, while in coordinating solvents decreasing the temperature tends to decrease the aggregation. Finally, carbon "anions" and oxygen anions apparently prefer to bond to a face of three lithium atoms, while nitrogen anions tend to bridge between two lithium atoms.[40]

The aggregation state of organolithium compounds has typically been assigned in solution based on colligative properties. These include: cryoscopy,[10,41] ebullioscopy,[42-43] vapor pressure osmometry,[44-45] and differential vapor pressure measurements using isopiestic[46-47] and barometric[48-49] methods. However, these are not always convenient or possible, particularly for complex mixtures of compounds, nor does the structure necessarily follow from a knowledge of the aggregation state.

A number of NMR methods are currently available which can aid in the determination of the structure of organolithium compounds. Those to be considered first depend upon scalar coupling between lithium and directly bonded nuclei. These methods, all of which require ^{6}Li-enriched compounds, include (1) multiplicity of the resonances for nuclei coupled to ^{6}Li, (2) magnitude of ^{6}Li coupling, (3) spin echo spectra, (4) 2D-J-resolved spectra, and (5) one- and two-dimensional polarization transfer experiments.

(B) Multiplicity of NMR resonances

Fraenkel[21] and Seebach[22] pioneered the use of scalar spin-spin coupling for the assignment of aggregation states in solution. In particular, they noted that the multiplicity of the ^{13}C resonance for the carbon alpha to lithium will vary depending on the aggregation state and on the exchange properties of the aggregate.

To observe coupling to lithium, two criteria must be met. First, there must be no, or only slow, interaggregate exchange

which breaks the X-^{6}Li bond and randomly exchanges nuclei.
Species undergoing rapid interaggregate exchange will exhibit no
coupling, while species undergoing only fluxional exchange or no
exchange may show coupling. The rate of interaggregate exchange
which causes collapse of the multiplet depends on the magnitude of
the coupling, but is typically quite small. Despite this,
interaggregate exchange can often be slowed sufficiently by
lowering the temperature.

It is often presumed that there also must be some covalent
Secondly, the relaxation of the lithium nucleus must be slow.
This is most easily achieved by isotopically enriching in ^{6}Li.
Although coupling to ^{7}Li has been observed for some compounds, the
quadrupolar relaxation of the ^{7}Li nucleus is usually too rapid to
resolve coupling, particularly at the low temperatures required to
slow interaggregate exchange. Unless otherwise specified,
discussions in this chapter will assume ^{6}Li-isotopic enrichment.

It is often presumed that there also must be some covalent
character to the lithium-carbon (or lithium-X) bond to observe
coupling. Indeed, the lack of observable coupling has been used
as evidence for "ionic" bonding. However, Streitweiser[50] noted
that spin polarization (or "through-space" coupling) could produce
scalar coupling without covalent bonding.

Lithium-lithium scalar coupling is not resolvable, although
under favorable conditions it can be used to establish correlation
between non-equivalent lithium atoms (see Sect. IIID). Multiple-
bond scalar couplings are likewise generally not resolvable;
essentially all of the scalar couplings observable as splitting of
NMR resonances are due to atoms directly bonded to lithium. (For
an exception, see Sec. IIC2.)

The multiplicity observed for nuclei bonded to lithium depends
on the number of equivalently coupled ^{6}Li nuclei. The number of
equivalently coupled ^{6}Li nuclei depends on both the structure of
the aggregate and any possible fluxional exchange. The
multiplicity is equal to 2mI+1, where m is the number of
equivalently coupled ^{6}Li nuclei and I = 1, the spin of ^{6}Li. Each
α-carbon in a dimer or static monocyclic aggregate is coupled to
two lithium atoms. In a static tetrahedral tetramer or D_{3d}
hexamer, each α carbon is coupled to three lithium atoms. For
fluxional aggregates, the α-carbon is coupled to all of the
lithium nuclei in the aggregate. The ^{13}C multiplicities expected
for a variety of static and dynamic aggregates (excluding larger

cyclic structures) are summarized in Table 3. Bonding in the octamers and nonamers is expected to be similar to that for the hexamer, but no structures are known. Therefore no values are given for these non-fluxional higher aggregates.

TABLE 3
Multiplicity of resonances showing coupling to ^{6}Li as a function of aggregation state and fluxional exchange

	RLi	$(RLi)_2$	$(RLi)_3$		$T_d-(RLi)_4$		$D_{3d}-(RLi)_6$		$(RLi)_8$		$(RLi)_9$
n	1	2	3		4		6		8		9
m	1	2	2	3	3	4	3	6	− 8	−	9
mult	3	5	5	7	7	9	7	13	− 17	−	19

The relative intensity of the observed lines within a multiplet is a function only of the number of coupled ^{6}Li nuclei. The first-order multiplets that arise from coupling to m magnetically equivalent ^{6}Li nuclei (I=1) are given in Table 4.

TABLE 4
Coupling patterns for nuclei coupled to m equivalent ^{6}Li nuclei

m	Number of Lines	Relative Intensities
1	3	1:1:1
2	5	1:2:3:2:1
3	7	1:3:6:7:6:3:1
4	9	1:4:10:16:19:16:10:4:1
5	11	1:5:15:30:45:51:45:30:15:5:1
6	13	1:6:21:50:90:126:141:126:90:50:21:6:1
7	15	1:7:28:77:161:266:357:393:357:266:161:77:28:7:1
8	17	1:8:36:112:266:504:784:1016:1107:1016:784:504:266:112:36:8:1
9	19	1:9:45:156:414:882:1554:2304:2907:3139:2907:2304:1554:882:414:156:45:9:1

(C) J values

It is clear from Table 4 that many of the outermost peaks of the larger multiplets will not be observed due to their low relative intensities. An alternate source of information concerning the aggregation state is the magnitude of the coupling, J. Not only does the multiplicity of the ^{13}C resonance change for fluxional aggregates, but the magnitude of the observed peak separation (which we will call observed $^{13}C-^6Li$ coupling, J_{obs}) also changes. For fluxional carbons bonded to a triangular face of lithium atoms:

$$J_{obs} = [3J + (n-3)J']/n \tag{2}$$

where J is the one bond coupling, J' is the multiple bond coupling and n is the aggregation state. Since J' is nearly or identically equal to zero, equation 2 reduces to

$$J_{obs} = 3J/n \tag{3}$$

Fraenkel used this relationship to assign n for a series of n-propyllithium aggregates in hydrocarbon solution.[21] In this case, J was assumed to be constant. More recently, Schleyer and co-workers[51,52] have suggested that J is nearly constant for *all* organolithium aggregates except monomers. This includes dimers, which would not have bonding to a triangular face. They suggest the general formula

$$J_{obs} = (17 \pm 2 \text{ Hz})/m \tag{4}$$

where m is the number of equivalently coupled ^6Li nuclei.

(1) Carbon-lithium coupling. Carbon-lithium scalar coupling has been observed for a range of organolithium compounds. Examples are collected in Table 5. The values are generally derived by observing the splitting of the peaks in the ^{13}C spectra. Coupling can also be measured by observing the ^{13}C satellites of the ^6Li resonance. This latter method is preferred if very accurate J values are desired, since the ^6Li spectra typically have a higher data point density than the ^{13}C spectra. The organolithium compounds are arranged in Table 5 by the type

TABLE 5

Examples of ^{13}C-^{6}Li Coupling

Li Compound	Solvent*	Temp (°C)	^{13}C mult	m	n	J_{obs}** (Hz)	Ref
ALKYLS							
MeLi	THF, Et$_2$O, Et$_3$N	-60	–	–	4	[≈5.6]	53,54
n-PrLi	cyclopentane	-56	13	6	6	3.3	21
"	"	-56	17	8	8	2.5	21
"	"	-56	19	9	9	2.2	21
i-PrLi	"	-15	7	3	4	6.1	55
"	"	-15	13	6	6	3.3	55
n-BuLi	Et$_2$O	-70	–	–	4	[5.3]	56
"	THF	-100	5	2	2	7.8	22
iso-BuLi	cyclopentane	<-20	13	6	6	3.11	55
sec-BuLi	"	-41	7	3	4	6.1	57
"	"	-41	–	6	6	3.25	57
"	"	-1	–	4	4	4.6	57
"	"	–	5	2	2	2.5	57
"	THF/1eq PMDTA	-96	3	1	1	14.0	51
t-BuLi	cyclohexane	RT	–	–	4	[4.2]	56
"	toluene	RT	–	–	4	[3.8]	56
"	cyclopentane	>-5	9	4	4	4.10	58
"	"	<-22	7	3	4	5.44	58
"	cyclopentane/Et$_2$O	-80	5	2	2	7.8	59
"	cyclopentane/Me$_3$SiOMe	-90	5	2	2	7.8	60
"	THF-d$_8$	-90	3	1	1	11.9	51
"	Et$_2$O	-64	–	–	2	[7.6]	51
2-Me-butylLi	"	-20	–	6	6	3.3	55
sec-pentylLi	"	5	9	4	4	4.3	55
"	"	<-25	7	3	4	5.7	55
"	"	-20	–	–	6	≈3.2	55
t-pentylLi	"	all	9	4	4	3.98	55
neo-pentylLi	Et$_2$O	all	5	2	2	8.9	61
"	Et$_2$O/6 eq PMDTA	-107	3	1	1	14.7	61
2-Et-butylLi	cyclopentane	<0	13	6	6	3.1	55
MIXED ALKYL/ALKOXIDES							
(iso-Pr)$_a$(iso-PrO)$_b$Li$_6$	"	-40	13	6	6	3.4	62
(t-Bu)$_3$(t-BuO)Li$_4$	"	25	9	4	4	3.8	62
"	"	-30	5x3	2/1	4	4.9/5.4	62
(t-Bu)$_2$(t-BuO)$_4$Li$_6$	"	25	13	6	6	2.5	62
SUBSTITUTED ALKYLS							
LiCHCl$_2$	THF	-100	3	1	1	16.3	22
LiCHBr$_2$	"	-100	3	1	1	16.3	22
LiCHI$_2$	"	-100	3	1	1	16.6	22
LiCCl$_3$	"	-100	3	1	1	17.0	22
LiCBr$_3$	"	-100	3	1	1	16.3	22
LiCI$_3$	"	-100	3	1	1	17.0	22
MeCBr$_2$Li	"	-100	3	1	1	17.0	22
LiCH$_2$SiMe$_2$CH$_2$OMe	cyclopentane	7	7	3	4	4.7	60
Li$_2$C(SiMe$_3$)SO$_2$Ph	THF-d$_8$	-103	various peaks			1.9-9.2	63
Li$_2$C(CPh$_2$)$_2$	THF	-116	5	2	1	9	76

Li Compound		Solvent*	Temp (°C)	^{13}C mult	m	n	J_{obs}** (Hz)	Ref
ALKENES								
1	C^1	THF-d_8	-90	3	1	1	10.6	28
2	C^8	"	-71	5	2	1	5.9	28
"	C^8	"	"	5	2	1	7.6	28
"	C^1	"	"	5x3	2/1	2	4.1/7.7	28
"	C^8	"	"	5x3	2/1	2	4.3/6.1	28
3		THF	-127	3	1	1	12.0	64
4		"	-113	3	1	1	11.0	64,65
5		"	-106	3	1	1	12.0	64,65
6		"	-93	3	1	1	10.7	65
7		"	-100	3	1	1	10.0	65
8		2-MeTHF	-115	5	2	2	9.0	22
9		THF	-100	3	1	1	16.8	22
CYCLO- AND BICYCLOALKYLS								
10		THF	-100	3	1	1	17.3	22
11		"	"	3	1	1	17.0	22
12		"	"	3	1	1	17.0	22
13		"	"	3	1	1	17.2	22
14		"	"	3	1	1	17.0	22
15		"	"	3	1	1	17.0	22
16		"	"	5	2	2	9.7	22
17		"	"	5	2	2	8.5	22
18		"	"	5	2	2	10.3	22
19		"	"	3	1	1	15.3	66
20		THF-d_8	-78	3	1	1	6.9	67
21		THF	-76	5	2	2	9.4-10.3	68
22		Et$_2$O	-142	5	2	2	7.6	69
23		"	"	3	1	1	16.0	69
24		"	"	5	2	1	8.9	69
ALKYNES								
t-BuCCLi		THF	-100	7	3	4	5.7	70,71
"		"	"	7	3	4	5.8	70,71
"		"	"	5	2	2	8.3	70,71
PhCCLi		"	-95	5	2	2	8.2	24
ARYLS								
PhLi		THF-d_8	-118	5	2	2	8	22
"		THF/PMDTA	-103	3	1	1	15.6	51
"		TMEDA/Et$_2$O	-75	5	2	2	8	30
"		DMS	--	9	4	4	--	72
PhLi/PhI		THF/HMPA	-105	3	1	1	13	73
mesitylLi		THF-d_8/ 1 eq PMDTA	-100	3	1	1	15.3	51
supermesitylLi		THF-d_8	-81	3	1	1	16.1	51
25		toluene-d_8	-85 to +50	5	2	3	6.5	74
$(2,6\text{-}(MeO)_2Ph)Li$		"	-90	5x3	2/1	4	3.5/4.9	77

Li Compound	Solvent*	Temp (°C)	^{13}C mult	m	n	J_{obs}** (Hz)	Ref
<u>HETEROARYLS</u>							
26	toluene-d_8/TMEDA	-90	5	2	2	7.7	75
27	"	"	5	2	2	8.2	75
"	"	"	5	2	2	7.8	75

*Abbreviations are explained at the end of the chapter.
**Square brackets denote $^{13}C-^{6}Li$ coupling calculated from observed $^{13}C-^{7}Li$ coupling (see text).

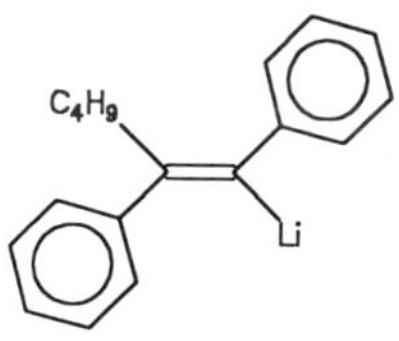

1

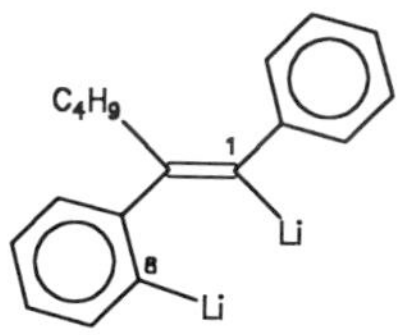

2

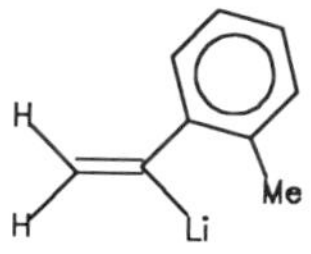

3

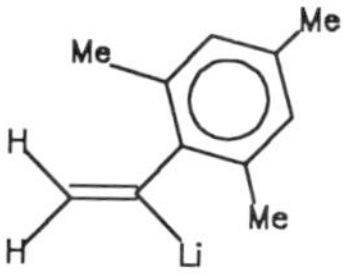

4

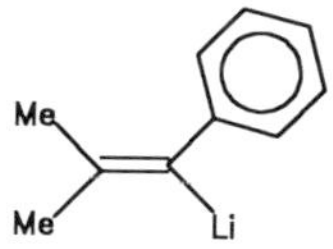

5

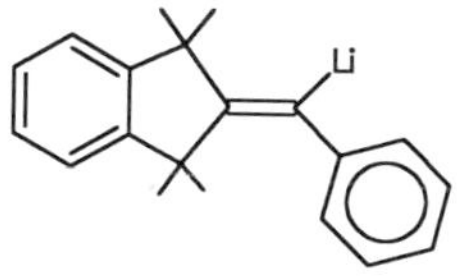

6

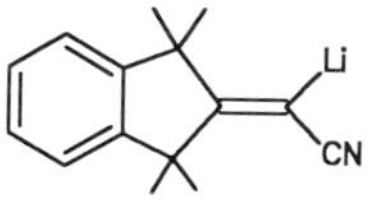

7

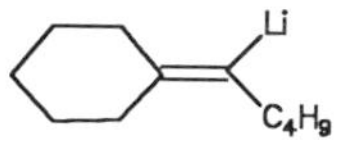

8

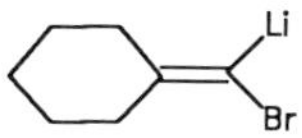

9

Fig. 2. Selected lithium compounds from Table 5 showing carbon-lithium coupling. (con't next page)

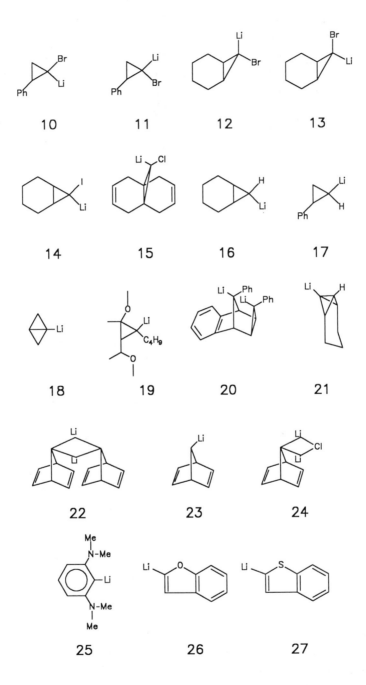

Fig. 2. Selected lithium compounds from Table 5 showing carbon-lithium coupling. (con't from previous page)

of organic group. Since the observation of carbon-lithium coupling is often heavily dependent upon solvent and temperature, these are also included. The observed coupling and multiplicity, along with the inferred number of equivalently coupled ^{6}Li nuclei, m, and the aggregation state, n, are listed in each case.

In a few instances, ^{13}C-^{6}Li coupling has not been reported, although ^{13}C-^{7}Li coupling is known. The corresponding ^{13}C-^{6}Li J_{obs} is expected to be 37.9% of ^{13}C-^{7}Li J_{obs}, as derived from the ratio of magnetogyric ratios for ^{6}Li and ^{7}Li (see Table 1). For these cases, the calculated value for ^{13}C-^{6}Li is listed in square brackets.

As seen in Table 5, J_{obs} is relatively constant for a given m, but decreases with increasing m. For m = 3,4, and 6, J_{obs} values range from 5.4 to 6.2, 4.0 to 4.7, and 3.1 to 3.4 Hz, respectively. These compare with predicted values from eq. 4 of 5.7, 4.25, and 2.8 Hz, respectively. Only one example each of coupling for m = 8 and m = 9 is known. These are 2.5 and 2.2 Hz, respectively, compared to the predicted values of 2.125 and 1.9 Hz. Although the number of examples are limited, the magnitude of J apparently provides a fairly good indication of the aggregation state for m ≥ 3, although the observed values tend to be larger than that predicted by eq. 4. As noted in the table, alkyl groups in mixed aggregates containing alkoxides have smaller J_{obs} values than for the corresponding alkyllithium aggregate alone.

For m = 1 or 2, the observed values for J cover a much broader range. For m = 2, values range from 2.5 to 10.3 Hz; for m = 1, the values range from 6.1 to 17.3 Hz. These values compare with predicted values of 8.5 and 17 Hz, respectively. It is interesting to note that many of the compounds with unusually small J_{obs} values are dilithium compounds.

(2) <u>Coupling to other nuclei</u>. Coupling between ^{6}Li and other directly bonded atoms has also proven to be a valuable tool, particularly for determining lithium-nitrogen connectivities. This method has been used extensively by Jackman[77-79] and by Collum.[34,80-83]

Unlike most of the examples cited above for ^{6}Li-^{13}C coupling, ^{6}Li-^{15}N coupling has generally been observed in *doubly* labeled compounds. As noted in Table 1, ^{15}N has a very low sensitivity, due to its very low natural abundance (0.37%) and its small magnetogyric ratio. This usually precludes observing ^{6}Li coupling using natural abundance ^{15}N NMR. However, enriching in both ^{6}Li and ^{15}N not only enables the observation of ^{6}Li coupling with ^{15}N NMR, but also makes possible the observation of ^{15}N coupling using ^{6}Li NMR. While observation of the multiplicity of the ^{15}N

TABLE 6
Selected examples of ^{6}Li coupling to nuclei other than carbon

Li Compound	Solvent*	Temp (°C)	^{6}Li mult	equiv X	X mult	equiv Li	n	J_{obs}** (Hz)	Ref
X = ^{15}N									
Ph$_2$NLi	toluene-d$_8$/ 10 eq THF	-90	3	2	-	-	2	3.12	81
"	3 eq THF	-90	-	-	5	2	2	3.34	81
Ph$_2$N(Br)Li$_2$	toluene-d$_8$/ 2.5 eq THF	-90	2	1	-	-	2	3.24	81
Ph(Me)NLi	Et$_2$O	-100	-	-	5	2	2	3.8	80
Ph(i-Pr)NLi	THF	-80	-	-	3	1	1	7.5	80
"	Et$_2$O/4 eq HMPT	-85	-	-	3	1	1	6.4	79
"	"	-90	2	1	-	-	1	6.6	79
"	"	-85	-	-	3	1	1***	5.9	79
"	"	-90	3	2	-	-	1***	6.0	79
Ph(c-hexenyl)NLi	toluene-d$_8$/								
"	2 eq THF	-94	3	2	5	2	2	≈3.3	84
"	20 eq THF	-94	2	1	-	-	1	6.3	84
"	8 eq THF	-94	-	-	3	1	1	6.1	84
i-Pr(c-hexyl)NLi	toluene-d$_8$/THF	-90	3	2	-	-	2	5.1	82
"	"	-90	-	-	5	2	2	4.8	82
"	"	-90	-	-	5	2	2	5.3	82
(i-Pr)$_2$NLi	"	-90	3	2	5	2	2	5.1	83
Li-2,3,3-Me$_3$- indolenide	Et$_2$O	-80	-	-	5	2	2	3.5	78
X = ^{29}Si									
Ph$_3$SiLi	2-MeTHF	-100	-	-	3	1	1	17	85
Ph$_2$(Me)SiLi	THF	-115	-	-	3	1	1	15	85
"	2-MeTHF	-80	-	-	3	1	1	16	85
Ph(Me)$_2$SiLi	THF	-100	-	-	3	1	1	18	85
"	2-MeTHF	-80	-	-	3	1	1	18	85
X = ^{31}P									
Ph$_2$PLi	Et$_2$O	-73	-	2	-	2	2	[17]	86
((Me$_3$Si)$_2$CH)$_2$PLi	"	25	-	-	-	2	2	[30]	87
((Me$_3$Si)$_2$CH)$_2$PLi	Et$_2$O/TMEDA	25	-	-	-	1	1	[46]	87

*Abbreviations are explained at the end of the chapter.
**Square brackets denote ^{6}Li-X coupling calculated from observed ^{7}Li-X coupling (see text).
***Triple ion salt.

resonance can determine the number of directly bonded lithium atoms, the combined results can further distinguish between monomeric, cyclic oligomeric, and ion triplet forms of lithium-nitrogen compounds. Examples of $^{15}N-^{6}Li$ coupling are shown in Table 6.

Also listed in Table 6 are examples of one bond $^{6}Li-^{29}Si$ and $^{6}Li-^{31}P$ coupling. Coupling between lithium and phosphorus was originally observed as $^{7}Li-^{31}P$ coupling in natural abundance samples.[86-88] The observed J values have been converted here to the corresponding $^{6}Li-^{31}P$ coupling, as predicted by the ratio of the magnetogyric ratios of ^{6}Li and ^{7}Li.

Two bond $^{7}Li-^{31}P$ coupling has been observed for HMPA solvated Li^{+} ions, $Li^{+} \cdot (HMPA)_n$.[73,89] In this case the multiplicity of the lithium resonance clearly shows the number of coordinated HMPA molecules. The $^{7}Li-^{31}P$ coupling decreased from 11.2 Hz for the mono-solvated cation to 7.5 Hz for the tetra-solvated ion. These correspond to $^{6}Li-^{31}P$ couplings of 4.2 to 2.8 Hz.

Lithium-proton coupling, either $^{7}Li-^{1}H$ or $^{6}Li-^{1}H$, is not resolved in simple alkyl- and aryllithium compounds. However, $^{7}Li-^{1}H$ coupling of 6.4 and 8.4 Hz (corresponding to $^{6}Li-^{1}H$ coupling of 2.4 and 3.2 Hz, respectively) has been observed for some transition-metal lithium complexes.[90]

(D) Spin-echo experiments.

Although the determination of the number of coupled ^{6}Li nuclei from the multiplicity of the resonance or from the value of the coupling often works well for individual compounds, it is not well suited for complex reaction mixtures. Splitting of the α-carbon resonances into multiple lines leads to poor signal to noise for species in low concentrations and to poor resolution of overlapping multiplets. A possible alternative solution is the use of multiplicity selection and spectral editing techniques using spin-echo spectra.

(1) 1D-J-modulated ^{6}Li spin-echo spectra. Gated ^{13}C decoupling can be used to produce phase selection of the ^{6}Li resonances, similar to the now common APT (attached proton test) experiment[91] used in ^{13}C NMR. The pulse sequence (see Sequence 1) is exactly the same as the standard APT pulse sequence except the X-nucleus decoupler is gated, rather than ^{1}H. The general pulse sequence shown in Sequence 1 contains a second 180° refocusing pulse. This modification[92] allows the first pulse to be set to any angle, θ,

necessary to optimize sensitivity and save time in the acquisition of multiple transients. This may be either the Ernst angle[93] or some other angle.[94] The time τ determines the nature of the final spectrum. The time Δ is merely a short time to permit the receiver to recover from the final pulse before data acquisition begins.

PULSE SEQUENCE 1
J-modulated spin-echo

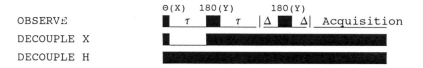

For ^{6}Li spin-echo spectra, the observe nucleus is ^{6}Li and the X-nucleus is ^{13}C (or some other nucleus coupled to lithium). When $\tau=1/(2J_{obs})$, the ^{6}Li signals are either positive or negative, depending on whether the ^{6}Li nuclei are coupled to an odd or even number of ^{13}C nuclei. This experiment requires ^{13}C decoupling capabilities and ^{13}C enrichment of the carbons bonded to lithium.

This technique was used to establish the multiplicities of the various ^{6}Li signals from the different aggregates of mono- and di-lithiumtrimethyl((phenylsulfonyl)methyl)silane.[63,19,95]

(2) <u>1D-J-modulated ^{13}C spin-echo spectra.</u> Phase selection can also be done using the same pulse sequence, but with ^{13}C observation and ^{6}Li decoupling. This is similar to spin echo techniques developed for ^{13}C, ^{2}H spin systems.[96-98] Unlike spin-echo experiments in which a spin-1/2 nucleus is decoupled, the intensities of the ^{13}C signals with gated decoupling of a spin-1 nucleus are greatly reduced. CX_n signals (X = spin-1 nucleus) with an evolution time $\tau = 1/(2J_{obs})$ are proportional to $(-1/3)^m$. This means that for the common m values for organolithium compounds of 2, 3, 4, and 6, the intensity of the ^{13}C signal is only 11, 3.7, 1.2, and 0.1%, respectively, of a ^{6}Li decoupled ^{13}C signal. As noted in an earlier review,[19] the very low signal intensities make this experiment impractical in this form for most applications, although Günther has shown that the experiment does work as predicted.[99]

Selective observation of ^{13}C signals for different organolithium aggregates can be achieved by using evolution times

to totally refocus the magnetization at $\tau = 1/J_{obs}$. Since, as
noted in Section IIC, J_{obs} is different for different aggregates,
subspectra can be produced which contain only selected aggregates.
The subspectra are produced with $\tau = 1/J_{obs} \approx k/(3J)$, where k is
an integer $\geq$ m and $3J \approx 17$ Hz.[31] For example, dimer signals
refocus at multiples of $2/(3J)$, static tetramers at multiples of
$3/(3J)$, and fluxional tetramers at multiples of $4/(3J)$. An
example of this experiment for t-butyllithium in cyclopentane
containing diethyl ether is shown in Figure 3.

Due to the long evolution times of 0.2 - 2 sec, the method is
susceptible to significant signal loss via T_2 relaxation during
the evolution time. This is a greater problem for larger
aggregates, since longer evolution times are required. The method
is most useful for branched chain alkyllithium compounds since
they generally exist as smaller aggregates and are less likely to
undergo interaggregate exchange.

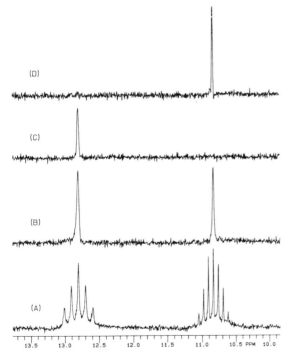

Fig. 3. 75 MHz ^{13}C J-modulated spin-echo spectra of the α-carbon
region of t-butyllithium-^{6}Li in cyclopentane/Et$_2$O at -60° C.[31] A,
normal ^{13}C$\{^{1}$H$\}$, 256 acquisitions (total exp. time = 2.5 hr). B,
$\tau=0$, ^{13}C$\{^{1}$H,^{6}Li$\}$, 16 acquisitions (total exp. time = 9 min). C,
$\tau=0.128$s (optimized for dimers), 16 acquisitions (total exp. time
= 9 min). D, $\tau=0.184$ s (optimized for static tetramers), 16
acquisitions (total exp. time = 9 min).

(3) <u>2D-J-resolved spectra</u>. The spin-echo experiments described above depend upon the value of J_{obs}. If the J values are not known or if a wide range of J values are expected, an alternative approach is the use of 2D-J-resolved NMR spectroscopy.[100-102] Both of the spin-echo experiments listed above can easily be expanded to achieve 2D-J-resolved spectra by making the delay time, τ, a variable. In these experiments the chemical shift information is retained for the observe nucleus in one dimension, while each chemically shifted peak appears as the appropriate multiplet in the second dimension. These experiments can greatly simplify overlapping multiplets, enabling the determination of both the multiplicity of the resonance and the magnitude of the coupling. This method (when used with [6]Li decoupling) also is a potential means of separating coupling to lithium, which only appears in the f_1 dimension, from coupling to other nuclei, which appears only in the f_2 dimension.

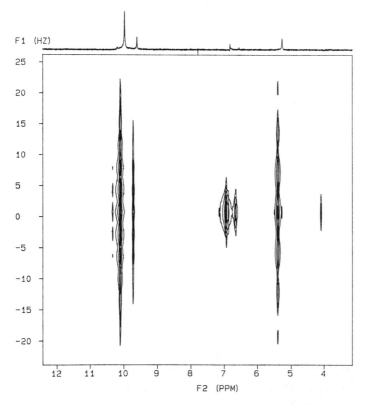

Fig. 4. α-carbon region of 2D-J-resolved ^{13}C NMR spectrum of isopropyllithium-[6]Li/[6]Li-lithium isopropoxide in cyclopentane at -40° C.

Both ^{6}Li-2D-J-resolved experiments[19,63] and ^{13}C-2D-J-resolved experiments[103] have been demonstrated for organolithium compounds. Figure 4 is a ^{13}C-2D-J resolved spectrum of the α-carbon region of a sample of isopropyllithium/lithium isopropoxide. The 2D spectrum clearly shows the multiplicity and coupling of each of the ^{13}C resonances. Even the four center peaks of the very small downfield peak, which are totally obscured by overlap with the large adjacent multiplet in a normal ^{13}C{^{1}H} experiment, are clearly visible. With this configuration (i.e. decoupling of ^{6}Li) the peaks in the f_1 dimension are separated by J_{obs}, not $J_{obs}/2$ as in most such experiments in which a spin-1/2 nucleus is decoupled.

(E) Polarization transfer experiments

(1) One-dimensional experiments. An additional set of experiments which can be used for spectral editing is based on polarization transfer between coupled nuclei. A number of experiments such as INEPT,[104,105] DEPT,[106-108] and their many variations[109,110] were originally developed as a general means of sensitivity enhancement by transferring magnetization from protons to a less sensitive nucleus. The observation of some insensitive nuclei, with negative magnetogyric ratios such as ^{29}Si or ^{15}N still make major use of these techniques.[111-113] By appropriate choice of delay times, these experiments can be used for spectral editing and multiplicity selection.

Polarization transfer normally requires resolvable spin-spin coupling. ^{6}Li is thus among the relatively few quadrupolar nuclei for which these experiments may be applicable, although not generally with protons. Analysis and examples of INEPT and DEPT experiments for a number of quadrupolar nuclei have appeared.[91,114-116] However, we are unaware of any examples of one-dimensional experiments using transfer of polarization from spin-1/2 nuclei to ^{6}Li.

So-called "reverse" INEPT, i.e. polarization transfer from quadrupolar nuclei to spin-1/2 nuclei is also known. This has been demonstrated for transfer from ^{6}Li to ^{13}C.[103] This experiment suffers from sensitivity loss since the signal intensities are proportional to $\gamma(^6$Li$)/\gamma(^{13}$C$) = 0.58$. The experiment also suffers from the long relaxation times of ^{6}Li. It is therefore unlikely that this experiment, in contrast to the reverse INEPT experiments used for other quadrupolar nuclei[117-120] (which take advantage of rapid the relaxation rates of these nuclei) will have widespread utility.

(2) <u>Two-dimensional experiments</u>. Although the one dimensional polarization transfer experiments have not found widespread use for organolithium compounds, they do form the basis for the important method of 2D-heteronuclear spin correlation.[121-122] Such experiments are of great importance in the identification of ^6Li, ^{13}C spin pairs.

One pulse sequence which has been used to achieve heteronuclear shift correlated spectra in organolithium compounds is shown in Sequence 2. In this pulse sequence, D is a delay for the

PULSE SEQUENCE 2

2D-heteronuclear shift correlation via polarization transfer

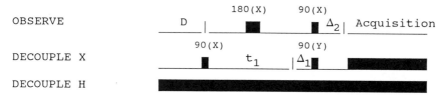

relaxation of the decoupled nucleus; Δ_1 depends on the magnitude of the observed coupling and the spin of the decoupled nucleus; Δ_2, which allows for refocusing of the observe nucleus and thus decoupled signals in the f_2 dimension, depends on the multiplicity of the observe signal and the spin of the decoupled nucleus.

Attempts to use this sequence with ^6Li observation and ^{13}C decoupling for the study of a dilithiosulfone were unsuccessful.[123] This was likely due to the rapid transverse relaxation of the carbon resonances. To avoid this problem, Ellington[103] carried out the reverse experiment in which ^{13}C is observed and ^6Li is decoupled. This takes advantage of the slow relaxation of the ^6Li nucleus. For this configuration (i.e. ^{13}C observe and ^6Li decouple) D is set to 1.5 times the ^6Li T_1 time; Δ_1 is set to $1/(2J_{obs})$; Δ_2 is set to 25 ms. The optimal refocusing time, Δ_2, depends on the multiplicity of the ^{13}C signal. However, the use of the average time of 25 ms leads to relatively small losses in signal intensity for a wide range of organolithium compounds.[103]

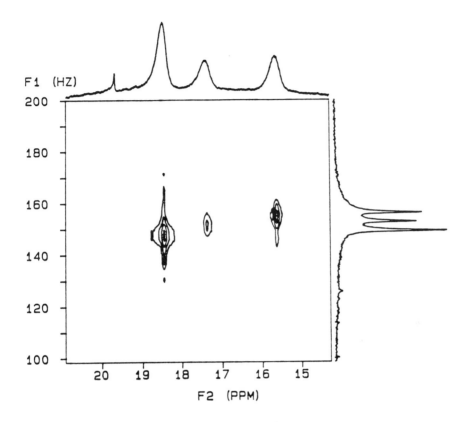

Fig. 5. ^{13}C-^{6}Li 2D HETCOR spectrum of 2F n-propyllithium in cyclopentane at -80° C. 64 transients for each of 32 increments. D=10s, Δ_1=0.180s, Δ_2=0.025s (Total exp. time = 6 hr).

Figure 5 is a 2D-heteronuclear correlated NMR spectrum of n-propyllithium-^{6}Li using Pulse Sequence 2. This unambiguously shows the correlations of the ^{13}C and ^{6}Li resonancs previously assigned to the hexameric, octameric, and nonameric aggregates.[21] In this case the correlations are clearly evident even though the ^{13}C-^{6}Li coupling is not resolved in the one-dimensional ^{13}C spectrum. Despite reduced sensitivity due to the "reverse" nature of the polarization transfer, this technique has proven useful for a variety of organolithium compounds.[103]

Heteronuclear shift correlations can also be achieved using double-quantum spectroscopy.[124,125] The pulse sequence for this experiment is shown in Pulse Sequence 3. This technique, where

PULSE SEQUENCE 3
2D-heteronuclear shift correlation via double quantum coherence

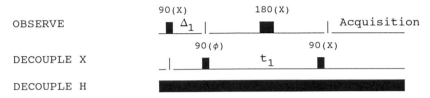

the observe nucleus is ^{6}Li and the decoupled nucleus is ^{13}C, has been used successfully to obtain 2D-heteronuclear correlations between ^{13}C and ^{6}Li.[63,123] For this configuration $\Delta_1 = 1/(2J_{obs})$.

III. ^{6}LI NMR

(A) Practical considerations

The use of ^{6}Li-enriched compounds has not only made possible the experiments discussed above based on coupling, but has also resulted in the widespread use of ^{6}Li NMR. Although ^{6}Li-enrichment is not required for the observation of ^{6}Li NMR, the additional sensitivity makes this an easy nucleus to observe in such compounds. Spectra can often be observed with only one pulse.

Due to long relaxation times discussed below, ^{6}Li resonance lines are normally very narrow; line widths of less than 1 Hz are not uncommon. Despite a shift range (in Hz) of only 38% of that for ^{7}Li, the very narrow line widths of the ^{6}Li resonances typically result in an increase in effective resolution. Individual organolithium aggregates can usually be resolved, at least at low temperature, and individual stereoisomers have been observed in some cases.[57,126] The long relaxation times can however be a problem, making some experiments prohibitively long.

^{6}Li spectra are typically run with ^{1}H decoupling. Although coupling is not resolved, unresolved coupling can broaden lines.[53,127] More importantly, ^{1}H decoupling can lead to significant sensitivity gains through NOE. However, since the amount of the Overhauser enhancement varies dramatically, such enhancement should be avoided for quantitative work through gated decoupling techniques.[128,129]

Although ^{6}Li NMR has excellent sensitivity and resolution, it is often difficult to assign the spectra based merely on chemical shifts. There are relatively few studies of lithium chemical shifts in organolithium compounds,[130] and lithium chemical shifts remain relatively uninformative compared to the chemical shifts of other nuclei in these compounds.[131] The entire chemical shift range for lithium is only 5-6 ppm, with a range of only 2-3 ppm for organolithium compounds. Despite these problems, lithium chemical shifts can be particularly informative within a closely related set of compounds.[186]

Understanding lithium chemical shifts is further hindered by the difficulty in comparing chemical shift data from different laboratories. There is no generally accepted reference for lithium chemical shifts. Most shifts given in the literature are relative to an external standard, typically a solution of a lithium salt. However, the chemical shifts of the lithium salts, as well as the organolithium compounds themselves, are concentration, temperature, and solvent dependent.[17]

Another significant problem, although often less appreciated, is the lack of correction for magnetic susceptibility between the sample and the external reference. Due to the small chemical shift range of lithium this can be quite large, and is most severe when comparing data from an iron magnet with data from a superconducting magnet.[132,133] For example, in a superconducting magnet, t-butyllithium in cyclopentane has a chemical shift of 0.2 ppm relative to external 1 M LiClO$_4$ in d$_6$-acetone. The identical sample has a chemical shift of 1.07 ppm in an iron magnet.[134] Since the entire chemical shift range of organolithium compounds is only 2-3 ppm, shifts of 0.9 ppm due to magnetic susceptibility differences are huge.

Given these problems, ^{6}Li spectra are best assigned by correlation to other nuclei within the compound. This can be via 2D heteronuclear correlated spectra (Section IIE2), by selective decoupling experiments, or merely by observation of coupling in the ^{6}Li spectrum. The latter method may be with doubly-labeled compounds, as discussed in Section IIC2, or by observation of satellites in the ^{6}Li spectrum. For example, isopropyllithium exists as a mixture of hexamers and tetramers in hydrocarbon solvent. The two [6]Li resonances were unambiguously assigned by observation of ^{13}C satellites for one of the peaks with coupling appropriate for the tetramer.[55]

(B) [6]Li spin-lattice relaxation

The great utility of [6]Li-enriched organolithium compounds comes from its unusual relaxation properties. Despite a spin of 1, [6]Li has very inefficient quadrupolar relaxation, even in unsymmetrical environments. The result is narrow lines in [6]Li NMR and resolvable coupling for many nuclei bonded to lithium.

In spite of the central role that [6]Li relaxation plays in the properties of the lithium nucleus, there have been very few studies of [6]Li relaxation.[15,103,135] Only one of these[103] includes [6]Li-enriched compounds. A knowledge of relaxation times is critical to the proper choice of delay times in many of the pulse sequences. [6]Li T_1 relaxation times and NOE data are listed in Table 7 for a few representative compounds.

TABLE 7

[6]Li spin-lattice relaxation times and NOE values for selected organolithium compounds

| Li Compound | Solvent | Natural Abundance [6]Li | | | | 95% [6]Li | | | |
		Temp (°C)	T_1 (sec)	η	Ref	Temp (°C)	T_1 (sec)	η	Ref
MeLi	Et$_2$O	28	72	0.7	15	--	--	--	--
n-PrLi	cyclopentane	30	20	0.8	135	29	46	1.4	103
" (hexamer)	"	-85	2	0.7	135	-85	4	0.5	103
" (octamer)	"	"	3	0.9	135	"	4	0.7	103
" (nonamer)	"	"	3	1.0	135	"	4	0.9	103
i-PrLi	"	30	61	2.5	135	30	135	2.2	103
" (tetramer)	"	-85	6	2.3	135	-85	10	2.3	103
" (hexamer)	"	"	3	1.8	103	"	4	1.8	103
n-BuLi	n-hexane	28	20	1.2	15	--	--	--	--
"	C$_6$H$_6$	"	12	0.9	15	--	--	--	--
"	C$_6$D$_6$	"	9	0.8	15	--	--	--	--
t-BuLi	cyclopentane	30	97	2.8	103	30	202	2.9	103
"	"	--	--	--	--	41	228	2.8	103
"	"	-85	23	2.8	103	-85	31	2.8	103
"	2,2-DMB	--	--	--	--	31	122	2.4	103
"	"	--	--	--	--	41	17	2.1	103
"	"	--	--	--	--	-70	25	2.6	103
PhLi	C$_6$H$_6$/Et$_2$O	28	13	0.6	15	--	--	--	--

As seen in Table 7, ^{6}Li T_1 times depend on R-group, solvent, temperature, and lithium isotopic abundance. The relaxation times for organolithium compounds range from a few seconds to hundreds of seconds, but are all less than lithium cation in D_2O solution. In this symmetrical environment, and in the absence of protons, relaxation times are as great as 1000 sec.[15]

(1) Relaxation mechanisms. The observed relaxation rate, $1/T_1$, is the sum of relaxation rates from each of a number of relaxation mechanisms. The mechanisms making major contributions to ^{6}Li relaxation are the quadrupolar, ^{6}Li-^{1}H dipole-dipole, and ^{6}Li-^{7}Li dipole-dipole mechanisms.

$$1/T_1 = \Sigma(1/T_1{}^i) = 1/T_1{}^Q + 1/T_1{}^{DD}(^6Li-^1H)$$
$$+ 1/T_1{}^{DD}(^6Li-^7Li) \qquad (5)$$

Chemical shift anisotropy does not contribute to the relaxation due to the small chemical shift of the ^{6}Li nucleus, while scalar relaxation has yet to be found important for ^{6}Li. The spin-rotation mechanism is only important in a few cases, and only at elevated temperatures. The observed ^{6}Li T_1 times are best understood by considering each of the possible relaxation mechanisms individually.

(2) ^{6}Li-^{1}H dipole-dipole relaxation. Wehrli very early determined[15] that unlike the relaxation of most other quadrupolar nuclei, ^{6}Li relaxation contained a significant component from ^{1}H dipolar interactions. The actual contribution from ^{6}Li-^{1}H dipole-dipole relaxation can be determined by measuring the nuclear Overhauser enhancement factor, η ($\equiv(I-I_0)/I_0$).[136] The maximum value for η is $\gamma(H-1)/2\gamma(Li-6) = 3.4$. In essentially all cases studied thus far there is a contribution from ^{6}Li-^{1}H dipole-dipole relaxation, but this contribution varies widely. The values in Table 7 range from 0.5 to 2.9, corresponding to contributions from ^{6}Li-^{1}H dipole-dipole relaxation of from 14.7 to 85.3%. Due to the presence of this mechanism, sensitivity increases for ^{6}Li resonances if ^{1}H decoupling is used. However, the amount of the enhancement varies greatly. This mechanism is also the basis for the one- and two-dimensional NOE experiments discussed in Section IIIC.

(3) Quadrupolar relaxation. Due to its larger quadrupole moment, the ^{7}Li nucleus relaxes almost entirely by the quadrupolar

mechanism.[137] T_1 values for [7]Li typically range from 0.01 - 10 sec. However, [6]Li quadrupolar relaxation is over 600 times less efficient.[15] That is, [6]Li quadrupolar relaxation times are of the order of 6-6000 sec. This represents roughly half of the relaxation for hexameric, [6]Li-enriched alkyllithium compounds, but is less efficient for more symmetrical structures.[103]

(4) [6]Li-[7]Li dipole-dipole relaxation. As noted in Table 7, there is a significant increase in the relaxation times with [6]Li-enrichment. This difference in the relaxation properties upon isotopic substitution is apparently due to [6]Li-[7]Li dipolar relaxation in the natural abundance samples. This mechanism accounts for 30-60% of the relaxation in these compounds, as much as twice the contribution from [6]Li-^{1}H dipole-dipole relaxation.

If the rotational correlation time governing the [6]Li-[7]Li dipole-dipole interaction can be determined independently, the relaxation rate can be used to determine the lithium-lithium bond distance. Using this method, the lithium-lithium internuclear distance of tetrameric t-butyllithium at -85° C was found to be 2.7 Å.[103] A similar approach, based on the [13]C-[7]Li dipole-dipole relaxation time, has been used to measure the carbon-lithium internuclear distance in phenyllithium.[138]

(5) Temperature effects. As expected for relaxation dependent upon rotational or translational correlation times (which includes dipole-dipole and quadrupolar relaxation mechanisms),[139] the T_1 times generally decrease with a decrease in temperature. However, in compounds without nearby protons such as t-butyllithium, the T_1 times remain relatively long even at low temperatures.

(6) Solvent effects. Relatively little work has been done on the effect of various solvents on the [6]Li relaxation time. It is likely that different solvents will greatly affect the relaxation rate. It should be noted that even different hydrocarbon solvents can lead to dramatically different relaxation times. It was observed that small amounts of 2,2-dimethylbutane, present as an impurity in some commercial cyclopentane, contributed specifically to the relaxation of t-butyllithium above room temperature. Subsequent studies in 2,2-dimethylbutane solvent showed this was a specific interaction between the methyl protons of the hydrocarbon and lithium atoms in a less aggregated state which is only present at higher temperatures.[103] Such interactions can produce dramatic reductions in the observed T_1 times. For example, the T_1 time of

t-butyllithium in cyclopentane increases from 202 sec to 228 sec in going from 30° to 40° C. However, in 2,2-dimethylbutane, the T_1 time decreases from 122 sec to 17 sec!

(C) $^6Li\{^1H\}$ NOE experiments

As noted above, the ^6Li-^1H dipole-dipole mechanism contributes to the relaxation of most ^6Li nuclei. The presence of this relaxation mechanism can be used experimentally to identify short proton-lithium separations. The amount any given proton contributes to dipole-dipole relaxation is inversely proportional to r^6, where r is the lithium-proton internuclear distance.[140] For intermolecular interactions, the expression for dipolar relaxation is different,[139,140] but the relaxation still depends on the distance between the lithium nucleus and the proton. Since the amount of NOE observed (and the rate at which it builds up) is governed by T_1^{DD}, the observation of NOE caused by any given proton identifies is it as close to the ^6Li nucleus.

(1) 1D $^6Li\{^1H\}$ NOE experiments. One of the first applications of $^6Li\{^1H\}$ NOE to structural problems was a study of alkyltrihydro-metalates.[141] A strong enhancement of the ^6Li resonance when the hydrido hydrogens were irradiated showed their close proximity to the lithium ion. The method was also effective in detecting weak broad resonances in the ^1H spectrum.

$^6Li\{^1H\}$ NOE experiments have also shown that lithium is closest to the methylene protons anti to the methyl group in 2-methylallyllithium, and more distant from the methyl protons.[141] In this case, all of the protons contributed to ^6Li NOE, but the enhancement varied from only 12% for irradiation of the methyl protons to 29.5% for irradiation of the anti protons. A similar set of experiments indicated that lithium is near the C_1 and C_3 carbons of [1-(trimethylsilyl)allyl]lithium, but far from C_2.[160]

(2) 2D $^6Li\{^1H\}$ NOE experiments -- HOESY. An alternative method to sequentially irradiating each individual proton resonance is two dimensional heteronuclear NOE, HOESY. In these spectra the observe nucleus chemical shift (in this case ^6Li) is on one axis and the ^1H chemical shift is on the other axis. Correlations occur between ^6Li nuclei and any protons which contribute to the ^6Li nucleus's T_1^{DD} relaxation, i.e. those protons which are physically close to that particular ^6Li nucleus. First applied to $^{13}C\{^1H\}$ systems,[143-145] $^6Li\{^1H\}$ HOESY has been developed by Schleyer and coworkers for the detection of short

(<ca. 3.5Å) ^{6}Li-^{1}H separations.[51,52,146-152]

The basic pulse sequence for ^{6}Li{^{1}H}HOESY is shown as Pulse Sequence 4, where τ_m is the mixing time during which cross-

PULSE SEQUENCE 4

2D-heteronuclear NOE

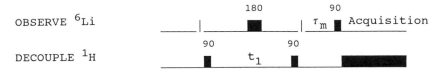

relaxation between ^{1}H and ^{6}Li occurs. The optimal value for τ_m has been discussed in detail for ^{13}C{^{1}H} systems.[153,154] Optimal values for the τ_m in ^{6}Li{^{1}H} HOESY have generally been ca. 2 sec for intramolecular interactions, although longer mixing times (3-4 sec or longer) are needed for some longer range or intermolecular interactions.

Using the originally suggested phase cycling,[144] the suppression of axial peaks (peaks at $f_1=0$, the center of the ^{1}H chemical shift range) and other artifacts is poor. Bauer and Schleyer[147] found that phase cycling of the lithium refocusing pulse (leading to a 64-step cycle) or using a 4-step CYCLOPS[94,155,157] phase cycling solves this problem. These workers also found[147] a further improvement in the 2D spectra by acquiring data in the pure absorption (phase sensitive) mode using the method reported by States.[156] As with other phase sensitive 2D experiments,[157] resolution was improved by removal of the strong tailing present in magnitude mode spectra. Phase sensitive spectra have significantly improved resolution, although magnitude mode spectra may be sufficient for many applications. Use of pseudo-steady-state 2D HOESY[158] produced no further improvement in the quality of the spectra.[147]

The ^{6}Li{^{1}H} HOESY experiment has been useful in the study of the mechanism of organolithium reactions by identifying short ^{1}H-^{6}Li separations.[52,148,151] More recently, the technique has been used as a direct means of studying ion pairs.[150] HOESY correlations were observed for contact ion pairs, but were not observed for solvent separated ion pairs. ^{6}Li{^{1}H} HOESY has also been important in establishing connectivity between resonances arising from nuclei within different portions of the same aggregate when methods dependent on scalar coupling (see Section II) can not be used. This method has been used successfully for lithium enolates[159] and lithium alkoxides,[62] as well as for oxygen and nitrogen bases coordinated to lithium.[51,52,147,150-152] For example, Figure 6 is the 2D ^{6}Li{^{1}H} HOESY spectrum of a mixture of (t-Bu)$_a$(tBuO)$_b$Li$_n$ aggregates where n=4 and 6. The connectivities established from the correlations in the HOESY spectrum enable the assignment of five different aggregates.

It should be noted that ^{6}Li-enrichment is not required for ^{6}Li{^{1}H} HOESY experiments. In fact, the first demonstration of the experiment was with a compound containing natural abundance lithium.[146] However, ^{6}Li-enrichment does significantly improve the sensitivity. This is due not only to the greater number of ^{6}Li nuclei, but also to the increased percent contribution of ^{6}Li-^{1}H dipole-dipole relaxation to the overall relaxation of ^{6}Li in the absence of ^{7}Li.

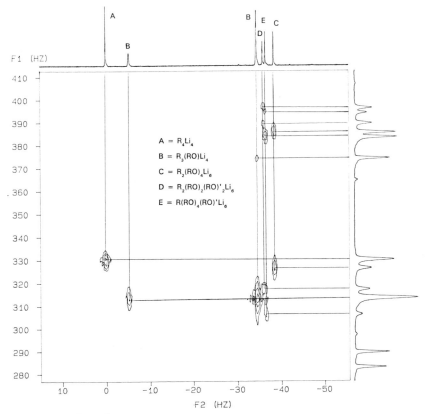

Fig. 6. ^{6}Li{^{1}H} 2D HOESY spectrum of (t-Bu)$_a$(t-BuO)$_b$Li$_n$ aggregates in cyclopentane at -15°C.

(D) Correlation between non-equivalent lithium nuclei--^{6}Li-^{6}Li COSY

Most organolithium aggregates contain magnetically equivalent lithium nuclei. This is due either to the symmetry of the aggregate or to rapid fluxional exchange which averages the chemical shifts of magnetically distinct lithium atoms. There are however exceptions. For these cases it is often critical to unambiguously establish that magnetically different lithium atoms belong to the same aggregate. Although there is likely some coupling between lithium nuclei, it is very small[161] and has never been resolved. However, correlations can be established even between lithium nuclei which show no resolvable coupling by using ^{6}Li-^{6}Li 2D homonuclear correlated spectroscopy, COSY.[162,52]

2D COSY NMR is one of the most widely used 2D NMR experiments.[91,157] Although primarily used for ^{1}H NMR, it has also been applied to other spin-1/2 (e.g. ^{31}P[163]) and to quadrupolar (e.g. ^{11}B,[164] ^{51}V,[165] ^{2}H[166]) nuclei.

The basic COSY pulse sequence could presumably be used for ^{6}Li, but it would require very large t_1 values due to the very small ^{6}Li-^{6}Li coupling. The preferable approach is the addition of fixed delays, Δ, as in Sequence 5 to produce longer evolution periods without having to increase the number of t_1 increments.[167,157]

PULSE SEQUENCE 5
2D homonuclear correlated spectroscopy, COSY, with fixed delays for optimization with small coupling

                        90   90
OBSERVE ^{6}Li   ■ Δ | t_1 ■ Δ | Acquisition
DECOUPLE ^{1}H   ▬▬▬▬▬▬▬▬▬▬▬▬▬▬▬▬▬▬▬▬▬▬▬▬

This pulse sequence, or the variation in which the second pulse is 45°, has been used successfully to observe ^{6}Li-^{6}Li correlations in several lithium compounds.[162,52,19] The choice of Δ in this experiment depends on the magnitude of the ^{6}Li-^{6}Li coupling and on the T_2 relaxation,[19] but is generally 0.1 - 0.4 sec.

(E) 2D-Exchange NMR

An additional NMR experiment which has had only limited use for organolithium compounds thus far is 2D-exchange spectroscopy, EXSY.[168-171] The pulse sequence for this experiment is the normal

NOESY pulse sequence shown in Pulse Sequence 6. In the resulting 2D spectrum, both f_1 and f_2 axes are identical chemical shift axes. Off-diagonal correlations occur when chemical exchange occurs between magnetically different sites during the mixing time, τ_m. The choice of τ_m depends upon the exchange rate and the T_1 relaxation times. The best sensitivity is typically achieved for $T_1/2 < \tau_m < 3T_1/2$.[172]

PULSE SEQUENCE 6
2D NOESY for use as 2D-EXSY NMR

This experiment is complicated in ^{1}H NMR due to homonuclear correlations (so-called J peaks), in addition to chemical exchange. However, this is not a problem for dilute spins or for nuclei which lack homonuclear dipolar interactions. It has been successfully used for a variety of other nuclei (^{13}C,[173] ^{119}Sn,[174,175] ^{31}P,[176] ^{195}Pt,[177] ^{51}V[172]) including ^{6}Li.[19,52] It is likely this technique will increase in importance as a means of qualitatively mapping exchange networks in organolithium compounds.

The exchange can also be treated quantitatively using the intensity of the off-diagonal peaks. This is best done with data acquired in the pure absorption (phase sensitive) mode.[178] A detailed error analysis of quantitative 2D EXSY has appeared.[177] For accurate results, it is often necessary to record a series of 2D exchange spectra with different τ_m values. Alternatively, multiple mixing times can effectively be introduced into one 2D experiment using accordion spectroscopy.[179] In this experiment the mixing time is equal to kt_1, where k is a constant, and the mixing time is incremented along with t_1. For this experiment, information on rates is present in the lineshapes of the cross peaks.

It should be noted that these techniques are not restricted to ^{6}Li NMR. In fact, for organolithium compounds with particularly long ^{6}Li T_1 times (and hence prohibitively long experiment times), ^{7}Li and/or ^{13}C{^{6}Li} 2D-EXSY is preferable.

IV.   OTHER TECHNIQUES

(A)   Chemical shifts of remote nuclei

Other experiments in this chapter were either based on lithium
NMR or depended upon coupling to lithium.   For those reasons, they
would be most affected by lithium isotopic substitution.   There
are some other techniques, based on nuclei remote from the lithium
atom, which can provide a significant amount of indirect or
secondary information.   For example, ^{13}C chemical shifts of
alkyllithium compounds, compared to the corresponding
hydrocarbons, give an indication of the degree of aggregation.[131]
Likewise, the ^{13}C chemical shifts of remote carbon atoms have been
used to study lithium phenolates,[48] enolates,[180] and phenyllithium.[181]   Since their application to [6]Li-enriched organolithium
compounds is essentially identical to that for compounds
containing natural abundance lithium, they will not be discussed
further here.

(B)   Solid state NMR

All of the techniques discussed in this chapter have been for
organolithium compounds in solution.   However, much of the
structural information for organolithium compounds still comes
from single crystal x-ray structures.   One possible means of
linking these experiments is by solid state NMR.

Solid state ^{13}C NMR using cross-polarization and magic angle
spinning (CPMAS)[182] has been used for several organolithium
compounds.[183-185]   The first such spectra were of samples of $LiCH_3$
and Li_2CH_2 containing either [6]Li-enriched or natural abundance
lithium.[183,184]   The resonances were quite broad for the natural
abundance samples due to dipolar coupling to [7]Li.   However,
isotopic substitution of lithium caused a significant decrease in
the line width.   Further narrowing, to ca. 80 Hz at half height,
was produced using [6]Li decoupling.

For more ionic compounds, lithium isotopic substitution
apparently has much less effect on the ^{13}C linewidths, as seen for
a ^{13}C CPMAS study of lithium fluorenide.[185]   The detailed
structural information for this latter compound suggests ^{13}C CPMAS
NMR studies may become an important tool for studying such
compounds.

ABBREVIATIONS, ACRONYMS, AND SYMBOLS

2-MeTHF	2-methyl-tetrahydrofuran
2,2-DMB	2,2-dimethylbutane
APT	attached proton test
Bu	butyl
COSY	2-D homonuclear correlated NMR spectroscopy
CPMAS	cross-polarization magic angle spinning
D_{3d}	3-fold symmetry point group
D_{4h}	4-fold symmetry point group
DEPT	distortionless enhancement by polarization transfer
DMS	dimethyl sulfide
eq	equivalents
Et	ethyl
Et_2O	diethyl ether
EXSY	2D exchange spectroscopy
f_1	second frequency domain in a 2D spectrum
f_2	observe frequency domain in a 2D spectrum
HETCOR	2-D heteronuclear correlated NMR spectroscopy
HMPA	hexamethylphosphoramide
HMPT	hexamethylphosphorictriamide
HOESY	2-D heteronuclear NOE correlated NMR spectroscopy
I	spin quantum no.
INEPT	insensitive nucleus enhancement by polarization transfer
J_{obs}	observed coupling constant
Me	methyl
mesityl	2,4,6-trimethylphenyl
NOE	nuclear Overhauser effect
NOESY	2-D homonuclear NOE correlated NMR spectroscopy
P	percent isotopic abundance
Ph	phenyl
PMDTA	N,N,N',N",N"-pentamethyldiethylenedtriamine
Pr	propyl
supermesityl	2,4,6-tri-(t-butyl)phenyl
T_1	spin-lattice relaxation time
T_2	spin-spin relaxation time

T_d	tetrahedral symmetry point group
THF	tetrahydrofuran
$THF-d_8$	deuterated tetrahydrofuran
TMEDA	N,N,N',N'-tetramethylethylenediamine
WALTZ	wideband alternating phase low-power technique for zero residue splitting
X	any nucleus other than carbon
γ	magnetogyric ratio
η	nuclear Overhauser enhancement factor

ACKNOWLEDGMENTS

Sincere thanks goes to my present and former co-workers who have contributed to the organolithium portion of our research efforts: Tim Bates, Matt Clarke, George DeLong, Randy Jensen, Don Ellington, Jon Longlet, Danny Pannell, and Corby Young. I also thank the Robert A. Welch Foundation for partial financial support of this work.

REFERENCES

1. B. J. Wakefield, The Chemistry of Organolithium Compounds, Pergamon Press, Oxford (1974).

2. E. Buncel, Carbanions: Mechanistic and Isotopic Aspects, Elsevier, Amsterdam, (1975).

3. J. C. Stowell, Carbanions in Organic Synthesis, Wiley, New York (1979).

4. M. Schlosser, Struktur und Reaktivität polarer Organometalle, Springer, Berlin (1973).

5. B. J. Wakefield, Comprehensive Organometallic Chemistry, G. Wilkinson, F. G. A. Stone, E. W. Abel, Eds., 7 (1982) 1.

6. P. von R. Schleyer, Pure Appl. Chem., 56 (1984) 151.

7. W. N. Setzer and P. von R. Schleyer, Adv. Organomet. Chem, 24 (1985) 353.

8. G. Boche, Angew. Chem. Int. Ed. Engl., 28 (1989) 277.

9. D. Seebach, Angew. Chem. Int. Ed. Engl. 27 (1988) 1624.

10. T. L. Brown, Acc. Chem. Res., 1 (1968) 23.

11. T. L. Brown, Adv. Organomet. Chem., 3 (1965) 365.

12. T. L. Brown, Pure Appl. Chem., 23 (1970) 447.

13. F. W. Wehrli, J. Magn. Reson., 23 (1976) 527.

14. F. W. Wehrli, Org. Magn. Reson., 11 (1978) 106.

15. F. W. Wehrli, J. Magn. Reson., 30 (1978), 193.

16. F. W. Wehrli, Ann. Rep. NMR Spec., 9 (1979) 125.

17. B. Lindman and S. Forsén, in NMR and the Periodic Table, R. K. Harris and B. E. Mann, Eds., Academic Press, London (1978) 129.

18. J. J. Dechter, Prog. Inorg. Chem., 29 (1982) 285.

19. H. Günther, D. Moskau, P. Bast, and D. Schmalz, Angew. Chem. Int. Ed. Engl., 26 (1987) 1212.

20. G. Fraenkel, A. M. Fraenkel, M. J. Geckle, and F. Schloss, J. Am. Chem. Soc., 101 (1979) 4745.

21. G. Fraenkel, M. Henrichs, J. M. Hewitt, B. M. Su, and M. J. Geckle, J. Am. Chem. Soc., 102 (1980) 3345.

22. D. Seebach, R. Hässig, and J. Gabriel, Helv. Chim. Acta, 66 (1983) 308.

23. D. Seebach, H. Siegel, J. Gabriel, and R. Hässig, Helv. Chim. Acta, 63 (1980) 2046.

24. R. Hässig and D. Seebach, Helv. Chim. Acta, 66 (1983) 2269.

25. N. N. Greenwood and A. Earnshaw, Chemistry of the Elements, Pergamon, Oxford (1984) 21.

26. K. J. Franklin, J. D. Halliday, L. M. Plante, and E. A. Symons, J. Magn. Reson., 67 (1986) 162.

27. B. J. Wakefield, Organolithium Methods, Academic Press, London (1988).

28. W. Bauer, M. Feigel, G. Müller, and P. von R. Schleyer, J. Am. Chem. Soc., 110 (1988) 6033.

29. D. Moskau, F. Brauers, H. Günther, and A. Maercker, J. Am. Chem. Soc., 109 (1987) 5532.

30. G. Fraenkel, H. Hsu, and B. M. Su, Lithium, Current Applications in Science, Medicine, and Technology, R. O. Bach, Ed., Wiley, New York (1985) 273.

31. R. D. Thomas and D. H. Ellington, Magn. Reson. Chem., 27 (1989) 628.

32. L. Muller, J. Am. Chem. Soc., 101 (1979) 4481.

33. M. F. Summers, L. G. Marzilli, and A. Bax, J. Am. Chem. Soc., 108 (1986) 4285.

34. J. H. Gilchrist, A. T. Harrison, D. J. Fuller, and D. B. Collum, J. Am. Chem. Soc., 112 (1990) 4070.

35. J. L. Wardell, Comprehensive Organometallic Chemistry, G. Wilkinson, F. G. A. Stone, E. W. Abel, Eds., 1 (1982) 43.

36. D. Seebach, Proc. Robert A. Welch Found. Conf. Chem. Res., 27 (1984) 93.

404

37. J. F. McGarrity and C. A. Ogle, J. Am. Chem. Soc., 107 (1985) 1805.

38. J. F. McGarrity, C. A. Ogle, Z. Brich, H.-R. Loosli, J. Am. Chem. Soc., 107 (1985) 1810.

39. P. Beak and A. I. Myers, Acc. Chem. Res., 11 (1986) 356.

40. A.-M. Sapse, K. Raghavachari, P. von R. Schleyer, and E. Kaufmann, J. Am. Chem. Soc., 107 (1985) 6483.

41. W. Bauer and D. Seebach, Helv. Chim. Acta, 67 (1984) 1972.

42. G. van Koten and J. G. Noltes, J. Organomet. Chem., 174 (1979) 367.

43. M. Weiner, C. Vogel, and R. West, Inorg. Chem., 1 (1962) 654.

44. P. West and R. Waack, J. Am. Chem. Soc., 89 (1967) 4395.

45. P. West, R. Waack, and J. I. Purmont, J. Am. Chem. Soc., 92 (1970) 840.

46. B. Y. Kimura and T. L. Brown, J. Organomet. Chem., 26 (1971) 57.

47. D. Margerison and J. P. Newport, Trans. Faraday Soc., 59 (1963) 2058.

48. L. M. Jackman and C. W. DeBrosse, J. Am. Chem. Soc., 105 (1983) 4177.

49. G. Fraenkel, W. E. Beckenbaugh, P. P. Yang, J. Am. Chem. Soc., 98 (1976) 6878.

50. A. Streitwieser, Jr., J. E. Williams, S. Alexandratos, and J. M. McKelvey, J. Am. Chem. Soc., 98 (1976) 4778.

51. W. Bauer, W. R. Winchester, and P. von R. Schleyer, Organometallics, 6 (1987) 2371.

52. W. Bauer, M. Feigel, G. Müller, P. von R. Schleyer, J. Am. Chem. Soc., 110 (1988) 6033.

53. L. D. McKeever, R. Waack, M. A. Doran, and E. B. Baker, J. Am. Chem. Soc., 90 (1968) 3244.

54. L. D. McKeever, R. Waack, M. A. Doran, and E. B. Baker, J. Am. Chem. Soc., 91 (1969) 1057.

55. R. D. Thomas, R. M. Jensen, and T. C. Young, Organometallics, 6 (1987) 565.

56. L. D. McKeever and R. Waack, J. Chem. Soc. Chem. Commun. (1969) 750.

57. G. Fraenkel, M. Henrichs, M. Hewitt and B. M. Su, J. Am. Chem. Soc., 106 (1984) 255.

58. R. D. Thomas, M. T. Clarke, and R. M. Jensen, Organometallics, 5 (1986) 1851.

59. T. F. Bates, M. T. Clarke, and R. D. Thomas, J. Am. Chem. Soc., 110 (1988) 5109.

60. T. F. Bates and R. D. Thomas, J. Organometal. Chem., 359 (1989), 285.

61. G. Fraenkel and W. R. Winchester, J. Am. Chem. Soc., 110 (1988) 8720.

62. R. D. Thomas and G. T. DeLong, unpublished results.

63. H-J Gais, J. Vollhardt, H. Günther, D. Moskau, H. J. Lindner, and S. Braun, J. Am. Chem. Soc., 110 (1988) 978.

64. R. Knorr, T. von Roman and V. von Roman, footnote 34 of ref. 28.

65. R. Knorr, T. von Roman and V. von Roman, footnote 32 of ref. 51.

66. D. Brendel and P. Warner, footnote 31 of ref. 51.

67. D. Hoell, J. Lex, K. Müllen, J. Am. Chem. Soc., 108 (1986) 5983.

68. A.-D. Schlüter, H. Huber, G. Szeimies, Angew. Chem. Int. Ed. Engl., 24 (1985) 404.

69. M. J. Goldstein, T. T. Wenzel, Helv. Chim. Acta, 67 (1984) 2029.

70. G. Fraenkel, Recent Advances in Anionic Polymerization, T. E. Hogen-Esch and J. Smid, Eds., Elsevier, New York (1987) 23.

71. G. Fraenkel, and P. Pramanik, J. Chem. Soc. Chem. Commun. 1983) 1527.

72. Footnote 13 in S. H. Bertz and G. Dabbagh, J. Am. Chem. Soc., 110 (1988) 3668.

73. H. J. Reich, D. P. Green, and N. H. Phillips, J. Am. Chem. Soc., 111 (1989) 3444.

74. S. Harder, J. Boersma, L. Brandsma, J. A. Kanters, W. Bauer, and P. von R. Schleyer, Organometallics, 8 (1989) 1696.

75. S. Harder, J. Boersma, L. Brandsma, J. A. Kanters, W. Bauer, R. Pi, P. von R. Schleyer, and H. Schöllhorn, and U. Thewalt, Organometallics, 8 (1989) 1688.

76. A. Rajca and L. M. Tolbert, J. Am. Chem. Soc., 107 (1985) 2969.

77. S. Harder, J. Boersma, L. Brandsma, A. von Heteren, J. A. Kanters, W. Bauer, P. von R. Schleyer, J. Am. Chem. Soc., 110 (1988) 7802.

78. L. M. Jackman, L. M. Scarmoutzos, B.D. Smith, and P. G. Williard, J. Am. Chem. Soc., 110 (1988) 6058.

406

79. L. M. Jackman, L. M. Scarmoutzos, and W. Porter, J. Am. Chem. Soc., 109 (1987) 6524.

80. L. M. Jackman and L. M. Scarmoutzos, J. Am. Chem. Soc., 109 (1987) 5348.

81. J. S. DePue and D. B. Collum, J. Am. Chem. Soc., 110 (1988) 5518.

82. A. S. Galiano-Roth, E. M. Michaelides, and D. B. Collum, J. Am. Chem. Soc., 110 (1988) 2658.

83. A. S. Galiano-Roth and D. B. Collum, J. Am. Chem. Soc., 111 (1989) 6772.

84. N. Kallman and D. B. Collum, J. Am. Chem. Soc., 109 (1987) 7466.

85. U. Edlund, T. Lejon, T. K. Venkatachalam, E. Buncel, 107 (1985) 6408.

86. I. J. Colquhoun, H. C. E. McFarlane, and W. McFarlane, J. Chem. Soc., Chem. Commun. (1982) 220.

87. P. B. Hitchcock, M. F. Lappert, P. P. Power, and S. J. Smith, J. Chem. Soc., Chem. Commun. (1984) 1669.

88. I. J. Colquhoun, H. C. E. McFarlane, W. McFarlane, Phosphorus Sulfur, 18 (1983) 61.

89. H. J. Reich and D. P. Green, J. Am. Chem. Soc., 111 (1989) 8729.

90. T. M. Gilbert and R. G. Bergman, J. Am. Chem. Soc., 107 (1985) 6391.

91. W. S. Brey, Pulse Methods in 1D and 2D Liquid-Phase NMR, Academic Press, San Diego (1988) 2.

92. S. L. Patt and J. N. Shoolery, J. Magn. Reson., 46 (1982) 535.

93. R. R. Ernst and W. A. Anderson, Rev. Sci. Instrum., 37 (1966) 93.

94. R. Freeman, A Handbook of Nuclear Magnetic Resonance, Longman Scientific, Essex (1988).

95. J. Vollhardt, H.-J. Gais, K. L. Lukas, Angew. Chem. Int. Ed. Engl., 24 (1985) 696.

96. P. Schmitt, J. R. Wesener, and H. Günther, J. Magn. Reson., 52 (1983) 511.

97. J. R. Wesener, H. Günther, Org. Magn. Reson., 21 (1983) 433.

98. J. R. Wesener, P. Schmitt, and H. Günther, J. Am. Chem. Soc., 106 (1984) 10.

99. O. Eppers and H. Günther, Tet. Lett., 30 (1989) 6155.

100. L. Müller, A. Kumar, and R. R. Ernst, J. Chem. Phys., 63 (1975) 5490.

101. H. Kessler, M. Gehrke, and C. Griesinger, Angew. Chem. Int. Ed. Engl., 27 (1988) 490.

102. R. Benn and Günther, Angew. Chem. Int. Ed. Engl., 22 (1983) 350.

103. D. H. Ellington, Ph.D. Dissertation, University of North Texas, Denton, TX (1990).

104. G. A. Morris and R. Freeman, J. Am. Chem. Soc., 101 (1979) 760.

105. D. P. Burum and R. R. Ernst, J. Magn. Reson., 39 (1980) 163.

106. D. M. Doddrell, D. T. Pegg, and M. R. Bendall, J. Magn. Reson., 48 (1982) 323.

107. D. T. Pegg, D. M. Doddrell, and M. R. Bendall, J. Chem. Phys., 77 (1982) 2745.

108. M. R. Bendall, D. M. Doddrell, D. T. Pegg, J. Am. Chem. Soc., 103 (1981) 4603.

109. O. W. Sørensen and R. R. Ernst, J. Magn. Reson., 51 (1983) 477.

110. C. J. Turner, Prog. NMR Spec., 16 (1984) 311.

111. G. A. Morris, J. Am. Chem. Soc., 102 (1980) 428.

112. J. Kowalewski and G. A. Morris, J. Magn. Reson., 47 (1982) 331.

113. T. A. Blinka, B. J. Helmer, and R. West, Adv. Organometal. Chem., 23 (1984) 193.

114. D. M. Doddrell, D. T. Pegg, M. R. Bendall, W. M. Brooks, and D. M. Thomas, J. Magn. Reson., 41 (1980) 492.

115. D. T. Pegg, D. M. Doddrell, W. M. Brooks, and M. R. Bendall, J. Magn. Reson., 44 (1981) 32.

116. N. Chandrakumar, J. Magn. Reson., 60 (1984) 28.

117. P. L. Rinaldi and N. L. Baldwin, J. Magn. Reson., 61 (1985) 165.

118. P. L. Rinaldi and N. L. Baldwin, J. Am. Chem. Soc., 104 (1982) 5791.

119. P. L. Rinaldi and N. L. Baldwin, J. Am. Chem. Soc., 105 (1983) 7523.

120. T. T. Nakashima, R. E. D. McClung, and B. K. John, J. Magn. Reson., 58 (1984) 27.

121. A. A. Maudsley, L. Müller, and R. R. Ernst, J. Magn. Reson., 28 (1977) 463.

408

122. G. Bodenhausen, and R. Freeman, J. Magn. Reson., 28 (1977) 471

123. D. Moskau, F. Brauers, H. Günther, and A. Maercker, J. Am. Chem. Soc., 109 (1987) 5532.

124. A. Bax, R. H. Griffey, and B. L. Hawkins, J. Magn. Reson., 55 (1983) 301.

125. L. Müller, J. Am. Chem. Soc., 101 (1979) 4481.

126. A. S. Galiano-Roth and D. B. Collum, J. Am. Chem. Soc., 110 (1988) 3546.

127. T. L. Brown and J. A. Ladd, J. Organomet. Chem., 2 (1964) 373.

128. R. Freeman, J. Chem. Phys., 53 (1970) 457.

129. K. Müllen and P. S. Pregosin, Fourier Transform NMR Techniques: A Practical Approach, Academic Press, New York (1976) 37.

130. P. A. Scherr, R. J. Hogan, and J. P. Oliver, J. Am. Chem. Soc., 96 (1974) 6055.

131. R. D. Thomas, M. T. Clarke, T. C. Young, J. Organomet. Chem., 328 (1987) 239.

132. E. D. Becker, High Resolution NMR, Theory and Chemical Applications, 2nd Ed., Academic Press, New York (1980) 59.

133. J. K. Becconsall, G. D. Daves, Jr., and W. R. Anderson, Jr., J. Am. Chem. Soc., 92 (1970) 430.

134. R. D. Thomas and M. T. Clarke, unpublished results.

135. R. D. Thomas, Ph.D. Dissertation, Wayne State University, Detroit, MI (1981).

136. J. H. Noggle and R. E. Schirmer, The Nuclear Overhauser Effect, Academic Press, New York (1971).

137. G. E. Hartwell and A. Allerhand, J. Am. Chem. Soc., 93 (1971) 4415.

138. L. M. Jackman and L. M. Scarmoutzos, J. Am. Chem. Soc., 106 (1984) 4627.

139. R. K. Harris, Nuclear Magnetic Resonance Spectroscopy, Pitman, London (1983).

140. T. C. Farrar and E. D. Becker, Pulse and Fourier Transform NMR, Academic, New York (1973).

141. A. G. Avent, C. Eaborn, M. N. A. El-Kheli, M. E. Molla, J. D. Smith, and A. C. Sullivan, J. Am. Chem. Soc., 108 (1986) 3854.

142. G. Fraenkel and W. R. Winchester, J. Am. Chem. Soc., 111 (1989) 3794.

143. P. L. Rinaldi, J. Am. Chem. Soc., 105 (1983) 5167.

144. C. Yu and G. C. Levy, J. Am. Chem. Soc., 105 (1983) 6694.

145. C. Yu and G. C. Levy, J. Am. Chem. Soc., 106 (1984) 6533.

146. W. Bauer, G. Müller, R. Pi, and P. von R. Schleyer, Angew. Chem. Int. Ed. Engl., 25 (1986) 1103.

147. W. Bauer and P. von R. Schleyer, Magn. Reson. Chem., 26 (1988) 827.

148. W. Bauer, T. Clark, and P. von R. Schleyer, j. Am. Chem. Soc., 109 (1987) 970.

149. W. Bauer, P. A. A. Klusener, S. Harder, J. A. Kanters, A. J. M. Duisenberg, L. Brandsma, and P. von R. Schleyer, Organometallics, 7 (1988) 552.

150. D. Hoffmann, W. Bauer, and P. von R. Schleyer, J. Chem. Soc., Chem. Commun, (1990) 208.

151. W. Bauer and P. von R. Schleyer, J. Am. Chem. Soc., 111 (1989) 7191.

152. K. Gregory, M. Bremer, W. Bauer, P. von R. Schleyer, N. P. Lorenzen, J. Kopf, and E. Weiss, Organometallics, 9 (1990) 1485.

153. K. Kövér and G. Batta, J. Magn. Reson., 69 (1986) 344.

154. K. Kövér and G. Batta, Prog. Nucl. Magn. Reson. Spec., 19 (1987) 223.

155. D. I. Hoult and R. E. Richards, Proc. R. Soc. London, Ser. A, 344 (1975) 311.

156. D. J. States, R. A. Haberkorn, and D. J. Ruben, J. Magn. Reson., 48 (1982) 286.

157. G. A. Morris, Magn. Reson. Chem., 24 (1986) 371.

158. P. Bigler, Helv. Chim. Acta, 71 (1988) 446.

159. E. M. Arnett, F. J. Fisher, M. A. Nichols, and A. A. Ribeiro, J. Am. Chem. Soc., 111 (1989) 748.

160. G. Fraenkel, A. Chow, and W. R. Winchester, J. Am. Chem. Soc., 112 (1990) 2582.

161. T. L. Brown, L. M. Seitz, B. Y. Kimura, J. Am. Chem. Soc., 90 (1968) 3245.

162. H. Günther, D. Moskau, R. Dujardin, and A. Maercker, Tet. Lett., 27 (1986) 2251.

163. R. J. Crowte and J. Evans, J. Chem. Soc., Chem. Commun., (1984) 1332.

164. T. L. Venable, W. C. Hutton, R. N. Grimes, J. Am. Chem. Soc., 106 (1984) 29.

165. P. J. Domaille, J. Am. Chem. Soc., 106 (1984) 7677.

410

166. D. Moskau and H. Günther, Angew. Chem. Int. Ed. Engl., 99 (1987) 156.

167. A. Bax and R. Freeman, J. Magn. Reson., 44 (1981) 542.

168. J. Jeener, B. H. Meier, P. Bachmann, and R. R. Ernst, J. Chem. Phys., 71 (1979)4546.

169. B. H. Meier and R. R. Ernst, J. Am. Chem. Soc., 101 (1979) 6441.

170. S. Macura and R. R. Ernst, Mol. Phys., 41 (1980) 95.

171. Y. Huang, S. Macura, and R. R. Ernst, J. Am. Chem. Soc., 103 (1981) 5327.

172. D. C. Crans, C. D. Rithner, and L. A. Theisen, J. Am. Chem. Soc., 112 (1990) 2901.

173. S. Macura, Y. Huang, D. Suter, and R. R. Ernst, J. Magn. Reson., 43 (1981) 259.

174. C. Wynants, G. Van Binst, C. Mugge, K. Jurkschat, A. Tzschach, H. Pepermans, M. Gielen, and R. Willem, Organometallics, 4, (1985) 1906.

175. R. Ramachandran, C. T. G. Knight, R. J. Kirkpatrick, E. J. Oldfield, J. Magn. Reson., 65 (1985) 136.

176. A. A. Ismail, F. Sauriol, S. Jacqueline, and I. S. Butler, Organometallics, 4 (1985) 1914.

177. E. W. Abel, T. P. J. Coston, K. G. Orrell, V. Sik, D. Stephenson, J. Magn. Reson., 70 (1986) 34.

178. C. L. Perrin and R. K. Gipe, J. Am. Chem. Soc., 106 (1984) 4036.

179. G. Bodenhausen and R. R. Ernst, J.Am. Chem. Soc., 104 (1982) 1304.

180. L. M. Jackman, N. M. Szeveneyi, J. Am. Chem. Soc., 99 (1977) 4954.

181. L. M. Jackman, L. M. Scarmoutzos, J. Am. Chem. Soc., 106 (1984) 4627.

182. C. S. Yannoni, Acc. Chem. Res., 15 (1982) 201.

183. J. A. Gurak, J. W. Chinn, Jr., R. J. Lagow, H. Steifink, C. S. Yannoni, Inorg. Chem., 23 (1984) 3717.

184. J. A. Gurak, J. W. Chinn, Jr., R. J. Lagow, R. D. Kendrick, C. S. Yannoni, Inorg. Chim. Acta, 96 (1985) L75.

185. D. Johnels and U. Edlund, J. Am. Chem. Soc., 112 (1990) 1647.

186. U. Edlund, T. Lejon, P. Pyykkö, T. K. Venkatachalam, and E. Buncel, J. Am. Chem. Soc., 109 (1987) 5982.

Isotopes in the Physical and Biomedical Science, Vol. 2, edited by E. Buncel and J.R. Jones   411
© 1991 Elsevier Science Publishers B.V., Amsterdam

Chapter 8

NMR AS A QUANTITATIVE ANALYTICAL TOOL IN CHEMICAL APPLICATIONS OF
ISOTOPES

ALEX D. BAIN and LYDIA LAO
Department of Chemistry, McMaster University, Hamilton, Ontario, Canada

CONTENTS

I. SCOPE OF THIS REVIEW

Nuclear magnetic resonance (NMR) is one of the most powerful methods of chemical analysis and is, by definition, isotope-specific. It has long been used in many types of chemical and isotope research, but the past few years have shown an explosive development in NMR.[1-4] In 1977, it was possible to review much of the quantitative work in ^{13}C NMR,[5] but this would fill many volumes now.[6] Rather, the purpose of this review is to acquaint workers in the field of isotope research with the power of basic quantitative NMR as well as with the new developments in the field.

The growth of the technique has been coupled with dramatic improvements in both the quality and availability of NMR instruments, to the point where now all serious chemical research laboratories have access to a multinuclear pulse Fourier transform spectrometer equipped with a superconducting magnet. This means that newer quantitative NMR methods are being developed and also that the standard analyses are being applied to a much wider range of isotopes and nuclei.[7,8] Because of this wide applicability, it is important to consider NMR as a quantitative analytical method.[9-13] This review will cover the various techniques that are used and the precautions that are needed for good results, and then illustrate them with representative examples.

The review is divided into three parts. The first section lays out the reasons to use NMR as an analytical tool. The advantages and disadvantages of NMR are discussed, compared to alternatives such as mass spectrometry and the use of radioactive isotopes. The second section covers the techniques and methods required to get good quantitative data out of an NMR spectrometer. Much of current NMR spectrometry is concerned more with qualitative rather than quantitative analysis; the identification of a new molecule or system rather than careful measurement of the amount of each species. In order to get reliable quantitation, modifications must be made to the way that NMR spectra are acquired. Finally, the third section covers applications and examples of the sorts of measurements that can be done.

II. THE CHOICE OF NMR AS AN ANALYTICAL TECHNIQUE

The basic phenomenon of NMR is the interaction of the magnetic moment of a nucleus with an external magnetic field. The nucleus is permitted only certain orientations in space, and we observe transitions between them. This obscure physical phenomenon has blossomed into one of the main methods for chemical analysis. For some elements, such as hydrogen, fluorine or phosphorus, the major isotope is magnetic and hence much NMR has been applied to their study. Although ^{13}C is only 1.1% abundant in natural samples, the pervasive presence of carbon has made its study very important.[5,6] In fact, there are certain advantages to the fact that ^{13}C is relatively dilute in a large amount of ^{12}C, a non-magnetic nucleus. Almost all elements in the periodic table have at least one isotope which is NMR-active and in some cases there is more than one (e.g. ^6Li, ^7Li, or ^{10}B, ^{11}B) which opens further possibilities. At least in principle, NMR can be applied to almost any system both in liquids and in the solid state.

In practice, the vast majority of NMR is done on molecules in solution. This is because in solution the lines in the NMR spectrum are remarkably narrow[14,15] (compared to other forms of spectroscopy) and so it is possible to observe very subtle differences in chemical environment for a nucleus. It is common to see lines that are less than 0.5 Hz wide in a 500 MHz proton spectrum, giving a precision of 1 part in 10^9. This is often taken for granted, but it is one of the reasons for the tremendous analytical power of NMR. The two main features of an NMR spectrum, the chemical shift and the spin-spin coupling patterns, give an almost complete picture of the structure of a molecule. The chemical shift reflects the electronic environment of the nucleus and the coupling patterns give the connectivity of the nuclei. In favorable cases, the full structure of a molecule can be deduced from this information. This is the use of NMR for qualitative analysis - identification of an unknown substance. As will be seen in this review, NMR also has important quantitative applications as well. For these reasons, NMR has become an indispensable tool for chemical analysis.

The real power of NMR in doing isotope research comes from the fact that not only does it signal the presence of an isotope in the sample, but also it often gives its exact location in the molecule. If a particular carbon is enriched in ^{13}C, this would show up as an unusually intense (M+1) peak in a mass spectrum, but it may not be possible to localize that enrichment without chemical degradation of the molecule. A ^{13}C NMR spectrum would show precisely where the enrichment is in the molecule. By means of these single label isotope experiments, the fate of an atom in a molecule can be studied.

This argument can be carried further to include double labelling experiments which follow the fate of groups of atoms comprising a chemical bond.[16-20] Two NMR-active nuclei which are bonded together almost always show a scalar spin-spin coupling. If these are rare isotopes, such as ^{13}C or ^{15}N, then the probability of the isotopomer with them directly bonded is very small. If a label is introduced with this pair of isotopes bonded together, then the observation of the scalar coupling in the final product is proof that the bond between them has not been broken. This is vital information for unraveling the mechanism of the reaction, and cannot be obtained in any other way.

The major disadvantage of NMR is its lack of absolute sensitivity. To begin with, only the difference in population between the energy levels of the nucleus in the magnetic field is observed in an NMR experiment. For typical cases, this difference is of the order of 10^{-5} of the total number of nuclei; in other words, only one nucleus in 10^5 is visible by NMR. This problem is made worse by the fact that many of the NMR-active isotopes are of relatively low natural abundance. Finally, NMR is a very low-frequency form of spectroscopy, so the transitions are much more difficult to detect than high-energy particles or photons. Compared to mass spectrometry or radioisotope work, where individual events can be counted, NMR is many orders of magnitude less sensitive and requires relatively large amounts of material (of the order of milligrams).

In spite of this serious disadvantage, there are several reasons for choosing NMR for doing isotope studies. To begin with, it is non-destructive. Although relatively large amounts of material are required, the sample is contained in a glass tube and can be fully recovered for further analysis. This non-destructive nature partially compensates for the lack of sensitivity of the technique, since signal averaging can recover some of the sensitivity. The signal/noise ratio in an NMR spectrum is proportional to the square root of the number of scans accumulated, so if a single scan takes a second, an hour's accumulation will increase the signal/noise sixty-fold.

The samples need only be soluble, so preparation and handling are minimal. Some radioactive isotopes (tritium, technicium)[21,22,23] are studied by NMR, but generally speaking the usual precautions of handling chemicals suffice. Because the sample is contained, there are no cross-contamination problems, such as occur in the vacuum system of a mass spectrometer. The samples can be changed (even automatically) as rapidly as the measurement of the spectrum permits. Solubility places some restriction on molecular weight as does the rigidity of the molecule, but the state of the art at the moment permits the observation and full assignment of the proton NMR spectrum of proteins with more than 100 amino acids[24,25] and there is considerable quantitative work done on synthetic polymers.[26-29]

For quantitative work, NMR has the considerable advantage of simplicity. For those that are familiar (or not familiar enough!) with a sophisticated modern spectrometer and the multitude of available experiments this may sound paradoxical, but the basic physics behind an NMR measurement is very simple. At equilibrium, a nuclear spin will assume the Boltzmann distribution amongst its various spin states. The NMR signal that is measured is directly proportional to the difference in level population between these states. Since the various chemical environments all have very similar energy level differences (chemical shifts are measured in parts per million), this population difference will be the same for all the different chemical environments of a given isotope. The NMR signal then should be directly proportional to the number of spins, regardless of the chemical environment.

The pulse NMR experiment proceeds as follows. In classical terms, this Boltzmann population difference creates a macroscopic magnetization along the magnetic field (which is conventionally called the z axis), as shown in figure 1(a). This is like a bar magnet which is aligned with the magnetic field. The essence of the pulse NMR experiment is that this magnetization is flipped down into the x-y plane by a pulse of radiofrequency power. This magnetization precesses at its resonant frequency around the z axis, and induces a current in coils which are placed in the x-y plane. If there are several lines, they precess independently. The current is detected to give the free induction decay (FID) which is then Fourier transformed to give the spectrum. Therefore, the connection between the number of spins and the detected signal is very direct: the magnetization is flipped down and it induces a measurable current. There is an overall sensitivity factor due to the gain of amplifiers, etc. but within a spectrum line intensities are directly comparable. There are, at least in principle, no extinction coefficients or calibrations necessary. The signal that is observed is directly proportional to the number of spins.

This is the simplest NMR experiment. In practice, for routine work more complex experiments such as DEPT[30,31] (for liquids) or cross-polarization[32] (for solids) are used to increase the signal/noise ratio in the spectrum. Furthermore, they can sort different types of nuclei, such as CH groups from CH_2 or CH_3 for separate quantitation. However, these experiments lose quantitative accuracy since the intensity now also depends on several experimental parameters such as pulse flip angles, coupling constants or cross-polarization rates.[32] For accurate work the simple experiments are the most reliable.

In order to get good quantitation, therefore, certain precautions are needed to ensure that the spin system is in equilibrium before it is observed, and that the collection and Fourier transformation of the FID are faithful. The next section covers the details of what these precautions are.

III. THE PRACTICE OF QUANTITATIVE NMR

Since the physics of the NMR experiment is so straightforward and simple, it is easy to understand the reasons behind the experimental techniques that are recommended for good quantitative measurements. These can be classified according to whether they ensure a valid initial state before the measurement, whether they mean that the different parts of the spectrum are measured uniformly, or whether the data processing has in any way distorted the data. It is taken for granted that good analytical practice has been followed in obtaining the sample and preparing it for the NMR measurement, so this section will concentrate on methods that are required in order to obtain a faithful measurement of the contents of the NMR tube.

To begin with, the simplest way of ensuring that a spin system is in equilibrium is to wait between acquisitions at least five times the longest spin-lattice relaxation time in the system.[33] After 5 time constants, an exponential relaxation is 99.3% complete, so this would give an error of no more than 0.7% in the quantitation. For careful work, a measurement of the spin-lattice relaxation time, T_1, should be done at least to establish an upper limit. Typical values of T_1 for most nuclei in solution are less than a few seconds, although pathological cases of relaxation times of 60 seconds or more do exist in liquids. For solid-state NMR samples, this can be a serious problem, since relaxation times in solids can be hours long. If long T_1 values mean that the time required to obtain a spectrum is excessive, then paramagnetic reagents can be added to the solution to reduce them. For organic solvents, the usual reagent is $Cr(acac)_3$ at approximately 5-50 mM concentration, and in aqueous solution $Cu(II)$ or $Mn(II)$ salts can be used. The purpose of these reagents is to relax the nuclei by interaction with the magnetic fields of the unpaired electrons, as they tumble in solution. In many cases the relaxation reagents affect all the nuclei similarly, but there also can be specific effects if there is association with the paramagnetic species.

If five T_1's are not left between pulses, then the spin system does not relax fully before the next pulse. After a few pulses, a steady state population difference is set up, which depends on T_1, the flip angle of the pulse, α, and the delay, τ, between pulses. If the x-y magnetizations are ignored, the effect of the pulse is to take the z magnetization M_z into $\cos(\alpha)M_z$. If M_{eq} is the equilibrium magnetization, then equation 1 gives the full expression for the steady state magnetization M_{ss}.

$$M_{ss} = M_{eq} \frac{(1 - e^{-\tau/T_1})}{(1 - \cos(\alpha)\, e^{-\tau/T_1})} \tag{1}$$

Therefore, if the relaxation times are similar, then partially saturated spectra can be used for rough quantitation, but for careful work, complete equilibration is necessary.

If proton decoupling is used during the acquisition of the spectrum, as is the usual case in ^{13}C work, this can also perturb the equilibrium state. If two nuclei, A and X, relax each other through the dipole-dipole mechanism and if A is irradiated the equilibrium population of X will be changed by the nuclear Overhauser effect (nOe).[34] If the original X magnetization was 1 then the steady state magnetization under irradiation of the A nucleus is given by equation 2.

$$M = 1 + \frac{\gamma_A \; R_{1\,dd}}{2\gamma_X \; R_1} \qquad (2)$$

where R_{1dd} is the dipolar contribution to the relaxation rate $(1/T_1)$ of the X spin, R_1 is the total relaxation rate of the X spin and the γ's are the magnetogyric ratios of A and X. Note that the magnetogyric ratio has a sign associated with it, so the second contribution in (2) may be positive or negative. When A is ^{1}H (the usual case in NMR) and X is ^{13}C, then the second part of (2) has a value of approximately 2 if dipolar relaxation dominates. This considerably increases the strength of the signal. However, for nuclei such as ^{15}N and ^{29}Si, the factor is negative so it is possible for the signal to vanish. For quantitative work, these Overhauser effects must be avoided by only decoupling the spectrum during the acquisition. During the delay between pulses the decoupler should be gated off and only turned on during the acquisition. This produces a decoupled spectrum, but with the intensities faithfully reflecting the equilibrium populations.

The next step in the NMR measurement is to excite the spectrum with an rf pulse. For multi-line spectra, all the nuclei should feel the same effect from the pulse. If a line is directly on-resonance with the rf radiation, then the magnetization is flipped down into the x-y plane as was shown in figure 1. If it is off-resonance, then the flipping is around an effective magnetic field which is the vector resultant of the rf magnetic field, B_1, along the x axis and the offset from resonance expressed as a magnetic field $(\omega - \omega_0)/\gamma$, where γ is the

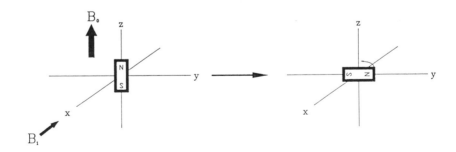

Figure 1.   Basic principle of the NMR measurement. The spins create a magnetization along the static field B_0, which is flipped into the x-y plane by the radiofrequency field B_1. This magnetization precesses around the z axis and induces a current in the receiver coil.

magnetogyric ratio of the nucleus. The rf magnetic field is often measured by quoting the 90^0 pulse width, which can easily be converted to a frequency, γB_1 in order to estimate these off-resonance effects. A 90^0 pulse width of 10 μs corresponds to the spins going through 360^0 or one cycle in 40 μs. This corresponds to a frequency of $1/(40 \times 10^{-6}) = 25$ kHz. The exact effect of this off-resonance excitation can be calculated, but a rule of thumb is that the offset should be less than γB_1. If the offset is equal to γB_1, then the intensity is still 98% of its on-resonance value,[35] but the phase is substantially distorted.

In between the end of the pulse and the start of the acquisition there is always a short deadtime or ringdown time to allow the electronics to recover from the strong rf pulse. If the lines in the spectrum are of very broad (i.e. the FID decays quickly), it is possible that this can affect the quantitation.[36] For instance, the $Al_{13}O_{40}$ cation has the Keggin structure with one tetrahedral aluminum at the centre and 12 surrounding aluminum sites which are approximately octahedral. The central site is very symmetrical, so the resonance is relatively sharp, but the slight asymmetry of the octahedral site causes strong relaxation through the nuclear quadrupole interaction. For a typical spectrum,[36] the sweep width would be about 30 kHz, and a normal value for the ringdown time would be about 20 μs. If the octahedral aluminum line is 4 kHz broad, then the time constant for the decay of the FID is $1/(\pi \times 4000) = 80$ μs. In the deadtime, this broad signal has decayed by $e^{-(20/80)} = 0.78$, whereas the sharp signal has hardly been affected. This is a particularly dramatic example, but it illustrates one of the problems with dealing with broad lines. This can be avoided partly by triggering the acquisition more quickly after the pulse, but there is a minimum value imposed by the electronics itself.

Once the FID has been acquired, it is almost always multiplied by a window or apodization function. The simplest and most popular is the exponential window function, which has the advantage of being an approximation to a matched filter. In this case, signal/noise is improved at the expense of some broadening of the lines. Since the intensity information in an FID tends to be near the beginning of the FID (the first point of the FID is directly proportional to the integral of the whole spectrum), the exponential window is the least distorting of quantitative information. Many other window functions are routinely used to enhance resolution,[37,38] but these will enhance the intensity of narrow lines relative to broad ones. If the lines in the spectrum are all of similar width (as is often the case in ^{13}C spectroscopy) then the choice of window functions is not crucial. On the other hand, if linewidths are quite different, serious bias can be introduced by the use of the wrong window function. Since Fourier transformation is done after the acquisition is finished, the effects of apodization can be estimated, since the same raw data can be re-apodized and re-transformed with a series of window functions. The results can then be compared and the best one selected.

The final step is of course the measurement of the intensity in the spectrum itself. In principle, the integral over the spectral lines corresponding to a given isotopic species is

418

directly proportional to the number of spins. In practice, getting good integrals (accurate to a few percent) can bring out some problems.[39,40] For example, figure 2 represents a typical integral. To begin with, the baseline of a Fourier transform spectrum is not directly related to the instrument, as it was in a CW NMR spectrometer. Often the baseline is substantially different from zero and since the integral is very sensitive to the baseline, it must be corrected. Most software packages include routines to run a smooth curve through selected points which are designated as baseline. This curve is then subtracted from the spectrum. Alternatively, the slope and bias of the integral itself can often be corrected, but this is equivalent to a correction of the baseline. Once the baseline is corrected the spectrum can then be numerically integrated.

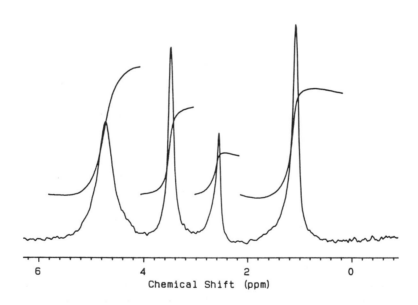

DEUTERIUM SPECTRUM, ETHANOL + STANDARD

Chemical Shift (ppm)

Figure 2.   Deuterium NMR spectrum of ethanol at natural abundance, run at 11.7 Tesla (500 MHz for ^{1}H). The peak at 2.3 ppm is due to tetramethylurea, an external intensity standard.

Instead of integrating the spectrum itself, it is sometimes better, particularly when peaks overlap, to fit a sum of calculated lines to the spectrum.[41-48] The individual components of the fitted spectrum can be integrated and so the overlapping lines can be separated. This procedure eliminates some of the operator biases in performing an integration, but it is limited by the power of the fitting procedure. Lines in an NMR spectrum are rarely pure Lorentzian or pure Gaussian, so some bias creeps in when these lineshapes are used, but unless the lines are severely overlapped this procedure works well.

The final way to extract quantitative information from an NMR spectrum is to use peak heights. This is a common practice in relaxation time measurements,[49] where the spectra are directly comparable, but it is also more widely applicable.[50] Peak heights are much less sensitive to operator biases and are more robust[51-53] than integrals, but they have the disadvantage that they destroy the direct connection to the number of spins. For an integral, no calibration factor is needed, but the peak height measurement requires a peak width to give accurate information. In practice, peak heights are often used for rough quantitation, and for accurate work the calibration can be done relatively easily. The integral is more closely related to the quantitative information that is required, but it is more sensitive to biases and systematic errors. The peak height is much easier to measure, but requires some calibration to recover the accuracy that is needed.

Several groups have investigated alternatives to the Fourier transform for interpreting the free induction decay.[54-56] The FID is a sum of decaying sine waves, and although an FT produces a spectrum from this, it is not the only way of separating all the frequencies that make up the sum. Essentially this sum can be modeled as a sum of frequencies with given intensity and phase, and then a best fit, according to some criterion, can be calculated. Two popular methods are the maximum entropy method (MEM)[54] and the linear prediction technique (LPZ).[55] These techniques require much more computation than a standard FT, and also work best when there is some a priori knowledge about the system. In some cases, spectacular improvements in the spectrum over a normal FT are obtained, but the methods are still not in common use. For accurate quantitative work, these methods may turn out to be important, since the a priori information about the number of peaks in the spectrum and their positions is often already available. Probably the lack of readily-available software is holding up a wide application of these techniques.

With these precautions, good quantitative NMR data can be acquired. The system should be in equilibrium before each acquisition, the acquisition should be done with as simple a pulse sequence as possible, and some care should be taken in the Fourier transform and other data processing to ensure the information is not distorted. These conditions are not always feasible, but the majority of the applications that follow illustrate their importance.

IV. APPLICATIONS OF QUANTITATIVE NMR

It is, of course, impossible to cover all the applications of quantitative NMR in isotope research, so this section will review some recent work that illustrate the various methods. There are the classic analytical applications of isotope ratios and the use of quantitative NMR to determine structures. Some examples will be the use of isotopic labels to follow the kinetics and mechanism of chemical reactions in both chemical and biosynthetic systems. One of the roles of the isotopic label is to follow the fate of an atom or a chemical bond through a complex series of reactions. In these reactions, there are kinetic isotope effects which can provide vital information on rate-determining steps in a mechanism, and can sometimes help distinguish between different possible schemes. The discovery of substantial differences in the natural abundance of deuterium, due to these isotope effects has opened new and powerful analytical procedures for some natural products. Another fast-growing area is the study of isotopes in living systems with in-vivo spectroscopy. Since NMR is non-destructive, either naturally-occurring species or labels can be studied with minimal perturbation of the living system. These results illustrate the tremendous power of NMR in such systems.

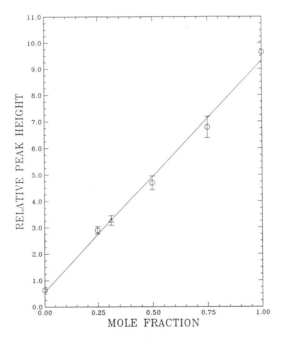

Figure 3.   Relative peak height of the HOD peak in a deuterium NMR spectrum of water. The samples were mixtures of water with 9.3 ppm ^{2}H and water with 148 ppm ^{2}H, and the x axis represents the mole fraction of 148 ppm water.

A. Hydrogen Isotopes

The simplest way to measure the amount of an isotope is to observe it directly against a standard. Zhang studied critically both the internal standard and external standard methods in his study of deuterium kinetic isotope effects by NMR methods.[57] In the internal standard method, one of the signals within the solution was used as a reference, whereas the external standard was in a separate sealed capillary contained inside the sample. Figure 3 shows the results of some similar work from our laboratory. The work involved measuring deuterium content of the water in water/DMSO mixtures, and deuterium NMR was chosen because mass spectrometry would require a separation of the DMSO whereas NMR could be applied directly to the raw sample. Figure 3 represents a calibration curve and test of the method. A series of calibration standards were made up by mixing waters of known enrichment, and then measuring the deuterium NMR intensity against a standard sample of DMSO in a separate concentric NMR tube. The circles in Figure 3 show the calibration points obtained this way, with their error bars ($\pm$ 2σ). The square on the graph represents a test sample of known concentration, and it fits well within the calibration curve.

Perhaps the most developed analytical methodology in the quantitation of hydrogen isotopes by NMR has been the work of Martin and his group[58-62] on the site-specific natural isotope fractionation (SNIF) of deuterium.[63,64] Because of kinetic isotope effects, the natural abundance of deuterium at various sites within a molecule can vary quite dramatically: $\pm$ 20 % or more.[58,63] This pattern of natural abundances of deuterium can be quite characteristic of the source of the material, and Martin has developed the technique to the stage where differences in deuterium content of less than 0.5% can be detected.[50,60,62] Unfortunately, this powerful technique is restricted to relatively small mobile molecules. Because of rapid quadrupolar relaxation, deuterium NMR lines become very broad (>50 Hz) for larger systems, and the chemical shift range for deuterium is very narrow (10 ppm = 767 Hz, at 11.7 Tesla, 500 MHz for ^{1}H). This leads to considerable overlap in the spectra. However, in the analysis of foodstuffs, where small molecules such as ethanol play an important role, the technique has provided one of the few ways to detect adulteration.

Because of the relatively high NMR sensitivity of protons, they can be used to detect quite low levels of impurities in large samples. With long acquisitions, impurities approaching the trace level can often be detected.[65] For a modern high-field spectrometer (500 or 600 MHz) proton signal/noise on a sample of 0.1% ethylbenzene of 500:1 is readily achievable. This is in a single acquisition, so with a thousand acquisitions (approximately one hour of spectrometer time) one ppm of material can be detected. For a typical small molecule, this represents less than a microgram of material.

The power of proton NMR spectroscopy has led to its being employed in quantitating many reactions and other chemical phenomena. Proton NMR was used to follow a photochemical reaction that was performed in the NMR tube.[66] Many other groups have used similar techniques to follow reactions[67] and to probe their mechanisms via the isotope effects.[39,40,68]   One elegant use of NMR has been the determination by Anet, Floss and co-workers of the absolute configuration of a chiral methyl group[23] by tritium NMR and the measurement of the enantiomeric purity of a mixture. The methyl group was incorporated into 2-methylpiperidine of known configuration, and the resulting diastereomers could be distinguished in the tritium NMR spectrum. McInnes and his group[20] used deuterium NMR to distinguish between a single-chain polyketide pathway in the biosynthesis of mollisin over a double-chain route. In more complex systems, Nicholson[69] has been applying chemometric and pattern recognition methods to the quantification of the proton NMR of biological fluids such as urine. With in-vivo NMR, it is now possible to do the quantitation in a living system as well, particularly for drugs or ethanol.[70,71]   These studies were all concerned with direct detection of the various isotopes. In favorable cases, both deuterium and protons can be measured simultaneously through the splittings observed in the ^{31}P NMR spectrum.[72]

B. Alkali Metals and Alkaline Earths

In some cases, such as with alkali metal ions, the standard and the unknown should be measured simultaneously, so something must be done to separate the two signals. In a measurement[73] of the ratio of ^6Li to ^7Li, solvent effects were used to achieve this. A coaxial NMR tube with a known solution of the two isotopes was used as an internal standard, and the standard signal was separated from the unknown by using different solvents. The authors found the results good to about $\pm$ 3% (2σ) and certainly competitive with thermionic mass spectrometry. Especially for ionic solutions, NMR shieldings can be quite solvent dependent, and in this case the difference was about 3 ppm - enough to ensure baseline separation between the peaks for both isotopes.

Where solvent effects are not large enough, chemical shift reagents can also be used. In aqueous solution, paramagnetic complexes[74,75] such as $Dy(EDTA)^{1-}$, where EDTA = ethylenediaminetetraacetate, can be used and the common shift reagent in non-aqueous solutions is $Eu(FOD)_3$.[76]   These complexes can also affect the relaxation times in the systems, since they are paramagnetic, and so while the NMR lines are shifted in these solutions, they may also be broadened. Careful attention to concentration can usually produce a solution where the unknown and the standard peaks are nicely separated.

A major application of this sort of quantitation is in biological systems. Because NMR is usually applied to species in solution and it is non-destructive, it provides a unique way of studying biological processes and following their time course. A simple example is the measurement of cell volume in a living species. One approach is to use ^{35}Cl NMR.[77-79] Inside the cell, the ^{35}Cl signal is very broad, because of rapid quadrupolar relaxation, so the signal from a cell suspension gives the extracellular volume. If the cells are then removed and the pure extracellular fluid measured, the ratio of the two intensities gives the cell volume. Similarly, cobalt complexes[80] are often impermeable to cell membranes and give a good strong ^{59}Co NMR signal. The signal strength in a cell suspension and a cell-free extract again gives cell volume.

If a molecule does permeate the cell membrane, then an impermeable shift reagent can distinguish the "inside" and "outside" species. These techniques have been widely used to detect and measure the relative concentrations of ions and their transport across cell membranes. In the natural state of the cell, quantitative NMR gives a non-invasive probe of both Na^+ and K^+.[81] Some care must be taken, however, since both [23]Na and [39]K are quadrupolar and in the presence of species that bind to the ions and slow their motions, part of the lineshape may broaden out.[82] This is due to spin relaxation[83] and part of the sodium may become "invisible". On the other hand, the quadrupolar nature of these nuclei can be exploited by performing a double-quantum filtered experiment[84] to detect and analyze these isotopes. Because these double-quantum experiments are more complex than the simple one-pulse acquisitions,[1] quantitative accuracy will suffer, but the experiments will be useful.

As well as simply measuring the ratio of ions inside and outside a cell, quantitative NMR can be used to study transport across a cell membrane. Sodium and potassium are of particular importance in normal metabolism,[74] and both lithium and rubidium have been studied in connection with antibiotics that alter the membrane properties of cells.[85-87] With the use of shift reagents, the ions inside and outside are distinguished, and if they permeate the membrane then the two sites are usually in the NMR slow exchange limit. Under these circumstances, the classical NMR methods for studying chemical exchange can be applied and the permeability of the membrane can be directly measured.

The alkali metals all have isotopes that are reasonably easy to study by NMR methods, but the same can not be said for the alkaline earths. In particular, [43]Ca and [25]Mg are both relatively difficult to detect directly. However, for both these isotopes indirect detection schemes are possible. In the ^{19}F NMR spectrum of several fluorinated chelating ligands, the position of the Ca-bound and the free ligand are well separated.[88] Since ^{19}F is 100% abundant and is almost as sensitive as protons, the Ca concentration can be measured quite easily at concentrations which would be impossible for direct [43]Ca NMR.[89] Similarly, fluorinated ligands have been used for magnesium, and in other biological investigations.[90,91]

C. Carbon

[13]C is probably the most used isotope for labelling and tracer studies, since it is a rare isotope but carbon forms part of almost all the systems that are studied.[5,6,16,17] There are too many applications of quantitative ^{13}C NMR to review completely, but they fall into a few classes. Allerhand has used his ultra-high resolution NMR[14,15] techniques to detect the small proportion of the enol form of acetone directly. In order to elucidate the photochemistry of cyclopropenes, Fahie and Leigh labelled one of the carbons and then analyzed the product mixture by quantitative NMR to see where the label went.[92] Vederas' group[16,19,93] has used stable isotopes and NMR extensively in his biosynthetic studies. In these cases, products could usually be isolated and purified for individual analysis, so that the fate of the label (or labels) could be measured. This is the classical labelling strategy.

One field of quantitative ^{13}C NMR that has developed recently is the extension to much more complex mixtures and systems, such as fossil fuels and living systems.[18,30,31] These represent extremely complex chemical mixtures, and the aim of the quantitative NMR in this case is to separate the substance to be analyzed from the background, as well as measure its level. Unless it constitutes 1% or more of the mixture, a sample which is 100% enriched in ^{13}C will still be invisible against the natural abundance carbon in all the other components. With biological systems,[18,94-97] there are further problems of simply getting the system into the spectrometer and keeping it stable for the course of the measurement. Furthermore, the NMR spectra must be localized in the particular organ or tissue under study. The explosive development of magnetic resonance imaging and related techniques has made this sort of work possible. Starting with labelled glucose or acetate, the time course of the incorporation into the biochemistry of the organism can be studied, since NMR is non-destructive of the sample. This can be done on perfused organs inside a relatively standard spectrometer,[97] or on whole animals.[94-96]

Carbon-13, in particular, is also amenable to many spectroscopic tricks to help in sorting the different types of environments prior to quantitation. Carbons can be sorted according to the number of directly bonded protons attached[3,4,30,98] and in the case of systems with multiple labels,[31] pulse sequences can be applied that select doubly- or multiply labelled species. Again, since these experiments rely on pulse flip angles and delays, the quantitative accuracy is not as good as a single-pulse experiment but the sorting can be extremely helpful.

D. Other Applications

In the study of chiral compounds, one of the crucial parameters is the enantiomeric purity of the sample, either before or after a reaction. This has, of course, important scientific implications, but is also of vital importance to the pharmaceutical industry.[99] The NMR of a variety of isotopes has been used to investigate this. As mentioned above,[23] tritium has been used in chiral methyl groups, but other work has been concerned with carbon,[100] fluorine,[101] phosphorus[102] and most recently, selenium.[103] In the selenium work, the authors were able to detect chiral groups five or six bonds away from the ^{77}Se nucleus.

An unusual application of quantitative NMR has been the measurement of empty space.[104-107] Ripmeester[104] discovered that the shielding of the ^{129}Xe nucleus is very sensitive to its physical environment - a physical shift, rather than a chemical one. The NMR spectrum of Xe adsorbed into the cavities of zeolites clearly shows the various environments and their relative proportions. Fraissard[105,106] and his group have taken this work and applied it to a wide variety of zeolites, and Brownstein has seen similar effects in polymers.[107]

Isotopes and their NMR have also played a role in determining the structure and properties of some inorganic solids. Zeolites are the classic case where large crystals are almost impossible to grow, so ^{27}Al and ^{29}Si NMR have been vital to the elucidation of these structures. In particular, aluminum and silicon atoms can be exchanged within the same structure, and these two elements are essentially indistinguishable by X-ray methods. NMR provides the only way of establishing where the elements are and in what proportion.[108] Similarly, in the study of solid electrolytes (lithium containing solids, in particular) NMR has been vital.[109]

V. CONCLUSIONS

NMR has proven to be an extremely useful method for chemical analysis in isotope research and will continue to grow. The technique is simple, since little sample preparation is needed and calibrations are not usually needed. Provided a few precautions are taken, the NMR experiment can give precise and accurate measurements of not only the total amount of an isotope, but also its chemical environment. This is the principal advantage of NMR: the provision of chemical information along with the quantitation. Although it lacks absolute sensitivity, NMR is non-destructive so the sample can be recovered completely. Now, since multinuclear spectrometers are readily available and the methods have been developed, we will see an increase in the use of quantitative NMR in isotope research.

ACKNOWLEDGEMENTS

We would very much like to thank the Natural Sciences and Engineering Research Council of Canada and Esso Petroleum Canada for their financial support, and IBM Canada for the provision of computers under the IBM - McMaster development project. Many of our colleagues made useful and constructive suggestions on the contents of this paper, and we would like to thank them all.

REFERENCES

1.   R.R. Ernst, G. Bodenhausen and A. Wokaun, Principles of Nuclear Magnetic Resonance in One and Two Dimensions, Clarendon Press, Oxford, 1987.

2.   M-L. Martin, G.J. Martin and J.J. Delpuech, Practical NMR Spectroscopy, Heyden, London 1980.

3.   J.K.M. Sanders and B.K. Hunter, Modern NMR Spectroscopy. A Guide for Chemists, Oxford University Press, Oxford, 1987.

4.   A.E. Derome, Modern NMR Techniques for Chemistry Research, Pergamon Press, Oxford, 1987.

5.   J. Hinton, M.Oka and A. Fry, in Isotopes in Organic Chemistry, volume 3, E. Buncel and C.C. Lee, editors, Elsevier, Amsterdam, 1977.

6.   H-O. Kalinowski, S. Berger and S. Braun, Carbon-13 NMR Spectroscopy, J. Wiley, New York, 1988.

426

7. R.K. Harris and B.E. Mann, NMR and the Periodic Table, Academic Press, New York, 1978.

8. P. Lazlo, NMR of Newly Accesible Nuclei, Volumes 1 and 2, Academic Press, New York, 1983 and 1984.

9. J.N. Shoolery, Prog. Nucl. Magn. Reson. Spectrosc., 11 (1977) 79.

10. D.E. Leyden and R.H. Cox, Analytical Applications of NMR, J. Wiley, New York, 1977.

11. L.D. Field and S. Sternhell, Analytical NMR, J. Wiley, New York, 1989.

12. D.L. Rabenstein and T.T. Nakashima, in Trace Analysis, Spectroscopic Methods for Molecules, G.D. Christian and J.B. Callis eds., J. Wiley, New York, 1986.

13. B. Thomas and A. Portzel, Zeitschrift für Chemie, 26 (1986) 59.

14. S.R. Maple and A. Allerhand, J. Magn. Reson., 80 (1988) 394.

15. S.R. Maple and A. Allerhand, J. Am. Chem. Soc., 109 (1987) 6609.

● 16. J.C. Vederas, Natural Products Reports, 4 (1987) 227.

17. T.J. Simpson, Chem. Soc. Rev., 16 (1987) 123.

18. R.E. London, Prog. Nucl. Magn. Reson. Spectrosc., 20 (1988) 337.

● 19. B.J. Rawlings, P.B. Reese, S.E. Ramer and J.C. Vederas, J. Am. Chem. Soc., 111 (1989) 3382.

20. A.G. McInnes, J.A. Walter, J.L.C. Wright, L.C. Vining, N.Ranade, R. Bentley and E.P. McGovern, Can. J. Chem., 68 (1990) 1.

21. E.A. Evans, D.C. Warrell, J.A. Elvidge and J.R. Jones, Handbook of Tritium NMR Spectroscopy and Applications, J. Wiley, New York, 1985.

22. E. Buncel and C.C. Lee, Editors, Tritium in Organic Chemistry, volume 4 in Isotopes in Organic Chemistry, Elsevier, Amsterdam, 1978.

23. F.A.L. Anet, D.J. O'Leary, J.M. Beale and H.G. Floss, J. Am. Chem. Soc., 111 (1989) 8935.

24. G.M. Clore and A.M. Gronenborn, Crit. Rev. Biochem. Mol. Biol., 24 (1989) 479.

25. J. Weng, A.P. Hinck, S.N. Loh and J.L. Markley, Biochemistry, 29 (1990) 4242.

26. D.R. Eaton, M. Mlekuz, B.G. Sayer, A.E. Hamielec and L.K. Kostanski, Polymer, 30 (1989) 514.

27. R. Chujo, K. Hatada, R.K. Kitamura, T. Kitayama, H. Sato and Y. Tanaka, Polymer J., 19 (1987) 413.

28. R. Chujo, Y. Tanaka, Y. Terawaki, K. Hatada, T. Kitayama, R. Kitamura, F. Horii and H. Sato, Polymer J., 20 (1988) 627.

29. B.J. Burger, M.E. Thompson, W.D. Cotter and J.E. Bercaw, J. Am. Chem. Soc., 112 (1990) 1566.

30. D.A. Netzel, Anal. Chem., 59 (1987) 1775.

31. D. Baudot, J. Brondeau, D. Canet and F. Martin, FEBS Letters, 231 (1988) 11.

32. C.E. Snape, P.Tekely, B.C. Gerstein, D.E. Axelson, M. Pruski, M.A. Wilson, R.E Botto, J.J. Delpeuch and G.E. Maciel, Fuel, 68 (1989) 547.

33.   R.K. Harris and R.H. Newman, J. Magn. Reson., 24 (1976) 449.

34.   D. Neuhaus and M.P. Williamson. The Nuclear Overhauser Effect in Stereochemical and Conformational Analysis, VCH Publishers, New York, 1989.

35.   P. Meakin and J.P. Jesson, J. Magn. Reson., 10 (1973) 290.

36.   A.R. Thompson, A.C. Kunwar, H.S. Gutowsky and E. Oldfield, J. Chem. Soc. Dalton, (1987) 2317.

37.   A.G. Ferrige and J.C. Lindon, J. Magn. Reson., 31 (1978) 337.

38.   D.D. Traficante and D. Ziessow, J. Magn. Reson., 66 (1986) 182.

39.   F.A.L. Anet and M. Kopelevich, J. Am. Chem. Soc., 108 (1986) 1355.

40.   F.A.L. Anet and D.J. O'Leary, Tetrahedron Lett., 30 (1989) 1059.

41.   C. Giancasparo and M.B. Comisarow, Applied Spectroscopy, 37 (1983) 153.

42.   K. McLeod and M.B. Comisarow, J. Magn. Reson., 84 (1989) 490.

43.   F. Ni and H. Scheraga, J. Magn. Reson., 82 (1989) 413.

44.   M. Schmidt and D. Ziessow, Ber. Bunsen. Gesellschaft für Phys. Chem. 91 (1987) 1110.

45.   F. Abildgaard, H. Gesmar and J.J. Led, J. Magn. Reson., 79 (1988) 78.

46.   T. Kupka and J.O. Dziegielewski, Magn. Reson. Chem., 26 (1988) 353.

47.   R.E. Hoffman and G.C. Levy, J. Magn. Reson., 83 (1989) 411.

48.   S.J. Nelson and T.R. Brown, J. Magn. Reson., 84 (1989) 95.

49.   I.M. Armitage, H. Huber, D.H. Live, H. Pearson and J.D. Roberts, J. Magn. Reson., 15 (1974) 142.

50.   C. Guillou, M. Trierweiler and G.J. Martin, Magn. Reson. Chem., 26 (1988) 491.

51.   G.E.P. Box, Biometrika, 40 (1953) 318.

52.   P.J. Huber, Annals of Mathematical Statistics, 43 (1972) 1041.

53.   A.D. Bain, J. Magn. Reson., 77 (1988) 125.

54.   S. Sibisi, J. Skilling, R.G. Brereton, E.D. Laue and J. Staunton, Nature, 311 (1984) 446.

55.   J. Tang and J.R. Norris, J. Magn. Reson., 69 (1986) 180.

56.   H. Barkuijsen, R. deBeer and D. VanOrmondt, J. Magn. Reson., 73 (1987) 553.

57.   B.L. Zhang, Magn. Reson. Chem., 26 (1988) 955.

58.   G.J. Martin and M-L. Martin, Tetrahedron Lett., 22 (1981) 3525.

59.   G.J. Martin, B.L. Zhang, N. Naulet and M-L. Martin, J. Am. Chem. Soc., 108 (1986) 5116.

60.   G.J. Martin and N.Naulet, Fresenius Z. Anal. Chem., 332 (1988) 648.

61.   G.J. Martin, C. Guillou, M-L. Martin, M-T. Cabanis, Y. Tep and J. Aerny, J. Ag. Food Chem. 36 (1988) 316.

62.   G.J. Martin, M-L. Martin, F. Mabon and M.J. Michon, Anal. Chem. 54 (1982) 2380.

63.   P.Rauschenbach, H. Simon, W. Stichler and H. Moser, Z. Naturfor., 34C (1979) 1.

428

64.   D.M. Grant, J. Curtis, W.R. Croasmun, D.K. Dalling, F.W. Wehrli and S. Wehrli, J. Am. Chem. Soc., 104 (1982) 4492.

65.   P. Fux, Analyst, 115 (1990) 179.

66.   R.F. Childs and G.S. Shaw, J. Am. Chem. Soc., 110 (1988) 3013.

67.   W. Smadja, Tetrahedron Lett., 29 (1988) 1283.

68.   R.A. Pascal, M.W. Baum, D.S. Huang, L.R. Rodgers and C.K. Wagner, J. Am. Chem. Soc., 108 (1986) 6477.

69.   J.K. Nicholson and I.D. Wilson, Prog. Nucl. Magn. Reson. Spectrosc., 21 (1989) 449.

70.   J.L. Evelhoch, B.P. Giri and C.L. McCoy, Magn. Reson. Medicine, 9 (1989) 402.

71.   O.A.C. Petroff, E.J. Novotny, T. Ogino, M. Avison and J.W. Prichard, J. Neurochem., 54 (1990) 1188.

72.   J.A. Valley and R.F. Evila, Spectroscopy Lett. 22 (1989) 541.

73.   K.J. Franklin, J.D. Halliday, L.M. Plante and E.A. Symons, J. Magn. Reson., 67 (1986) 162.

74.   C.S. Springer, Annual Review of Biophysics and Biophysical Chemistry, 16 (1987) 375.

75.   J. Szklaruk, J. Marecek, A.L. Springer and C.S. Springer, Inorg. Chem., 29 (1990) 660.

76.   G.M. Whitesides and D.W. Lewis, J. Am. Chem. Soc., 92 (1970) 6979.

77.   B.M. Rayson and R.K. Gupta, J. Biol. Chem., 260 (1985) 7276.

78.   D. Hoffman and R.K. Gupta, J. Magn. Reson., 70 (1986) 481.

79.   D. Hoffman, A.M. Kumar, A. Spitzer and R.K. Gupta, Biochem. Biophys. Acta, 889 (1986) 355.

80.   H. Shinar and G. Navon, FEBS Letters, 193 (1985) 75.

81.   T.L. Knubovets, U. Eichoff, A. Revazov and L.A. Sibeldina, Magn. Reson. Medicine, 9 (1989) 261.

82.   A. Delville, C. Detellier and P. Laszlo, J. Magn. Reson., 34 (1979) 301.

83.   S. Padmanabhan, M.T. Record, B. Richey and C.F. Anderson, Biochemistry, 27 (1988) 4367.

84.   Y. Seo, M. Murakami, E. Suzuki, S. Kuki, K. Nagayama and H. Watari, Biochemistry, 29 (1990) 601.

85.   M.P. Williamson, FEBS Letters, 254 (1989) 171.

86.   F. Riddell, Biochem. Biophys. Acta, 984 (1989) 6.

87.   A. Nakamura, S. Nagal, T.Ueda, J. Sakakibara, Y. Hotta and K. Takeya, Chem. Pharm. Bull., 37 (1989) 2330.

88.   G.A. Smith, R.T. Hesketh, J.C. Metcalfe, J. Feeney and P.G. Morris, Proc. Nat. Acad. Sci., 80 (1983) 7178.

89.   F. Marban, Y. Koretsune, M. Corretti, V.P. Chacko and H. Kusuoka, Circulation, 80 (suppl. IV) (1989) 17.

90. L.A. Levy, R.E. London, E. Murphy and B. Raju, Biochemistry, 27 (1988) 4041.

91. R.E. London, and S.A. Gabel, Biochemistry, 28 (1989) 2378.

92. B.J. Fahie and W.J. Leigh, Can. J. Chem., 67 (1989) 1859.

93. C.R. McIntyre, F.E. Scott, T.J. Simpson, L.A. Trimble and J.C. Vederas, Tetrahedron, 45 (1989) 2307.

94. R.S. Badar-Goffer, H.S. Bachelard and P.G. Morris, Biochem. J., 266 (1990) 133.

95. J.R. Brainard, R.E. London, D.M. Bier and R.S. Downey, Analytical Biochemistry, 176 (1989) 307.

96. T. Jue, D.L. Rothman, G.I. Shulman, B.A. Tavitian, R.A. DeFronzo and R.G. Shulman, Biophysics, 86 (1989) 4489.

97. S.M. Cohen, Biochemistry, 26 (1987) 581.

98. R. Radeglia, J. für Praktische Chemie, 331 (1989) 705.

99. H.Y. Aboul-Enein, Analytical Letters, 21 (1988) 2155.

100. C. Rabiller and F. Maze, Magn. Reson. Chem., 27 (1989) 582.

101. Z. Glowacki, M. Topolski, E. Matczak-Jon and M. Hoffman, Magn. Reson. Chem. 27 (1989) 922.

102. D. Cuvinot, P. Mangeney, A. Alexakis, J-F. Normant and J-P. Lellouche, J. Org. Chem. 54 (1989) 2420.

103. L.A. Silks, R.B. Dunlap and J.D. Odom, J. Am. Chem. Soc., 112 (1990) 4979.

104. J.A. Ripmeester, J. Am. Chem. Soc., 104 (1982) 289.

105. J. Fraissard and T. Ito, Zeolites, 8 (1988) 350.

106. T. Ito, J. Fraissard and M.A. Springuelhuet, Zeolites, 9 (1989) 68.

107. S.K. Brownstein, J.E.L. Roovers and D.J. Worsfold, Magn. Reson. Chem., 26 (1988) 392.

108. P.P. Man and J. Klinowski, J. Chem. Soc. Chem. Commun., (1988) 129.

109. H. Eckert, J.H. Kennedy and Z.M. Zhang, J. Non-Cryst. Solids, 107 (1989) 271.

Isotopes in the Physical and Biomedical Science, Vol. 2, edited by E. Buncel and J.R. Jones
© 1991 Elsevier Science Publishers B.V., Amsterdam

431

Chapter 9

APPLICATION OF STABLE ISOTOPE LABELLING AND MULTINUCLEAR NMR TO BIOSYNTHETIC STUDIES

T. J. SIMPSON
School of Chemistry, University of Bristol, England.

CONTENTS

I.   INTRODUCTION

The study of biosynthetic pathways received a major new impetus in the early 1970s with the development of pulsed Fourier-transform n.m.r. spectrometers which greatly facilitated the routine determination of ^{13}C n.m.r. spectra of realistically available amounts of natural products. In addition, precursors highly enriched with ^{13}C and other stable isotopes became more available. These were opportune developments as structures were increasingly being determined by physical methods so that biosynthetic studies using radioisotopes which necessitated extensive degradations to locate the position of incorporation of label became correspondingly difficult. This was further exacerbated by the increasing complexity of many of the molecules of current interest, e.g. vitamin B_{12}. The development of n.m.r. based methodology again provided a biosynthetic technique complementary to the methods used for structural work. Early work in this area has been reviewed.[1]

Subsequent and equally important developments have involved the use of precursors doubly labelled with ^{13}C together with ^{18}O, ^{2}H or ^{15}N. These enable the biosynthetic origins and metabolic fate of hydrogen, oxygen and nitrogen to be determined by observation of isotope-induced shifts (or spin coupling) in ^{13}C n.m.r. spectra. The incorporation of label from ^{2}H or ^{15}N-enriched precursors may also be observed directly by ^{2}H or ^{15}N n.m.r. Some success has been achieved in the use of ^{17}O and ^{3}H n.m.r., and ^{31}P n.m.r. has been particularly valuable in the study of phosphate metabolism.

In this chapter the general principles involved in applying ^{13}C n.m.r. and n.m.r. of other isotopes to the study of secondary metabolism will be described. For convenience, many of these techniques will be illustrated by reference to studies with micro-organisms in the author's own laboratories.[2] Due to the high efficiency of incorporation of label usually required, many n.m.r. based studies have been carried out on microbial metabolites or using purified enzyme systems. However, many successful studies have been achieved using higher plants.[3] The overall aim will be to describe the basic requirements for carrying out labelling studies with n.m.r. observation of isotope incorporation, to give some idea of the strengths and limitations of these techniques, and general considerations in interpreting the results obtained.

II.   ASSIGNMENT OF SPECTRA

An essential prerequisite of n.m.r. studies is the complete, unambiguous assignment of all the resonances in the n.m.r. spectrum. In the case of large molecules with many carbons and hydrogens, this may constitute the major part of the biosynthetic study. While the number of techniques for this are increasing rapidly and some general principles apply, each molecule may pose its own particular challenges. The traditional methods include the use of compilations of known chemical shifts and tables of substituent chemical shift effects, off-resonance and specific proton decoupling, lanthanide and solvent-induced shift studies, comparisons of chemical shifts in model compounds and in related series of derivatives, synthesis of specifically isotopically-labelled compounds, and incorporation of labelled precursors. These have been adequately discussed elsewhere[1,3] so they will not be considered further here. In recent years the task has been greatly assisted by the development of a wide range of multi-pulse sequences which allow selective decoupling and observation of resonances, and of two-dimensional spectroscopic methods which allow correlation of resonances through bonds and also through space. Most of these techniques are, of course, of general applicability although the rigorous requirements of biosynthetic work often demand their most extensive use. A number of recent books and reviews should be consulted for clear and concise applications of these techniques.[4-8] The application of several of these techniques is illustrated in spectral and structural arrangement studies of the fungal metabolites LL-D253α (1),[9] terretonin (2)[10] and astellatol (3).[11]

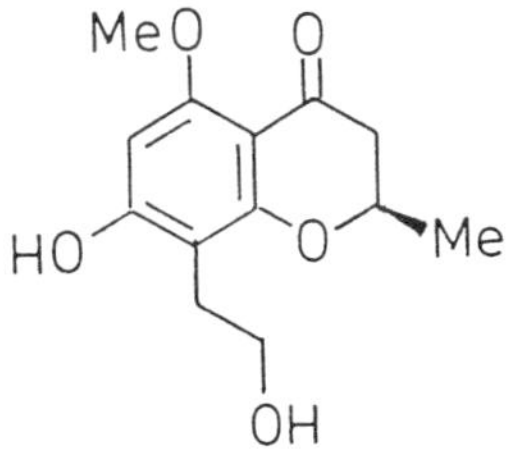

(1)

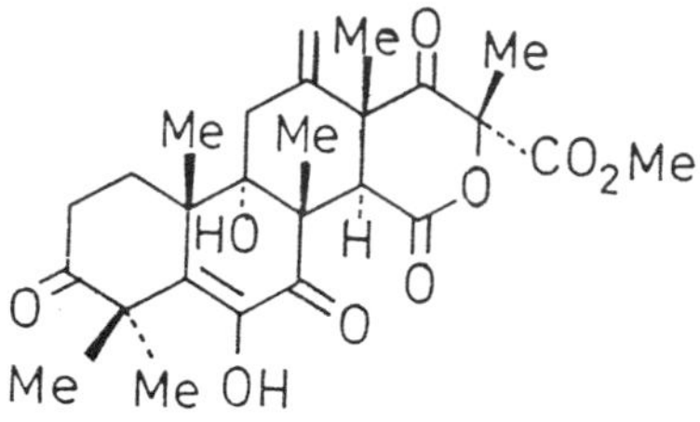

(2)

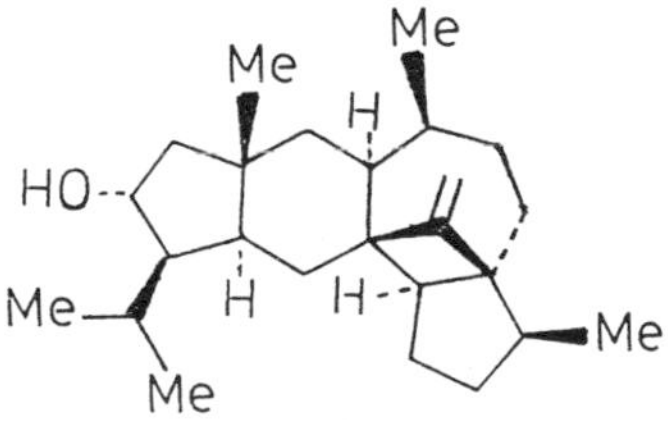

(3)

III.   PRECURSOR INCORPORATION[1]

Before carrying out experiments with stable isotopes it is necessary to carry out extensive experiments to optimise conditions for feeding of labelled substrates. It is routine procedure to perform these preliminary experiments with ^{14}C-labelled precursors. Precursor efficiency may be assessed in several ways in biosynthetic experiments but for ^{13}C studies the important criterion is *dilution* of added label. For ^{14}C, dilution per labelled site is given by:-

$$\text{Dilution} = \frac{\text{specific activity of precursor}}{\text{specific activity of product}} \text{ x number of labelled sites}$$

434

To obtain unequivocal results in ^{13}C studies using typically 90% enriched precursors, dilutions per labelled site of *ca.*100 or less are therefore required.

Three main parameters require studying: time of precursor addition, period of growth after addition, and amount of precursor required. Maximum incorporation usually occurs when precursors are added at the start of maximum metabolite production, e.g. the start of the idiophase in the case of microbial fermentations. This necessitates determination of cell growth and metabolite production curves. The period of growth after addition of precursor may be critical. If the turnover of exogenous precursor is rapid, then prolonged growth results merely in increased dilution of labelled metabolite by further metabolite production from unlabelled endogenous substrate. Thus it is important to note that neither maximum yield of metabolite nor maximum total incorporation of label is the important factor, the prime consideration being to obtain the *minimum dilution* of label given a *sufficient yield* of metabolite for the ^{13}C n.m.r. spectrum to be determined. In general, dilution decreases with increasing amounts of precursor. Thus mass versus incorporation studies will determine the minimum amount of labelled precursor that must be added to obtain a satisfactory enrichment.

The amount of labelled precursor that can be used is limited both by the actual expense of the isotopic label and also due to the adverse effects that high levels of added precursor may have on the metabolism of the organism and may in fact change the biosynthetic pathway under study. Toxicity effects may be overcome by pulsed addition of precursor or by continuous slow addition over a prolonged period by use of e.g. peristaltic pumps. Although this disadvantage of elevated amounts of precursors is often cited, it is rarely encountered. Indeed, the use of elevated amounts of precursor may provide distinct advantages over the tracer amounts used in most studies with radioactive isotopes. By flooding metabolic pools with exogenous precursor, enzymes which cause randomisation of label, e.g. via the tricarboxylic acid cycle may be overloaded and so more specific incorporation of label may be observed with stable isotopes than in corresponding experiments with radioisotopes.

IV. ^{13}C ENRICHMENT STUDIES

The labelling pattern resulting from incorporation of ^{13}C-enriched precursors is determined by obtaining the proton noise decoupled (p.n.d.) ^{13}C n.m.r. spectrum of the labelled metabolite and comparing it with the spectrum of the unlabelled metabolite. A large number of different types of experiments are now available.

A.   Single ^{13}C-Labelling

This was the first method to be developed and remains the simplest. It is best understood by examining the formal model system shown in Figure 1.

Use of ^{13}C-nmr in Isotope Labelling Experiments*

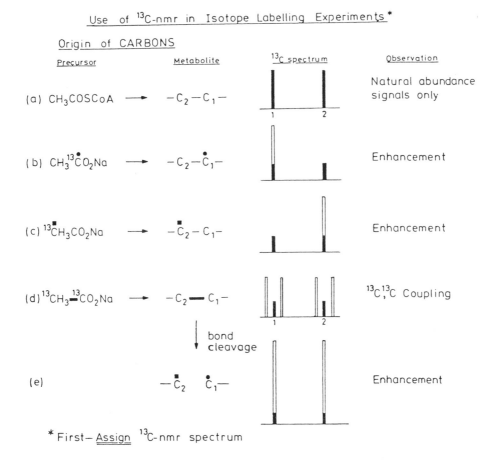

Origin of CARBONS

Precursor	Metabolite	^{13}C spectrum	Observation

(a) $CH_3COSCoA \longrightarrow -C_2-C_1-$ — Natural abundance signals only

(b) $CH_3{}^{13}\overset{\bullet}{C}O_2Na \longrightarrow -C_2-\overset{\bullet}{C}_1-$ — Enhancement

(c) $^{13}\overset{\blacksquare}{C}H_3CO_2Na \longrightarrow -\overset{\blacksquare}{C}_2-C_1-$ — Enhancement

(d) $^{13}CH_3{}^{13}CO_2Na \longrightarrow -C_2\!\!-\!\!C_1-$ — $^{13}C,^{13}C$ Coupling

bond cleavage

(e) $-\overset{\blacksquare}{C}_2 \quad \overset{\bullet}{C}_1-$ — Enhancement

*First – Assign ^{13}C-nmr spectrum

Figure 1   Simulated p.n.d. ^{13}C n.m.r. spectra of a polyketide-derived moiety (a) at natural abundance, (b) enriched from [1-^{13}C]acetate, (c) enriched from [2-^{13}C]acetate, (d) enriched from [1,2-^{13}C$_2$]acetate, and (e) enriched from [1,2-^{13}C$_2$]acetate after bond cleavage.

If we consider any four contiguous carbons in a polyketide derived molecule, normally these carbons will be derived from endogenous acetyl CoA produced by the cell's normal metabolism and will contain only natural abundance ^{13}C (1.1%) and so in the p.n.d. ^{13}C n.m.r. spectrum each carbon will give rise to one sharp line of more or less equal intensity, Figure 1a. Now, if sodium acetate in which the carboxyl carbon is highly enriched (*ca* 95%) with ^{13}C, ([1-^{13}C]acetate) is added, then this exogenous acetate will be diluted to a greater or lesser extent by the endogenous acetate pool, but some of it will be incorporated into the metabolite and so those carbons, say C-2 and C-4, which were originally derived from the carboxyl carbon of acetate will contain extra ^{13}C and this will manifest itself as an increase in the appropriate signal intensities in the ^{13}C n.m.r. spectrum of the enriched metabolite, Figure 1b. If acetate in which the methyl carbon is enriched

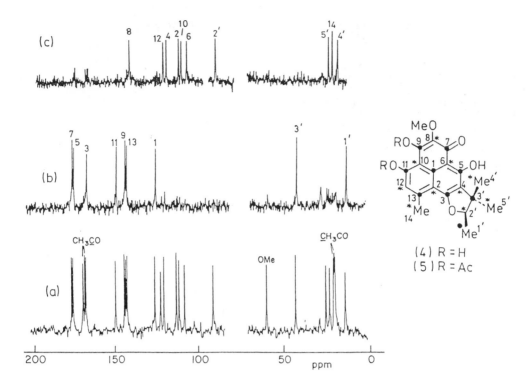

Figure 2 15.04 MHz p.n.d. ^{13}C n.m.r. spectra of deoxyherqueinone diacetate (a) at natural abundance, and enriched (b) from [1-^{13}C]acetate (●), and (c) from [2-^{13}C]acetate (*).

([2-^{13}C]acetate) is now added, then enhancement of the alternate signals is obtained, Figure 1c. Thus, simply by feeding ^{13}C-labelled precursor and determining the ^{13}C n.m.r. spectrum, the signals which show enchanced intensities indicate the sites of enrichment provided the ^{13}C n.m.r spectrum has been assigned unambiguously. It is important to note that this does not give further information than (in theory at least) could have been obtained by classical ^{14}C techniques, though in practice it is easier and usually more comprehensive. As may be seen, a severe limitation to this type of experiment is the existing ^{13}C natural abundance which as a rule necessitates precursor incorporation efficiencies which will produce a doubling of the ^{13}C content if enrichment is to be reliably observed. This means that the maximum permissible *dilution* of label from precursor into product is *ca.* 100. These dilutions are more easily obtained in micro-organisms than plants and it is for this reason that a more significant number of studies have been in microorganisms.

An example of single labelling studies with acetate is the incorporation of [1-^{13}C]- and [2-^{13}C]-acetates into deoxyherqueinone (4), a metabolite of *Penicillium herquei*. In this study[12] the ^{13}C n.m.r. spectrum of deoxyherqueinone diacetate (5) was fully assigned by analysis of long range ^{1}H-^{13}C couplings in the fully ^{1}H coupled ^{13}C n.m.r. spectrum, a technique which remains one of the best, but under utilised methods for both structure elucidation and spectral assignment studies in aromatic systems. The resultant spectra are shown in Figure 2. If high enrichments are obtained, as in this case, the labelled sites are readily apparent from visual inspection of the spectrum. However in general, enrichments are lower and identification of enriched sites with certainty can be more difficult. This is discussed below.

B.   Quantitative ^{13}C measurements

In principle at least, quantitative values for the degree of enrichment by labelled precursors and intermediates can be determined by comparing the integrated intensities of different carbon resonances in the p.n.d. ^{13}C n.m.r spectrum. The percentage of ^{13}C isotope above natural abundance has been defined as:

$$1.1 \times \frac{\text{(integrated intensity at labelled centre)}}{\text{(integrated intensity at unlabelled centre)}} -1.1$$

In practice, however, comparing signal height or integrals is problematic and the complications have been widely discussed.[1,13] Due to the wide range of spin-lattice relaxation times (T_1's) encountered in ^{13}C n.m.r. spectroscopy and the variable effects of the nuclear Overhauser enhancement (n.O.e.) induced by proton decoupling, the line intensities in a ^{13}C spectrum usually show wide variation even in a natural abundance spectrum. Fourier transform n.m.r. spectrometers exacerbate this problem using normal pulse sequences and add another one: the digitization of data often results in a poorly defined peak shape. The first problem can be circumvented to an extent by comparing unenriched and enriched resonances of carbons in a similar environment, but a better solution is to obtain natural abundance and enriched spectra under indentical conditions, normalise both spectra to a standard which is known to be unlabelled, and compare the intensities of the lines in the two spectra directly. This method was first used in the study[14] of the incorporation of [1-^{13}C]- and [2-^{13}C] acetates into tajixanthone (6), a metabolite of *Aspergillus variecolor*. In this case the ^{13}C n.m.r. spectrum was assigned by comparison of chemical shifts among a series of closely related derivatives and co-metabolites. Subsequent analysis of fully ^{1}H coupled ^{13}C n.m.r. spectra necessitated interchanging the assignments of two resonances but fortunately this does not affect the biosynthetic conclusions. An alternative approach is to add a paramagnetic relaxation agent such as chromium tris-acetoacetonate which suppresses the n.O.e. and shortens the T_1's. It may also be necessary to use "inverse gated" proton decoupling in which the n.O.e. is reduced by only decoupling during acquisition. Both these techniques were used to obtain the spectra shown for deoxyherqueinone in Figure 2 and as can be seen, near integral values for ^{13}C resonances can be obtained. Further alleviation of the effects of long T_1's may be achieved by the use of a long delay between pulses.

The second problem, that of data handling, manifests itself in random variations in intensities in the lines of spectra obtained from the same sample under the same experimental parameters. These variations can be reduced by using more data points to define the spectrum or by acquiring the spectrum several times and averaging the results. The spectra resulting from incorporation of [2-^{13}C]malonate into phomazarin (7), an azaanthraquinone metabolite of *Pyrenochaeta terrestris* were analysed in this way.[15] This analysis revealed that C-15 showed a lower level of enrichment (1.3%) when compared to other enriched carbons (1.6%) consistent with the derivation of phomazarin *via* a single nonaketide (1 acetate + 9 malonates) chain as shown in Scheme 1. The price to be paid in instrument time, in return for reliable enrichment data, is considerable and some compromise is generally required.

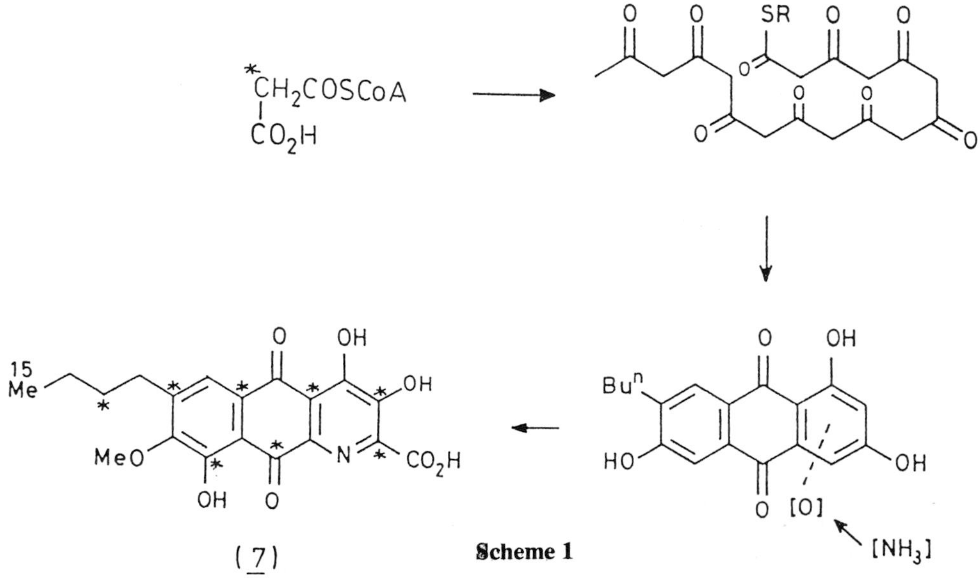

Scheme 1

C.   Single ^{13}C-labelling in plants[3]

Due to the reasons discussed above ^{13}C-labelling studies are much more difficult to carry out on higher plant metabolites compared to microbial metabolites. Some of the problems associated with precursor incorporation and transport, and dilution by endogenous pools of precursors and metabolites can be overcome by use of cell-free enzyme systems and in particular by use of tissue cultures. Obviously both of these require much effort to develop workable systems. However there have been a number of successful applications, particularly in the alkaloid field and in the important area of phytoalexin metabolites. A brief description of a few studies serves to illustrate these ideas.

^{14}C-Labelling studies indicated that the lactam (9) was an intermediate in the biosynthesis of camptothecin (10) in *Camptotheca acuminata* (Scheme 2). Owing to the absence of suitable degradations, ^{13}C n.m.r. was used to prove specificity of incorporation. Thus, [1-^{13}C]tryptamine (8) was synthesised from K^{13}CN and converted to the [5-^{13}C]lactam (9), 38 mg of which was wick-fed to intact plants. After 2 days growth, 20 mg of comptothecin was isolated which showed a *ca* 55% enhancement of the C-5 resonance only.[16]

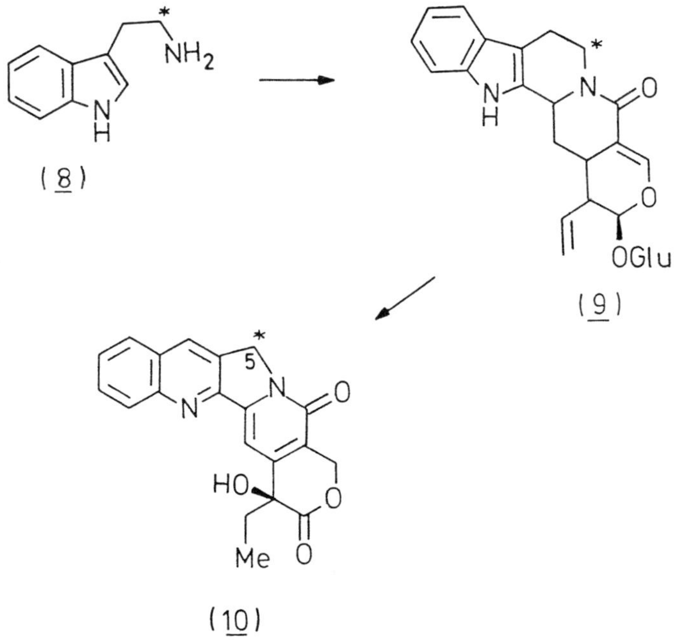

Scheme 2

440

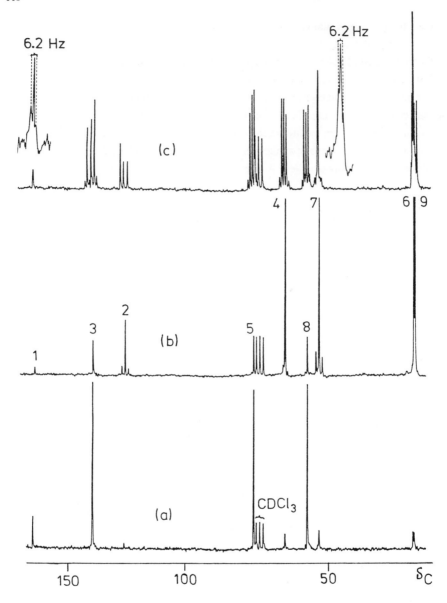

Figure 3   25.2 MHz p.n.d. ^{13}C n.m.r. spectra of aspyrone enriched from (a) [1-^{13}C]acetate, (b) [2-^{13}C]acetate, and (c) [1,2-$^{13}C_2$]acetate.

[1-^{13}C]Autamnaline (11) was synthesised and injected as the hydrochloride (300 mg) into seed capsules (1 mg per capsule) of *Colchicum autumnale*. After 2 weeks growth 1.24 g colchicine (12) was isolated (Scheme 3). The resultant ^{13}C n.m.r. spectrum showed a 2.5 fold enhancement of the C-7 signal only.[17]

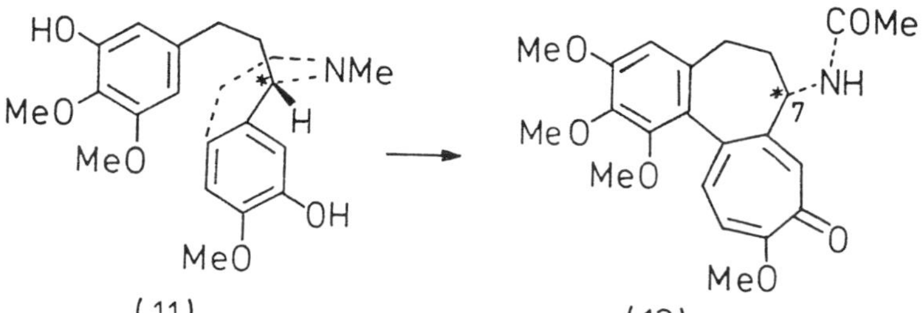

Scheme 3

V. ^{13}C-^{13}C SPIN-SPIN COUPLING

In natural abundance ^{13}C n.m.r. spectra ^{13}C-^{13}C spin-spin coupling is not normally observed as the probability of ^{13}C nuclei being adjacent is equal to the square of the natural abundance, giving satellites of 0.55% of the intensity of the main signal. They can be observed by resorting to the double quantum resonance technique, INADEQUATE, which allow observation *only* of coupled signals,[18] but this requires very large samples and sophisticated spectrometers. However in enriched metabolites, the probability of having adjacent ^{13}C nuclei is much higher and the detection of a ^{13}C-^{13}C coupling can provide conclusive evidence that two labels have been incorporated into adjacent positions in a molecule. This is normally done by using doubly-labelled precursors but ^{13}C-^{13}C couplings can also arise from administration of singly labelled precursors.

A.   Singly ^{13}C-labelled precursors

Molecular rearrangement of a biosynthetic intermediate may give rise to a ^{13}C-^{13}C coupling. Aspyrone (13), a metabolite of *Aspergillus melleus*,[19] when enriched from [2-^{13}C]acetate shows a coupling of 61 Hz between C-2 and C-7 (Figure 3b). This coupling arises due to an intramolecular rearrangement of the precursor pentaketide chain as shown in Scheme 4.

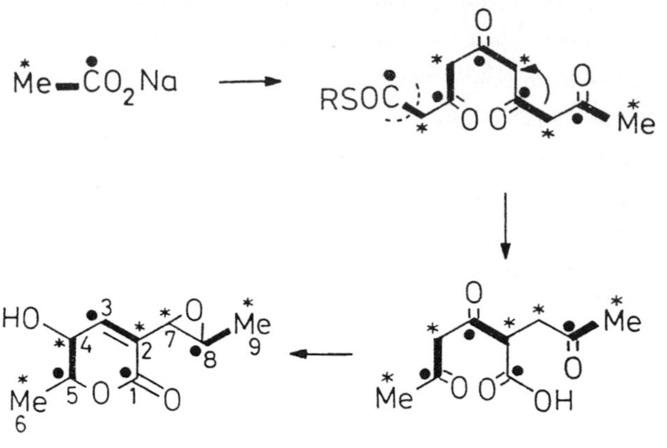

Scheme 4

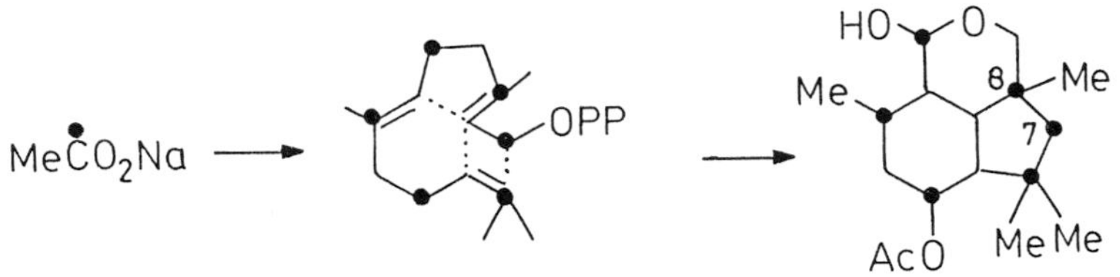

Scheme 5

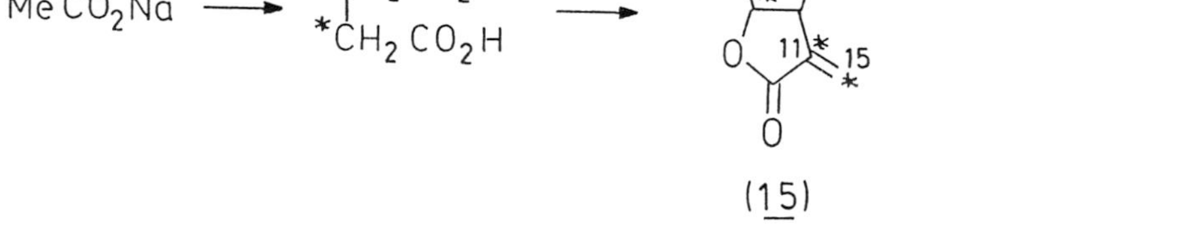

Scheme 6

Incorporation of [1-^{13}C]acetate into the sesquiterpene dihydrobotrydial (14) in *Botrytis cinerea* resulted in a coupling being observed between C-7 and C-8 (Scheme 5).[20] Conversion of [2-^{13}C]-acetate into succinate in the Krebs cycle results in ^{13}C-^{13}C coupling between C-11 and C-15 in avenaciolide (15) (Scheme 6).[21] Similar metabolic transformations give rise to ^{13}C-^{13}C couplings when [2-^{13}C]glycine is incorporated *via* serine into prodigiosin (16) (Scheme 7)[22] and [5-^{13}C]-δ-aminolaevulinic acid (17) is incorporated into vitamin B$_{12}$ (19) *via* porphobilinogen (18) (Scheme 8).[23] However it is important to note that all of these were in effect fortuitous observations resulting from high specific incorporation of labelled precursors rather than by experimental design.

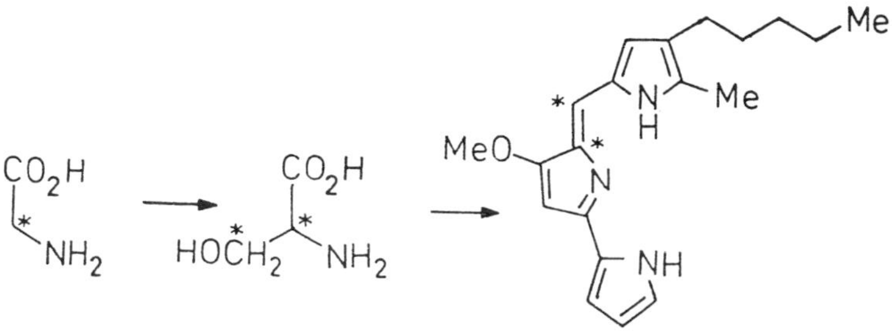

(16)

Scheme 7

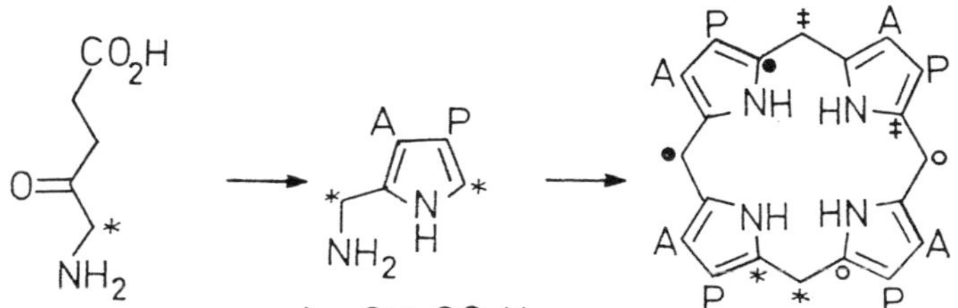

A = CH$_2$CO$_2$H

B = CH$_2$CH$_2$CO$_2$H

(17)   (18)   (19)

Scheme 8

B.   Doubly ^{13}C-labelled precursors

This has been one of the major developments in biosynthetic methodology and permits information to be obtained which would have been impossible or at best extremely difficult to obtain by classical radio-isotope labelling techniques. Again, the basic concept can be illustrated by consideration of the model polyketide system shown in Figure 1.

If we consider a molecule of acetate in which both carbons are entirely ^{13}C, ([1,2-^{13}C$_2$]acetate), it contains two adjacent nuclei of spin 1/2 and so they will couple to each other. If this acetate molecule is incorporated intact into a metabolite, then in any individual molecule, those pairs of carbons derived from an originally intact acetate unit must necessarily both be enriched simultaneously and so will show a mutual ^{13}C-^{13}C coupling. Thus if C-1 is enriched then C-2 must also be enriched. In the resultant ^{13}C n.m.r spectrum, the natural abundance signal is flanked by ^{13}C-^{13}C coupling satellites (Figure 1d). By analysing the coupling patterns, information is obtained on the way in which the precursor molecules are assembled on the enzyme surface, and on the way the, in this case, polyketide, chain folds up prior to condensation and cyclisation. If at any stage in the biosynthesis, the bond between two carbons originally derived from an intact acetate unit is broken, then the ^{13}C-^{13}C coupling is lost and these carbons then appear simply as enhanced singlets as shown for C-3 and C-4 in Figure 1e. In this way bond cleavage and rearrangement processes occurring during biosynthesis can be detected. The introduction of doubly labelled precursors results in the concept of "bond labelling". This concept will be extended to bonds other than carbon-carbon bonds in much of the ensuing discussion.

^{13}C-^{13}C Couplings are generally between 30 and 80 Hz and increase in proportion to the amount of "s" character of the atoms in the bond. Hence sp^3-sp^3 bonds are typically 35 Hz; sp^2-sp^3 are 45Hz, and sp^2-sp^2 are 60 Hz.[24] Substitution of one or both atoms by oxygen increases the size of the coupling. Usually the coupling will be sufficiently different in magnitude to enable the pairs of coupled carbons to be matched up. In addition, if the spectrum has been unambiguously assigned it will be apparent which carbons are mutually coupled. Couplings may also be confirmed by selective homonuclear ^{13}C-^{13}C decoupling[25] or by a 2D INADEQUATE spectrum.[26] A useful compilation of ^{13}C-^{13}C coupling constants observed in biosynthetic experiments has appeared.[27]

The contiguous double labelling technique has been used extensively. This is partly because it extends the permissible dilution factor to at least 2000, as small ^{13}C-^{13}C coupling satellites can be observed with more certainty than the corresponding enrichment from a singly-labelled precursor. This is apparent in the ^{13}C n.m.r. spectrum which results from incorporation of [1,2-^{13}C$_2$]acetate into the sesterterpene metabolite, astellatol (3) by cultures of *Aspergillus variecolor*.[28] In this study only very low incorporation levels were achieved. However, the high signal to noise ratio available at 100 MHz allowed the ready observation of ^{13}C-^{13}C coupling satellites with as little as 5% of the intensity of the natural abundance signals as shown in Figure 4. However its success is largely because of the extra information obtainable in respect of bond cleavages, rearrangements and symmetrical elements of intermediates.

The ^{13}C n.m.r. spectrum resulting from incorporation of [1,2-^{13}C$_2$]acetate into aspyrone (13) is shown in Figure 3c. ^{13}C-^{13}C couplings of 68, 41 and 44 Hz are observed between C-2 and C-3, C-4 and C-5, C-8 and C-9 respectively, indicating their derivation from originally intact acetate units. C-1,C-6 and C-7 however appear as enriched singlets indicating their derivation from cleaved acetate units as indicated above in Scheme 4. In the study of deoxyherqueinone (4), incorporation of [1,2-^{13}C$_2$]acetate was used to distinguish among the 3 possible foldings of the

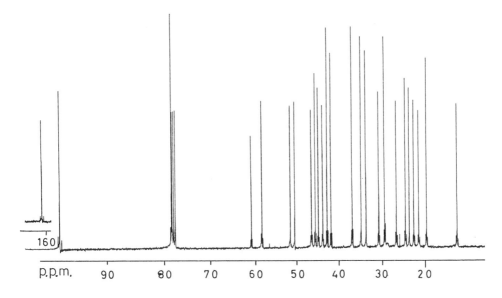

Figure 4   90.50 MHz p.n.d. ^{13}C n.m.r. spectrum of astellatol enriched from [1,2-^{13}C$_2$]acetate.

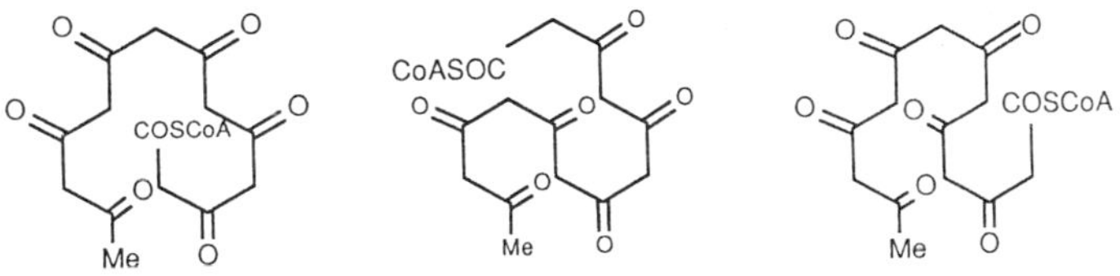

Alternative polyketide derivations of the phenalenone
ring system

Scheme 9

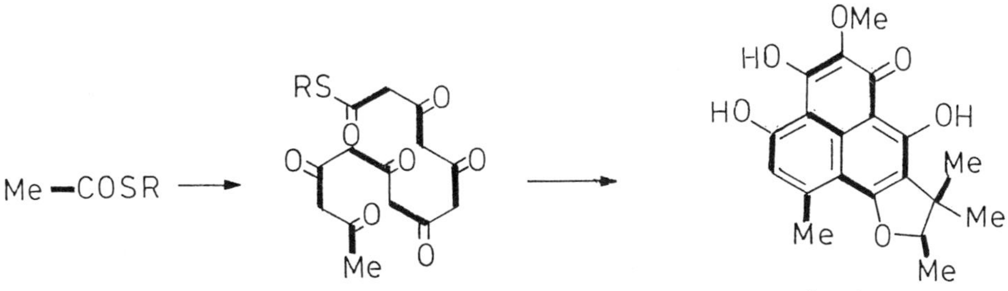

Scheme 10

precursor heptaketide chain (Scheme 9). The pattern of ^{13}C-^{13}C coupling observed indicated the incorporation pattern of intact acetate units. This defined the mode of folding shown in Scheme 10.

Due to its extra sensitivity the technique has been applied to many higher plant metabolites. Some of these studies are discussed briefly below to indicate the type of information these studies can provide.

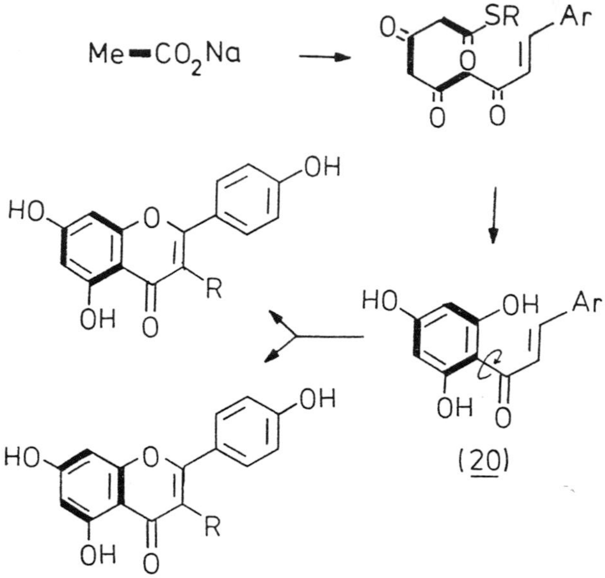

(20)

(21) R = H
(22) R = OH

Scheme 11

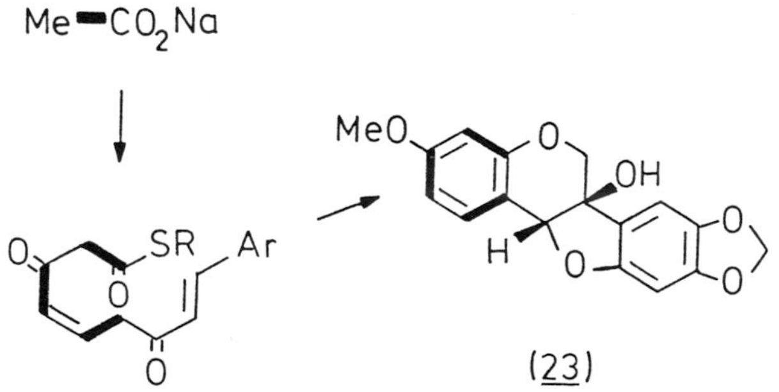

(23)

Scheme 12

[1,2-^{13}C$_2$]Acetate was incorporated into the flavone, apigenin (21), and the flavonol (22) by cell suspension cultures of parsley, *Pteroselinum hortense,* with randomisation of ^{13}C-^{13}C couplings in ring A.[29] Each carbon shows two sets of coupling satellites due to coupling to *both* adjacent carbons. This means that a symmetrical intermediate, presumably the chalcone (20) in which ring A is free to rotate, is an intermediate in the biosynthesis of (21) and (22) as shown in Scheme 11. Detection of symmetrical intermediates by this type of randomisation of coupling is one of the more important applications of double ^{13}C-labelling experiments. In contrast, [1,2-^{13}C$_2$]acetate is incorporated into ring A of the phytoalexin pisatin (23) in *Pisum sativum* without randomisation, showing that deoxygenation of the polyketide precursor must occur *before* cyclisation and aromatisation (Scheme 12).[30] The labelling pattern of paniculide (24) derived from [1,2-^{13}C$_2$]acetate in callus tissue cultures *Andrographis paniculata* demonstrates[31] that the manner of folding of the farnesyl pyrophosphate in the biosynthesis of (24) must be as shown in Scheme 13. [4,5-^{13}C$_2$] Lysine was incorporated into anabasine (25) by *Nicotania glauca.* The ^{13}C labels were incorporated into C-4 and C-5 only (Scheme 14).[32]

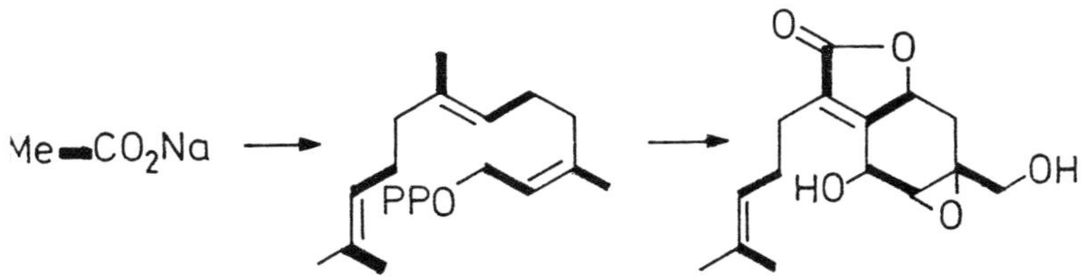

Me—CO$_2$Na

(24)

Scheme 13

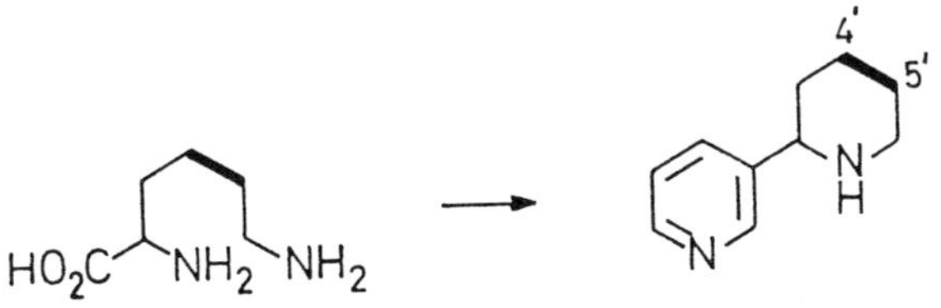

(25)

Scheme 14

In contrast, [2,3-^{13}C$_2$]ornithine was incorporated[33] into both C-2/3 and C-4/5 of nicotine (26) to indicate the involvement of a symmetrical intermediate, presumably putrescine (27) (Scheme 15). Incorporation of [2,3-^{13}C$_2$]putrescine into retronecine (28) by seedlings of *Senecio isatideus* and observation of coupling satellites on C-1/2 and C-6/7 established the symmetrical labelling of the two halves of the pyrrolizidine alkaloids (Scheme 16).[34]

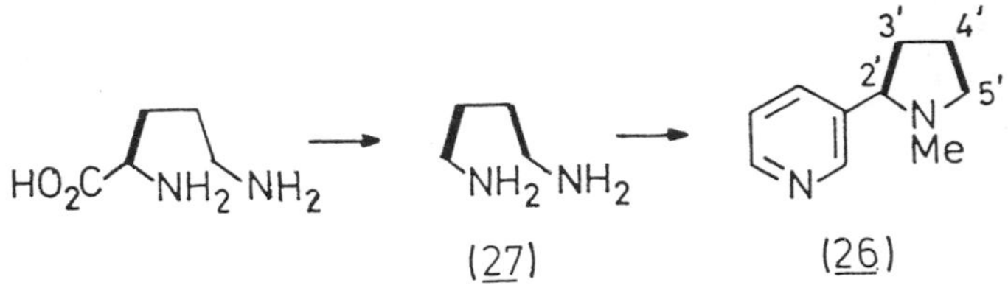

(27)   (26)

Scheme 15

(28)

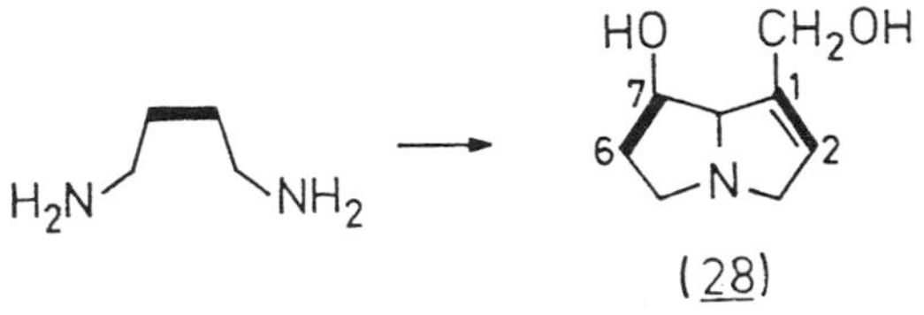

Scheme 16

C.   Non-contiguous double ^{13}C-labelling

All the above examples have involved precursors in which there is a one-bond coupling between ^{13}C labels in both precursors and final metabolites. Other studies yielding valuable biosynthetic information have involved precursors with contiguous labels which become contiguous, or even non-contiguous labels which remain non-contiguous but detectable during biosynthesis.

1.   Two-bond ^{13}C-^{13}C couplings from contiguous ^{13}C$_2$-labelled precursors

When two ^{13}C-labelled atoms are in a two-bonded relationship to each other they may or may not display a mutual ^{13}C-^{13}C coupling. These couplings, when observed, are much smaller than one-bond couplings and typically range from 1 to 15 Hz.[24] However with careful spectral determination they can be observed.

In the study of aspyrone (13) discussed above, the rearrangement proposed in Scheme 4 involves cleavage of an intact acetate unit during an intramolecular rearrangement. Thus C-1 and C-7 which are in a 2-bonded relationship in aspyrone arise from the same acetate unit. This was confirmed by redetermination of the spectrum of [1,2-^{13}C$_2$]acetate-enriched aspyrone under conditions of higher resolution, whereupon a mutual ^{13}C-^{13}C coupling of 6.2 Hz was observed between C-1 and C-7, Figure 3c. Similar 2-bond couplings arising from the same acetate unit have also been observed for vulgamycin (29)[35] and the tetrahydrofuran (30) (Scheme 17).[36]

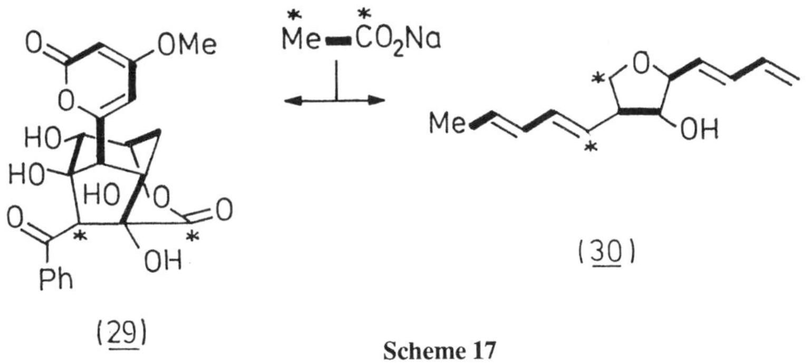

(29)

(30)

Scheme 17

2.   One-bond ^{13}C-^{13}C couplings from non-contiguous $^{13}C_2$-labelled precursors

This is an especially valuable method for elucidating the direction of intramolecular rearrangements and also for establishing the intact incorporation of precursors. It is in essence the reversal of the process described above and is best illustrated by examples.

Tropic acid (32) as found in scopolamine (33) is formed from phenylalanine with the side-chain rearrangement involving an intramolecular 1,2-shift of the carboxy-group. This was established[37] by synthesis of [1,3-$^{13}C_2$]phenylalanine (31) and its incorporation into (32) and (33) in *Datura inoxia*. In the ^{13}C n.m.r. of the resultant alkaloids, a ^{13}C-^{13}C coupling of 38 Hz was observed between C-2 and C-3 to prove that the rearrangement is an intramolecular and not intermolecular process (Scheme 18). Similar studies carried out on tenellin (34) showed a ^{13}C-^{13}C coupling of 45 Hz between C-5 and C-6 indicating that these two carbons have migrated together from C-2 and C-3 of phenylalanine during biosynthesis (Scheme 18).[38]

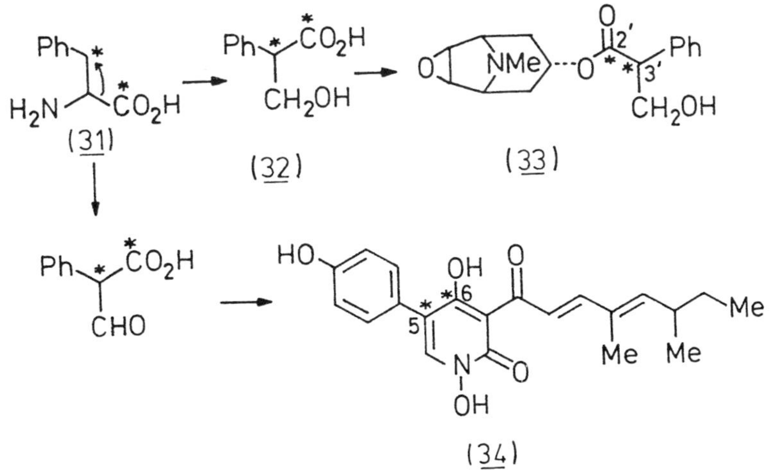

(31)

(32)

(33)

(34)

Scheme 18

In a study[39] of cholesterol (36) biosynthesis [4,6-^{13}C$_2$]mevalonic acid was fed to a cell-free enzyme preparation from rat-liver, C-13 and C-18 showed a ^{13}C-^{13}C coupling of 35 Hz proving that they were derived from the same mevalonate unit of squalene and confirming that C-18 becomes bonded to C-13 by a 1,2-methyl migration from C-14 of the initial cyclisation product (35) rather than by a 1,3 migration from C-8, as indicated in Scheme 19.

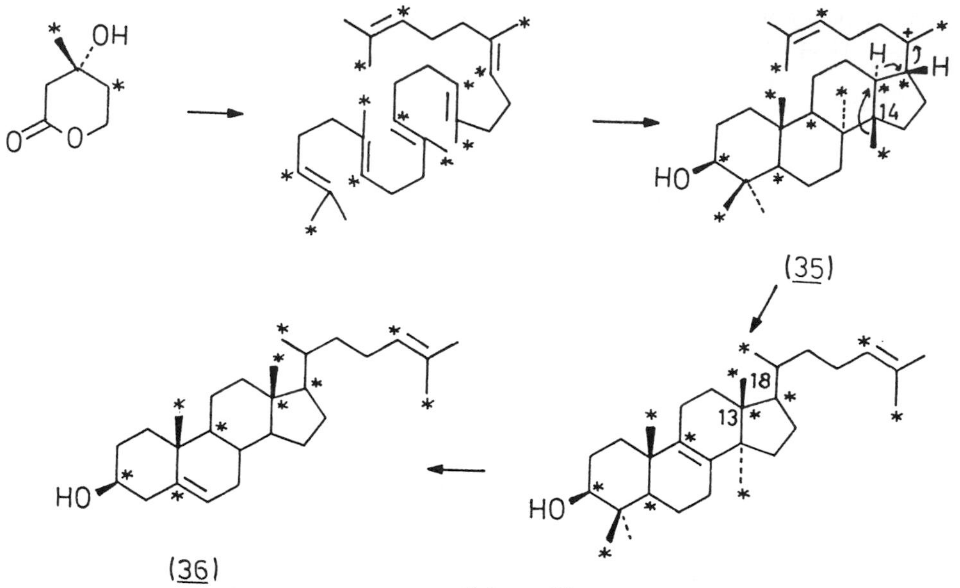

Scheme 19

One of the first and most elegant examples of this approach was in Battersby's study of uroporphyrinogen III (38) biosynthesis. [2,11-^{13}C$_2$]Porphobilinogen (37) was synthesised and showed a long-range coupling of 4 Hz. This was incorporated into (38) using a cell-free system from avian blood. Analysis of the resultant ^{13}C n.m.r. spectrum showed three doublets each of 5 Hz splitting corresponding to the α, β and δ carbons, and one doublet of 72 Hz for the γ carbon, indicating that PBG unit D must undergo an intramolecular rearrangement with respect to itself during biosynthesis (Scheme 20).[40]

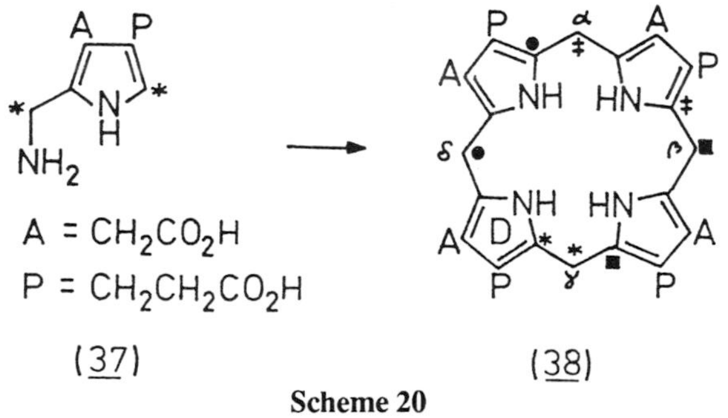

A = CH$_2$CO$_2$H

P = CH$_2$CH$_2$CO$_2$H

(37)   (38)

Scheme 20

3. Two-bond ^{13}C-^{13}C couplings from non-contiguous $^{13}C_2$-labelled precursors

The value of this method lies in its ability to display intact incorporation of complex biosynthetic intermediates. A powerful illustration of the method is the demonstrated intact incorporation of homospermidine into the alkaloid retronecine. [1,9-$^{13}C_2$]Homospermidine (39) was synthesised and fed to *Senecio isatideus*. The ^{13}C n.m.r. spectrum of the derived retronecine (40) showed doublets for C-8 and C-9 that were superimposed on the natural abundance singlets (Scheme 21).[41]

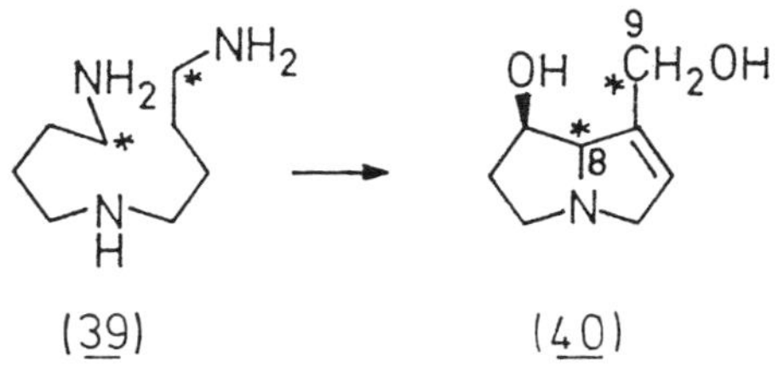

Scheme 21

VI. ^{13}C-^{15}N DOUBLY LABELLED PRECURSORS

As discussed above ^{15}N has spin 1/2 and so will couple to ^{13}C. Thus, observation of a ^{13}C-^{15}N coupling on incorporation of a precursor doubly labelled with ^{13}C-^{15}N can be used to establish intact incorporation of precursor or to probe whether the integrity of a particular carbon-nitrogen bond is maintained during a biosynthetic sequence. ^{13}C-^{15}N Couplings are in general smaller and less predictable than ^{13}C-^{13}C couplings and in certain cases they may not be observable at all so results must be interpreted with caution.[42]

As with doubly ^{13}C-labelled precursors we can use precursors in which the ^{13}C and ^{15}N are already in a one-bonded relationship or we can use precursors labelled so that the ^{13}C and ^{15}N become bonded during biosynthesis. An elegant series of experiments encompassing both approaches was carried out during studies on the biosynthesis of the antibiotic streptothricin F (41). As shown in Scheme 22 a series of arginines specifically labelled with ^{13}C and ^{15}N were synthesised and fed to *Streptomyces* L-1689-23. The streptothricin F isolated in each case was analysed by ^{13}C n.m.r. and the ^{13}C-^{15}N couplings indicated by heavy lines in Scheme 22 were observed.[43]

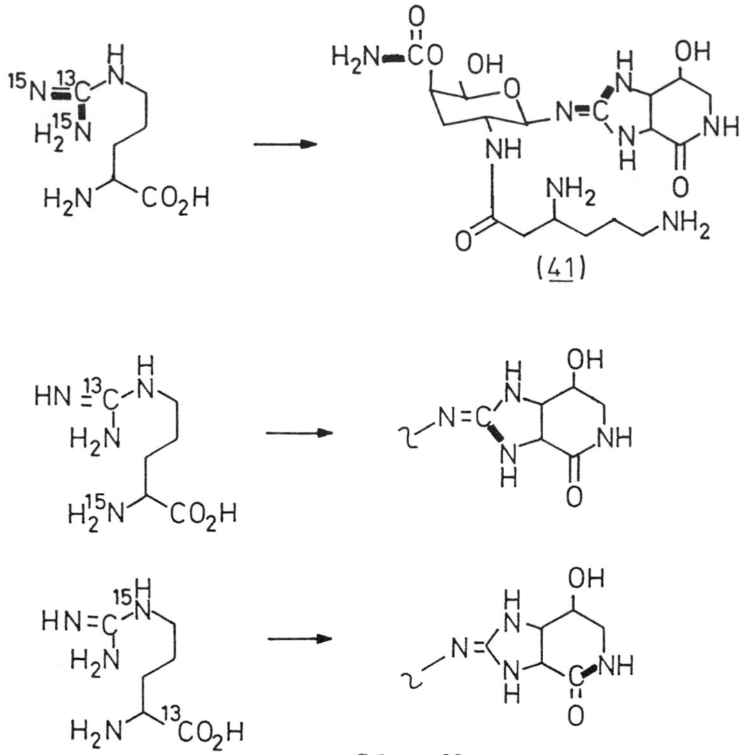

Scheme 22

A particularly elegant application to penicillin biosynthesis has been reported.[44] δ-(α-L-Amino-adipoyl)-L-[3-^{13}C]cysteinyl-D-[^{15}N]valine (42), the "Arnstein tripeptide", was efficiently incorporated into isopenicillin N (43) by a cell-free enzyme system prepared from cells of *Cephalosporium acremonium* (Scheme 23). The ^{13}C n.m.r. spectrum showed a ^{13}C-^{15}N coupling of 4.4 Hz for the enriched C-5 resonance of (43).

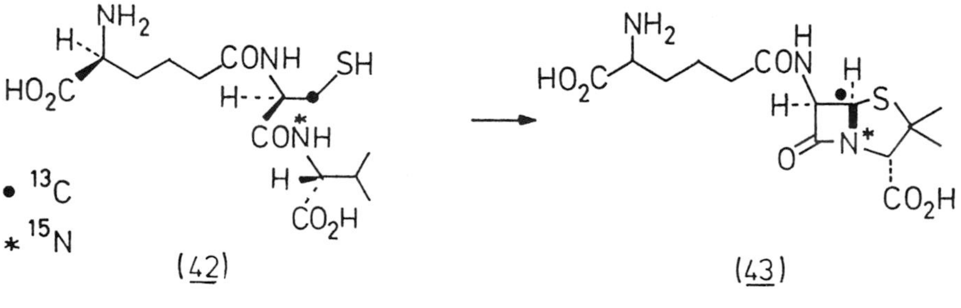

Scheme 23

The technique has also been applied in a number of higher plant alkaloids. [1-^{13}C,*methylamino*-^{15}N]-N-methyl putrescine has been incorporated into scopolamine (44) and nicotine (45). In both cases C-5 show [13]C-^{15}N coupling satellites due to the intact incorporation of the contiguous ^{13}C and ^{15}N atoms as shown (Scheme 24).[45] [1-^{13}C,1-^{15}N]Putrescine has been incorporated into retronecine (47). The resultant [13]C n.m.r. spectrum had enhanced ^{13}C signals due to C-3,C-5, C-8 and C-9. In addition both C-3 and C-5 showed [13]C-^{15}N coupling satellites (4.1 and 3.8 Hz respectively) indicating the intermediacy of a symmetrical intermediate (46) in the biosynthesis of retronecine (Scheme 25).[46]

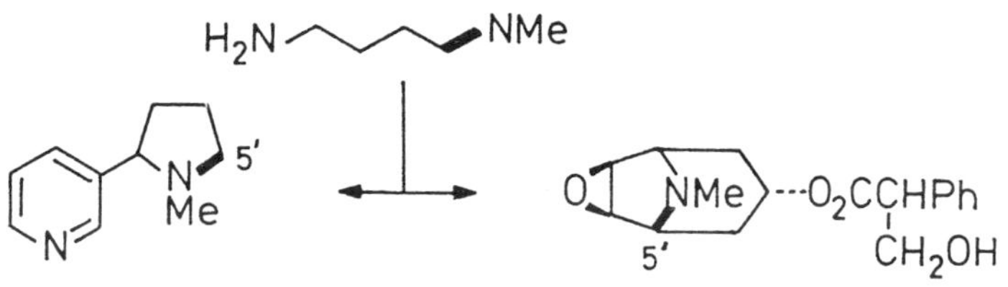

Scheme 24

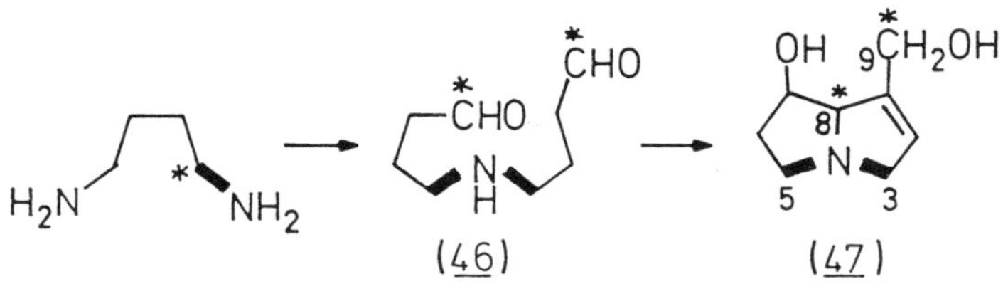

Scheme 25

VII. ISOTOPE-INDUCED SHIFTS IN [13]C N.M.R.

In studying the nature of the intermediates on a biosynthetic pathway and in particular elucidating the detailed mechanisms of their interconversions it is essential to determine the biosynthetic origins and fate of the hydrogen and oxygen atoms. Hydrogen and oxygen can be monitored directly by n.m.r. by using the n.m.r active isotopes ^{2}H, ^{3}H or [17]O (see below). However all of these have limitations which can be overcome by the use of isotope-induced shifts in [13]C n.m.r.

A.   The deuterium alpha-shift technique

The use of ^{13}C as a reporter nucleus of both hydrogen and oxygen represented a major advance in biosynthetic studies with stable isotopes and makes use of the observation that substitution of a proton *alpha* or *beta* to a ^{13}C by deuterium causes a change (usually upfield) in the ^{13}C chemical

Use of ^{13}C-nmr in Isotope Labelling Experiments

Origin of HYDROGENS and OXYGENS

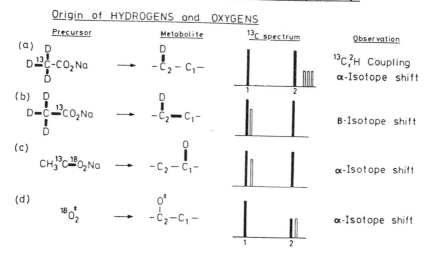

Precursor	Metabolite	^{13}C spectrum	Observation

Figure 5   Simulated p.n.d. ^{13}C n.m.r. spectra of a polyketide-derived moiety enriched from (a) [2-^{13}C, ^{2}H$_3$]acetate, (b) [1-^{13}C, ^{2}H$_3$]acetate, (c) [1-^{13}C, ^{18}O$_2$]acetate, and (d) ^{18}O$_2$ gas.

shift. Similarly the presence of ^{18}O *alpha* to a ^{13}C atom can be detected by an upfield shift in the ^{13}C n.m.r. spectrum. These effects are summarized in Figure 5. When the deuterium label is directly attached to a ^{13}C nucleus in the precursor molecule, the p.n.d. ^{13}C spectrum of the enriched metabolite shows, for carbons which have retained deuterium label, a series of signals upfield of the normal resonance. The presence of each deuterium shifts the centre of the resonance by 0.3–0.6 p.p.m. and spin-spin coupling ($^{1}J_{CD}$) produces a characteristic multiplet, hence CD appears (Figure 5a) as a triplet whereas CD$_2$ and CD$_3$ would give respectively a quintet and septet. Shifted signals arising from carbons which bear no hydrogen suffer reduced signal-to-noise ratio caused by poor relaxation and lack of n.O.e. enhancement, a disadvantage of the method which is compounded by the multiplicities due to coupling. Deuterium noise decoupling can assist in this by removing the ^{13}C-^{2}H coupling. However, information not obtainable by direct ^{2}H n.m.r. spectroscopy, such as the distribution of label as CH$_2$D, CHD$_2$ and CD$_3$ and the integrity of carbon-hydrogen bonds during biosynthesis, may be gained.

One of the first applications of this technique was to terrein (48), a cyclopentenone metabolite of *Aspergillus terreus*.[47] In the p.n.d. decoupled ^{13}C n.m.r. of terrein (48) enriched from [2-^{13}C,^{2}H$_3$]-acetate, the signals for isotopically shifted resonances were not clearly observed. However, on redetermining the spectrum with deuterium decoupling, a signal at 17.95ppm, 0.8 ppm upfield of the normal C-1 signal was observed indicating the presence of molecules triply labelled with deuterium at C-1. In addition a weak doublet signal (J = 123 Hz) centred at 18.2

ppm indicated the presence of molecules labelled as CHD$_2$. The detection of molecules labelled with three deuteriums at C-1 confirmed that the methyl is derived intact from the methyl of the acetate "starter" unit of the precursor polyketide chain (Scheme 26).

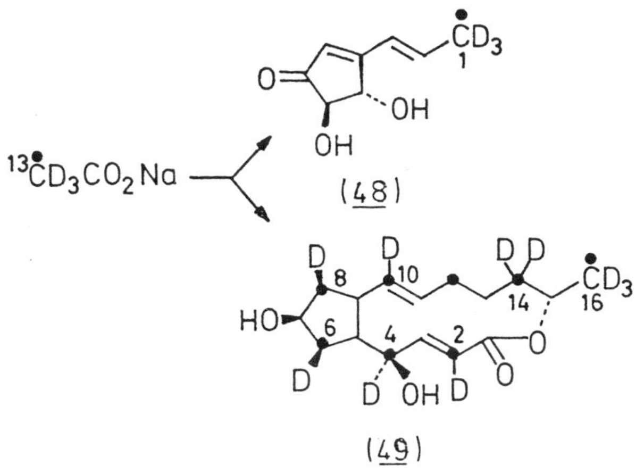

Scheme 26

The method has been used most successfully when the ^{13}C n.m.r.spectrum of the enriched metabolite can be determined with simultaneous proton *and* deuterium decoupling. Thus on incorporation of [2-^{13}C,^{2}H$_3$]acetate into brefeldin A (49) by cultures of *Penicillium brefeldianum,* the resultant proton and deuterium decoupled ^{13}C n.m.r. spectrum showed isotopically shifted signals indicating incorporation of up to three deuteriums on C-16, two on C-14 and one on carbons 2,4,6,8 and 10.[48] Incorporation of [2-^{13}C,^{2}H$_3$]acetate into aspyrone(13) gave the spectra shown in Figure 6. As may be seen deuterium noise decoupling leads to a great simplification of the otherwise uninterpretable results for carbons 7 and 10. Thus one deuterium atom is incorporated at C-4, one and two at C-6 and up to three at C-9. These along with oxygen-18 labelling results enabled a biosynthetic pathway involving epoxide-mediated rearrangement and ring-closure reactions (see Scheme 33 below) to be proposed.[49]

A pulse sequence has been described[50] which allows the selective observation of deuterated ^{13}C signals by selective suppression of signals from protonated carbons. This makes the technique more sensitive, but like the simultaneous proton and deuterium decoupling method requires instrumentation and expertise which are not widely available.

A continuing problem in biosynthetic labelling studies is the determination of stereospecificity as well as regiospecificity of isotope labelling. A new combination of methods namely incorporation of ^{13}C,^{2}H doubly labelled precursors and analysis by ^{2}H decoupled ^{1}H,^{13}C heteronuclear shift correlation n.m.r spectroscopy allows the stereochemistry of incorporation of deuterium at diastereotopic methylene positions to be determined.[51] This has been applied to

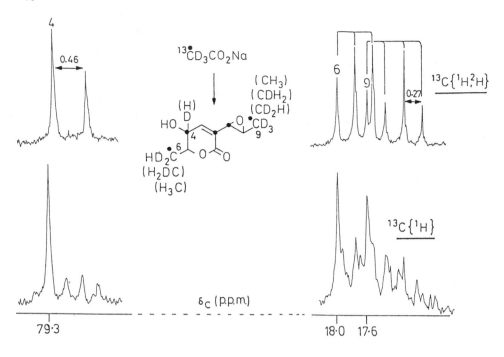

Figure 6 Alpha ^{2}H isotope-induced shifts observed in the 100.6 MHz and ^{1}H and ^{2}H noise decoupled ^{13}C n.m.r. spectrum of aspyrone labelled from [2-^{13}C, ^{2}H$_3$]acetate.

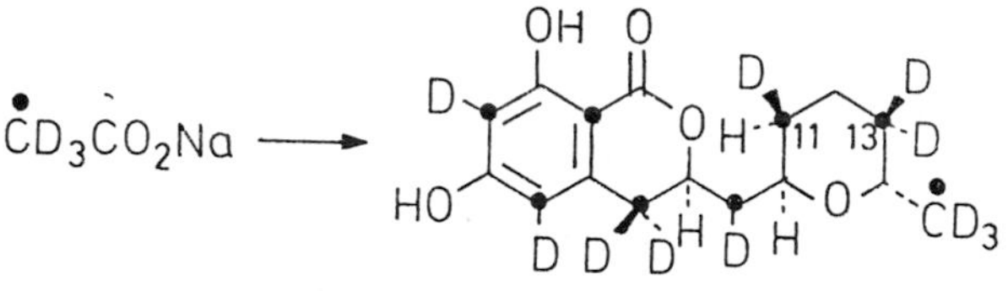

(50)

Scheme 27

elucidate the stereochemistry of incorporation of deuterium from [2-^{13}C,^{2}H$_3$]acetate into the fungal metabolite cladosporin (50) in *Cladosporium cladosporoides* (Scheme 27).[51] The results are shown in Figure 7 which shows that the hydrogen of the CHD group at C-11 occupies the downfield pro R position. Thus the deuterium has been incorporated into the pro-S (upfield position), a result with important implications for the stereochemistry of related reductive steps in polyketide and fatty acid biosynthesis.

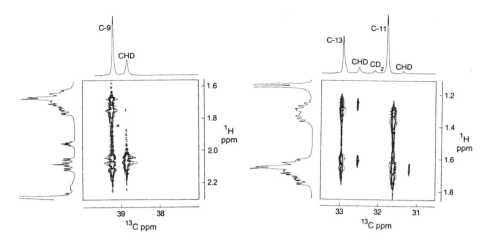

Figure 7 ^{2}H decoupled ^{1}H, ^{13}C chemical shift correlation plots of cladosporin diacetate enriched from [2-^{13}C, ^{2}H$_3$]acetate. Upfield shifts show that C-9 and C-11 are stereospecifically labelled but C-13 is not. (Reprinted with permission from *J.Am.Chem.Soc.,* 110 (1988) 316. Copyright 1988 American Chemical Society).

B.   The deuterium beta-shift technique

Many of the problems associated with directly attached deuterium are avoided by placing the deuterium label two bonds away from the ^{13}C reporter nucleus. The isotope shift, although reduced, is still observable, and as β-hydrogens only contribute markedly to the relaxation of non-protonated ^{13}C nuclei, the shifted signals otherwise retain any n.O.e. effect also experienced by the unshifted signals on proton decoupling. As geminal carbon-proton coupling constants are generally small anyway, and carbon-deuteron couplings are over six times smaller again, the shifted signals are effectively singlets (Figure 5b), even without deuterium decoupling, and this gives a further increase in the signal to noise ratio compared to the corresponding α-shift experiment.

The method was first applied to biosynthesis in a study of 6-methysalicylic acid (51). [1-^{13}C,^{2}H$_3$]Acetate was fed to cultures of *Penicillium griseofulvum*. The p.n.d. ^{13}C n.m.r. spectrum of the methyl ester of the resulting 6-MSA showed shifted signals for C-2, C-4,and C-6, corresponding to the presence of deuterium label at positions 3, 5, and 7 respectively (Scheme 28).[52] Thus the integrity of an acetate unit (heavy lines) can be established in certain cases without recourse to a double ^{13}C-labelled experiment.

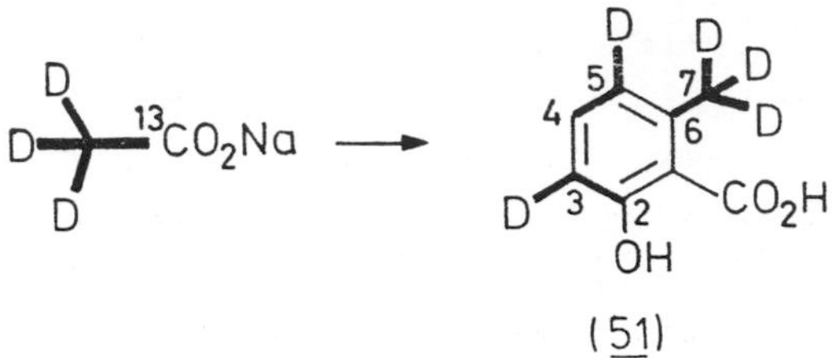

Scheme 28

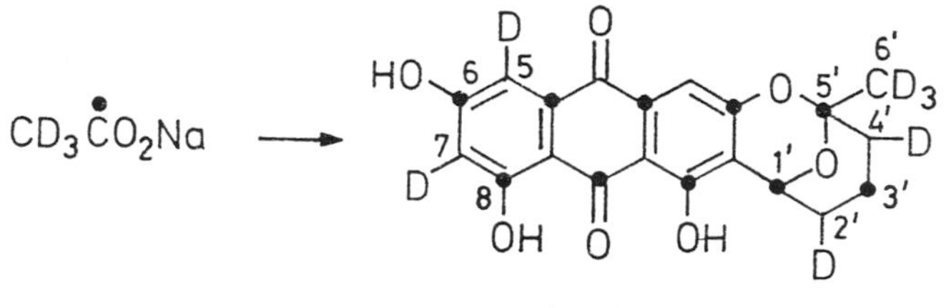

Scheme 29

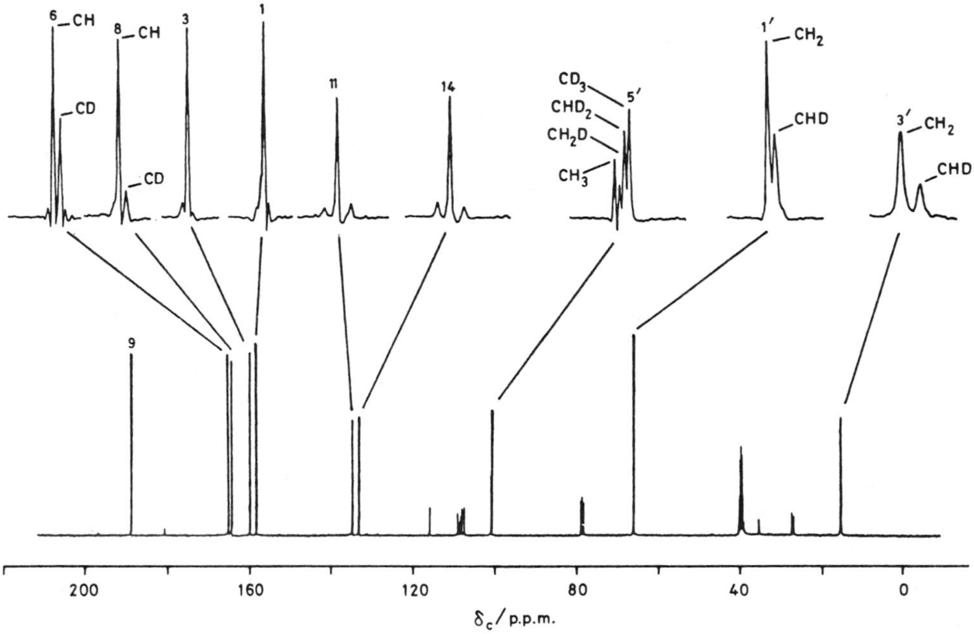

Figure 8   90.56 MHz p.n.d. ^{13}C n.m.r. spectrum of averufin labelled from [1-^{13}C, ^{2}H$_3$]acetate.

In a similar study[53] on the aflatoxin intermediate averufin (52) (Scheme 29), the regiospecificity of incorporation of ^{2}H from [1-^{13}C,^{2}H$_3$]acetate into (52) by cultures of *Aspergillus toxicarius* was determined using the β-^{2}H isotope shifts observed in the p.n.d. ^{13}C n.m.r. spectrum which is shown in Figure 8. This shows isotopically shifted signals for C-5 (the reporter nucleus) indicative of the incorporation of 1-3 deuteriums on C-6,and one deuterium at C-1, C-4, C-5 and C-7 respectively. (The stereochemistry of incorporation of the single deuterium label at C-2 and C-3 has been determined subsequently[54] to be as shown by incorporation of [2-^{13}C,^{2}H$_3$]acetate and by ^{1}H,^{13}C shift correlation spectroscopy as discussed above).

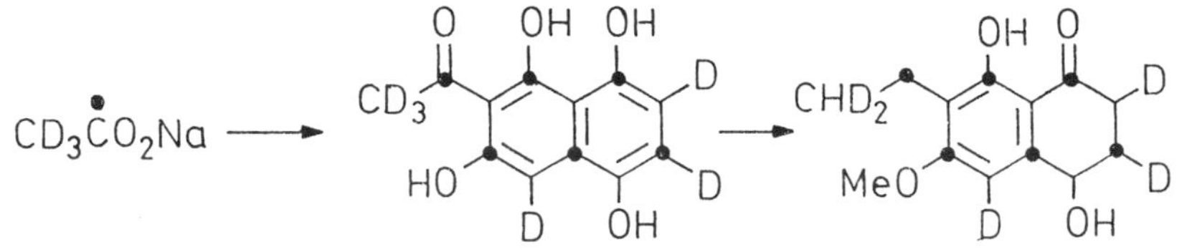

Scheme 30

Unlike the α-shifts, the β-shifts show a marked dependence on both stereochemistry of the deuterium label and the functionality of the ^{13}C reporter nucleus. Incorporation of [1-^{13}C,^{2}H$_3$]acetate into asparvenone (53) by cultures of *Aspergillus parvulus* (Scheme 30) and analysis of the ^{2}H β-isotope shifts in the resultant ^{13}C n.m.r. spectrum showed that one hydrogen was lost from the C-10 methyl to indicate formation of the ethyl moiety by a reduction-elimination-reduction sequence on the corresponding acetyl group.[55] The magnitude and direction of the β-isotope shifts were observed to depend markedly on the functionality of the reporter ^{13}C nucleus and surprisingly on the stereospecificity of ^{2}H incorporation. This was confirmed by an *in vitro* experiment when the C-2 methylene hydrogens were exchanged in equimolar MeOH and MeO^{2}H to give the spectrum shown in Figure 9. For carbonyl groups the observed shifts are downfield in contrast to the usually observed upfield shifts. Incorporation[56] of [1-^{13}C,^{2}H$_3$]acetate into colletodiol (54), Scheme 31, resulted in no observable ^{2}H isotope shift at either C-1 or C-1 despite the fact that ^{2}H n.m.r. analysis showed a high level of incorporation at both C-2 and C-2. These results indicate the need for caution in the interpretation of results when carbonyl groups are involved. ^{2}H n.m.r. analysis of the enriched asparvenone indicated that the ^{2}H label at C-2 was exclusively incorporated into the axial position. This analysis also revealed the presence of ^{2}H label at the 3-axial position. This label could only have arrived at C-3 by migration from C-2, presumably due to an NIH shift induced by hydroxylation of C-2 biosynthesis as indicated in Scheme 30.

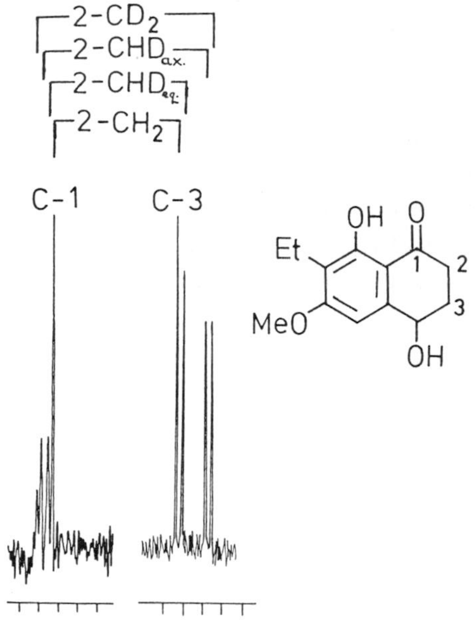

460

Figure 9   Signals from the 90.56 MHz p.n.d. ^{13}C n.m.r. spectrum of $\underline{O}$-methylasparvenone partially deuteriated at C-2. Resonances are for (a) C-1 and (b) C-3.

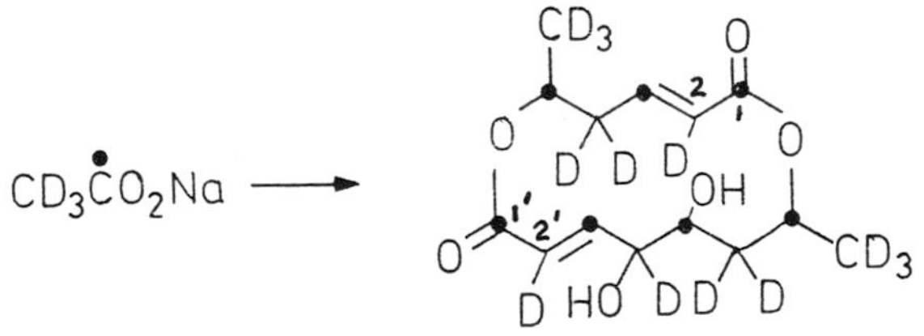

Scheme 31

C. ^{18}O Isotope-induced shifts

Until recently, the biosynthetic origin or fate of oxygen was determined almost exclusively by mass spectrometry, and locating the position of a label, present perhaps in only low concentrations, was difficult. It was known that the heaviest oxygen isotope, ^{18}O, (natural abundance - 0.204%) was capable of inducing resolvable isotope shifts in the n.m.r. spectra of certain other elements and this had been exploited in biological studies, *e.g.* the mechanisms and kinetics of enzymatic phosphoryl group transfer had been investigated via the ^{18}O isotope shift in the ^{31}P n.m.r. spectrum.[57] The detection of such an isotope shift in the ^{13}C spectrum provided a more general technique for biosynthetic research.[58] As shown in Figure 5, the ^{18}O may be conveniently introduced *via* a doubly labelled precursor or by growth in an ^{18}O$_2$ atmosphere. The resulting shifts are generally not much larger than 0.05 ppm. These are very small effects and are the same general size as β-^{2}H isotope shifts and are only readily observed with high field spectrometers. These techniques were first employed in a study of averufin (55) (Scheme 32). *Aspergillus parasiticus* when grown under an atmosphere containing ^{18}O$_2$ gas, produced averufin, the ^{13}C n.m.r. spectrum of which showed an isotopically shifted resonance for C-10 only.[59] In an experiment feeding with [1-^{13}C,^{18}O$_2$]acetate, carbons 1,3,6,8,9 and 1´ all showed prominent shifted signals. Thus, at the incorporation levels typically achieved with early precursors, carbon-oxygen bonds which have been preserved intact throughout the course of biosynthesis can be distinguished from those which have arisen between precursor units.

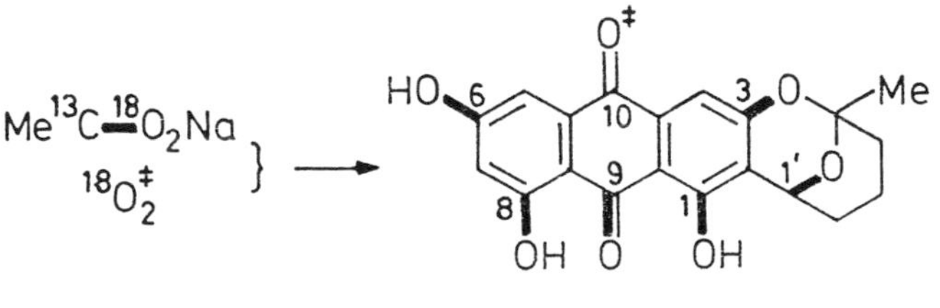

Scheme 32

462

Experiments with [1-^{13}C,^{18}O$_2$]acetate and ^{18}O$_2$ revealed the surprising result that *none* of the oxygens in aspyrone were derived from acetate, three being derived from the atmosphere and one from the medium. The pathway proposed to account for this is shown in Scheme 33. As indicated in Figure 10, the lactone carbonyl C-1 showed shifts due to the presence of ^{18}O in either the doubly or singly bonded oxygen.[48]

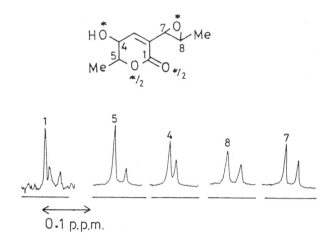

Figure 10 ^{18}O Isotope-induced shifts observed in the 100.6 MHz p.n.d. ^{13}C n.m.r. spectrum of aspyrone labelled by ^{18}O$_2$ gas.

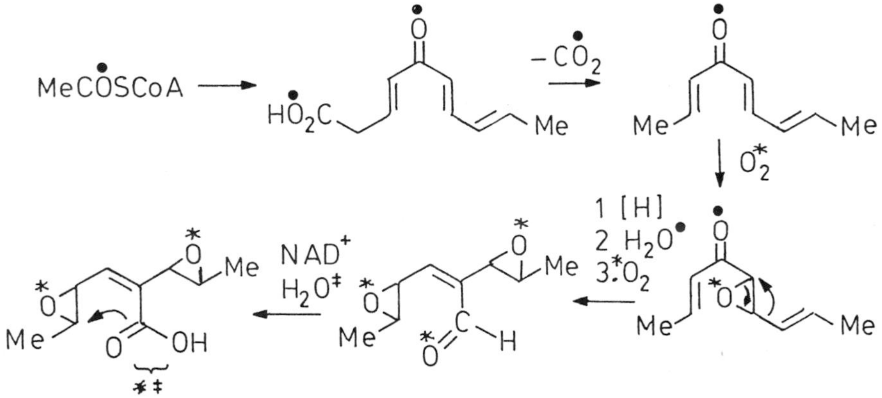

Scheme 33

In an application using an advanced precursor, ethyl 3,5-dimethyl orsellinate (56) was synthesised with double ^{13}C,^{18}O labelling in both the carbonyl of the ester and at the C-6 position as shown. This was incorporated with high efficiency into the mycotoxin austin (57) (Scheme 34). In the spectrum of the enriched metabolite both the C-8 carbonyl and C-6 tertiary alcohol carbon showed isotopically shifted signals[60] (Figure 11).

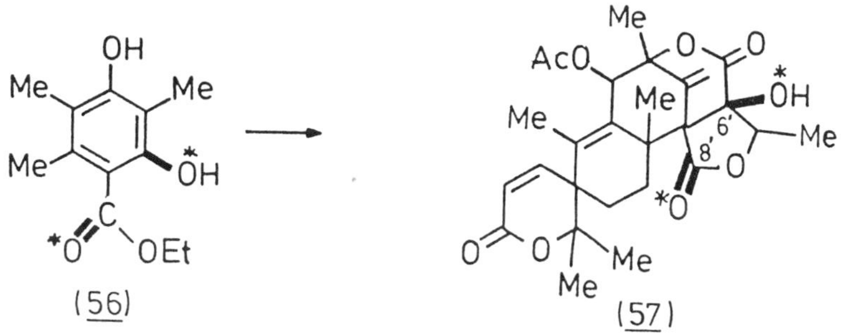

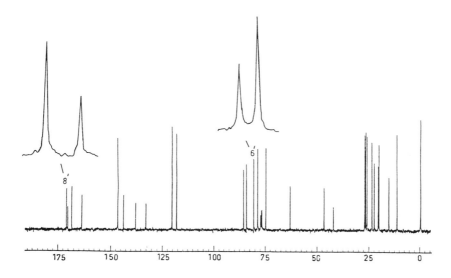

Scheme 34

Figure 11 ^{18}O Isotopically shifted resonances in the 100.6 MHz p.n.d. ^{13}C n.m.r. spectrum of austin enriched by ethyl [^{13}C, ^{18}O]-3,5-dimethylorsellinate.

VIII.   DEUTERIUM AND TRITIUM N.M.R. SPECTROSCOPY

The incorporation of isotopically-labelled hydrogen (^{2}H or ^{3}H) can also be monitored directly by ^{2}H n.m.r. or ^{3}H n.m.r.[61,62]

A. ^{2}H N.m.r.

^{2}H N.m.r. despite several inherent disadvantages has been the nucleus of choice in many biosynthetic studies. Its major limitations are mainly as a consequence of the low magnetogyric ratio, and the relaxation behaviour of the ^{2}H nucleus. Because it is a quadrupole nucleus (spin 1) and thus very efficiently relaxed, the spectral lines are rather broad and this, coupled with the low magnetogyric constant and the small chemical shift range for hydrogen nuclei, often results in poorly resolved spectra. However, the rapid relaxation and lack of any n.O.e. mean that accurate integration of ^{2}H n.m.r spectra is possible so that the relative enrichment at different sites in a metabolite can be accurately assessed. Another major advantage is that as a consequence of its

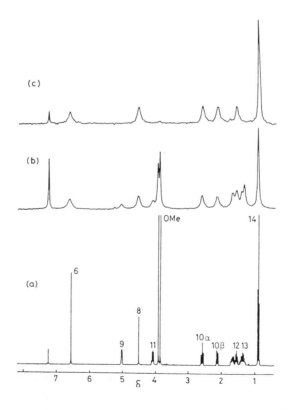

Figure 12   (a) 360 MHz ^{1}H n.m.r. spectrum of monocerin, (b) 55.28 MHz ^{2}H n.m.r. spectra of monocerin produced in a culture medium containing 5% ^{2}H$_2$O and, (c) after enrichment with [^{2}H$_3$]acetate.

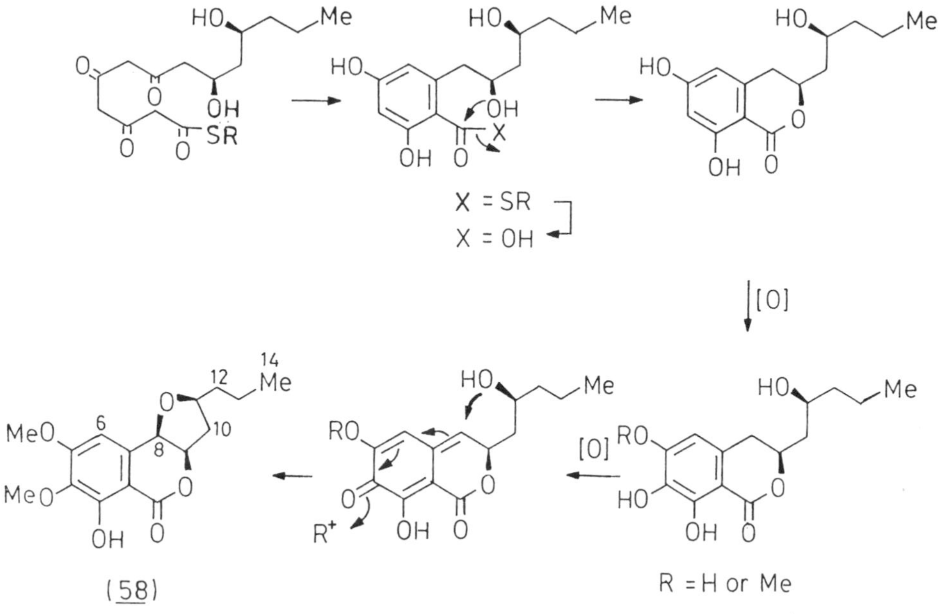

Scheme 35

low natural abundance (0.012%) much greater dilutions are tolerable than in the case of ^{13}C-labelling: a 100% [2]H-labelled precursor may be diluted 6,000 fold and still result in a doubling of intensity over the corresponding natural abundance signal. This makes [2]H-labelling particularly suitable for studying the incorporation of advanced intermediates on a biosynthetic pathway. The 55 MHz [2]H n.m.r. spectrum which results[63] on incorporation of [[2]H$_3$]acetate into monocerin (58) in cultures of *Dreschlera monoceras* is shown in Figure 12. This is compared with the 300 MHz [1]H n.m.r. spectrum and also the 55 MHz [2]H n.m.r. of uniformly [2]H-labelled monocerin, produced by the simple expedient of growing the organism in a medium containing 5% H_2O. This shows that all of the hydrogens including those of the diastereotopic methylenes at C-10, C-12 and C-13 are resolved. Figure 12 shows that acetate-derived ^{2}H has been incorporated at C-7, C-8 and that both diastereotopic hydrogens at C-10 are labelled by ony one of those at C-12. On feeding [1-^{13}C,[2]H$_3$]-acetate the β-shifts observed in the [13]C n.m.r.spectrum of the enriched metabolite showed that two acetate derived hydrogens were retained at C-10 allowing the pathway shown in Scheme 35 to be porposed. This illustrates the complimentary nature of the use of isotope-shifts (indirect) and direct observation by [2]H n.m.r. to monitor the metabolic fate of [2]H-labels. While [2]H n.m.r. is ideal for observing the stereospecificity of isotope labelling, only the isotope-shift method can confirm the number of labelled atoms at a given position.

The use of ^{2}H n.m.r. to monitor incorporation of advanced intermediates is illustrated in Figure 13. Tajixanthone (6), a metabolite of *Aspergillus variecolor* on the basis of ^{13}C-labelling studies, was believed to be biosynthesised *via* oxidative cleavage cleavage of anthraquinone inter-mediates. [*Methyl*-^{2}H$_3$]-chrysophanol (59) was synthesised and fed to cultures of *A.variecolor*.[64] The ^{2}H n.m.r. of the tajixanthone isolated from this experiment is shown in Figure 13 where it is compared with the uniformly labelled metabolite enriched as above by supplementing the medium with ^{2}H$_2$O. As required by the proposed biosynthetic pathway, ^{2}H label from chrysophanol is incorporated specifically into the 24-methyl group only of tajixanthone.

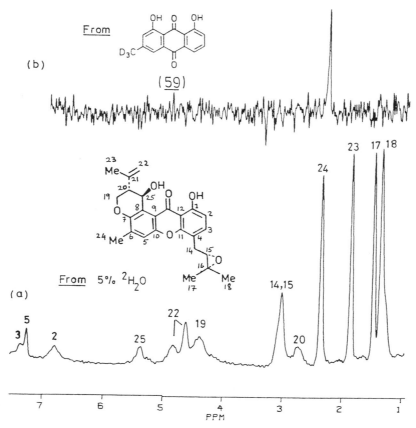

Figure 13   55.28 MHz ^{2}H n.m.r. spectra of tajixanthone (a) produced in a culture medium supplemented with 5% ^{2}H$_2$O, and (b) labelled from feeding [methyl-^{2}H$_3$]-chryso-phanol.

B.   **^{3}H N.m.r.**

Although best known as a radioactive tracer, the ^{3}H nucleus has ideal characteristics for use as an n.m.r. label.[61] As the natural abundance is practically zero and the magnetogyric ratio is the highest yet found for any nuclide, ^{3}H n.m.r. spectroscopy is uniquely sensitive, by the standards of other n.m.r. methods of tracing isotopes. The chemical shift values and coupling constants are very close to those of the corresponding ^1H n.m.r. spectrum so assignments can be made on this basis. Accurate integration is also possible.[65] The utility of the method in biosynthetic studies was first demonstrated in a study of penicillic acid (60), a metabolite of *Penicillium cyclopium* (Scheme 36).[66] Sodium [^3H]acetate of high specific activity was incorporated and the ^3H n.m.r. spectrum of the enriched metabolite showed the positions indicated in Figure 14 carried ^3H label. The signal due to the allylic methyl group appeared as a triplet, $J_{HT} = 15.5Hz$, due to coupling of the single ^3H with two protons. An interesting observation was the stereospecific retention of ^3H at the more upfield (*trans* to methyl) of the vinylic methylene hydrogens.

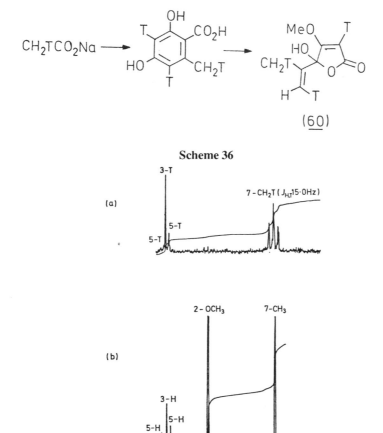

Scheme 36

Figure 14   N.m.r. spectra of penicillic acid in [$^2H_6$]acetone; (a) 96.02 MHz ^3H n.m.r. spectrum of [^3H]acetate enriched metabolite; (b) 90.02 MHz ^1H n.m.r. spectrum. (Reproduced with permission from *J.Chem.Soc., Chem.Commun.*, (1974) 220).

In an application making use of more advanced intermediates, the stereochemistry of the 1,3-proton loss from a chiral methyl group in the biosynthesis of cycloartenol has been determined by ^{3}H n.m.r.[67] Oxidosqualene (61) was synthesised containing a chiral methyl group as indicated and converted into ^{3}H-labelled cycloartenol (62) by incubation with a cell-free microsomal fraction of *Ochromonas malhamensis*. The ^{3}H-n.m.r. of the enriched metabolite showed resonances at δ 0.168, 0.438 and 0.456 in a ratio of 4.6: 1:1.5. These were assigned as follows: the resonance at δ 0.168 is due to an *exo* cyclopropyl tritium in molecules which also have an *endo* deuterium; the resonance at δ 0.456 is due to an *endo* cyclopropyl tritium in molecules which also have an *exo* deuterium; and the resonance at 0.456 is due to an *endo* cyclopropyl tritium in molecules which also have an *exo* proton. These results indicate that the conversion proceeds with retention of configuration as indicated in Scheme 37.

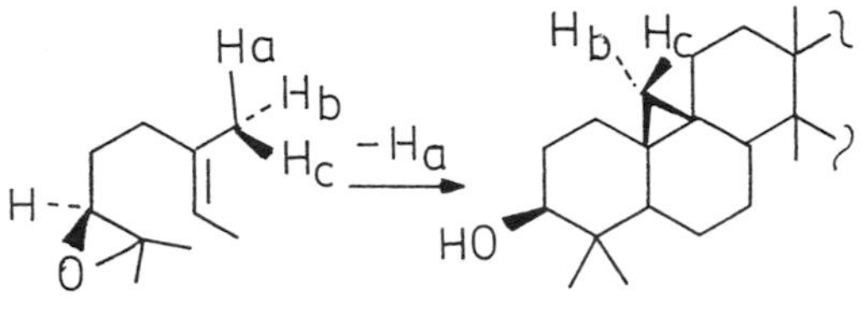

(61)   (62)

Scheme 37

There have been surprisingly few other biosynthetic applications. This is possibly due to the relatively high amounts of activity required. However, with modern high field spectrometers, it should be possible to observe signals due to as little as 100μCi of ^{3}H. Poor biosynthetic incorporation rates may, however, mean that considerably higher levels of activity may be required at the outset.

IX.   OXYGEN-17 N.M.R.

Of the isotopes of oxygen, only one, ^{17}O (natural abundance 0.037%), possesses nuclear spin ($\frac{1}{2}$) and thus gives rise to an n.m.r. spectrum.[68] Unfortunately it is an insensitive nucleus which gives broad lines due to quadrupolar relaxation. In addition acoustic ringing in the instrument probe and pulse breakthrough can cause serious interference with the recording of the signal from the sample and this may result in severe baseline roll which may obscure weak signals. However a number of biosynthetic experiments have appeared. The incorporation of acetate-derived oxygen into citrinin was followed by ^{17}O n.m.r.[69] and the technique was also used to demonstrate that no oxygen was lost from the L-α-aminoadipyl section of the tripeptide precursor in the biosynthesis of isopenicillin N.[70]

Evidence for the biosynthesis of hydroxymellein (64) via direct hydroxylation of the benzylic methylene of mellein (63) has been obtained by incorporation of [17]O-labelled acetate and oxygen followed by [17]O n.m.r. analysis of the enriched metabolites (Scheme 38).[71] The spectra arising from these studies are shown in Figure 15, which demonstrates the incorporation of oxygen from [[17]O] acetate into the lactone and phenolic oxygens. The benzylic hydroxyl was shown to be derived by a direct hydroxylation by growing the producing organism, *Aspergillus melleus* in closed system containing oxygen-17 gas.

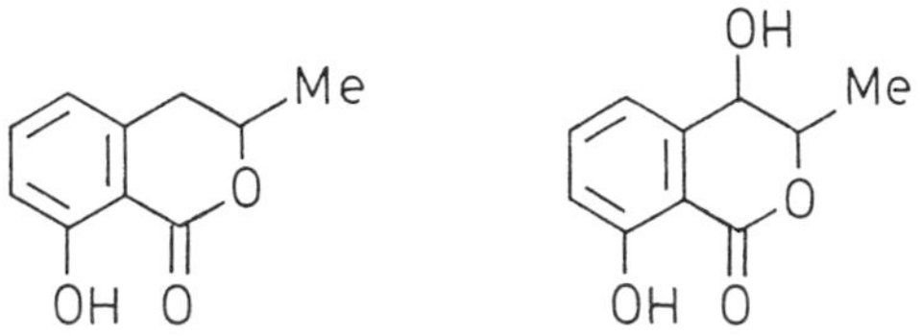

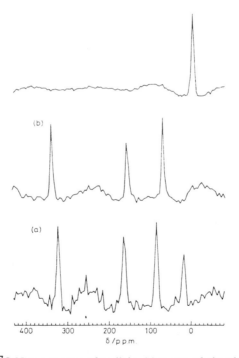

Scheme 38

Figure 15   54.2 MHz [17]O N.m.r. spectra of mellein, (a) at natural abundance, and after incorporation of label from (b) [[17]O]acetate and (c) [[17]O]O$_2$ gas. (Reproduced with permission from *J.Chem.Soc., Chem.Commun.*, (1987), 177).

In a related study[72] the quality of ^{17}O n.m.r. spectra has been greatly improved by application of the maximum entropy method. This obviates the problem of acoustic ringing and pulse breakthrough and produces vastly improved spectra as shown in Figure 16. It may be anticipated that this technique will encourage further use of ^{17}O-labelling in biosynthetic work.

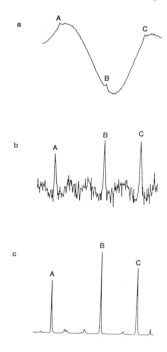

Figure 16 ^{17}O N.m.r. spectra of aspyrone derived from an FID corrupted by ring down. (a) Spectrum produced by conventional Fourier transform of the FID. (b) Spectrum produced by conventional Fourier transform with the first 10 points of the FID set to zero. (c) MEM spectrum with the first 10 points of the FID ignored. (Reproduced with permission from *J.Mag.Res.*, 72 (1987) 495).

X.   NITROGEN-15 N.M.R.

Since ^{15}N has a natural abundance of 0.36% and a number spin of $\frac{1}{2}$, ^{15}N-n.m.r. is possible.[73] Although most biosynthetic studies have detected biosynthetic incorporation of ^{15}N via its coupling to ^{13}C, the direct observation of ^{15}N incorporation is possible. One of the first applications was the incorporation of [^{15}N]-valine into penicillin G by *Penicillium chrysogenum*.[74] In a more recent application, ^{15}N-n.m.r. was used to demonstrate the incorporation of [^{15}N]-nitrosuccinate into the toxin, 3-nitropropanoic acid (65) in cultures of *Penicillium atrovenetum*.[75] This established that in the course of the conversion of aspartic acid to (65), complete oxidation of the amino group must precede decarboxylation as shown in Scheme 39. The same workers also established that both oxygen atoms of the nitro group are derived from molecular oxygen by growing *P. atrovenetum* under an atmosphere of $^{18}O_2$-$^{16}O_2$

(50:50) in a defined medium containing $^{15}NH_4Cl$ as the sole nitrogen source. The ^{15}N n.m.r. spectrum (Figure 17) of the resultant 3-nitropropanoic acid showed ^{18}O isotope induced shifts corresponding to the incorporation of one and two ^{18}O atoms into the nitro group.

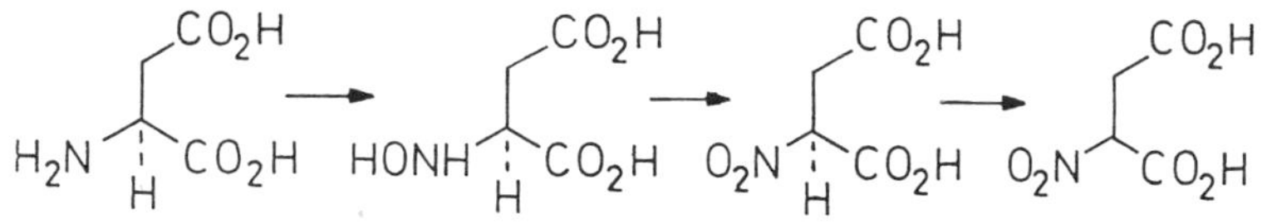

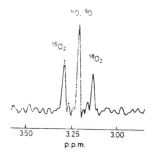

Scheme 39

Figure 17   36.5 MHz ^{15}N n.m.r. spectrum of biosynthetically ^{15}N, ^{18}O enriched 3-nitro-propanoic acid. Chemical shifts are expressed relative to nitromethane. The observed ^{18}O isotope shift is 0.08 ppm. (Reproduced with permission from *J.Chem.Soc., Chem.Commun.*, (1986) 176).

XI.   CONCLUSION

The development of ^{13}C n.m.r. and associated stable isotope labelling techniques has produced a major advance in biosynthetic and metabolic studies. ^{13}C, ^{15}N, ^{18}O, ^2H and ^3H labelling provides information on the origins of the carbon skeletons of metabolites and also the origins and fates of oxygen, nitrogen and hydrogen. This provides an invaluable insight into the nature of the intermediates on metabolic pathways and the detailed mechanisms of their interconversions that could not be otherwise obtained.

REFERENCES

1.   T.J. Simpson. *Chem.Soc.Review*, 4 (1975) 497.

2.   T.J. Simpson. *Chem.Soc.Review*, 16 (1987) 123.

3.   T.J. Simpson in: H.F. Linskens and J.F. Jackson (Eds.), Modern Methods of Plant Analysis, Volume 2, Nuclear Magnetic Resonance, Springer-Verlag, Berlin, Heidelberg, 1986, p.1.

4.   J.K.M. Saunders and B.K. Hunter. Modern NMR Spectroscopy, O.U.P., Oxford, 1987.

5.   A.E. Derome. Modern NMR Techniques for Chemistry Research, Pergamon, Oxford, 1987.

6.   I.H. Sadler. *Nat.Prod.Rep.,* 5 (1988) 101.

7.   A.E. Derome. *Nat.Prod.Rep.,* 6 (1989) 111.

8.   R. Brown and H. Gunther. *Angew. Chem. Int. Ed. Engl.,* 22 (1983) 350.

9.   C.R. McIntyre and T.J. Simpson. *J.Chem.Soc., Chem.Commun,* (1984) 704.

10.   C.R. McIntyre, D. Reed, I.H. Sadler, and T.J. Simpson. *J.Chem.Soc., Perkin Trans. 1* (1989)   1987.

11.   T.J. Simpson and I.H. Sadler. *J.Chem.Soc., Chem.Commun,* (1989) 1602.

12.   T.J. Simpson, *J.Chem.Soc., Perkin Trans. 1,* (1979) 1233.

13.   R.J. Abraham and P.Loftus. Proton and Carbon-13 NMR Spectroscopy, Heyden, London, 1978.

14.   J.S.E. Holker, R.D. Lapper, and T.J. Simpson. *J.Chem.Soc., Perkin Trans. 1,* (1974) 2135.

15.   A.J. Birch and T.J. Simpson. *J.Chem.Soc., Perkin Trans. 1,* (1979) 816.

16.   C.R. Hutchinson, A.H. Heckendorf, P.E. Doddona, E. Hageman, and E. Wenkert. *J.Am.Chem.Soc.,* 96 (1974) 5609.

17.   A.R. Battersby, P.W.S. Sheldrake, and J.A. Milner. *Tetrahedron Lett.,* (1974) 3315.

18.   R. Benn and H. Gunther. *Angew.Chem.Int.Ed.Engl.,* 22 (1983) 350.

19.   J.S.E. Holker and T.J. Simpson. *J.Chem.Soc., Perkin Trans. 1,* (1981) 1397.

20.   A.P. Bradshaw, J.R. Hanson, and M. Siverns. *J.Chem.Soc., Chem.Commun,* (1977) 319.

21.   M. Tanabe, T. Hamasaki, Y. Suzuki, and L.F. Johnson. *J.Chem.Soc., Chem.Commun.,* (1973)   212.

22.   H.H. Wasserman, R.J. Sykes, C.K. Shaw, and R.J. Cushley. *Tetrahedron Lett.,* (1974) 2787.

23.   C.E. Brown, J.J. Katz, and D. Shemin. *Proc.Nat.Acad.Sci., USA,* 69 (1972) 2585.

24.   J.L. Marshall. Carbon-Carbon and Carbon-Proton NMR Couplings, Verlag Chemie International, Deerfield Beach, 1983.

25.   A.G. McInnes, D.G. Smith, J.A. Walter, L.C. Vining and J.L.C. Wright, *J.Chem.Soc., Chem.Commun.,* (1975) 66.

26.   R.N. Moore, G. Bigam, J.K. Chan, A.M. Hogg, T.T. Nakashima and J.C. Vederas. *J.Am.Chem.Soc.,* 107 (1985) 3694.

27.   R.M. Horak, P.S. Steyn and R. Vleggaar. *Magn.Reson.Chem.,* 23 (1985) 995.

28.   T.J. Simpson. Unpublished results.

29.   R.J. Light and K. Hahlbrook. *Z. Naturforsch.,* 35C (1980) 717.

30. A. Stoessl and J.B. Stothers. *Z. Naturforsch.*, 34C (1979) 87.

31. K.H. Overton and D.J. Picken. *J.Chem.Soc., Chem.Commun.*, (1976) 105.

32. E. Leete. *J.Nat.Prod.*, 45 (1982) 197.

33. E. Leete and M.-L. Yu. *Phytochemistry*, 20 (1980) 1093.

34. H.A. Khan and D.J. Robins. *J.Chem.Soc., Chem. Commun.*, (1981) 146.

35. H. Seto, T. Sato, S. Urano, J. Uzawa, and H. Yonehara. *Tetrahedron Lett.*, (1976) 4367.

36. H. Seto, M. Saito, J. Uzawa, and H. Yonehara. *Heterocycles*, 13 (1979) 247.

37. E. Leete, N. Kowanko, and R.A. Newmark. *J.Am.Chem.Soc.*, 97 (1975) 6826.

38. E. Leete, N. Kowanko, R.A. Newmark, L.C. Vining, A.G. McInnes, and J.L.C. Wright. *Tetrahedron Lett.*, (1975) 4103.

39. G. Popjak, J. Edmond, F.Al. Anet, and N.R. Easton. *J.Am.Chem.Soc.*, 99 (1977) 931.

40. A.R. Battersby, E. Hunt, and E. McDonald. *J.Chem.Soc., Chem.Commun.*, (1973) 442.

41. J. Rana and D.J. Robins. *J.Chem.Res. (S)*, (1983) 146.

42. G.C. Levy and R.L. Lichter. Nitrogen-15 Nuclear Magnetic Resonance Spectroscopy. Wiley, New York, (1979).

43. K.J. Martinkus, C.-H. Tann, and S.J. Gould. Tetrahedron, 21 (1983) 3493.

44. R.L. Baxter, C.J. McGregor, G.A. Thomson, and A.I. Scott. *J.Chem.Soc., Perkin Trans 1*, (1985) 369.

45. E. Leete and J.A. McDonnell. *J.Am.Chem.Soc.*, 103 (1981) 608.

46. H.A. Khan and D.J. Robins. *J.Chem.Soc., Chem.Commun.*, 1981 146.

47. M.J. Garson, R.A. Hill, and J. Staunton. *J.Chem.Soc., Chem.Commun.*, (1977) 921.

48. C.R. Hutchinson, I. Kurobane, C.T. Mabuni, R.W. Kumola, A.G. McInnes, and J.A. Walter. *J.Am.Chem.Soc.*, 103 (1981) 2474.

49. S.A. Ahmed, T.J. Simpson, J. Staunton, A.C. Sutkowski, L.A. Trimble, and J.C. Vederas. *J.Chem.Soc., Chem.Commun.*, (1985) 1685.

50. D.M. Doddrell, J. Staunton, and E.D. Laue. *J.Chem.Soc., Chem.Commun.*, (1983) 602.

51. L.A. Trimble, P.B. Reese, and J.C. Vederas. *J.Am.Chem.Soc.*, 107 (1985) 2175; P.B. Reese, B.J. Rawlings, S.G. Ramer, and J.C. Vederas. *J.Am.Chem.Soc.*, 110 (1988) 316.

52. C. Abell and J. Staunton. *J.Chem.Soc., Chem.Commun.*, (1981) 856.

53. T.J. Simpson, A.E. de Jesus, P.S. Steyn, and R. Vleggaar. *J.Chem.Soc., Chem.Commun.*, (1982) 632.

54. C.A. Townsend, S.W. Brobst, S.C. Ramer, and J.C. Vederas. *J.Am.Chem.Soc.*, 110 (1988) 318.

55. T.J. Simpson and D.J. Stenzel. *J.Chem.Soc., Chem.Commun.*, (1982) 1074; (1981) 239.

56. T.J. Simpson and G.A. Stevenson. *J.Chem.Soc., Chem.Commun.*, (1985) 1822.

57. G. Lowe and B.G. Sproat. *J.Chem.Soc., Chem.Commun.*, (1978) 783.

58. J.H. Risley and R.L. Van Etten. *J.Am.Chem.Soc.*, 101 (1979) 252.

59. J.C. Vederas and T.T. Nakashima. *J.Chem.Soc., Chem.Commun.*, (1980) 183.

60. F.E. Scott, T.J. Simpson, L.A. Trimble, and J.C. Vederas. *J.Chem.Soc., Chem.Commun.*, (1986) 214.

61. M.J. Garson and J. Staunton. *Chem.Soc. Reviews,* 8 (1979) 539.

62. C. Abell in: H.F. Linskens and J.F. Jackson (Eds.), Modern Methods of Plant Analysis, Volume 2, Nuclear Magnetic Resonance, Springer-Verlag, Berlin, 1986, p 60.

63. F.E. Scott, T.J. Simpson, L.A. Trimble, and J.C. Vederas. *J.Chem.Soc., Chem.Commun.,* (1984) 756.

64. S.A. Ahmed, E. Bardshiri, and T.J. Simpson. *J.Chem.Soc., Chem.Commun.,* (1987) 883.

65. J.A. Elvidge in: J.A. Elvidge and J.R. Jones (Eds.), Isotopes: Essential Chemistry and Applications, The Chemical Society, London, (1980), p 123.

66. J.M.A. Al-Rawi, J.A. Elvidge, D.K. Jaiswal, J.R. Jones, and R. Thomas. *J.Chem.Soc., Chem.Commun.,* (1974) 220.

67. L.J. Attman, C.Y. Yan, A. Bertolino, G. Handy, D. Laungani, W. Muller, S. Schwartz, D. Shauker, W.H. de Wolf, and F. Yang. *J.Am.Chem.Soc.,* 100 (1978) 3235.

68. P. Diehl, E. Fluck, and R. Kosfield (Eds.), Oxygen-17 and Silicon-29 NMR, Basic Principles and Progress, Springer-Verlag, Berlin, (1981).

69. U. Sankawa, Y. Ebizuka, H. Noguchi, Y. Ishikawa, S. Kitigawa, T. Kobayashi, and H. Seto. *Heterocycles,* 16 (1981) 1115.

70. R.M. Adlington, R.T. Alpin, J.E. Baldwin, L.D. Field, E.-M.M. John, E.P. Abraham, and R.L. White. *J.Chem.Soc., Chem.Commun.,* (1982) 137.

71. C. Abell, A.C. Sutkowski, and J. Staunton. *J.Chem.Soc., Chem.Commun.,* (1987) 586.

72. E.D. Laue, K.O.B. Pollard, J. Skilling, J. Staunton, and A.C. Sutkowski. *J.Mag.Res.,* 72 (1987) 493.

73. T. Axenrod and G.A. Webb. Nuclear Magnetic Resonance Spectroscopy of Nuclei other than Protons: Wiley, New York, 1974.

74. H. Booth, B.W. Bycroft, C.M. Wels, K. Corbett, and A.P. Maloney. *J.Chem.Soc., Chem.Commun.* (1976) 110.

75. R.L. Baxter, A.B. Hanley, and H.W.-S. Chan. *J.Chem.Soc., Chem. Commun.,* (1988) 757.

76. R.L. Baxter and S.L. Greenwood. *J.Chem.Soc., Chem.Commun.* (1986) 175.

482